A Field Guide to the Common Trees and Shrubs of Sri Lanka

A Field Guide to the Common Trees and Shrubs of Sri Lanka

Mark S. Ashton, Savitri Gunatilleke, Neela de Zoysa,
M.D. Dassanayake, Nimal Gunatilleke and Siril Wijesundera

WHT Publications (Pvt.) Limited
1997

To all our teachers
for inspiring us
to appreciate and enjoy nature

A Field Guide to the Common Trees and Shrubs of Sri Lanka
(including an introduction to the flora, plant uses, and index).

Ashton, Mark S.; Gunatilleke, Savitri; de Zoysa, Neela;
Dassanayake, M.D.; Gunatilleke, Nimal and Wijesundera, Siril.

Published by WHT Publication (Pvt.) Ltd for the Wildlife Heritage
Trust of Sri Lanka,
95 Cotta Road, Colombo 8, Sri Lanka.

ISBN 955-9114-08-5.

Printed in Sri Lanka by Gunaratne Offset Limited.

Typesetting and colour-separation by
Iris Colour Graphics Limited, Colombo, Sri Lanka.

Contents

Preface

The stimulus for *The Field Guide to the common trees and shrubs of Sri Lanka* was the success of a similar guide for Puerto Rico compiled by one of us (MA). Begun as a modest effort in 1985, it was boosted in 1988 with grants from the WWF-US and from the Environmental Program of the MacArthur Foundation. The collective inertia made this comprehensive end-product possible.

This Guide is organized into three parts. The first part is a general introduction that deals with the geography, climate, geology, soils, natural vegetation and the history of landuse on the island. This section intends placing the tree and shrub species described in context with the biological and cultural history of the island.

The second part is the main body of the *Guide* which includes 95 families and 704 species of trees and shrubs, each with its own illustration. It is preceded by a key to identification and with an explanation on how to refer to the descriptions. All of the technical terms used are listed in a glossary. The information is organized in a form accessible to the practical user as well as the more advanced botanical reader.

The third part of the book is a reference to plant uses. It includes sections on timber, medicinal, as well as food and other uses of the trees and shrubs described. This section gives an overview to the general reader of the extent of traditional and contemporary uses of plants in this country and provides more precise information to the forester.

It is hoped that this book will serve a long-felt need by making available information on the fascinating plant life of Sri Lanka to a wide range of people. It is a pioneer effort, and the authors would welcome suggestions and criticisms from those who use the book. We also hope that in the coming few years we can compile a sequel to this book on the less common and rare trees and shrubs of Sri Lanka.

Acknowledgements

This work has been carried out over a period of eleven years somewhat intermittently by the authors in their spare time. It made it no less easy that they were scattered in several different places. We would like to thank the School of Forestry and Environmental Studies, Yale University for the use of computer printing facilities. We also thank Magdon Jayasuriya, Curator, National Herbarium, Peradeniya and D. B. Sumithraarachchi, Director of the Royal Botanic Gardens, Peradeniya for use of the herbarium and the helpful co-operation of the herbarium staff. We thank K. Vivekanandan, Chief Research Officer of the Forest Department, Sri Lanka, for use of the Department herbarium. We appreciate the important contributions made by Nishanta Rajakaruna by supervising and co-ordinating the guide during an important period and Jayantha Samarasinghe and Sunil Gamage by collating the plant-uses section. We also thank A. M. Gunapala for collection of plant specimens, and Milton Liyanage for the maps in the introduction.

We especially thank those who contributed the most to the illustrations: Malini Goonatilleke, L. S. B. Wadigamangawa, Chaminda Nagahapitiya, Sumedha Madawala, A.S.T.B. Wijetunga, Dharshini Goonatilleke and Jagath Kodituwakku. We also recognize the many others who helped with illustrations: Ruchindra Abeytunga, Maitri Jansz, Indula Silva, Shanta Jayaweera; and the numerous students of the Botany Department at Peradeniya: N. Ekanayake, C. de Silva, N. de Silva, K.D. Ratnayake, D. Welagedara (Botany Special Part III, 1989); Nilanthi Herath, Nilmini Kanthi, Champa Nalinie, Niranjala Perera, Hasantha Priyadharshani, Sunil Sarath Kumara, Asanga Uduwela, Malkanthi Wijepala (Botany Special Part I, 1990); Kamani Dambawinna, Deepthi Dissanayake, Priyani Kulatunga, Rohini Padmalatha, Renuka Premaratna, Mohamud Rizvi, Yojitha Seneviratne, Kushan Tennakoon, Dayani Tilakaratne, Thusitha Weerasekara (Botany Special Part II, 1990); Anoma Basnayake, D. Ekanayake, Anoma Perera, Udeni Pushpalatha, Geethani Rathnayake, Deepika Somaratne, Priyanka Weerasinghe, Hemanthi Wijeratne, Kapila Yakandawala (Botany Special Part III, 1990).

Our appreciation is also extended to Mr Rohan Pethiyagoda for the encouragement given to us during preparation of the manuscript, to Ruwan Herath and Miss Elita Atapattu for the long hours spent and their tireless efforts in typesetting the manuscript.

Part I:

Introduction to the Flora

INTRODUCTION

Since ancient times Sri Lanka has been well known for its great natural beauty. A rich and diverse flora is one of the country's important assets. This small island of 65,500 square kilometres has over 3,500 flowering plant species native to it. Over a quarter of these species are considered unique to the country. Today Sri Lanka is considered one of the most biodiverse areas in South Asia. Recent scientific evidence indicates that many of the plant species in the southwest of the country have a Deccan-Gondwana and ancestry.

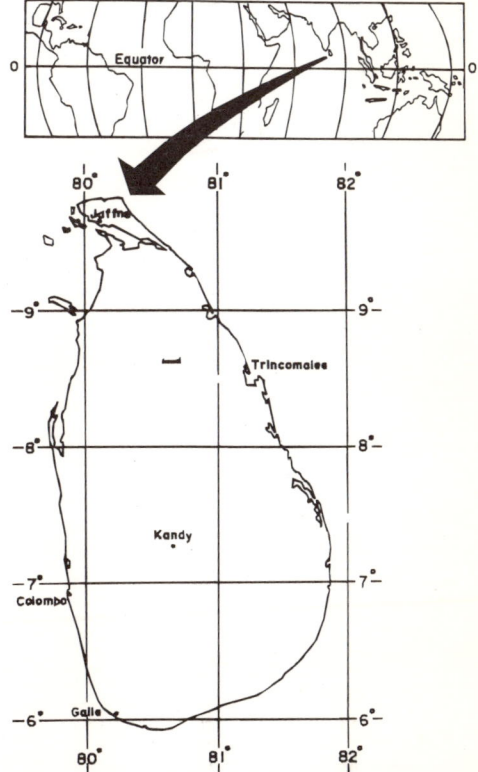

The island's natural flora has been enriched by the age-old Lankan tradition of cultivating spice, food, ornamental, and medicinal plants. Many of these were introduced from places as far off as China, the Mediterranean, Ethiopia, and Arabia, by travellers who visited the country during a period of at least two thousand years. The introduction of plantation crops such as tea and rubber during the 19th century has not only altered the landscape but the nation's economy as well.

This rich plant heritage is better appreciated if seen in the natural and historical context of the country. We therefore briefly describe factors such as geology, soils, climate and land form, and landuse trends that have influenced and shaped the plant life of Sri Lanka.

Figure 1. *Geographical location of Sri Lanka with respect to the tropical and subtropical regions of the world.*

Geography

Sri Lanka is situated close to the southeastern tip of India, north of the equator (Fig. 1). The two countries, although separated by the shallow Palk Strait, share the same continental plate and are linked by coral reefs and islets, popularly known as Adam's Bridge. Both the evolutionary history and the cultural traditions associated with the flora of this country have a strong Indian influence (see Box 1). The nearest land masses, other than India, are the Malayan region towards the east and the African continent to the west.

BOX 1. The origin of Sri Lanka's natural plant communities has become clearer with recent evidence on the geological past of the Asian region. It has been traced back a hundred million years to the early Cretaceous period when the ancient southern continent Gondwana began breaking up. The Deccan plate, a fragment consisting of India and Sri Lanka, drifted north and collided with mainland Asia nearly 55 million years later. After a lapse of another 20 million years, in the late Miocene, Sri Lanka was separated with the submersion of the land area between the two countries. The maps below show the geological past of Sri Lanka and the time scale indicating the different epochs, periods and eras with time of their initiation in millions of years before present (MYBP). Map stages show the Deccan plate, wedged between Africa and Australia, as part of Gondwana (I); as the Deccan plate separates from remaining Gondwana (II); and the isolation of Sri Lanka from India with rising sea levels (III).

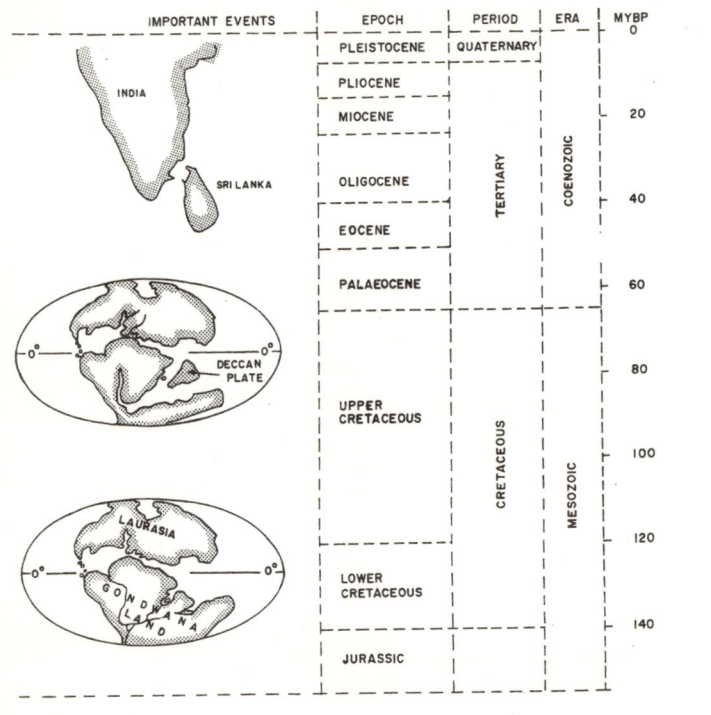

Land form

The present landform of Sri Lanka is the result of millions of years of weathering by rain and wind, as well as movements of the earth's crust. The topography of Sri Lanka is remarkably varied for its small area, with coastal plains, lowland hills and a mountainous interior (Fig. 2). This variation is reflected in the complexity of the island's diversity of natural plant communities and crops.

The coastal plains hardly exceed 100 m in elevation. The lowlands cover nearly three quarters of the country and are extensive in the north and east. Towards the south-central part, the land rises gently to about 600 m in elevation with low, rounded hills and crests of hard rock. Two large basins, the Kelani and Uva, characterize this area.

The central mountain area consists of a complex of plateaux, mountain chains and basins, much of it reaching elevations greater than 1800 m. Along the southern margin of the highlands are numerous waterfalls which have led to the speculation that the mountains were formed by upliftment during relatively recent geological times. The central mountains are steeply dissected particularly on the south and south-western faces. The highest part takes the shape of an anchor, marked by Adams Peak, Kirigalpoththa, Pidurutalagala and Namunukula, all major peaks reaching between 2000 m and 2524 m. These are interspersed with the plateaux of Horton Plains, Ambawela and Nuwara Eliya. To the northeast and southwest of the central mountains are two small massifs, the Knuckles mountains and the Rakwana hills respectively (Fig. 2).

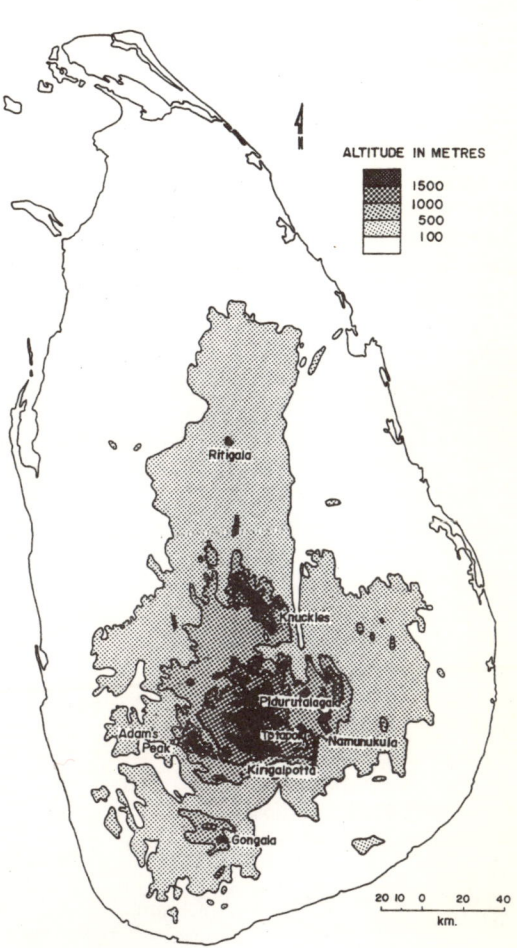

Figure 2. *Altitudinal variation within the island and its highest peaks.*

The drainage pattern of the country is almost entirely governed by the central highlands, with all the perennial water courses originating in the mountains and winding their way down to the plains below (Fig. 3).

Climate

Sri Lanka's equatorial position gives its lowlands a tropical climate, with year round temperatures of 27-28⁰C and a relatively constant day length. Rainfall is largely governed by monsoonal winds which occur during two seasons of the year. From mid-May to September, the monsoon blows from the southwest direction and brings in a greater amount of moisture than during December to February when the wind blows from the northeast. The distinct inter-monsoonal periods receive convectional rains and at times cyclones. During the southwest monsoon, the position and dramatic relief of the southwestern side of the central high-lands forces the moisture-laden air upwards. The rapidly cooled air condenses, causing pre-cipitation mostly on the windward slopes of the island's southwest. During this time the northeastern and south-eastern parts of the land hardly receive any rain. On the other hand, the northeast monsoon winds rise over the central highlands more gradually, and the rain shadow effect is not nearly so distinct, allow-ing precipitation to fall on the entire island. This has resulted in the division of the country into two major climatic regions; the **wet zone** which re-ceives rain from both

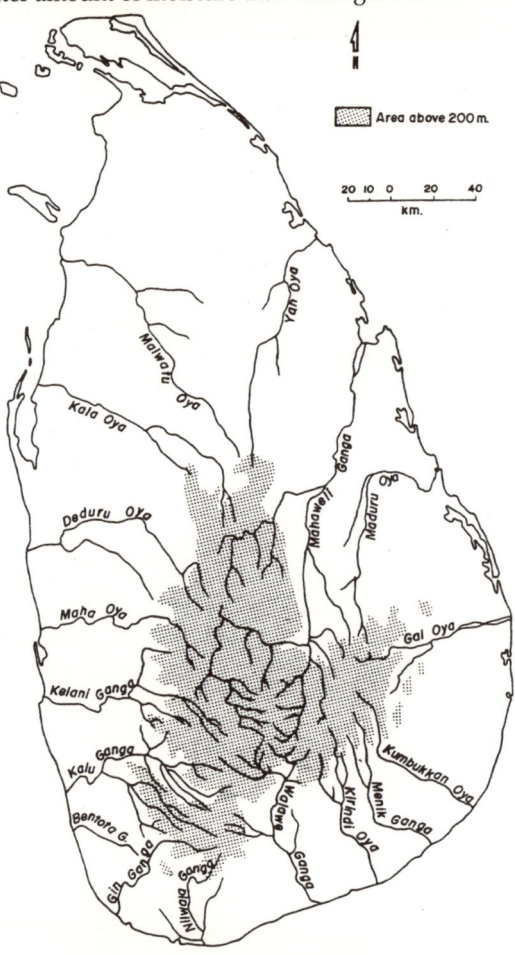

Figure 3. The main rivers showing their origin in the mountains and direction of flow to the sea.

monsoons, and the **dry zone** which receives rain from only one. The gradual change from the wet to dry zone allows an **intermediate zone** to exist. In addition, two small areas at the exteme northwest and southeast of the country have a very dry climate and are known as **arid zones** (Fig. 4).

Furthermore, owing to the small size of the island and its 'open' position in the Indian ocean, an oceanic climate dominates the coastal lowlands. This area is strongly influenced by local convectional winds and their associated thunderstorms. These local sea winds sometimes influence the interior of the country, often interacting with monsoonal winds.

Besides rainfall, temperature plays an important role in highland regions. For every 100 m increase in elevation, the mean temperature falls by 0.5⁰C. On the plateaux there is often ground frost in the lower lying areas between December and March.

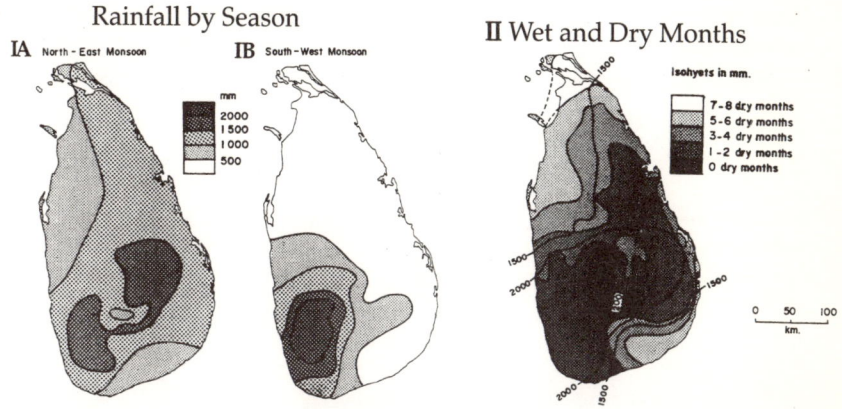

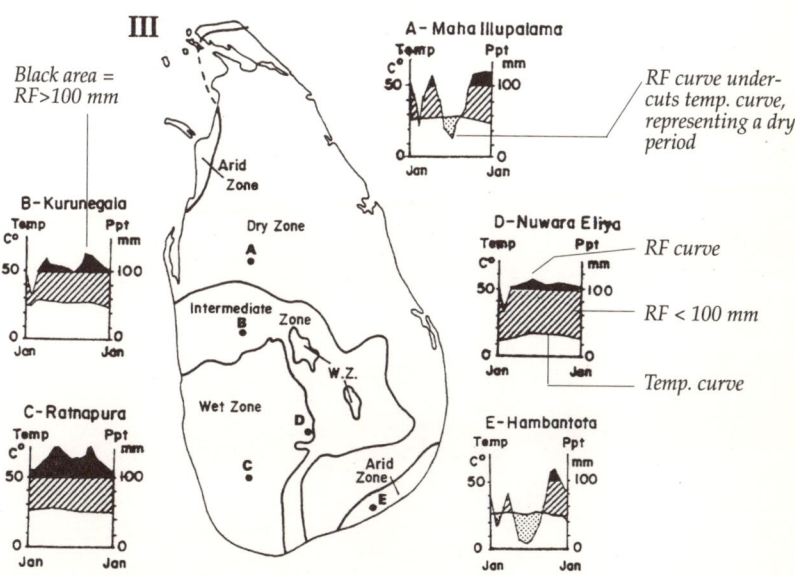

Figure 4. *Rainfall (RF) distribution during the northeast (IA) and southwest (IB) monsoons; duration of dry period and rainfall distribution in millimetres (II); climatic zones with representative climatic graphs for each zone in the country (III).*

Geology

Most of Sri Lanka is underlain by **Precambrian rocks** which are metamorphic. Two major groups of Precambrian rocks are recognized: i) the **Highland series** and ii) the **Vijayan complex**. The more recent rock types are of sedimentary origin, found mostly in the northwest of the country including the entire Jaffna peninsula. These are predominantly Miocene **limestone**. Small extents of Jurassic **sandstone, shale and grit** are found among the Vijayan rocks in the basins of Tabbowa and Andigama. **Alluvium, beach and dune sands, red earth and mottled gravel** of quaternary origin are found all over the island but predominantly near the coast (Fig. 5).

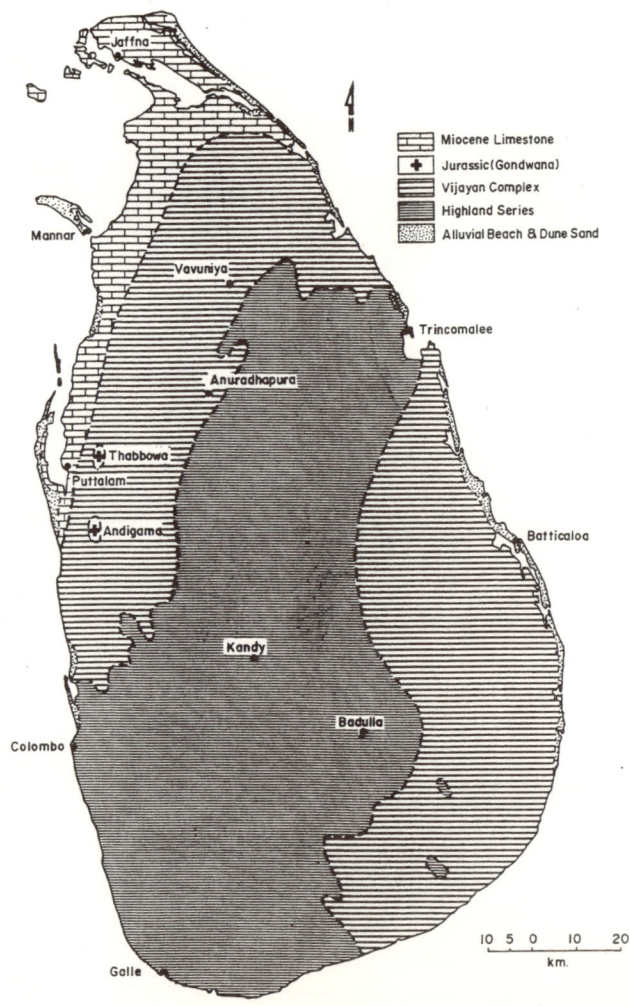

Figure 5. Major rock formations

Soils

Soil is formed by the weathering of parent rock material. Depending on the parent rock type, the weathering process, and climate, a variety of soils can form. In Sri Lanka fourteen major soil groups are recognized, but only two of them are found extensively (Fig. 6). Since the parent material of the Precambrian rocks is relatively uniform over the island mass, weathering by climate appears to be the main determinant of soil type. Soil groups therefore closely coincide with the climatic zones of the country.

Reddish brown earths occupy most of the dry and intermediate zones, over-lying the Vijayan rocks, and also on the Highland rocks between Polonnaruwa and Trincomalee. These soils are found on ridge crests, upper slopes and mid-slopes. In the valleys there are poorly-drained gleyed soils that are low in humus. The northern dry zone has two other main soil types found to smaller extents. These overlie sedimentary rock. **Red-yellow latosols,** considered to be the most ancient in the country, occur in a band from Puttalam and Mannar to Mullaitivu and parts of Jaffna. Southern Jaffna and Mannar districts have in addition soils known as **grumusols**. These are poorly drained, grey-black clay soils, which are sticky when wet and harden and crack when dry.

In association with reddish brown earths are **non-calcic brown soils.** They are most common in Kurunegala, Puttalam, Ampara, Batticaloa and Moneragala areas. These soils occur wherever the parent rock is deficient in iron and magnesium, and are particularly vulnerable to erosion. On the Kandy plateau and the montane regions of the intermediate and wet zones are **reddish brown latosols** (little developed). Also in this area are small areas of very young brown soils known as **immature brown loams**.

The soils in the lowland wet zone are predominantly **latosols.** These are underlain by the Highland series rocks. Two major groups are recognized here; those rich in alumina (well developed) and those near the coast which are mostly lateritic (very well developed). The lateritic soils harden on exposure, and severely restrict root penetration and infiltration.

The coastal and freshwater wetlands are characterised by young and unstable soils. In the northern and southern dry zones, Hambantota, Jaffna, Mannar and Puttalam areas, are found **solodized solonetz** and **solonchanks** (saline soils). These soils, found along the coast, tidal flats and flood plains in association with other soil types, are rather brownish or yellowish, saline and alkaline.

Acid swamp and bog soils are more characteristic of the wet zone. These poorly drained soils rich in organic matter overlie alluvial deposits. They are found on the flood plains, tidal marshes and filled-up lagoons of the Colombo, Kalutara, Galle, and Matara areas.

Introduction

Regosols found all over the country are recent deposits of deep, whitish, excessively drained sands. They are characteristic of coastlines with beaches and dunes. **Alluvial** soils are found adjacent to fresh water streams, rivers and flood plains. They are continually formed during frequent inundations with new deposits of sand, silt and clay brought down by water from the uplands.

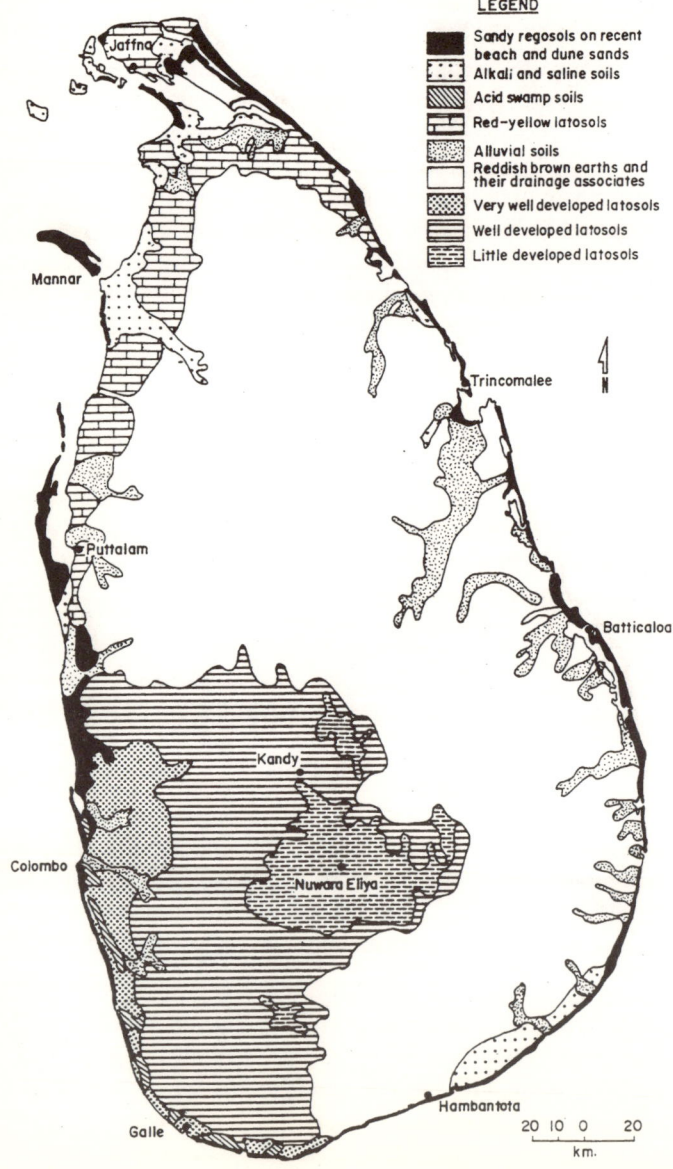

LEGEND

- Sandy regosols on recent beach and dune sands
- Alkali and saline soils
- Acid swamp soils
- Red-yellow latosols
- Alluvial soils
- Reddish brown earths and their drainage associates
- Very well developed latosols
- Well developed latosols
- Little developed latosols

20 10 0   20
km.

Figure 6. The major soil types

Vegetation types

Vegetation reflects the combined effect of topography, climate and soils. In Sri Lanka the natural vegetation is predominated by a diversity of forest types (Fig. 7). Only a small fraction of land is under non-tree-dominated vegetation. This is mainly grassland, and coastal and fresh water wetlands.

The most extensive type of forest in the island is the **dry mixed evergreen forest** found in the dry zone. Although deciduous species exist in these forests, their evergreen character is maintained by a few widespread species. Consequently these forests are also referred to as semi-evergreen forests. They receive rains only from the northeast monsoon. The strong seasonality in rainfall has prompted us to categorize them in this guide as **monsoonal forest**. In the arid zones of the northwest and southeast extremities of the country, **thorn scrub** predominates. This comprises small trees and thorny shrubs. Along the rivers, in both dry and arid zones, where there is no acute shortage of moisture, are impressive **riverine forests** (gallery forests). The tall, buttressed, spreading trees arch over waterways, and the cool, shady environment beneath them is in sharp contrast to the heat and dust away from the water.

In the intermediate zone, the vegetation gradually changes to **moist semi-evergreen forests.** Although these forests have a fair proportion of deciduous species, they are essentially evergreen. However, within the southeastern intermediate zone (Moneragala), mostly deciduous species constitute the forest canopy. Hence, these forests are more deciduous or semi-evergreen in character than those of the northern block in the Kurunegala area. In this guide, these forests have been lumped together as intermediate forests.

In the wet zone vegetation has been largely categorized by elevation with **wet-evergreen forests** or **rain forests** in the lowlands and hills, **lower montane forests** on the lower slopes of mountains between 1000 m and 1500 m, and **montane forests** above 1500 m. Although the wet zone is only a small area of 15,000 square km in extent, the combination of climate, topography and geological history has resulted in a diversity of species-rich associations in this zone, as compared to the vegetation types in the rest of the country. Consequently, at least nine floristic zones, based on species more or less restricted to each zone, have been recognized in this area (Fig. 8). The high proportion of endemic plant species (species found only in Sri Lanka), is another characteristic of these forests. These features of the Sri Lankan lowland rain forests and montane forests make them of critical importance in understanding the biogeography and floristic wealth of south and southeast Asia.

The non-forest vegetation types are mostly grasslands, found in small pockets in all climatic zones of the country. Most of these grasslands are secondary in origin and appear to be fire-maintained by humans. The dry zone also has **villu grasslands** which are associated with river flood plain systems such as those in the lower reaches of the Mahaveli. These seasonal marshes are inundated during rainy seasons. As the water recedes with the onset of the dry

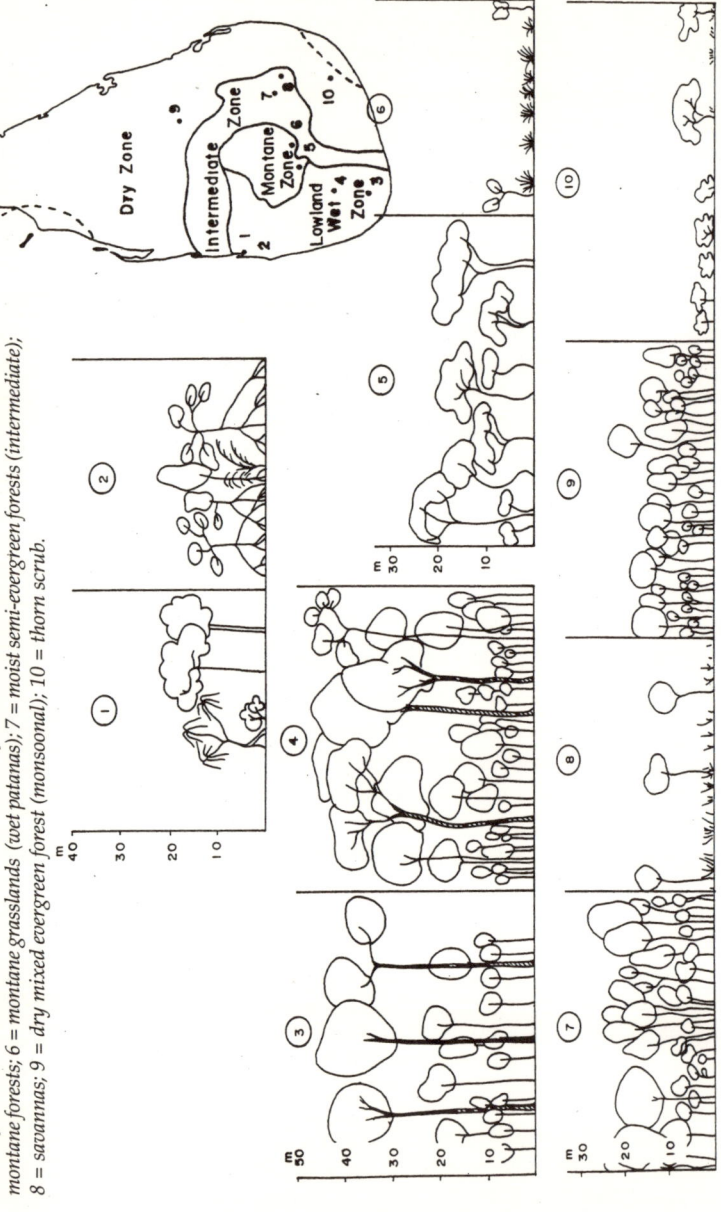

Figure 7. The climatic zones of Sri Lanka and the major natural vegetation types, indicated by numbers and vegetation profiles within each of these zones. 1 = coastal dune and beach scrub; 2 = mangroves; 3 = rain forests below 300 m elevation; 4 = hill rain forests between 300 and 1000 m elevation; 5 = montane forests; 6 = montane grasslands (wet patanas); 7 = moist semi-evergreen forests (intermediate); 8 = savannas; 9 = dry mixed evergreen forest (monsoonal); 10 = thorn scrub.

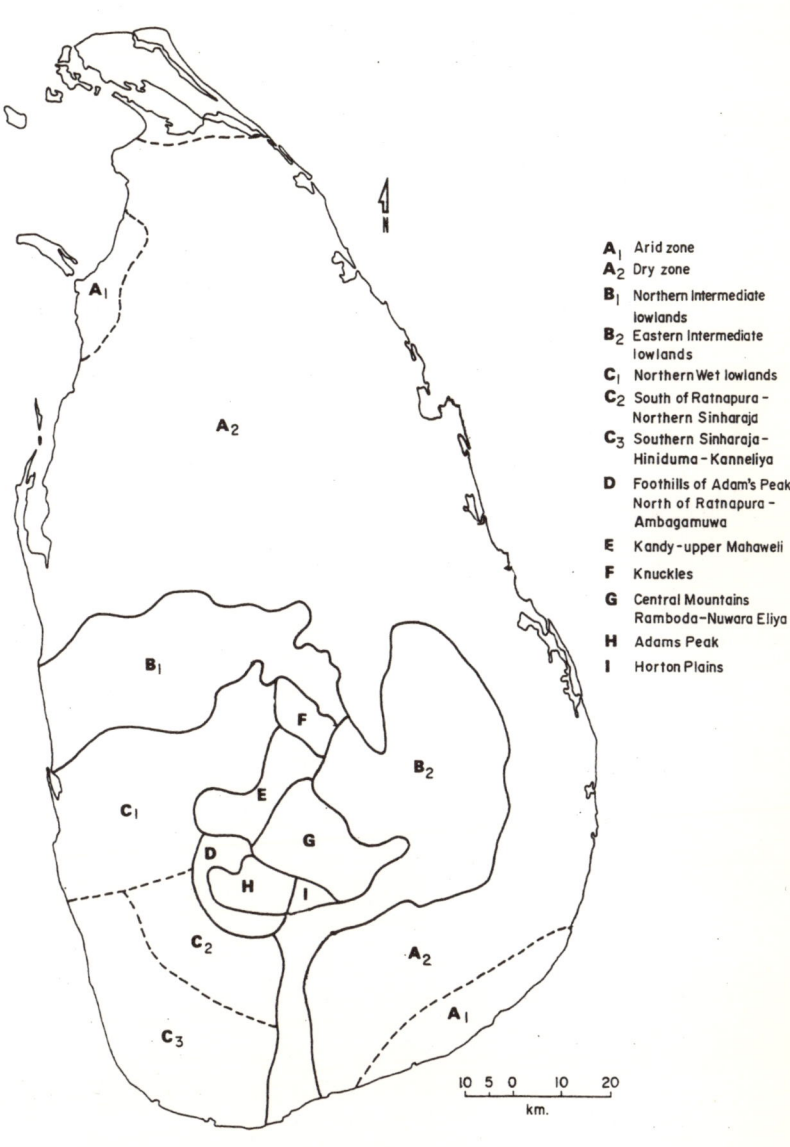

Figure 8. *The floristic zones*

A₁ Arid zone
A₂ Dry zone
B₁ Northern Intermediate lowlands
B₂ Eastern Intermediate lowlands
C₁ Northern Wet lowlands
C₂ South of Ratnapura – Northern Sinharaja
C₃ Southern Sinharaja – Hiniduma – Kanneliya
D   Foothills of Adam's Peak North of Ratnapura – Ambagamuwa
E   Kandy – upper Mahaweli
F   Knuckles
G   Central Mountains Ramboda – Nuwara Eliya
H   Adams Peak
I   Horton Plains

season, the land reverts to succulent grassland. On the eastern and southeastern side of the mountains in the Uva basin are the **savannas**. These are grasslands with a conspicuous tree component; the trees in them are scattered and often fire-tolerant. These savannas are thought to be maintained by fire following destruction of dry semi-evergreen forest.

There are fairly large tracts of grasslands in the Kandy, upper Mahaveli, and Ratnapura-Rakwana areas known as the **dry patanas**. These grasslands are of secondary origin and are maintained by annual burning which prevents their reversion to forest. In the wetter areas of the lowland wet zone are **fern** or **kekilla lands** which are of secondary origin as well. Here too, burning retains them in this state, preventing their being taken over by scrub. Perhaps the best known grasslands in the country are those of the highland plateaux: the Horton plains, Moon plains, Ambawela, and Nuwara Eliya. These picturesque montane grasslands are known as **wet patanas** and are interspersed with montane forest.

In the coastal areas **mangroves** and **salt marshes** colonize inundated bays, inlets and river estuaries with **scrub** vegetation invading sandy shores and dunes. Inland areas inundated by fresh water have **swamp** and **floodplain forests**. Most of these have been converted to paddy land.

Trends in landuse

Sri Lanka has been populated and cultivated for a long time. Although there is evidence of several prehistoric settlements, large scale settlements began after the 5th or 6th century BC. For 17 centuries thereafter, the dry zone was the major centre for economic activity (Fig. 9). The complex irrigation systems built during this time still continue to dominate life in these areas. The 3rd century AD saw the wet zone gradually colonised; especially Kegalle, the Kandy plateau, and the Kelani Ganga basin. By the 10th century a greater part of the lowland wet zone was inhabited. However, the hills above 600 m retained their forested landscape. With the decline of the dry zone civilization in the 13th century, the centre of activity shifted to the wet zone, where the last kingdoms of Gampola and Kandy were located. The early phase of Sri Lanka's history therefore suggests that nearly all the forests now in the dry zone are of secondary origin, having re-established over the last 7-8 centuries.

By the 16th century, Europeans had arrived in this country. The Portuguese occupied the maritime provinces and traded in spices and prize timbers. Later the Dutch replaced the Portuguese. In the early 19th century the British colonized the whole country. Much of the wet zone forest was cleared, initially for coffee, and then for tea and rubber.

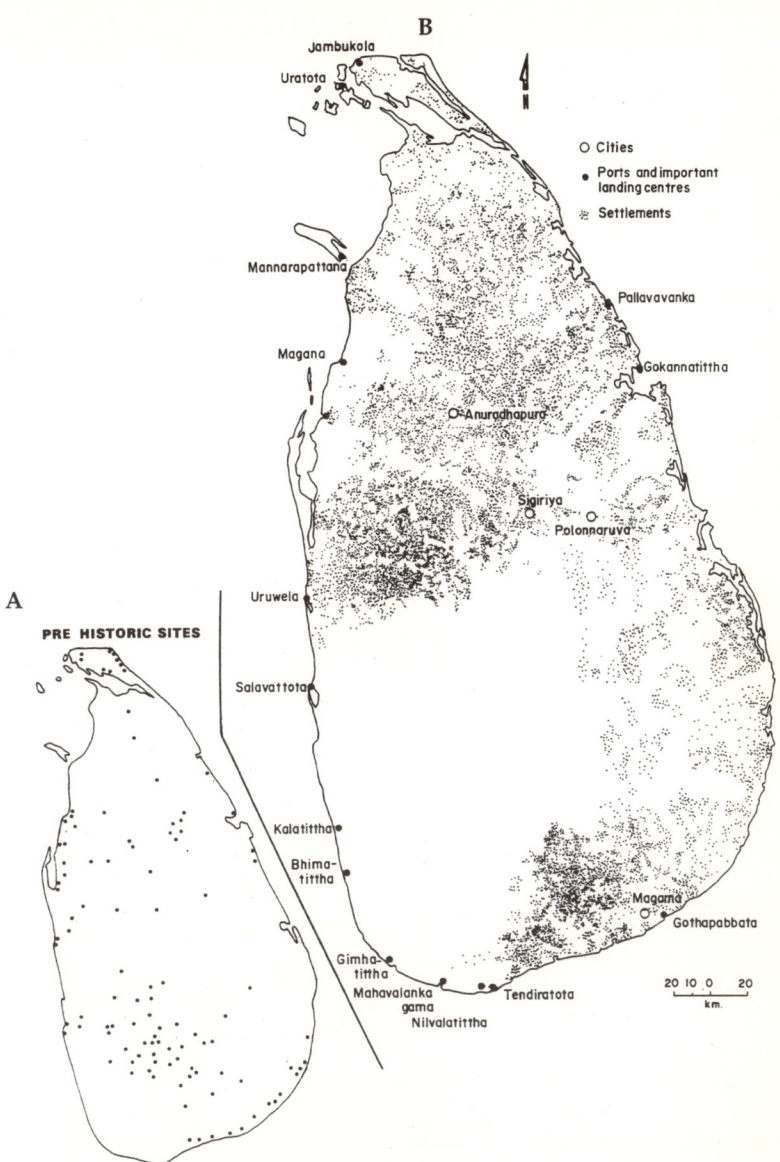

Figure 9. *Distribution of prehistoric sites (A) and the cities, ports and settlements before the 3rd century AD (B).*

BOX 2. The early landuse pattern usually included a small village settlement with paddy cultivation along the waterways. On higher ground tree and shrub crops were grown in home gardens. In the dry zone small reservoirs irrigated paddy fields. The catchment above the reservoirs and the area around the village was forest land subject to shifting cultivation and for gathering forest products not usually obtainable from the home gardens. Kandyan home-gardens are good examples of mixed-species tree cultivation. The plants in these gardens can comprise a spatial arrangement of vegetation that is complex. The paddy field and vegetable garden are single layered with fast growing arable crops. Vertical structure plays a more important role in the tree and shrub areas. The tallest trees of a garden form the framework within which are a heterogeneous mix of other subcanopy and groundstory plant species. Plant species can occupy a particular niche in both vertical and hori-zontal space. Plant species diversity is therefore high. In two gardens visited by Mendis *et al.* (1985) 105 tree species, 68 shrub species and 79 herb species were identified.

CANOPY–
Alstonia scholaris (timber)
Areca catechu (betel)
Artocarpus heterophyllus (jak)
Caryota urens (jaggery)
Cocos nucifera (coconut)
Mangifera indica (mango)

Gliricidia sepium

SUBCANOPY-
Myristica fragrans (nutmeg)
Garcinia mangostana (mangosteen)

UNDERSTORY-
Cinnamomum verum (cinnamon)
Coffea arabica (coffee)
Syzygium aromaticum (cloves)

VINES & SHRUBS-
Piper nigrum (pepper)
Elettaria cardamomum (cardamom)

Musa sp. (plantain & banana)
Carica papaya (pawpaw)

Cucurbitaceae (squash)
Solanaceae (tomatoes)
Leguminosae (pulses)

Lagerstroemia speciosa (ornamental)
Andropogon citratus (lemon grass)

Oryza sativa (rice)

Pandanus sp. (pandan)

SHRUB GARDEN

TREE GARDEN

VEGETABLE GARDEN

HOME SITE

ROAD

PADDY FIELD

FRINGE BOUNDARY

ALLUVIAL WATERWAYS

UPLANDS

Before the colonial period, the King was the traditional proprietor of land. The British followed this tradition with laws concerning 'crown land' and 'waste land'. All land without formal title was made the property of the government. The main reason for the British statutes was to obtain land for plantations. Food production took a secondary place and the importance of fallow for the rehabilitation of land was not appreciated.

From the beginning of the 20th century, and particularly since independence in 1948, the dry zone has been recolonized to augment food production. Ancient irrigation systems were restored and new ones established. By far the most important of these has been the Mahaveli development programme which was initially begun in 1970 and accelerated in 1977. It has provided irrigation for an additional 130,000 ha of dry zone land.

From a population that was 2 million in early colonial times, in the last fifty years it has burgeoned to 18 million. The wet zone and the Jaffna peninsula are the most densely populated areas with over 500 people per square kilometre. Since ancient times rice had been and continues to be the staple food crop, occupying 8% of the land area. Plantation crops, mostly introduced by the British, remain the basis of the island's economy and occupy 14% of the land. In the dry zone shifting cultivation is still practiced to a limited extent, while the wet zone provides most of the vegetables and fruits. Overall, about a quarter of the island's land is under cultivation (Fig. 10). Most cultivation is characterised by the coexistence of both traditional and modern agriculture. The latter involves improved varieties, especially of rice, and high inputs of fertilizer, pesticides and herbicides.

Many of the early food and medicinal plants were introduced from India - rice cultivars, medicinal plants of the ancient 'ayurvedic system', and temple trees such as the 'Bo tree'. During the British period many species of commercial importance as well as horticultural value were introduced. The three botanic gardens in the country, all of them over a hundred years old, have played a central role in these later introductions.

Since the end of the 19th century natural forests have dwindled to less than 25% of the total land area, due in large part to an array of encroachments (agricultural expansion, logging, urbanization, and shifting cultivation). Not more than 10% of this can now be classified as undisturbed. Fortunately though, at least half of the remaining natural forests in the country is now under some form of legal protection. Even so, a closer look reveals that the species-rich lowlands and mountains of the wet zone contain the least area in some form of natural forest. The proportions calculated in relation to each zone are about 8% in the lowland wet, 17% in the montane, and 47% in the dry (Fig. 11).

Since the extent of natural forest is very small, it is every person's repsonsibility to conserve it for future use. Current legislation concerning the protection of natural areas must be strengthened. Research should be accelerated to find out more about the status of plant species important to rural economies. Such species should be propagated and cultivated. Ancient rice cultivars and other valuable crop material should be conserved. Perhaps most importantly, cultivation practices should aim at innovative multi-species systems which are sustainable.

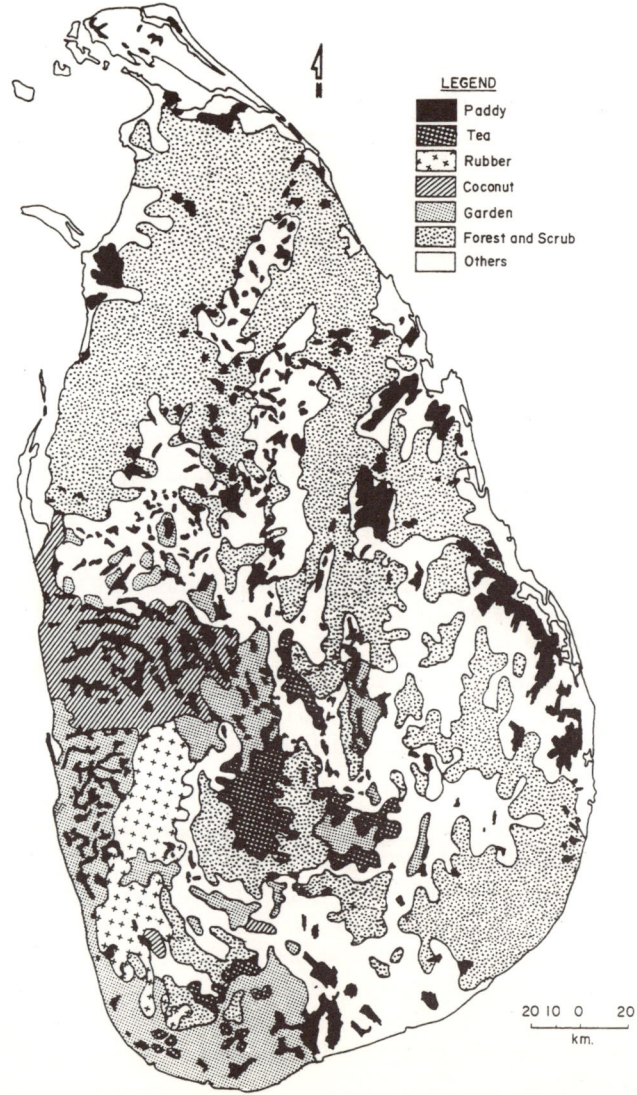

LEGEND

■ Paddy
▨ Tea
✶✶ Rubber
▨ Coconut
▨ Garden
▨ Forest and Scrub
☐ Others

20 10 0   20
km.

Figure 10. *Distribution of existing plantations, paddy, home gardens and remaining natural forest and scrub.*

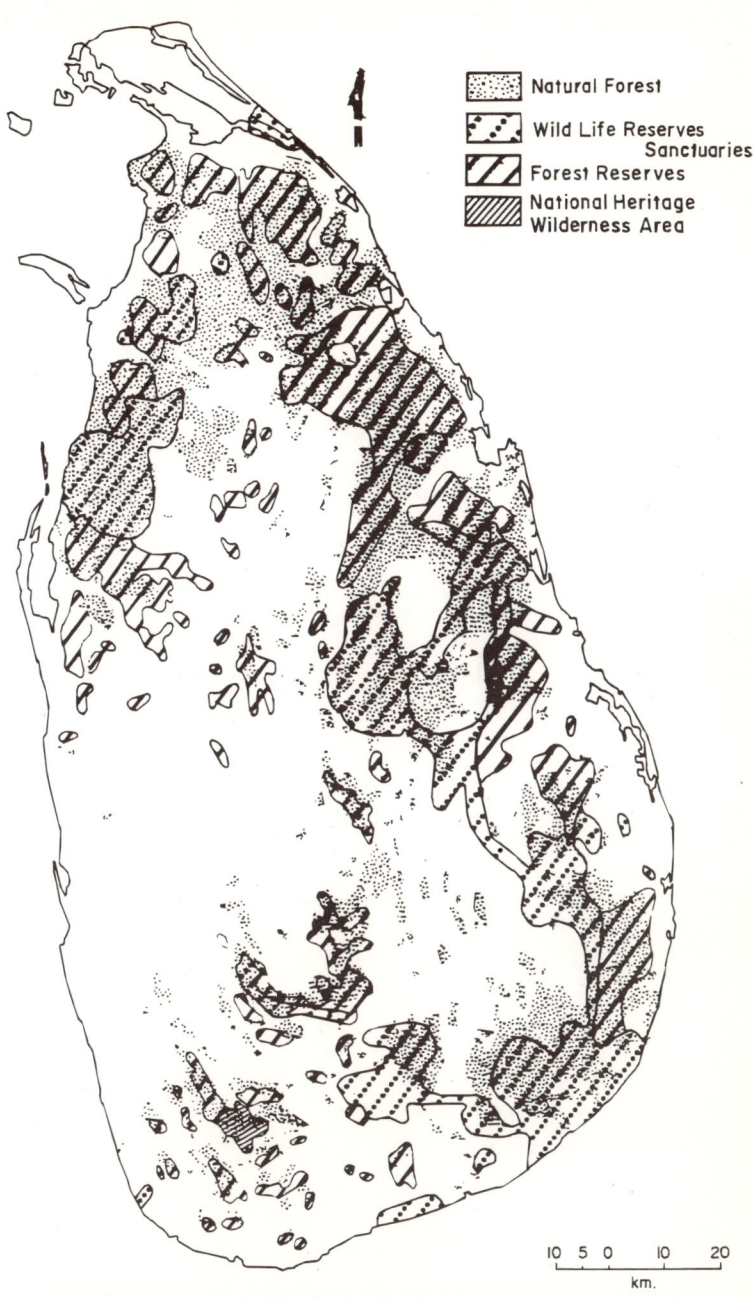

Figure 11. *Distribution of remaining natural forests and their administrative boundaries.*

Part II:

The field guide

An explanation of species and family descriptions

Family descriptions have been arranged within four groups of higher plants. In progressive order of the book they are: tree ferns and their allies, Gymnosperms, and Angiosperms (flowering plants) which have been further divided into monocotyledons and dicotyledons. Families belonging to each group have been arranged alphabetically. Family number codes are associated with the key. Species within each family have also been organized alphabetically at the generic level and numbered.

Families have been described to cater to individuals who have different levels of experience and interest. For those who are not interested in the details of floral parts, brief descriptions of each family are given immediately following the family name and code number that minimizes the use of technical language. For the more advanced user details of the flower parts, and in some cases fruit parts, that make different families taxonomically distinct from each other, are described and illustrated. This book has been designed in such a way that though floral information is useful, it is not necessary to understand it in order to fully utilize the guide and to identify tree and shrub species.

Species descriptions of trees and shrubs have been given and vegetative parts illustrated by simple black and white line drawings. Again an attempt has been made to minimize technical language. The description of each species has been organized with its scientific name and an associated **Headline**, below which further information is categorized under **Leaves**, **Trunk**, **Flowers**, **Fruits**, **Site**, and **Uses**. For the explanation of the technical terms within the species descriptions, the information has been arranged within each category as a series of questions that the user should become familar with when trying to identify a particular species under inspection. Definitions of technical terms have been listed separately in the **glossary** at the end.

The headline: AN EXAMPLE

Cryptocarya wightiana, gulumora (S), (T III:438), N, 30, tree

      A   B   C   D   E   F

A - **Scientific name** (always given in Latin).

B - **Common names** (E) = English; (S) = Sinhala; (T) = Tamil.

C - **Reference** (detailed description of a species by: author, volume, page no.).

        T = Trimen, H. (1893-1900) HANDBOOK TO THE FLORA OF CEYLON; Volumes I to V, and its updated Volume VI.

        DF = Dassanayake, M.D. and F.R. Fosberg. (1980 - ongoing). A REVISED HANDBOOK TO THE FLORA OF CEYLON; Volumes I to VIII. Amerind Publ., New Delhi.

D - **Origin** Endemic (E); Native (N); Introduced (I - country/region of origin) In certain cases where a species has recently been introduced into Sri Lanka a reference description has not been given.

E - **Tallest height** (in meters).

F - **Habit**: shrub (multi-stemmed, short woody plant), small tree (tree less than 5 m in height), tree.

Leaves: what is a leaf ?

Leaves are the organs of the plant which absorb atmospheric carbon dioxide and combine it with water taken in via the roots to produce sugars using light energy. This process is called photosynthesis. Sugars are simple carbohydrates that can be readily converted to other more complex substances. They can be thought of as the building blocks of plants. Different forms of carbohydrate are used for growth, reproduction, energy storage and physiological processes. The petioles (leaf stalks) support the leaf blade (lamina), uniting it with the main plant stem, and are responsible for leaf orientation. Veins facilitate leaf form, transport nutrients and water from the petiole to all areas of the leaf blade, and convey sugar synthesized in the leaf to other parts of the plant.

Leaf and stem parts

A - leaf tip
B - leaf margin
C - lamina
D - leaf base
E - petiole
F - stem
G - internode
H - node
I - tertiary veins
J - secondary veins
K - midrib

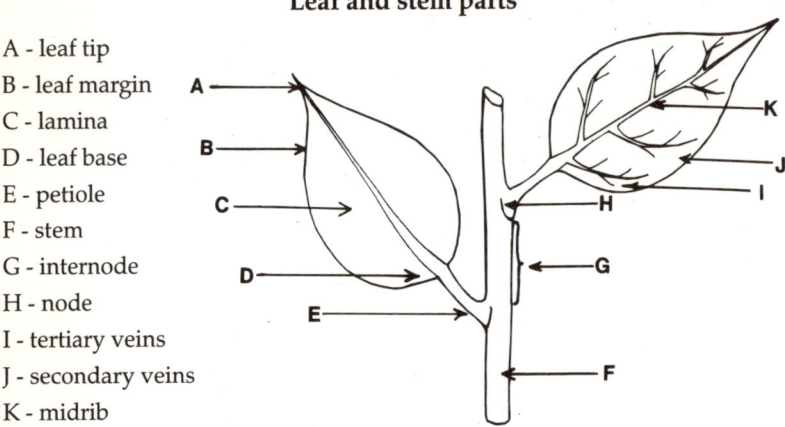

Questions about the leaf that you should ask yourself

1. **Is the leaf simple or compound? Compound leaves have several leaflets borne on an axis.**

LEAF TYPES

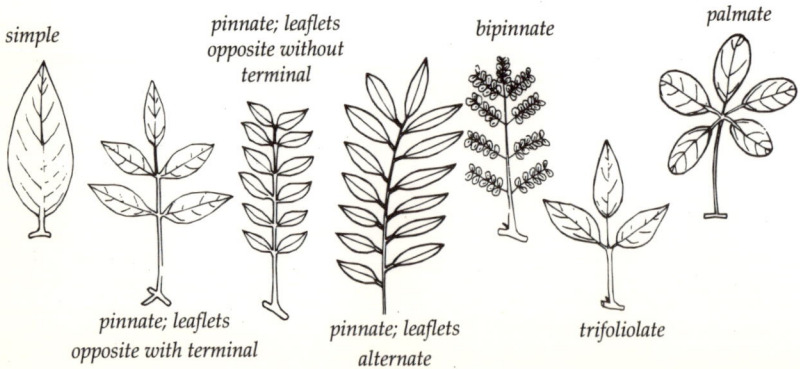

simple

pinnate; leaflets opposite without terminal

bipinnate

palmate

pinnate; leaflets opposite with terminal

pinnate; leaflets alternate

trifoliolate

2. How are the leaves arranged on the stem ?

LEAF ARRANGEMENT

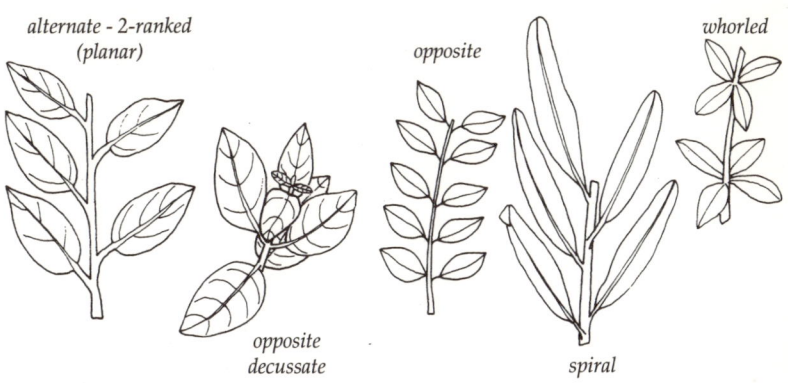

*alternate - 2-ranked
(planar)*

opposite

whorled

*opposite
decussate*

spiral

3. What shape is the leaf ?

LEAF SHAPE

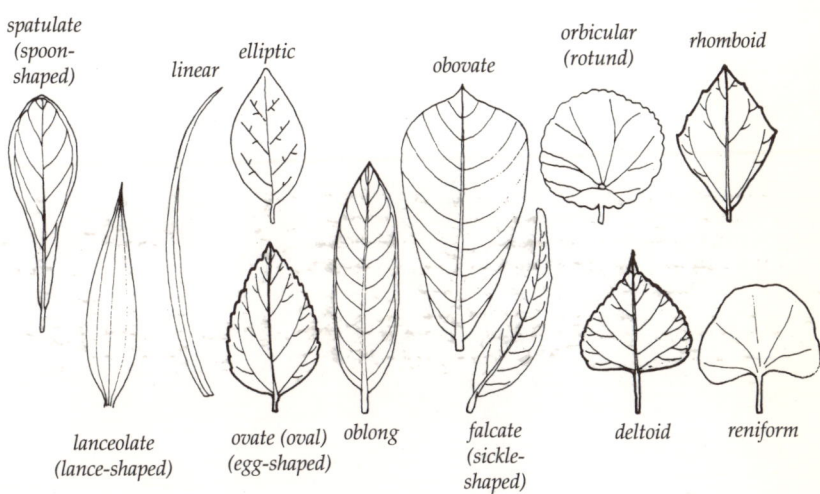

*spatulate
(spoon-
shaped)*

linear

elliptic

obovate

*orbicular
(rotund)*

rhomboid

*lanceolate
(lance-shaped)*

*ovate (oval)
(egg-shaped)*

oblong

*falcate
(sickle-
shaped)*

deltoid

reniform

4. What shape are the leaf apex and leaf base ?

LEAF ENDS

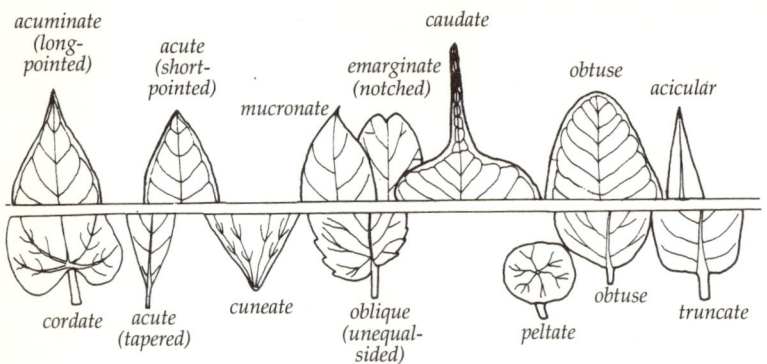

5. What kind of margin does the leaf have ?

LEAF MARGINS

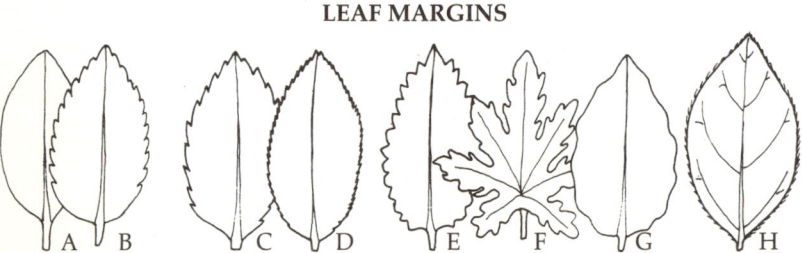

A-entire; B-crenate; C-serrate; D-serrulate; E-dentate; F-palmately lobed;
G-undulate; H-ciliate

6. What is the venation pattern on the leaf ? Veins are best seen on the leaf underside.

LEAF VENATION

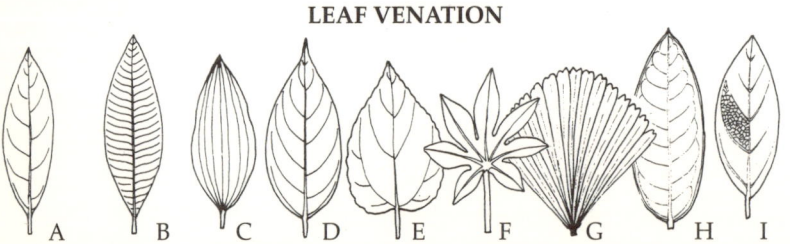

A,B - pinnate; C - parallel; D - strongly arched; E - basally 3-veined;
F - palmate; G - flabellate; H - intramarginal; I - reticulate.

7. What is the leaf surface like ?

LEAF SURFACE

Coriaceous - leather-like; **gland dotted -** with minute to clearly visible yellow dotted glands; **fulvous -** with reddish-yellow tawny hairs; **glabrous** - smooth, without hairs; **glaucous** - with waxy, whitish bloom; **hirsute** - with stiff, bristly hairs; **pubescent** - with soft, downy hairs; **puberulous** - very finely pubescent; **rugose** - wrinkled, with sunken veins; **scabrous** - with short, bristly hairs rough to the touch; **scaly** - covered with small plates, scurfy; **stellate** - star-shaped; **strigose** - with short, stiff hairs or scales; **tomentose** - with dense, matted pubescence; **villous** - with long, silky, straight hairs.

8. What is the petiole or rachis like ?

PETIOLE/RACHIS TYPES

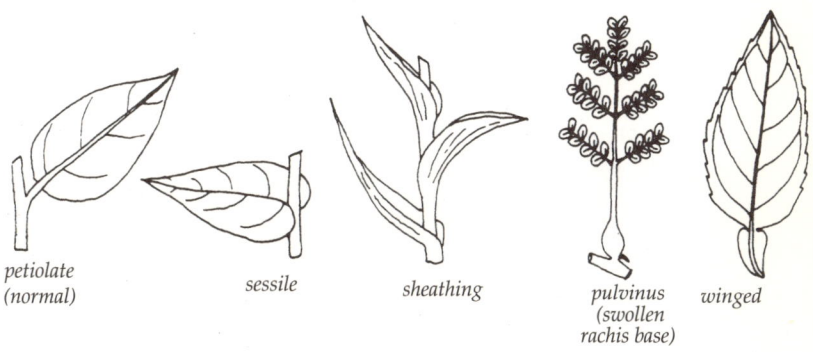

petiolate
(normal)   sessile   sheathing   pulvinus   winged
                                                                                    (swollen
                                                                                    rachis base)

9. Are stipules present ? If so, what are they like ? Are they deciduous leaving only a scar ? Stipules are small leaf-like structures near the base of the petiole.

STIPULE ARRANGEMENT

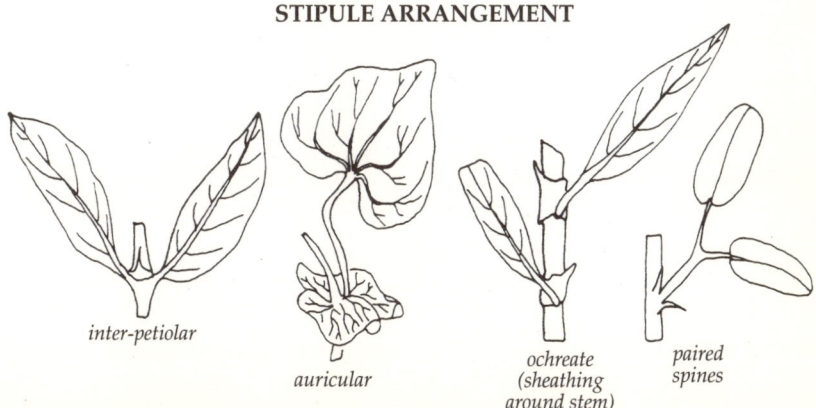

inter-petiolar

auricular   ochreate   paired
                             (sheathing   spines
                             around stem)

Trunk: stem - bark - branches - twigs - roots

The trunk and branches of a tree or shrub serve three functions, i) support the crown, ii) transport sugars, nutrients and water between roots and leaves and iii) store sugars for future use. The **bark (B)** is the outer dead skin of corky tissue that protects the tree from the elements. The texture, thickness and colour of the bark differ between species. The **inner bark (IB)** is the living portion that lies immediately beneath the dead bark.

The **wood (W)** is composed of **xylem** - tubular structures that conduct water and nutrients. The girth increase in a stem of a growing tree is largely due to new xylem formed from the cambium - a thin layer of actively dividing cells that exists as a sheath around the stem from which new tissues develop.

Sapwood (s) is the outer portion of the stem wood that has living xylem cells. **Heartwood (h)** is the central portion of the stem wood that has dead xylem. Heartwood is lignified and the xylem tubes are occluded with gums, resins and minerals. With age sapwood turns to heartwood. Heartwood is usually more resistant to rot than sapwood, it has a higher density and is also a stronger, more resilient structural timber.

Questions you should ask yourself

1. What is the colour and texture of the bark ?

BARK TEXTURES

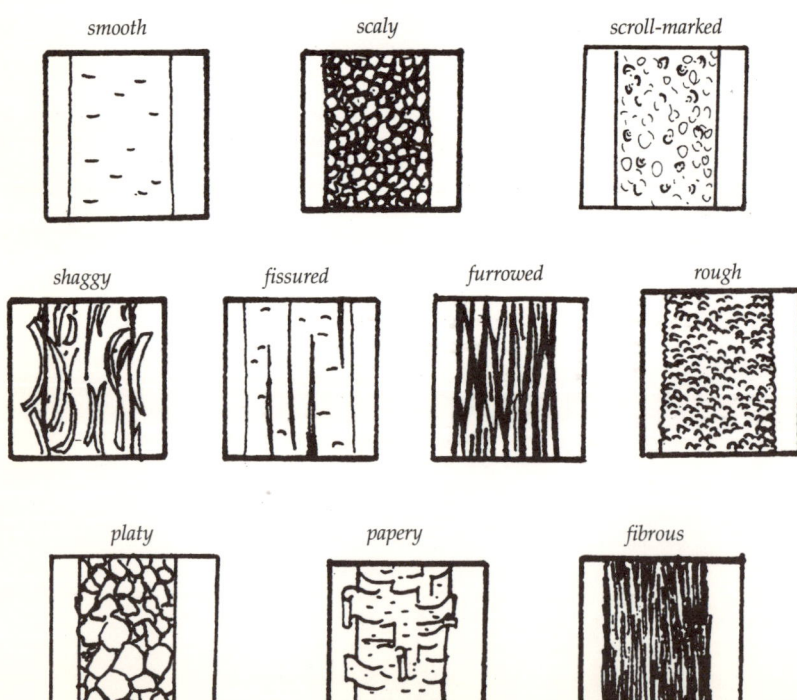

smooth   *scaly*   *scroll-marked*

shaggy   *fissured*   *furrowed*   *rough*

platy   *papery*   *fibrous*

2. How does the tree support itself ?

TRUNK

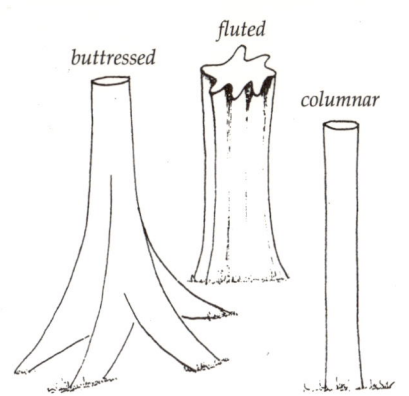

ROOTS

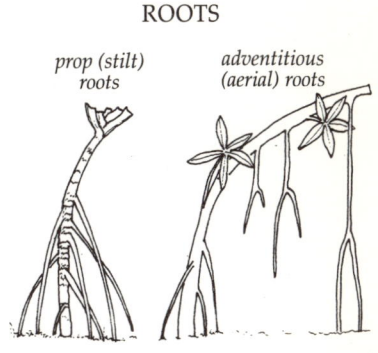

3. **What is the sap like ?** Sap is best observed by making a slash in the trunk of the tree and/or examining the freshly broken end of a twig or leaf. Is it a milky, yellow, or pink latex, or a clear sap that darkens on exposure ? Is it a mucilaginous, clear and watery, coloured and watery, or resinous sap ?

Flowers: what is a flower ?

The flower is the sexual reproductive structure of a plant. It comprises male parts - **androecium**; female parts - **gynoecium**; the flower base; and outer parts that protect the flower core from the elements and that attract animal visitation.

The androecium consists of a number of **stamens** that show a particular arrangement. They usually arise from the flower base, or are attached to the petals, and surround the gynoecium. The filament is the supporting stalk to which the pollen (male spore) producing **anther** is attached.

The gynoecium consists of a receptive end called the **stigma**, a stalk called the **style** which exposes the stigma to receiving pollen from other flowers, and a basal swelling called the **ovary**. The ovary is made of leaf-like structures called **carpels** which are usually joined together. The carpels enclose one or more chambers called **loculi** in which the **ovules** are found. The ovules are arranged in different ways within the loculi. For details refer to the figure illustrating **placentation**.

The floral base consists of the **receptacle**, upon which the androecium, gynoecium, and outer floral parts rest, and the stalk (**pedicel**), from which the flower obtains support and sustenance from the plant. Often plants that are pollinated by animals have sugar secretory glands (**nectaries**) at the base. Animals attracted to flowers unknowingly pollinate the stigma (s) with pollen carried on their bodies from previously visited flowers. Other forms of pollination can occur via wind and water.

User Notes

The outermost protective portions of the flower are called the **sepals**. Collectively they make up a whorl called the **calyx** and enclose the inner parts of the flower during development and before bloom. The next whorl of the flower comprises the **petals** that collectively form the **corolla**. They are usually large and showy to act as visual cues to animal visitors. Flowers that are wind and water pollinated have petals that are small and inconspicuous, or absent.

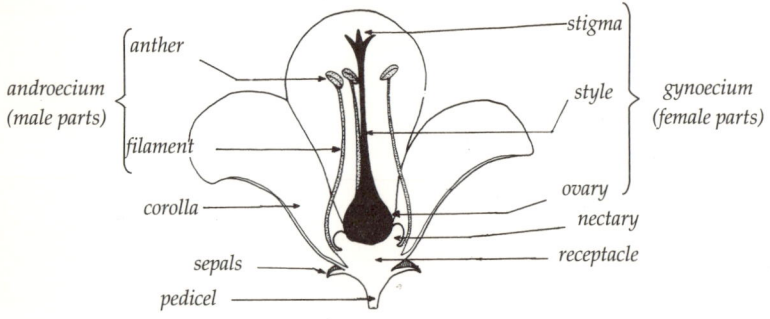

A diagram of the flower parts from a perfect flower.

Questions you should ask yourself

1. **What colour and shape is the corolla? Is it conspicuous? Is it actinomorphic** (radially symmetrical) **or zygomorphic** (asymmetrical)?

COROLLA FORMS

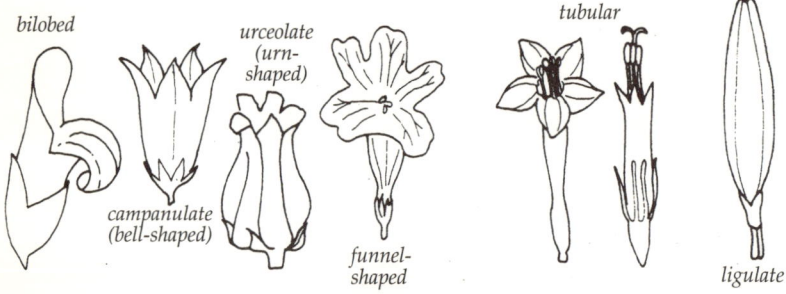

COROLLA SYMMETRY

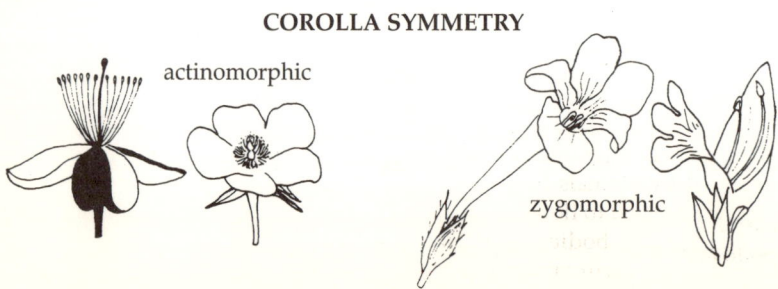

2. **How are the flowers arranged on the axis of the inflorescence ?** (An inflorescence is a flowering branch. Under **Flowers** in the species description I denotes inflorescence type)

INFLORESCENCE TYPES

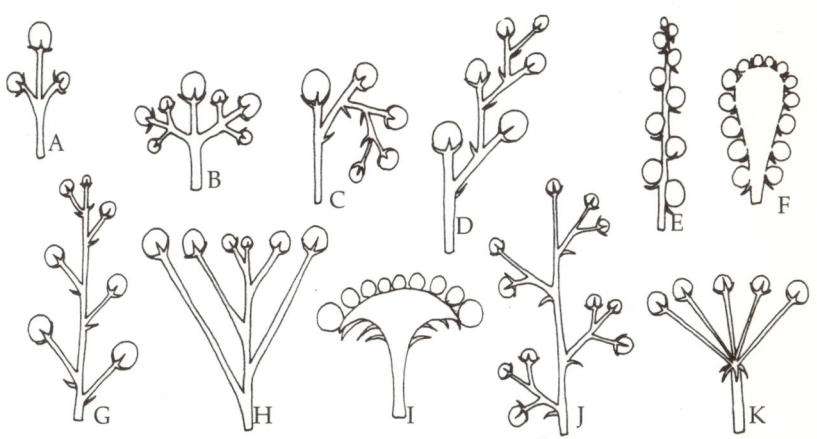

A - simple cyme; B - compound dichasial cyme; C - helicoid monochasial cyme; D - scorpioid monochasial cyme; E,F - spike; G - raceme; H - corymb; I - head or capitulum; J - compound raceme or panicle; K - umbel.

3. Many plants have evolved flowers that have only one functionally working sex, perhaps to avoid self-pollination. Flowers of different sex can occur on the same or different individuals. Some species have both, one sex and two-sex flowers on the same individual. The nature of the different flower sex combinations differs between families and genera and defines the breeding system of a plant. **What does the flower combination of the tree or shrub appear to be?**

BREEDING SYSTEMS

(♀=female, ♂=male, ⚥=bisexual flowers)

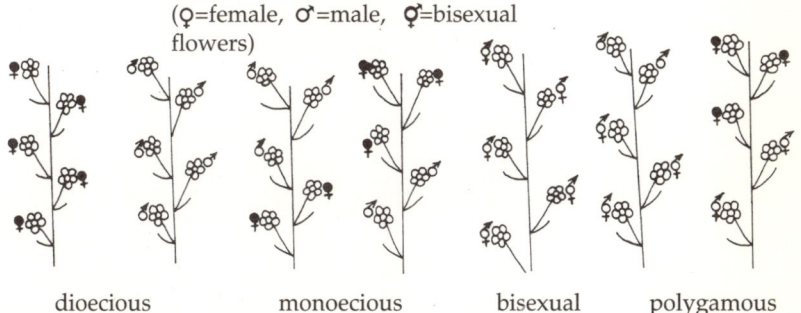

dioecious   monoecious   bisexual   polygamous

Dioecious - Unisexual. Male and female flowers borne on separate plants. **Monoecious** - Unisexual. Separate male and female flowers borne on the same plant. **Bisexual** - Male and female parts borne on the same flower. **Polygamous** - Unisexual and bisexual flowers borne on the same plant.

AN EXPLANATION OF THE FLORAL PARTS
FOR THE ADVANCED USER

The gynoecium and androecium comprise female and male parts of the flower respectively. In general the gynoecium is found at the centre of a flower and is surrounded by an androecium.

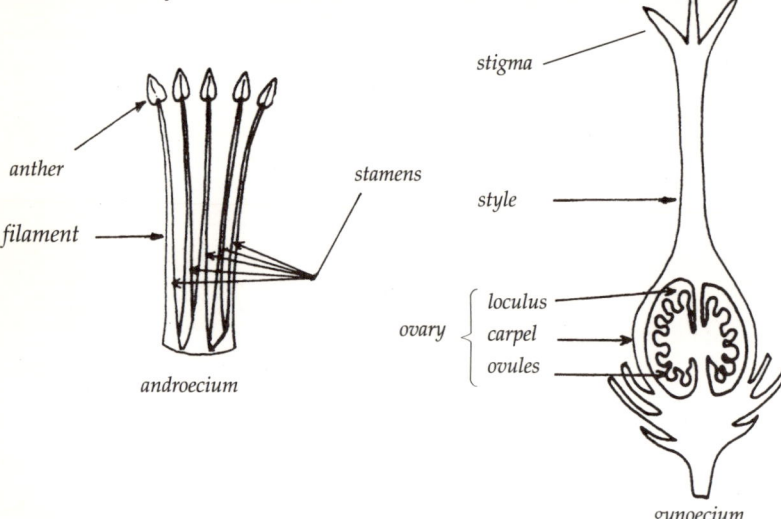

ANDROECIUM TYPES

1. The nature of stamen attachment and arrangement.

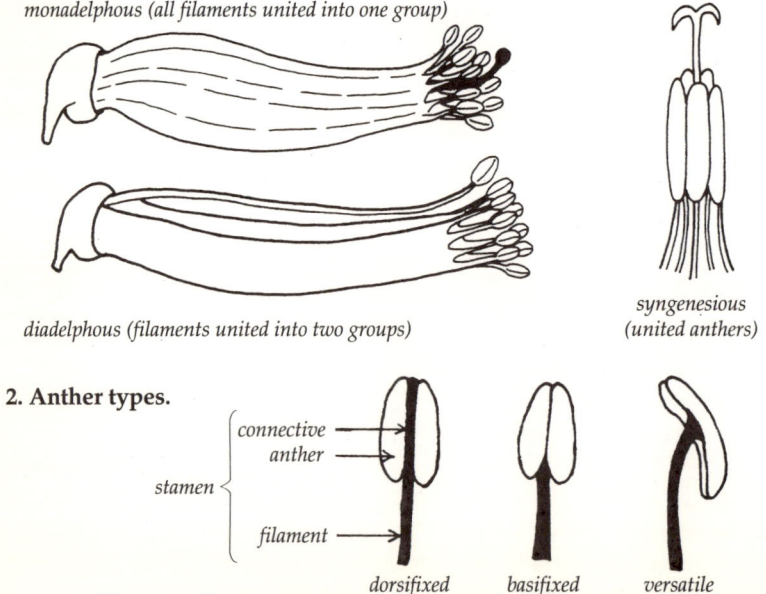

monadelphous (all filaments united into one group)

diadelphous (filaments united into two groups)

syngenesious (united anthers)

2. Anther types.

dorsifixed   *basifixed*   *versatile*

GYNOECIUM TYPES

1. **Calyx, corolla and androecium arrangement** in relation to the ovary, gives rise to different ovary types and flower types as shown below.

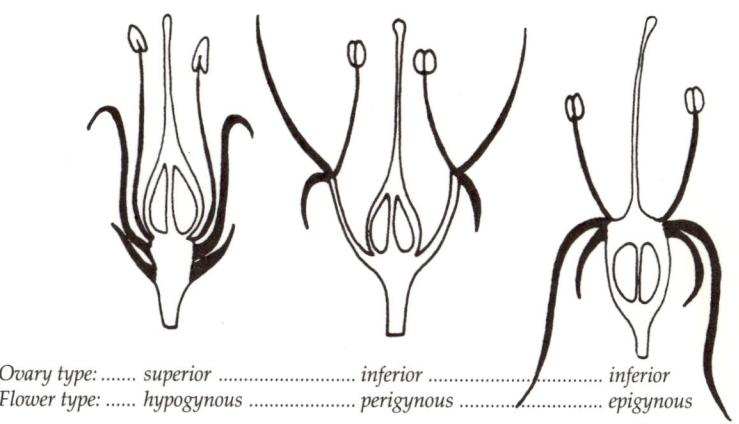

Ovary type: .... superior .... inferior .... inferior
Flower type: ...... hypogynous .... perigynous .... epigynous

2. Placentation (arrangement of ovules within the ovary).

Transverse (upper row) and longitudinal (lower row) sections of ovaries, showing different placentation types.

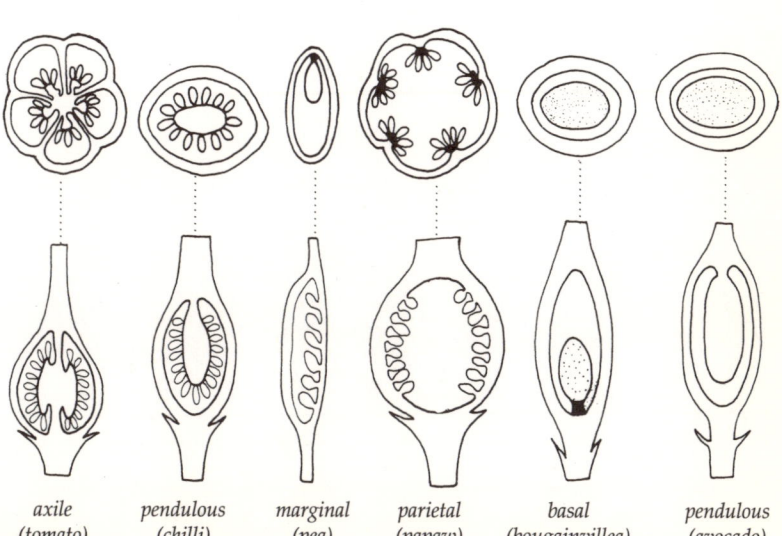

| *axile* | *pendulous* | *marginal* | *parietal* | *basal* | *pendulous* |
| *(tomato)* | *(chilli)* | *(pea)* | *(papaw)* | *(bougainvillea)* | *(avocado)* |

Fruit: what is a fruit ?

After fertilisation the ovary develops into the fruit and the ovule forms the seed. The seed consists of an embryo and a food store (cotyledons or endosperm). The fruit has a protective coat that can remain undifferentiated as a pericarp, or differentiate into a pericarp of several distinct layers (exocarp, mesocarp, and endocarp).

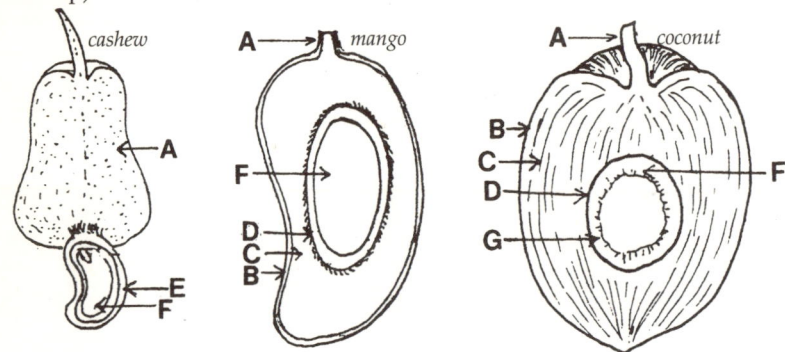

Parts of a simple fruit (one seeded): pedicel (A), fleshy in the cashew nut and normal in the mango and coconut. Pericarp or fruit wall uniform (E) or differentiated into an outer exocarp (B), middle mesocarp (C), and inner endocarp (D), in mango and coconut; embryo (F) and endosperm (G).

1. **What kind of fruit does it have ? Is it dry or fleshy ? Is it simple or composed of many together (composite) ?**

DRY FRUITS

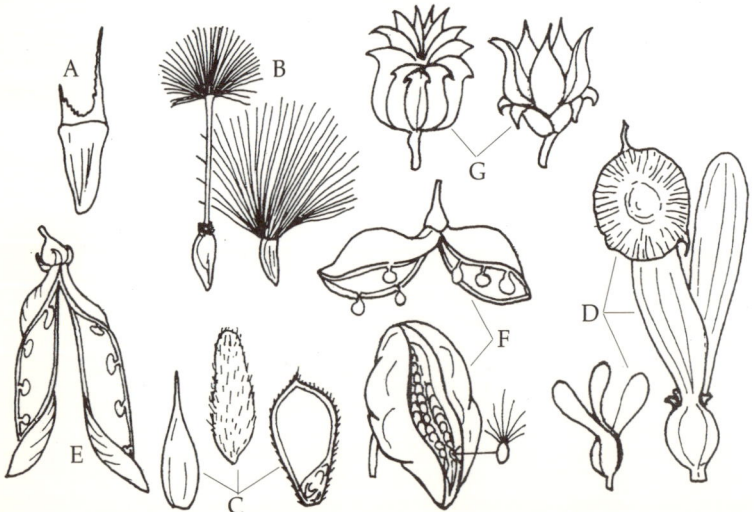

Indehiscent types: achene (A), cypselas (B), caryopses (C), samaras (D).
Dehiscent types: legume (E), follicles (F), and capsules (G).

FLESHY FRUITS

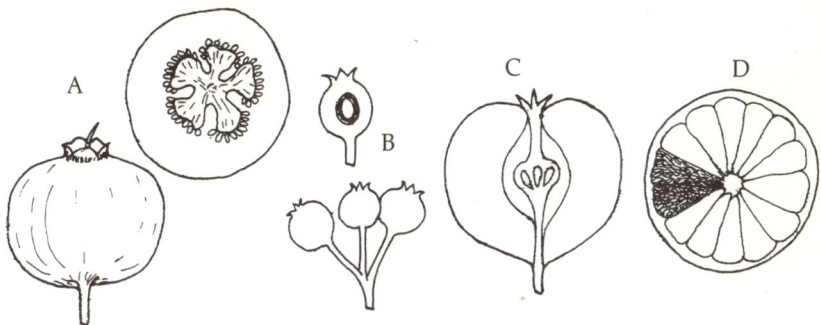

Different types of fleshy fruits where the pericarp is soft. Many seeded berry (A), single seeded berries (drupes) (B), pome (C) and hesperidium (D).

COMPOSITE FRUITS

Different types of composite fruits: fig (A); pineapple (B); and jak (C), all developed from an inflorescence.

AGGREGATE FRUITS

Different types of aggregate fruits that develop from a cluster of fruitlets formed from a single flower with an apocarpous ovary (ovary with free carpels). Each fruitlet develops from a single carpel. The fruitlets may be achenes, follicles (A), drupes (B) or berries (C); they may be separate from each other (A,B) or fused together (C).

Site: where is the tree or shrub growing ?

A **site** is a particular type of place where a plant lives. Each plant thrives under a set of climatic, soil and water conditions. This results in the establishment of a **plant community** composed of several species that are characteristic of that site. **Indicator** species are tree and shrub species that characterize a particular site and are not found elsewhere. **Generalists** are tree and shrub species that can grow on a wide range of different sites. A species that has a large population of individuals occupying a site is a **dominant** species.

Forest development is very slow because the life time of a tree is long. A forest may appear stable to the observer but in actual fact its species composition, growth and structure are changing. The term which describes these changes is called **succession**. It is a phenomenon where plant communities and the soils that they occupy pass through succeeding stages (**seral**) of composition and structure. The term **old growth** refers to the late seral stages of a forest's development.

Careful observation of the forest environment will inform you about where each tree or shrub species is found. Some questions that you should ask yourself include: what is the stature of the tree or shrub you are observing? Is it a large forest canopy tree or is it found in the forest understory? Can it be found in forest openings or is it on roadsides? Is the tree restricted to ridgetops, or swamps, or is it found along streams?

Answering some of these questions by observing where a plant grows will aid in its identification and provide you with an understanding of its site requirements.

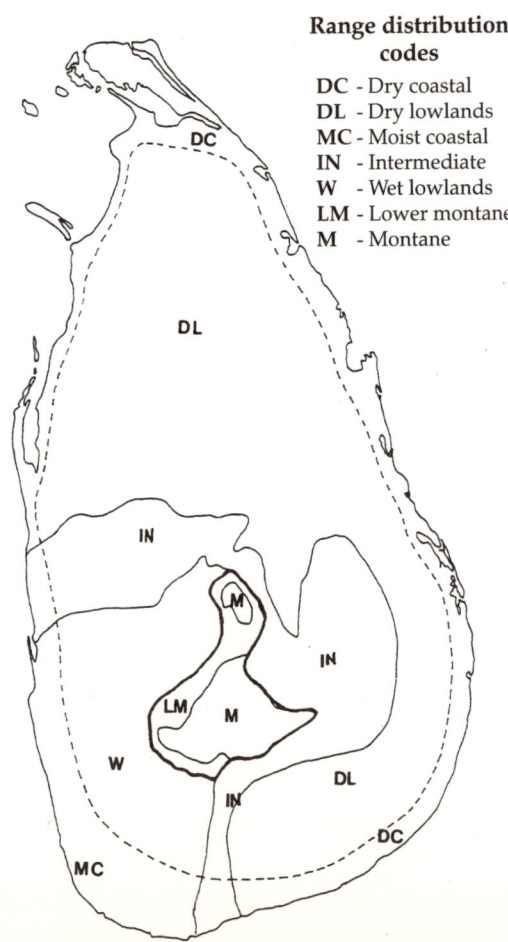

Range distribution codes

DC - Dry coastal
DL - Dry lowlands
MC - Moist coastal
IN - Intermediate
W - Wet lowlands
LM - Lower montane
M - Montane

GLOSSARY OF TERMS USED IN THIS GUIDE

Achene: a one-seeded dry indehiscent fruit from one carpel. Usually the seed is not fused to the fruit wall.

Acicular: needle-pointed.

Actinomorphic: a flower having radial symmetry; regular, more-or-less star-shaped.

Acumen: pointed end of a leaf.

Acuminate: leaf apex tapering gradually to a point (long-pointed).

Acute: sharply-pointed leaf apex (short-pointed).

Adnate: fusion of unlike parts, e.g. stamens with petals, or calyx with ovary.

Adpressed: flattened.

Adventitious roots: those arising from an unusual place other than at the base of the stem.

Aerial roots: those growing above ground, as in root climbers, trees with prop roots and strangling roots.

Aggregate fruit: fruit which is formed from an apocarpous ovary. Refers to an ovary of a single flower where the pistils are separate, each pistil gives rise to a fruit and they are attached to a single axis.

Alternate: leaves arranged singly at different heights on opposite sides of the stem in one plane.

Ampuliform: bladder-like.

Anatropous: inverted ovule in which opening (micropyle) is close to the funiculus (ovule stalk).

Androecium: collective term for male parts of the flower - each stamen is made up of filament, anther and connective.

Androgynophore: axis or stalk bearing androecium and gynoecium above the point of perianth (sepals and petals) attachment.

Annular: ring-like.

Anthesis: opening time of a flower or time when petals are unfolded.

Anterior: side of plant organ away from the central axis of the plant.

Apex: tip or free end of any plant part.

Apiculate: leaf apex abruptly and shortly pointed; or fruit terminated by a short sharp, flexible point.

Appendage: part of or expansion attached to the main organ.

Aril: appendage or outer covering of seed (growing out of the hilum or funiculus); sometimes appears as a pulpy covering.

Arillate: having an aril.

Aromatic: strong sweet or fragrant.

Asymmetrical: not divisible into equal halves.

Attenuate: gradually tapering to a point.

Auricle: an ear-shaped appendage, as in the top of a leaf-sheath in grasses.

Auriculate: with auricles.

Awn: a bristle-like part or appendage.

Axil: upper angle between twig or leaf and the stem, or between two veins.

Axile: placentation (arrangement of ovules within the ovary) where ovules are attached to the central axis formed by fusion of septa.

Axillary: floral buds (or branches) arising at the upper angle of the stem and leaf.

Bark: outer protective covering of stem consisting of inner live tissue (inner bark) and outer dead tissues (outer bark).

Basal: part of the flower where it is attached; or placentation (arrangement of ovules within the ovary) where the ovules are attached to the stalk side of the ovary.

Basifixed: attachment of anthers by their base to the filament.

Beaked: with long prominent point.

Berry: several seeded indehiscent fruit with a fleshy pericarp and without a hard layer surrounding the seeds. Usually formed from an ovary with axile placentation.

Bifarious: arranged in two rows in a single plane (two-rowed).

Bifid: divided in two or forked.

Bifurcate: forked.

Bilabiate: two lobed, often used to describe corolla and calyx.

Bilobed: two lobed.

Bipinnate: leaflets borne on secondary axes and not directly on the primary leaf axis.

Bisexual: both sexes represented in the same flower.

Blade: flattened or expanded part of the leaf (see lamina).

Bracteolate: bearing secondary bracts or bracteoles at the base of individual flowers. Each flower has two bracteoles.

Bracts: a much reduced or modified leaf in whose axil an inflorescence or flower arises.

Bristle: stiff, strong hair.

Bulbous: bulb-like.

Buttress: expanded base of stem, which provides additional support.

Caducous: falling off early or prematurely.

Callosity: leathery or hard thickening.

Calyptra: a hood or lid of an *Eucalyptus* fruit.

Calyx: outer whorl of floral parts (sepals) of a flower.

Cambium: the zone of cells that is responsible for forming the phloem and the xylem.

Campanulate: bell-shaped.

Capitate: in a head or aggregation of flowers into a very dense cluster.

Capitulum: dense inflorescence comprising an aggregate of usually sessile flowers.

Capsule: dry fruit formed from a syncarpous ovary, usually opening at maturity by one or more lines of dehiscence. The line of dehiscence may be along the septa (septicidal) or locules (loculicidal) or an apical pore (poricidal).

Carpellary: having a carpel or many carpels.

Caruncle: fleshy outgrowth of a seed.

Caryopsis: achene with testa (outer wall of seed) and pericarp (fruit wall) fused.

Caudate: with leaf apex drawn out into a long point (long-pointed).

Cauliflorous: condition where flowers are borne on the main stem.

Centrifugal: from the centre towards the periphery.

Ciliate: fringed with hairs.

Cinereous: ash-coloured; light grey.

Clavate: thickened towards free end or club-shaped.

Columnar: column-shaped stem, trunk, flower, or fruit, with little or no flaring at base.

Compound: made up of two or more similar units.

Connate: fusion or union of similar parts: sepal with sepal, petal with petal.

Connective: tissue between the two anther lobes in a stamen.

Cone: a coniferous fruit that consists of a series of overlapping bracts.

Conical: cone-like.

Contorted: mis-shaped, twisted.

Coppice shoot: a sprout from the stem base arising from dormant buds.

Cordate: leaf base heart-shaped.

Coriaceous: having a leathery texture.

Cork: soft, dead protective tissue.

Corm: short, solid, enlarged underground stem in which food is stored.

Corolla: inner whorl of floral segments (petals).

Corymb: a flat-topped racemose inflorescence, the main axis of which is elongated, but the pedicels of the older flowers are longer than those of the younger flowers.

Cotyledons: first leaf or leaves of a seed plant found in the seed. They may form the first photosynthetic leaves or remain below ground.

Crenate: scalloped leaf margin.

Crenulate: minutely scalloped leaf margin.

Culm: the stem of grasses and sedges, usually hollow except at the swollen nodes.

Cuneate: leaf base wedge-shaped or triangular with narrow end at point of attachment to petiole.

Cupular: having a cup-like structure.

Cupule: cup-like structure at the base of fruits, as in some palms, formed by the dry and enlarged floral envelopes.

Cyme: broad, more or less flat-topped, determinate flower cluster with central flowers opening first.

Cymose: sympodially (growth of axillary shoots after terminal has ceased growing) branched inflorescence.

Cypsela: inferior bicarpellary achene as in Compositae.

Deciduous: leaves, stipules or a plant part falling after a period of time; shedding.

Decurrent: a branching pattern that is not dominated by the main axis. Lateral branching is nearly or equally dominant to the main apical axis (see sympodial).

Decussate: opposite leaf arrangement where successive pairs of leaves are at right angles to each other, resulting in four rows of leaves along the stem.

Deflexed: to bend or turn downwards or aside.

Dehiscent: opening or splitting of a seed or fruit (or anther).

Deltoid: triangle-shaped.

Dentate: large saw-like toothed leaf margins.

Denticulate: finely dentate.

Dichasial: cymose inflorescence represented by a false dichotomy with the first flower to open situated between two lateral flowers.

Dichotomous: divided into two.

Diadelphous: stamens united into two sets.

Didynamous: four stamens, 2 long, 2 short.

Dilated: expanded or flattened.

Dimorphic: having two different forms.

Dioecious: having separate male and female flowers on different plants.

Disciform: disc-shaped or flat and circular.

Disc florets: tubular flowers in the centre of heads of most composites, as distinguished from the peripherally placed ray florets.

Distichous: leaves arranged singly at different heights on opposite sides of the stem in one plane (alternate 2-ranked).

Divaricate: spreading very far apart; forked.

Dorsifixed: stamen where the filament is attached to the back of the anther.

Dot glands: circular markings on the under surface indicating an internal secretory structure.

Drupe: a fleshy one-seeded indehiscent fruit with the seed enclosed in a stony endocarp.

Drupaceous: bearing drupes.

Elliptic: oval-shaped where the broadest part is equidistant from both ends.

Emarginate: the apex of a leaf that is indented (see notched).

Embryo: young plant in the early stages of development still within the seed.

Endocarp: inner-most layer of the pericarp, or fruit wall.

Endosperm: the tissue with stored food of a seed.

Entire: leaf margin continuous, without teeth or scallops.

Epigynous: having floral parts positioned above or on the ovary. Refers to floral parts when ovary is inferior.

Epipetalous: borne on or arising from the petals or the corolla.

Excurrent: the branching pattern of a tree that is dominated by a central apical axis. Often as in the case of palms this axis is unbranched (see monopodial).

Exocarp: outer-most layer of the pericarp, or fruit wall.

Extrorse: facing outward. Refers to dehiscence of anther when it opens on its outer length.

Exudate: mucus, gum or resinous substance that is exuded by the plant.

Falcate: sickle-shaped.

Fibrous: bark that is very finely fissured and stringy.

Filament: the stalk of the stamen.

Filiform: thread-like, long and slender.

Fissured: grooved or furrowed.

Flabellate: fan-shaped venation.

Flaccid: limp, floppy.

Flaky: broad parts of the bark shedding or coming off.

Fluted: fan-shaped base of a trunk, flower or fruit.

Follicle: dry dehiscent fruit formed of one carpel opening along one side (the dorsal side or that away from the axis).

Forked: divided.

Foveolate: having pits or shallow cavities.

Free central: refers to placentation where ovules are borne on a central axis that is free from the ovary wall.

Frond: leaf of a fern or palm.

Fruitlet: small fruit.

Funicular: of the stalk or funicle by which an ovule or seed is attached to the placenta of the ovary.

Fulvous: with deep yellow-coloured tawny hairs.

Furrowed: with narrow to wide parallel cracks, often describing a bark texture.

Fused: united.

Geniculate: bent like a knee.

Geocarpic: having fruits mature underground.

Geotropic: growth downwards towards the earth's centre.

Glabrous: having a smooth surface without hairs.

Gland: secretory structure.

Gland dotted: see dot gland.

Globose: globular or spherical.

Glomerate: in a dense compact cluster or clusters.

Glume: a membranous bract at the base of most grass inflorescences or spikelets.

Gum: various colloidal materials resulting from the breakdown of plant cells and exuding from the wounds of dicotyledonous angiosperms.

Gynoecium: collective term for the female part of the flower i.e., ovary, style and stigma.

Gynophore: stalk supporting the ovary. Elongation of the thalamus or receptacle between stamens and ovary.

Gynostegium: column formed by the fusion of stamens, styles and stigma.

Heartwood: dead tissue in the central portion of the stem, which does not conduct water.

Helicoid: in the shape of a helix, twisted.

Hesperidium: the fruit of an orange and other citrus plants.

Hilum: the scar or mark on the seed indicating its point of attachment to the placenta.

Hirsute: covered with rather rough or coarse hairs.

Hispid: covered with stiff bristle-like hairs.

Hypanthium: cup-like expansion of the receptacle formed from fusion of floral parts and androecium, on which the calyx, corolla and stamens are borne.

Hypogynous: borne beneath or below the gynoecium or ovary, referring to stamens, petals, sepals, and disk.

Imbricate: overlapping arrangement of sepals and petals like roof tiles.

Imparipinnate: pinnate with an odd terminal leaflet.

Indehiscent: fruits that do not release the seeds, instead the whole fruit is shed from the plant.

Indusium: an epidermal (outermost layer) outgrowth covering and protecting a sorus (collection of small stalked sporangia) as in ferns.

Inferior: position of ovary below the calyx, corolla and stamens.

Inflorescence: a flower cluster such as a spike, panicle or raceme.

Infructescence: a fruit cluster (originating as above).

Intercostals: tertiary veins set parallel to each other between two laterals.

Internode: part of the stem between successive nodes.

Intramarginal: vein running along the margin of a leaf.

Involucre: one or more whorls of small leaves or bracts arising close underneath a flower cluster.

Keel: boat-shaped structure formed by union of the two front petals in some Leguminosae; or ridge on the lower side of a leaf as in some Pandanaceae.

Lacerate: torn or cut into irregular lobes.

Laciniate: cut into narrow pointed lobes.

Lactiferous: plants that have lacticifers - cells that contain latex that are usually located in the phloem region.

Lamina: expanded portion of the leaf (see blade).

Lanceolate: lance-shaped or much longer than broad, wide at base and tapering to apex.

Laterals: main veins that arise from the midrib, or main secondary veins (see secondaries).

Latex: a milky, or clear, sometimes coloured juice or emulsion of diverse composition found in plants that have lacticifers - cells that contain latex that are usually located in the phloem region.

Leaflet: one small leaf-like part of a compound leaf.

Leaf scar: petiole scar or mark left on stem after a leaf has fallen.

Leaf sheath: lower part of leaf enveloping a stem or culm.

Leathery: (see coriaceous).

Legume: the pod of members of the Leguminosae. A fruit dehiscent along both sutures.

Lemma: lower of two membranous bracts enclosing a flower of grass. Upper is the palea.

Lenticel: ventilating pore on dead bark of stem which appears as a characteristic marking.

Ligule: a membranous outgrowth at the junction of blade and leaf sheath or petiole; or small scale on upper surface of leaf base; or a tongue-shaped corolla.

Ligulate: with a small expansion or outgrowth at the junction between the leaf blade and the leaf sheath.

Linear: long and narrow, the sides nearly parallel (see striate).

Lineolate: marked by fine lines or striate.

Lobe: any rounded projection of a floral part or leaf.

Glossary

Loculicidal: dehiscing more or less midway between the partitions or septa of the ovary.

Lodicules: scales at base of ovary in grasses, representing rudiments of the perianth.

Marginal: at or near the edge or border; placentation where ovules are borne on the side of an ovary made up of a single carpel e.g. beans.

Medullary rays: parenchyma arranged in ray form and that goes from the pith to the cortex.

Membranous: thin tissue, membrane-like.

Mesocarp: middle layer of the pericarp or fruit wall.

Midrib: main vein of the leaf that is continuous with the petiole.

Microsporophylls: leaves bearing the male reproductive structures in ferns and some gymnosperms.

Monadelphous: having stamens united by their filaments into one body.

Monochasial: cyme reduced to a single flower on each axis.

Monoecious: having separate male and female flowers on the same plant.

Monopodial: the branching pattern of a tree that is dominated by a central apical axis. Often as in the case of palms the axis is unbranched (see excurrent).

Mucilaginous: containing, or composed of mucilage. Widely occurring in plants, hard when dry but capable of absorbing water, swelling and becoming slimy.

Mucronate: with short, sharp pointed leaf apex.

Multiple: when applied to fruits, an aggregation or collection of fruits formed from the entire inflorescence. Fruits formed from each flower are attached to the swollen, fleshy inflorescence axis.

Nectariferous: producing nectar or having nectar secreting structures.

Nectary: nectar secreting gland, often appearing as a protuberance, scale or cap.

Node: part of the stem, usually thickened, where one or more leaves are attached.

Notched: a leaf apex that is indented (see emarginate).

Nut: an indehiscent, one-celled (locule or cavity), one-seeded fruit with a hard, woody pericarp.

Oblique: unequal-sided.

Oblong: longer than broad with nearly parallel sides.

Obovate: reverse of ovate. Leaf where basal part is narrower and broadest part is towards the apex.

Obtuse: blunt or rounded apex.

Ochrea: fused stipules sheathing around the stem at the node.

Operculum: lid, cap or covering flap; lid that protects the bud in *Eucalyptus* formed of sepals and/ or petals.

Opposite: when referring to leaves - two leaves per node, one on each side of stem.

Orbicular: circular or disc-shaped.

Ovary: part of a carpel or gynoecium containing the ovules.

Ovate: leaf where basal part is broader and narrower part is towards the apex.

Ovule: part of the ovary containing the egg. After fertilization ovule matures into seed.

Palea: the upper or inner of the bracts of the floret in grasses, often partly inclosed by the lemma.

Palmate: lobed, divided or ribbed as in the fingers of an out-stretched hand.

Panicle: a compound racemose inflorescence.

Papery: paper-like bark, peeling, thin and membranous.

Pappus: a circle or tuft of bristles, hairs or feathery processes modified from the calyx or corolla and persistent in the fruit.

Papilla: small fleshy projection.

Parietal: arrangement of ovules within the ovary where the ovules are borne on the walls of the uni-locular, one-to many-carpellary ovary.

Pedicel: stalk of a flower.

Peduncle: stalk of an inflorescence.

Peeling: narrow, long parts of the bark shedding or coming off.

Peltate: petiole attached to blade of leaf at its centre inside the margin (attached to the centre like the inside of an umbrella).

Pendant: drooping, hanging.

Pendulous: drooping, hanging downwards.

Perianth: corolla and calyx.

Pericarp: wall of ripened ovary or fruit wall.

Perigynous: borne or arising from around the ovary and not beneath or above it.

Persistent: remaining attached until mature or for an unusually long period.

Petal: one unit of the inner floral whorl or corolla.

Petiole: leaf stalk (petiolate - with a petiole).

Pilose: covered with soft hairs.

Pinna: single leaflet of compound leaf (plural, pinnae).

Pinnate: having leaflets on either side of a leaf axis.

Pisiform: pea-shaped.

Pistil: unit of gynoecium, comprising the ovary, style and stigma. When comprising only one carpel it is a simple pistil.

Pistillate: flowers with functional pistils and no functional stamens.

Placentation: arrangement of ovules within ovary (axile, marginal, parietal, free central, basal).

Planar: leaves arranged such that they are on either side of the stem and at one level.

Platy: a bark surface that comes off in small relatively circular pieces.

Plicate: folded like a fan.

Pneumatophores: roots that protrude vertically above the soil surface. Common in some mangroves.

Pod: a one-celled, one-to many-seeded fruit formed from a superior ovary with a single carpel and splitting into two valves at maturity.

Pollen: spores or grains borne by the anther, containing the male gamete.

Pollinia: mass of pollen organized as in orchids and asclepiads (milk weeds).

Polygamodioecious: species that are functionally dioecious, but have a few flowers of the opposite sex or a few bisexual flowers on all plants at flowering time.

Polygamous: bearing unisexual and bisexual flowers at the same time.

Pome: a fruit like an apple or pear, in which most of the edible part is the enlarged axis of the flower, rather than the ovary.

Pomiform: apple-shaped.

Prickles: rigid, stiff hairs.

Prop-roots: roots arising from the basal part of the stem providing additional support (see stilt roots).

Pseudostem: false stem made of sheathing bases.

Pubescent: covered with short soft hairs.

Puberulous: covered with fine hair, scarcely visible to the unaided eye.

Pulvinus: swollen petiole or rachis base which is responsive to vibrations and heat.

Punctate: with coloured or translucent dots or depressions or pits.

Pustular: blistery, usually minutely so.

Pyrene: a small stone of a drupe or similar fruit.

Pyriform: pear-shaped.

Quadrangular: four-sided.

Glossary

Raceme: flower cluster with the separate flowers attached by short equal stalks at equal distances along the central stem.

Rachis: axis bearing leaflets (or flowers). Petiole of a fern frond.

Rachilla: secondary axis bearing leaflets (or flowers) off of a rachis.

Ray florets: outer modified florets of some composites with an extended or strap-like part to the corolla.

Receptacle: the more or less elongated or enlarged end of flower axis or stalk on which the flower parts are borne.

Reduced: become smaller or degenerated.

Reflexed: bent over, directed back upon itself, recurved.

Reniform: kidney-shaped.

Resin: any of various substances with high molecular weight, including turpenes, which are found in mixture in plants. Often exuded from wounds and cuts (callus tissue), and by cells lining resin canals in the xylem, where it protects the plant from insects and fungi as it hardens to an amorphous vitreous solid.

Reticulate: patterned like a network.

Retuse: leaf apex rounded with a notch at the centre (see emarginate).

Revolute: margin rolled back towards under surface.

Rhizome: an elongated subterranean stem or branch; usually horizontal.

Rhomboid: diamond-shaped.

Rosette: a cluster of closely crowded radiating leaves arising from a very short stem near the surface of the ground.

Rotundate: round or circular.

Rough: uneven bark surface.

Rugose: wrinkled.

Sagittate: like an arrow head - triangular with basal lobes pointed downwards.

Salverform: flower shape when the fused corolla or petals form a slender tube basally and an abruptly expanded flat lip at the upper end.

Samara: a winged indehiscent fruit.

Sap: watery fluid in plants.

Sapwood: living outer part of the stem that conducts water and nutrients.

Sarcotesta: softer, fleshy outer portion of a testa.

Scabrid: somewhat rough.

Scabrous: rough or gritty to the touch with a covering of stiff hairs, scales, or points.

Scales: flat, small, plate-like structures usually found on the leaf under surface or on the bark.

Scalariform: ladder-shaped.

Scalloped: (see crenulate).

Scandent: climbing by stem, roots or tendrils.

Schizocarp: a dry fruit that is formed from an ovary with fused carpels which dehisces by splitting into two or more one seeded parts or mericarps.

Scorpioid: curled up at the end like a scopion tail.

Scroll-marked: hammer marked-like shallow depressions on the bark surface.

Scurfy: scaly skin; dried outer skin peeling off in scales. Scaly epidermal covering of some leaves.

Secondary: lateral veins off of the main midrib.

Sepals: separate part of the calyx, usually green and foliaceous (leaf-like).

Septa: a partition; referring especially to the partitions in a compound ovary.

Septicidal: dehiscing along the septa of the fruit.

Serrate: leaf margin toothed with teeth pointing forward.

Serrulate: minutely serrate.

Sessile: not stalked. Base of plant part directly resting on main axis (unpetiolate).

Shaggy: long, irregular pieces of bark that peel off.

Sheath: a protective covering. Lower part of leaf enveloping stem or culm.

Simple: not branched referring to stems; not compound, referring to leaves.

Sinuate: wavy margin.

Sori: clusters of sporangia on the back of fern fronds.

Spadix: racemose inflorescence with an elongated axis, sessile flowers, and an enveloping spathe.

Spathe: large, enveloping leaf-like structure, green or petaloid, protecting a flower cluster or spadix (spathaceous).

Spatulate: spoon-shaped, broad and rounded above the middle and tapering gradually to a narrow base.

Spicate: having the form of a spike.

Spike: an elongated inflorescence bearing sessile flowers.

Spikelet: a small spike; the unit of an inflorescence of grasses and sedges.

Spiny: thorny or prickly or with stiff, pointed projections.

Spiral: consecutive leaves arranged around the stem in all directions.

Sporangia: case, capsule, or cell in which spores are produced (e.g. megasporangia).

Spur: short, stubby branch with densely crowded leaves and leaf scars.

Spurred: having a tubular or sac-like projection from a petal or sepal, containing a nectar-secreting gland.

Stalk: a supporting structure as the peduncle (inflorescence stalk), pedicel (flower stalk), or petiole.

Stamen: unit of the androecium comprising a filament, anther lobes, and connective.

Staminate: flowers having stamens and no pistils - a male flower.

Staminode: sterile stamen, which may be reduced or expanded and petal-like.

Standard: upper and broad, more or less erect petal of some of the Leguminosae.

Stellate: star-like with radiating branches, or in star-like clusters.

Stem sheath: enveloping leaf base around the stem.

Sterile: lacks functional sex organs.

Stigma: receptive part of the pistil or gynoecium where pollen is received or trapped.

Stilt roots: (see prop roots) roots arising from basal part of stem and providing additional support.

Stipitate: borne on a stalk.

Stipule: appendage or leafy outgrowth at base of petiole.

Stoloniferous: capable of bearing many creeping stems or runners that develop rootlets and stem and ultimately a new individual.

Striated: (see linear) With fine longitudinal lines with channels or ridges.

Strigose: with sharp, adpressed straight hairs, stiff and often swollen at base.

Strobilus: synonymous with cone and best restricted to conifers.

Style: more or less elongated part between the ovary and stigma.

Subacute: nearly pointed leaf apex.

Subclavate: somewhat club-shaped.

Subtend: to be opposite or under.

Succulent: fleshy.

Superior: ovary free and separate from the calyx, usually lying above the attachment of the calyx to the receptacle.

Supra-axillary: above axils.

Sympodial: (see decurrent) A branching pattern that is not dominated by the main axis. Lateral branching is nearly or equally dominant to the main axis.

Syncarp: multiple or fleshy aggregate fruit.

Syngenesious: with united anthers.

Glossary

Tertiaries: small veins arising from the laterals.

Thryse: mixed inflorescence with main axis racemose, later axis cymose.

Tomentose: densely woolly or pubescent.

Transverse: lying across or between.

Trifoliate: three-leaved.

Trifoliolate: with three leaflets.

Tripinnate: compound leaf where leaflets are borne on the tertiary axess; divided pinnately three times.

Truncate: appearing as if cut off at the end; leaf base or apex nearly or quite straight across.

Tuber: a much thickened, usually short, subterranean stem, as in the potato, or root.

Tubercle: small rounded or knob-like extension from the stem or bark.

Tuberous: of the nature of a tuber.

Tubular: having the form of a tube.

Two-ranked: (see alternate, distichous, or planar).

Umbel: an umbrella-shaped inflorescence, in which the pedicels radiate from a common point at the summit of the peduncle.

Undulate: (see sinuate) leaf margins wavy.

Unifoliate: with one leaf.

Unisexual: bearing only one sex, either male or female.

Urceolate: urn-shaped.

Utricle: a small bladder-like structure; one seeded, usually indehiscent fruit.

Valvate: arrangement of petals and sepals where they meet without overlapping.

Variegated: marked with irregular patches of different colours.

Venation: arrangement of veins.

Verrucose: with minute warts or blunt projections.

Verticillate: with flowers arranged in whorls.

Versatile: attachment of filament to anthers near the middle, anther moving freely from side to side like a see-saw.

Vestigial: imperfectly developed or degenerate.

Villous: with long, silky, straight hairs.

Virgate: long, straight, slender, and rod-like.

Viviparous: germinating while still attached to the parent plant, like in some mangrove species.

Wart: firm or hard protuberance.

Whorled: with three or more leaves arising at a single node.

Wing: a flat expansion of petiole, fruit, or stem.

Woolly: having long, soft, more or less matted hairs.

Xylem: structures that conduct water and nutrients.

Zygomorphic: corollas which can be divided into equal halves in one plane only (see symmetrical).

Key to the identification of families
(based on vegetative characters)

Note. Numbers in parentheses denote family code number for species descriptions in the guide.

1 Plant a tree fern .... Cyatheaceae   (1)
   Plant not a tree fern 2

2 Reproductive structures are strobili,
   not flowers .... Gymnosperms (cycads & conifers)
   Reproductive structures are flowers ...... Angiosperms (flowering plants)

Key to gymnosperms (cycads & conifers)
1 Leaves large, pinnately compound, plants not resinous .... Cycadaceae   (2)
   Leaves not pinnately compound, plants resinous 2

2 Mature leaves needle-like and in bundles .... Pinaceae   (5)
   Mature leaves not needle-like 3

3 Branches whorled, growth strongly monopodial, leaves
   relatively large, often sharp-pointed, not scale-like .... Araucariaceae   (3)
   Branches not whorled, growth not strongly monopodial,
   leaves small, flat, scale-like .... Cupressaceae   (4)

Key to angiosperms (flowering plants)
1 Leaves usually parallel-veined .... Monocotyledons
   Leaves with tertiaries forming a network .... Dicotyledons

Key to monocotyledons
1 Leaf bases sheathing, forming a pseudostem .... Musaceae   (8)
   Pseudostem absent 2

2 Leaves large and compound, inflorescences large,
   enclosed in one or more large spathaceous bracts
   in bud, stems unbranched .... Palmae (Arecaceae[1])   (9)
   Leaves not compound, inflorescences not enclosed
   in such spathaceous bracts, stems usually branched 3

3 Leaves arranged in 3 rows, margins spinous,
   stilt roots present .... Pandanaceae   (10)
   Leaf arrangement and margins not so, stilt roots absent 4

4 Plants with many stems, conspicuous
   internodes and swollen nodes .... Gramineae (Poaceae[1])   (7)
   Internodes and nodes not so .... Agavaceae   (6)

---

[1]New family names.

Key to dicotyledons

Leaves reduced to scales, twigs green, slender,
jointed and finely ribbed or striate .... Casuarinaceae   (28)
Leaves normal, twigs not so—see Table 1: select appropriate group to continue

TABLE 1 - Key to dicotyledon groups

Leaf arrangement	Leaves simple		Leaves compound	
	Stipules Present	Stipules Absent	Stipules Present	Stipules Absent
Alternate or spiral	Group A	Group B	Group G	Group H
Opposite or Subopposite	Group C	Group D	Group I	Group J
Whorled	Group E	Group F	—	—

Key to Group A

Leaves simple, alternate or spiral , stipules present

1   Milky white sap present .... Moraceae, Euphorbiaceae
        (*Euphorbia, Exoecaria, Hevea, Manihot*)   (61; 46)
    Milky white sap absent 2

2   Reddish, orange or yellow sap present
    when stem is slashed .... Bixaceae   (21)
                                                        Cochlospermaceae   (32)

    Reddish, orange or yellow sap not present when stem is slashed 3

3   Stipules fused with petiole, enclosing young parts .... Dilleniaceae   (40)
    Stipules not fused with petiole 4

4   Leaves with distinct intramarginal veins .... Ochnaceae   (68)
    Leaves without intramarginal veins 5

5   Glandular pits at vein axils on leaf undersurface .... Elaeocarpaceae   (43)
    Glandular pits absent 6

6   Leaf margin serrate, dentate or crenate, not entire 7
    Leaf margin entire 16

7   Leaves 3-5 veined at base 8
    Leaves not 3-5 veined at base 9

8   Prickles present .... Rhamnaceae (*Zizyphus*)   (74)
    Prickles absent .... Rhamnaceae (*Colubrina*), Tiliaceae   (92)
                            (*Grewia, Muntingia*), Ulmaceae (*Celtis, Trema*),   (93)
                            Urticaceae (*Boehmeria, Debregeasia*)   (94)

9   Trunk and branchlets bearing spines .... Flacourtiaceae (*Flacourtia,*   (47)
                                        Scolopia), Rhamnaceae (*Scutia*)   (74)
     Trunk and branchlets unarmed .... 10 (see key to cluster 1)

Cluster 1—flower description

10   Stamens more than ten 11
     Stamens ten or less 14

11   Filaments free .... Flacourtiaceae (*Homalium, Hydnocarpus*)   · (47)
                                        Rosaceae (*Prunus, Pyrus*)   (76)

     Filaments united at base or along their length 12

12   Filaments united forming a tube .... Malvaceae (*Hibiscus*)   (57)
     Filaments united not forming a tube 13

13   Ovary inferior .... Lecythidaceae (*Careya*)   (52)
     Ovary superior .... Euphorbiaceae, Flacourtiaceae   (46; 47)

14   Stamens ten .... Sterculiaceae (*Helicteres, Pterospermum*)   (88)
     Stamens less than ten 15

15   Filaments free .... Urticaceae (*Villebrunea*)   (94)
     Filaments basally united ... Euphorbiaceae, Rhamnaceae (*Rhamnus*)   (46; 74)

16   Twigs with stipular thorns .... Capparidaceae (*Capparis*)   (26)
     Twigs without stipular thorns 17

17   Leaves heart-shaped .... Malvaceae (*Hibiscus, Malvaviscus*),   (57)
     .... Tiliaceae (*Berrya*)   (92)
     Leaves not heart-shaped 18

18   Leaves with golden yellow scales beneath .. Bombacaceae (*Cullenia*)   (22)
     Leaves without golden yellow scales 19

19   Lamina base or upper end of petiole with
     2 glandular dots .... Rosaceae (*Prunus*)   (76)
     Lamina and petiole without glandular dots 20

20   Bark resinous .... Dipterocarpaceae   (41)
     Bark not resinous .... 21 (see key to cluster 2)

Cluster 2—flower description

21   Stamens more than 15 22
     Stamens upto 15 25

22   Filaments free 23
     Filaments united basally 24

23   Carpels free.. Magnoliaceae   (56)
     Carpels united .... Flacourtiaceae (*Hydnocarpus*),   (47)
     .... Rosaceae (*Photinia*)   (76)

continued..

24	Ovary superior .... Euphorbiaceae, Flacourtiaceae (*Hydnocarpus*)	(46; 47)
	Ovary inferior .... Lecythidaceae (*Couroupita, Barringtonia*)	(52)
25	Filaments free .... Ulmaceae (*Gironniera, Holoptelia*)	(93)
	Filaments united at least partly 26	
26	Filaments united into a tube .... Erythroxylaceae	(45)
	....Sterculiaceae (*Heritiera, Sterculia*)	(88)
	Filaments united only basally into bundles ..... Celastraceae (*Bhesa*),	(29)
	Euphorbiaceae, Leguminosae (*Bauhinia, Crotalaria*)	(46; 53)

Key to Group B
Leaves simple, alternate or spiral, stipules absent

1	Milky or reddish orange watery sap present in stem 2	
	Such sap absent in stem 3	
2	Sap milky white .... Apocynaceae (*Cerbera, Plumeria,*	(14)
	Thevetia), Caricaceae, Sapotaceae	(27; 83)
	Sap reddish orange .... Myristicaceae	(63)
3	Leaf undersurface gland dotted (seen against the light	
	or with a hand lens), and when crushed smells citrus-like .... Rutaceae	(78)
	Leaf without gland dots and no citrus-like smell 4	
4	Young twigs with leaves of 2 sizes; large normal leaves alternating	
	with smaller deciduous leaves .... Rhizophoraceae (*Anisophyllea*)	(75)
	Leaves of one size only 5	
5	Leaves 3 veined at base 6	
	Leaves not 3 veined at base 7	
6	Leaves equal-sided at base .... Boraginaceae (*Cordia*)	(23)
	Buxaceae (*Sarcococca*), Hernandiaceae (*Gyrocarpus*)	(25; 49)
	Leaves not equal-sided at base .... Aristolochiaceae (*Apama*)	(17)
7	Leaves without petiole .... Goodeniaceae	(48)
	Leaves with petiole 8	
8	Leaf base distinctly cordate .... Datiscaceae (*Tetrameles*)	(38)
	Leaf base not cordate 9	
9	Leaf under surface white or densely hairy .... Compositae (*Vernonia*)	(34)
	Ericaceae (*Rhododendron*), Lauraceae (*Cryptocarya*)	(44; 51)
	Leaf under surface not white nor densely hairy 10	
10	Leaf when crushed smells of wintergreen .... Ericaceae (*Gaultheria*)	(44)
	Leaf without a wintergreen smell 11	
11	Leaf arrangement on stem planar .... Annonaceae	(13)
	Leaf arrangement not planar 12	

12 Leaf lamina with gland dots, dashes or glandular hairs ...... Myrsinaceae   (64)
    Leaf lamina without gland dots, dashes or glandular hairs 13

13 Irritant exudate present .... Anacardiaceae   (12)
    Irritant exudate absent 14

14 Leaf less than 3 cm long .... Aquifoliaceae (*Ilex*), Ebenaceae   (15; 42)
    Leaf more than 3 cm long .... 15 (see key to cluster 3)

Cluster 3—flower description

15 Ovary superior................. Combretaceae, Cornaceae, Symplocaceae (33; 36; 89)
    Ovary inferior 16

16 Corolla absent .... Daphniphyllaceae   (37)
    Corolla present.. 17

17 Corolla zygomorphic .... Sabiaceae, Sapindaceae   (79; 82)
    Corolla actinomorphic 18

18 Hypanthium present.. Thymelaeaceae   (91)
    Hypanthium absent 19

19 Stamens epipetalous ....Ebenaceae, Solanaceae   (42; 85)
    Stamens not epipetalous 20

20 Stamens more than 15 .... Theaceae   (90)
    Stamens less than 15 21

21 Anthers opening by valves .... Lauraceae   (51)
    Anthers not opening by valves 22

22 Disk or nectaries present ... Olacaceae, Sapindaceae, Simaroubaceae (70; 82;84)
    Disk or nectaries absent .... Icacinaceae, Pittosporaceae   (50; 71)

Key to Group C
Leaves simple, opposite or subopposite, stipules present

1   Plants growing only in mangrove habitats
    and showing vivipary .... Rhizophoraceae   (75)
    Plants showing no vivipary 2

2   Leaves with serrated margin .... Celastraceae (*Elaeodendron, Kokoona*),   (29)
                                                                    Chloranthaceae   (30)
    Leaves with entire margin .... Loganiaceae, Rubiaceae (*Gaertnera*)   (54; 77)

Key to Group D
Leaves simple, opposite or subopposite, stipules absent

1   Leaf margins serrate, dentate, crenate
    or ciliate, not entire .... Apocynaceae   (14)
                                                          Verbenaceae (some species)   (95)
    Leaf margin entire 2

Key

2   Milky, creamy, yellowish or pinkish sap present 3
     Milky, creamy, yellowish or pinkish sap absent 5

3   Depression at petiole base protecting
     young buds .... Clusiaceae (*Clusia, Garcinia*)   (31)
     Depression at petiole base absent 4

4   Secondary veins closely parallel
     to each other .... Clusiaceae (*Calophyllum, Mesua*)   (31)

     Secondary veins not parallel to each other .... Apocynaceae (*Carissa,*   (14)
                              Ervatamia, Pagiantha, Walidda, Wrightia),
                                        Asclepiadaceae (*Calotropis*)   (18)

5   Leaves with intramarginal vein 6
     Leaves without intramarginal vein 7

6   Undersurface of leaves gland-dotted
     (seen against the light or with a hand lens) .... Myrtaceae   (65)
     Undersurface of leaves not gland-dotted .... Melastomataceae   (58)

7   Stem spiny or leaves with prickly hairs .... Acanthaceae (*Barleria*),   (11)
                                    Nyctaginaceae, Punicaceae   (66; 73)
     Stem not spiny and leaves without prickly hairs 8

8   Young twigs quadrangular .... Lythraceae (*Lagerstroemia*),   (55)
           Melastomataceae (*Memecylon*), Sonneratiaceae, Verbenaceae   (58; 86; 95)
     Young twigs not quadrangular 9

9   Negatively geotropic roots present .... Avicenniaceae   (19)
     Negatively geotropic roots absent 10

10   Leaf arrangement planar .... Monimiaceae   (60)
     Leaf arrangement not planar .... Lauraceae (*Bielschmiedia,*   (51)
     .... *Cinnamomum*), Loganiaceae (*Fagraea, Strychnos*), Oleaceae,   (54; 70)
     .... Salvadoraceae, Thymelaeaceae   (80; 91)

Key to Group E
Leaves simple, whorled, stipules present .... see Rubiaceae   (77)

Key to Group F
Leaves simple, whorled, stipules absent .... see Apocynaceae (*Nerium*),   (14)
                          Lauraceae (*Actinodaphne, Alseodaphne, Neolitsea*)   (51)

Key to Group G
Leaves compound, alternate, stipules present

1   Leaflets deciduous, rachis and petiole
     modified to thorns .... Leguminosae (*Parkinsonia*)   (53)
     Leaflets not deciduous, thorns absent 2

2   Leaves trifoliolate .... Leguminosae (*Butea, Desmodium,*   (53)
                                        Flemingia, Crotalaria)

Leaves not trifoliolate but bipinnate, once pinnate or palmate 3

3   Leaves palmate with 4-9 leaflets 4
     Leaves not palmate 5

4   Stipule adnate to petiole .... Araliaceae (*Schefflera*)   (16)
     Stipule not adnate to petiole .... Bombacaceae (*Adansonia,*   (22)
                                   Bombax, Ceiba), Sterculiaceae (*Sterculia*)   (88)

5   Leaves bipinnate .... Leguminosae (*Caesalpinia, Cassia,*   (53)
                                                   Delonix, Peltophorum)
     Leaves once pinnate 6

6   Leaflets alternate .... Leguminosae (*Pericopsis, Pterocarpus, Sophora*)   (53)
     Leaflets paired 7

7   Leaves with a terminal leaflet .... Leguminosae (*Gliricidia, Pericopsis,*   (53)
                                        Pongamia), Burseraceae (*Canarium*)   (24)
     Leaves without a terminal leaflet .... Leguminosae (*Cassia, Humboldtia,*   (53)
                                        Saraca, Sesbania,Tamarindus)

Key to Group H
Leaves compound, alternate, stipules absent

1   Leaves tripinnate .... Moringaceae   (62)
     Leaves not tripinnate 2

2   Leaves bipinnate .... Meliaceae (*Melia*)   (59)
     Leaves not bipinnate 3

3   Leaves trifoliolate .... Capparidaceae (*Crateva*), Meliaceae (*Walsura*),   (26; 59)
     .... Sapindaceae (*Allophylus*)   (82)
     Leaves not trifoliolate 4

4   Leaves pinnate, leaflets alternate .... Meliaceae (*Dysoxylum*),   (59)
                              Proteaceae (*Grevillea*), Sapindaceae (*Harpullia*)   (72; 82)
     Leaves pinnate, leaflets paired 5

5   Leaves without a terminal leaflet .... Meliaceae (*Cedrela, Chukrasia,*   (59)
          Swietenia), Sapindaceae (*Nephelium, Pometia, Sapindus, Schleichera*),   (82)
                                                                         Simaroubaceae   (84)
     Leaves with a terminal leaflet 6

6   Leaflet margin serrate .... Anacardiaceae (*Spondias*), Meliaceae (*Aglaia,*   (12; 59)
                                              Azadirachta, Cipadessa)
     Leaflet margin entire .... Anacardiaceae (*Lannea*), Connaraceae   (12; 35)

Key to Group I
Leaves compound, opposite or subopposite, stipules present

     Leaves pinnate with terminal leaflet .... Staphyleaceae (*Turpinia*)   (87)
     Leaves palmately compound; 3-7 leaflets .... Verbenaceae (*Vitex*)   (95)

Key to Group J
Leaves compound, opposite or subopposite, stipules absent

1   Leaf undersurface gland dotted (seen against the light or
    with a hand lens), and when crushed smells citrus-like .... Rutaceae   (78)
    Leaf without gland dots and no citrus-like smell 2

2   Leaves palmately compound .... Bignoniaceae (*Tabebuia*)   (20)
    Leaves not palmately compound 3

3   Leaves bipinnate or tripinnate .... Bignoniaceae (*Jacaranda, Oroxylum*)   (20)
    Leaves not bipinnate nor tripinnate .... Bignoniaceae (*Dolichandrone,*   (20)
                                            Spathodea, Stereospermum)

1. CYATHEACEAE

FAMILY DESCRIPTION - **Habit**: tree ferns. Trunks mostly unbranched, covered with hairs or scales. **Leaves**: spiral, generally large, bi-, tri-, quadri-pinnate, rarely simple. Stalk densely covered with hairs or scales or spiny at the base. **Sori** (groups of sporangia): on the under surface of leaflets. Indusium present or absent.

1. *Cyathea crinita*, tree fern (E)/ginihota (S), N, 5, small tree.
    Leaves: very large, feathery, **tri- or quadri-pinnate**, crowded at end of stem, coriaceous, scaly or hairy beneath, margins minutely wavy-toothed.
    Trunk: slender, **unbranched**, covered with many adventitious roots; B-**scaly, blackish brown**, with large, oval leaf scars.
    Sori: small, brown, on veins beneath; **indusium absent**.
    Site: forest understory, gaps and fringes; along water courses and marshy sites; M.
    Uses: ornamental.

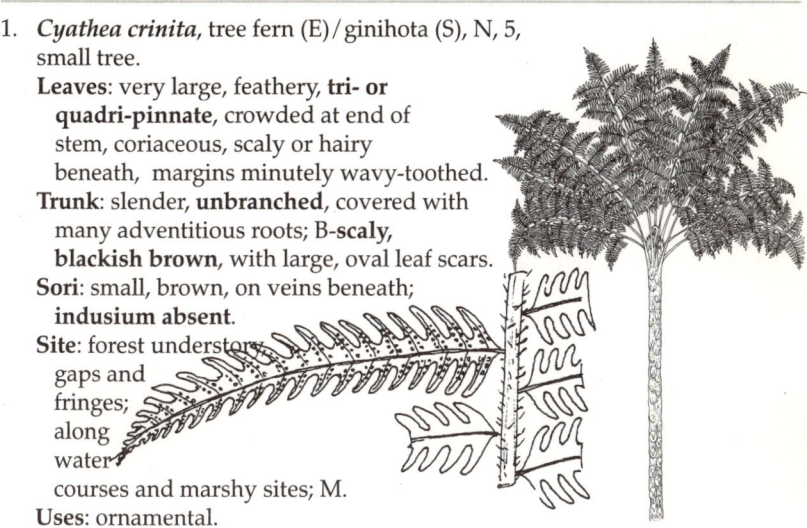

2. *Cyathea gigantea*, tree fern (E)/ginihota (S), N, 5, small tree.
    Leaves: very large, feathery, **bi- or tri-pinnate**, crowded at the end of stem; stalk purple to black, **glabrous** or with a **few scales** at base.
    Trunk: slender, unbranched, covered with many adventitious roots; B-scaly, brown with large, oval leaf scars.
    Sori: small, brown, on veins beneath; **indusium absent**.
    Site: forest understory, gaps and fringes; along water courses; M.
    Uses: ornamental.

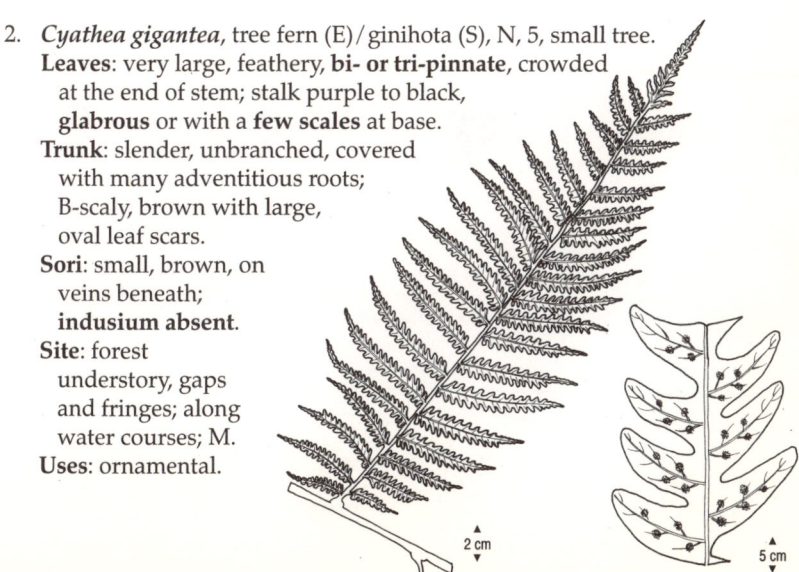

2 cm

5 cm

CYATHEACEAE

3. *Cyathea hookeri*, tree fern (E)/
   ginihota (S), E, 2, small tree.
   Leaves: feathery, **pinnate**, crowded at
   the end of stem, margins minutely
   wavy-toothed.
   Trunk: slender, **unbranched**; B-**scaly,
   brown,** large oval leaf scars.
   Sori: small, brown, on veins beneath;
   indusium present.
   Site: forest understory, gaps and fringes;
   along water courses; hybridises with
   C. sinuata; M, W.
   Uses: ornamental.

4. *Cyathea sinuata*, tree fern (E)/
   ginihota (S), E, 2, small tree.
   Leaves: **simple**, crowded at
   end of stem, margins
   minutely wavy-toothed.
   Trunk: slender, unbranched;
   B-scaly, brown, large oval leaf
   scars.
   Sori: small, brown, on veins
   beneath.
   Site: forest understory; in
   colonies along water courses;
   hybridises with *C. hookeri*;
   M, W.
   Uses: ornamental.

5. *Cyathea walkerae*, tree fern (E)/
   ginihota (S), N, 5, small tree.
   Leaves: very large, feathery,
   tri- or quadri-pinnate, crowded at
   end of stem, coriaceous, margins
   minutely wavy-toothed.
   Trunk: slender, **unbranched**;
   B-scaly, brown, large oval leaf
   scars.
   Sori: small, brown, on veins
   beneath; **indusium present**.
   Site: forest understory, gaps
   and fringes; along water
   courses and in marshy
   sites; W.
   Uses: ornamental.

2. CYCADACEAE

FAMILY DESCRIPTION - Habit: small, palm-like trees or shrubs. Stem thick, tuberous and partly underground, columnar and rarely branched. **Leaves**: pinnate, spiral, crowded at the stem apices. Leaf bases persist on trunk. **Cones**: unisexual, variable in size, plants dioecious. **Fruits**: drupe-like, sometimes brightly coloured.

REPRODUCTIVE PARTS - Male cones: microsporophylls spirally arranged, with abundant microsporangia. Female cones: megasporophylls usually in a cone (not in *Cycas*), and leaf-like, toothed or lobed. Megasporangia or ovules are large, marginal, naked.

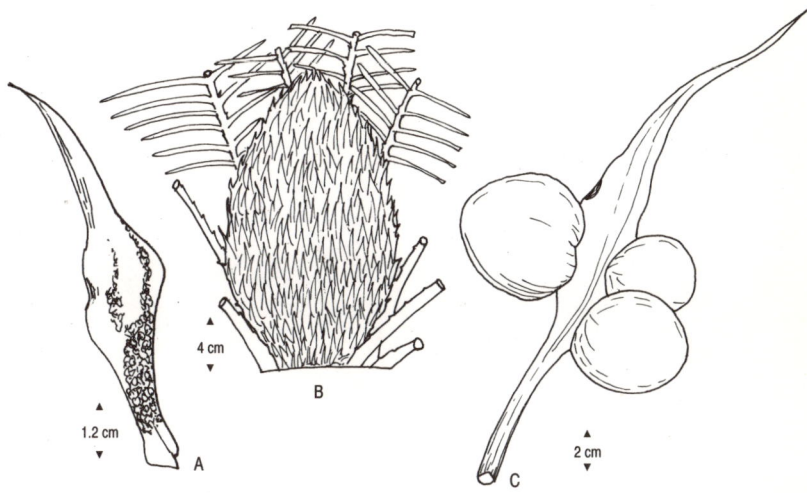

Key: a single microsporophyll (A), male cone (B) and megasporophyll (C) of *Cycas circinalis*.

1. ***Cycas circinalis***, madu (S), (T IV:121), N, 3, small tree.
   Leaves: pinnate; leaflets linear-lanceolate, pointed to sickle-shaped apex; stalk with **short deflexed spines** at base.
   Trunk: single or forked; B-glabrous, brown.
   Cones: male shortly stalked, cylindrical, ovoid; megasporophylls deltoid-obovate, spiny-tipped, buff tomentose; seed large, pale reddish yellow, ovoid.
   Site: savannahs, home gardens; IN.
   Uses: leaves, seeds-medicinal, vegetable; ornamental.

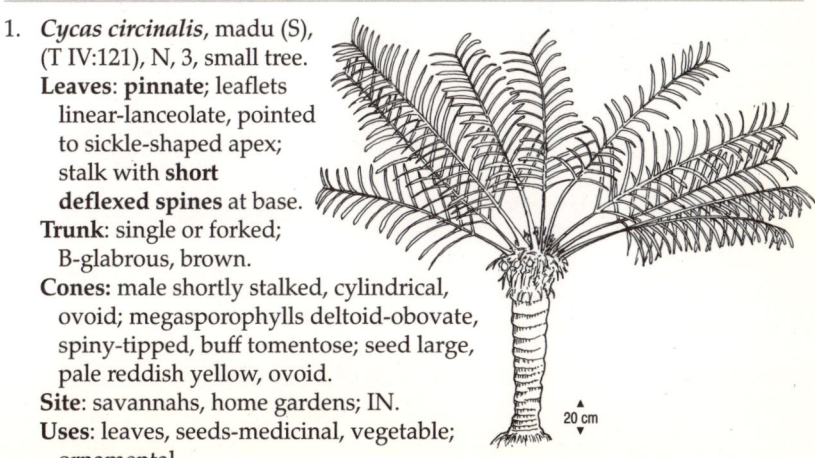

3. ARAUCARIACEAE

FAMILY DESCRIPTION - Habit: large trees, more or less symmetrical branches, often whorled. **Leaves**: alternate, often two ranked, broad to needle-like, sometimes sharply pointed. **Cones**: unisexual; plants dioecious or monoecious. Seeds large, winged or wingless.

REPRODUCTIVE PARTS - Male cones: large, cylindrical, axillary or terminal. Microsporophylls many, spiral. Microsporangia linear, many. Female cones: large, globose, woody, each megasporophyll with 1 ovule, cone disintegrating when seeds mature. Scale adnate to seed (*Araucaria*) or not (*Agathis*).

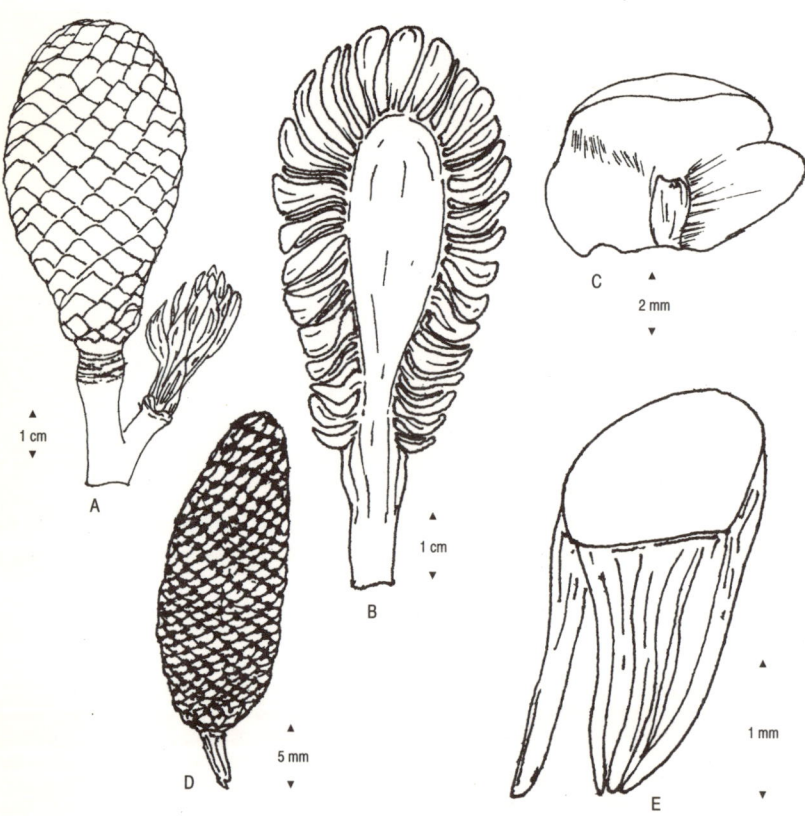

Key: female cone in full (A), and in longitudinal section (B), a single megasporophyll with seed (C), male cone (D) and a microsporophyll from the male cone showing the microsporangia (E) of *Araucaria* sp.

1. **Agathis robusta**, kauri pine (E),
   I (Australia), 25, tree.
   Leaves: **mostly opposite**, dark
   green, oblong to ovate, leathery,
   no visible veins.
   Trunk: B-smooth to scaly, grey-
   brown; IB-pink; branches
   whorled; resinous.
   Cones: male small; female ovoid to
   rounded, large, woody.
   Site: gardens and parks; M, W.
   Uses: ornamental; W-construction,
   pulp, resin.

   2.3 cm

2. **Araucaria bidwillii**, bunya-
   bunya pine (E), I (Australia),
   30, tree.
   Leaves: crowded, ovate to
   lanceolate, pointed apex,
   dark lustrous green,
   keeled beneath.
   Trunk: stout, basal
   half denuded of
   branches, branches
   whorled.
   Cones: male cone-like;
   female erect,
   subglobose, scale winged,
   seed wingless.
   Site: gardens; M, W.
   Uses: ornamental; W-construction.

   2 cm

   225 cm

3. **Araucaria cookii**, Cook's pine (E), I (Australia), 30, tree.
   Leaves: spirally crowded, laterally compressed, needle-shaped,
   sharp-pointed apex, bright green, midrib
   prominent.
   Trunk: crown columnar, branches whorled;
   B-rough, curling.
   Cones: male subcylindrical;
   female shortly-
   stalked, ovoid-
   globose, scales
   winged, seed
   wingless.
   Site: gardens; M, W.
   Uses: ornamental,
   W-construction, resin.

   1 cm

   230 cm

ARAUCARIACEAE

4. *Araucaria cunninghamii*, hoop
   pine (E), I (Australia), 30, tree.
   Leaves: crowded, broad
     base, tapering and
     pointed apex,
     laterally
     compressed;
     crown
     pyramidal.
   Trunk: branches
     whorled, spreading
     horizontally.
   Cones: similar to
     A. bidwillii but smaller.
   Site: gardens; M, W.
   Uses: ornamental,
     W-construction, resin.

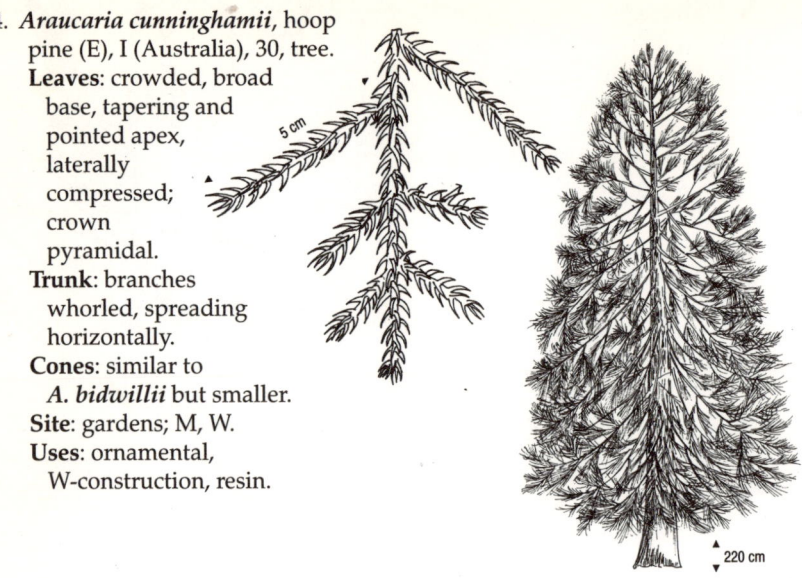

4. CUPRESSACEAE

FAMILY DESCRIPTION - **Habit**: trees or shrubs. **Leaves**: opposite or whorled, mostly small and scale-like or needle-like, sometimes of two different types; juvenile leaves being more slender than those of adult foliage. **Cones**: unisexual; plants dioecious or monoecious. Seeds often winged.

REPRODUCTIVE PARTS - MALE CONES: small, terminal or axillary, 2-24 microsporophylls each with 2-8 microsporangia borne on lower surface. FEMALE CONES: megasporophylls flattened and imbricate, peltate, woody or fleshy, united. Ovules erect, 1-12 per megasporophyll.

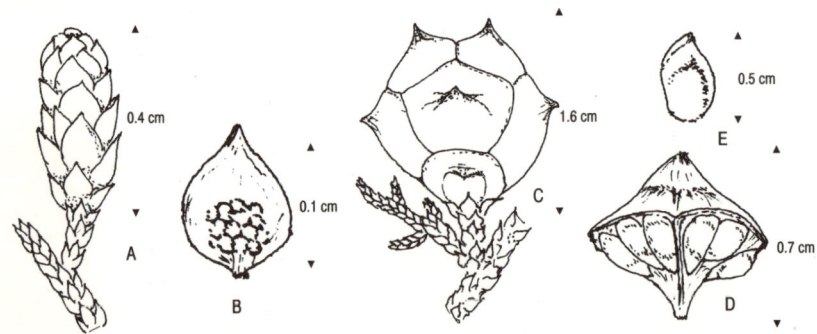

Key: male cone (A), microsporophyll with microsporangia (B), female cone (C), megasporophyll showing seeds (D) and a single seed (E) of *Cupressus* sp.

1. ***Cupressus macrocarpa***, Monterey cypress (E),
   I (SW North America), 25, tree.
   Leaves: closely appressed, swollen towards
      rounded apex, not glandular, dark green;
      crown broad and
      spreading.
   Trunk: B-fibrous, ridged,
      dark reddish brown;
      branches
      horizontal;
      branchlets stout,
      erect, stiff.
   Cones: male short,
      ridge-like, rounded;
      female globular to
      oblong, with 8-12
      woody scales.
   Site: mid- and up-country tea
      estates, timber plantations, home
      gardens; M, LM.
   Uses: christmas trees; ornamental; W-fuelwood, light construction.

2.5 cm

4

2. ***Cupressus torulosa***, Bhutan cypress (E),
   I (Himalayas & China), 20, tree.
   Leaves: appressed or slightly spreading at apex,
      small, scale-like, bright or bluish
      green; crown pyramidal.
   Trunk: B-dark brown; branches short,
      horizontal; branchlets slender, drooping.
   Cones: male short, woody, rounded;
      female with 8-12 scales.
   Site: mid- and up-country in tea estates,
      home gardens, timber plantations; M, LM.
   Uses: christmas trees, ornamental; W-fuelwood,
      light construction.

1 cm

5. PINACEAE

FAMILY DESCRIPTION - Habit: trees, rarely shrubs. Resinous. Branches usually opposite or whorled, rarely alternate. Plants monoecious. **Leaves**: spiral, needle-like in bundles. **Cones**: male small, not woody. Female usually woody, with spirally arranged scales. Seeds usually 2 per scale and usually winged.

REPRODUCTIVE PARTS - MALE CONES : sub-cylindrical, numerous, with spirally arranged microsporophylls. FEMALE CONES: paired scales of two types arranged spirally on a central axis, the upper bearing ovules. Mature cones woody. Seeds nut-like, winged.

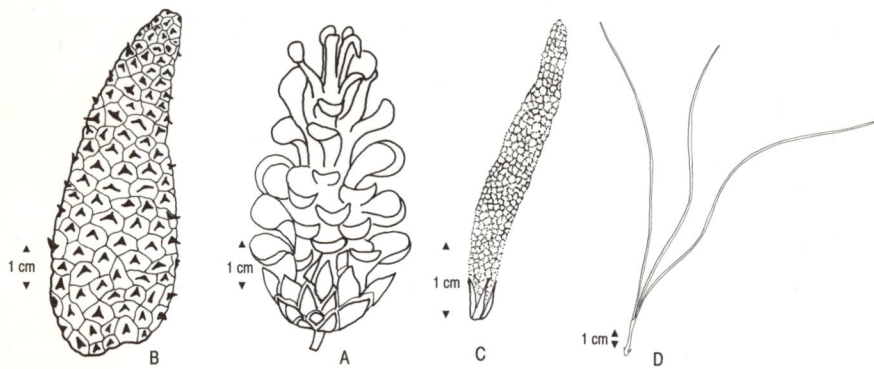

Key: dehisced (A), and undehisced (B) female cones, staminate cone (C), and a bundle of needle leaves (D) of *Pinus caribaea*.

1. ***Pinus caribaea***, Caribbean pine (E), I (Central Amer.), 20, tree.
    Leaves: needles, **3** in a **bundle**, dark green, glossy.
    Trunk: **resinous**; B-rough, scaly, thick, reddish grey; IB-reddish brown; W-reddish brown, light weight.
    Cones: male, red-brown, cylindrical, small; female, reddish brown, ovoid; seeds winged.
    Site: plantations; LM, UW, W.
    Uses: ornamental, reforestation; resin; W-pulpwood, posts, light construction.

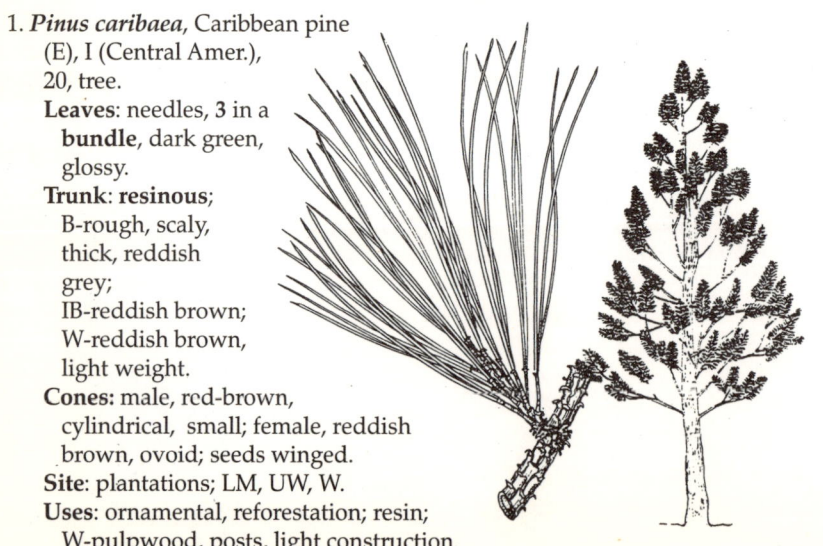

2. *Pinus patula*, patula pine (E),
   I (Central Amer.), 20, tree.
   Leaves: needles, **3 or 4** in a
     bundle.
   Trunk: resinous;
     B-rough, scaly,
     thick, grey;
     IB-reddish brown;
     W-reddish brown,
     light weight.
   Cones: male, red-brown, cylindrical,
     small; female, reddish brown, ovoid;
     seeds winged.
   Site: plantations; M, LM.
   Uses: reforestation; resin; W-posts,
     pulpwood, light construction.

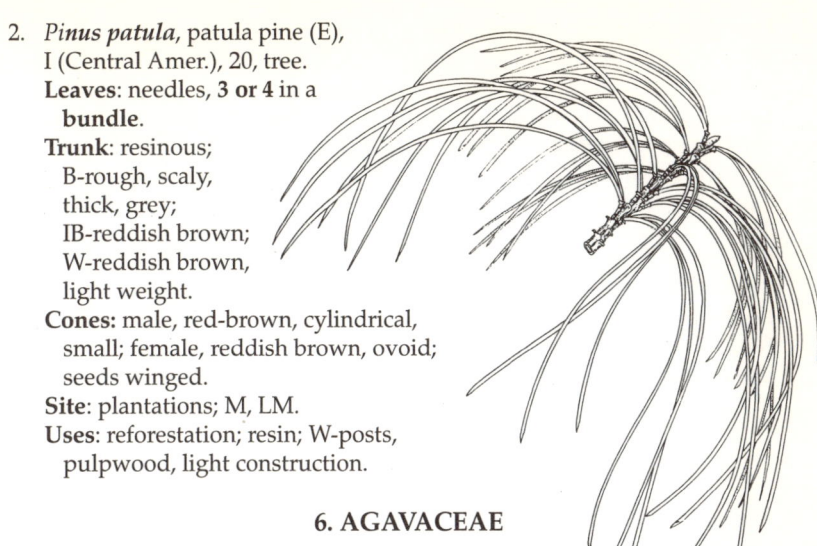

6. AGAVACEAE

FAMILY DESCRIPTION - Habit: coarse stemmed herbs to sparsely branched trees. **Leaves**: spiral, usually in rosettes, simple, narrow, parallel-veined, often prickly, sometimes succulent. **Stipules**: absent. **Flowers**: bisexual, actinomorphic, solitary or in spikes, racemes, panicles or umbels. **Fruits**: usually a capsule, rarely a berry, or nut.

FLOWER PARTS - PERIANTH: usually in 2 petaloid whorls, sometimes with a basal tube, sometimes a corona. Nectaries septal, or at base of perianth or androecium or absent. ANDROECIUM: stamens 6, hypogynous or adnate to perianth. GYNOECIUM: superior or inferior. Usually 3-carpellary, with axile or deeply intruded parietal placentation. Stigmas as many as carpels on 1-3 styles. Ovules numerous.

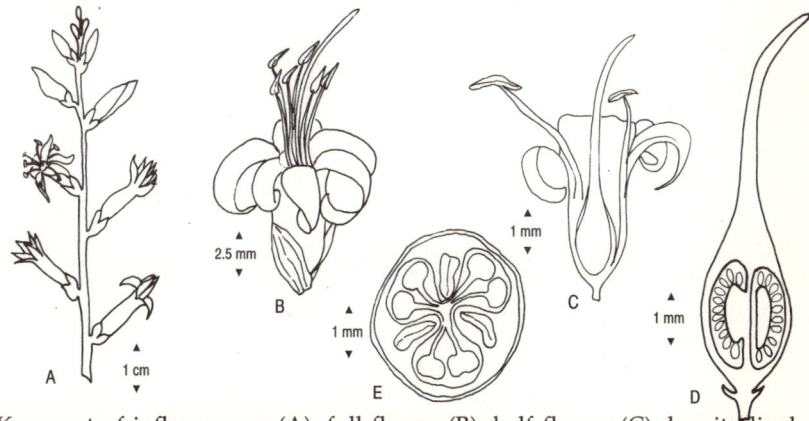

Key: part of inflorescence (A), full flower (B), half flower (C), longitudinal section of gynoecium (D) and transverse section of ovary (E) of *Cordyline fruticosa*.

1. **Dracaena fragrans**, dracaena (E)/
   botal gas (S), I (E. Africa), 3, shrub
   to small tree.
   Leaves: terminally **crowded** at
      top, long, lanceolate to
      narrowly oblong,
      parallel-veined.
   Trunk: **slender, unbranched**.
   Flowers: yellowish, funnel-shaped,
      crowded; I-branched heads.
   Fruits: orange, round berries.
   Site: home gardens; M, LM, W.
   Uses: ornamental.

2. **Dracaena reflexa**, song of India(E),
   I (Madagascar & India), 3, shrub to
   small tree.
   Leaves: densely crowded, elongate, narrow,
      leathery, deep glossy green with two whitish
      bands of yellow or cream at
      margin, without midrib,
      margin wavy, reflexed;
   Trunk: grey-brown; twigs
      crowded.
   Flowers: white.
   Fruits: clusters, rounded.
   Site: home gardens, parks; IN, W.
   Uses: ornamental.

7. GRAMINEAE

FAMILY DESCRIPTION - Habit: usually herbs, or rarely woody and tree-like.
Stems circular with hollow internodes, rarely solid. **Leaves**: composed of two
parts - a sheath enveloping the stem and overlapping; and a flat, linear to
lanceolate blade. At the junction of the leaf blade and sheath, on the inner side,
is the ligule, an appendage that is typically membranous, occasionally hyaline
or reduced to hairs. Flow**e**rs: bisexual or unisexual. The basic inflorescence unit
is a spikelet. Spikelets are aggregated terminally on primary culms or branches
as a spike, raceme or panicle. **Fruits**: caryopsis, rarely a nut, berry or utricle.

FLOWER PARTS - Spikelets consist of a pair of subopposite bracts (glumes)
and one to several distichous florets often on a zig-zag rachilla. Each floret
usually has 2 bracts (lemma and palea). The lemma often resembles the glumes.
It is greenish, keeled or rounded, nerved, awned or awnless, fertile (enveloping
a pistillate flower) or sterile (enveloping a staminate flower or none at all). The
palea lies between the flower and the rachilla, usually 2 nerved and 2 keeled,
enclosing the flower. Sometimes the palea is completely reduced. When bisexual,
perianth highly modified, represented by up to 6 or more lodicules in
Bambusoideae and 2 in other tribes. Lodicules are situated at the base of the

flower outside the stamens. At anthesis their increased turgidity parts the palea and lemma pushing the stamens and/or stigma out of the flower. ANDROECIUM: stamens usually 3-6, when 3 sometimes 1 or 2 suppressed with 2 fertile stamens or 1. Rarely stamens numerous. Anthers elongated, basifixed but deeply sagittate so as to appear versatile, with longitudinal slits. GYNOECIUM: superior, 3-carpellary, unilocular with 2-3 styles, which have often large, feathery stigmas.

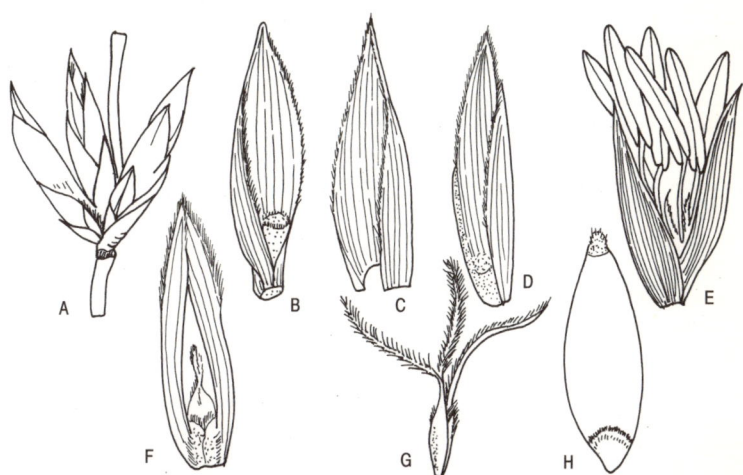

Key: cluster of pseudo-spikelets (A), floret (B), lemma (C), palea (D), lemma and palea separated to show six stamens (E), palea showing lodicules and ovary (F), gynoecium (G) and caryopsis (H) of *Bambusa bambos.*

1. ***Bambusa bambos***, spiny bamboo (E)/katu una (S)/moongil (T), (DF VIII:74), N, 30, tree-like clump.
   Leaves: linear-lanceolate, glabrous or hairy, veins parallel, margins **ciliate**; leaf sheath ends in **thick callus**, auricle very **short, bristly**, ligule short.
   Culms: many with tiny spiny branches from base; nodes prominent; culm sheath **orange-yellow** streaked **green-red** when young, margins ciliate, **golden** hairy.
   Flowers: I-large panicle, loose pale clusters.
   Fruits: oblong, beaked.
   Site: cultivated; ridgetops and along water courses; DL, IN, W.
   Uses: culms-basketry, light construction, handicrafts; hedges; seeds-edible; silicious deposits (tabashir) within internodes-medicinal.

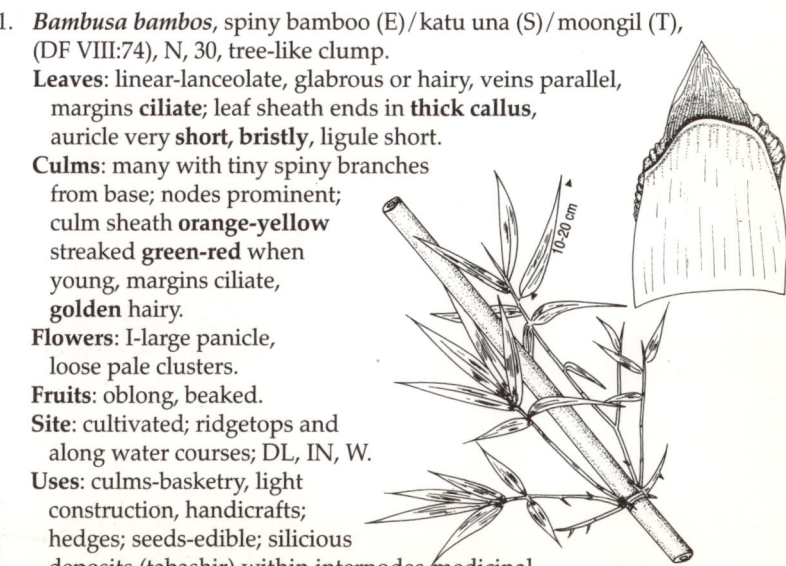

2. **Bambusa multiplex**, Chinese bamboo (E)/
   cheena bata (S), (DF VIII:76), I (S.E. Asia), 3,
   shrub.
   Leaves: lnear-lanceolate, veins parallel,
      whitish and softly hairy
      beneath; leaf sheaths tipped
      with callus, auricles **bristly**.
   Culms: many, branching from
      base; nodes thickened; culm
      sheath slightly narrowed to
      rounded apex, stiff, not hairy,
      green yellow; sheath blade long
      and narrow, base decurrent,
      glabrous, or hairy above and
      beneath, margins with hair that
      fall off early.
   Flowers: I-diffuse, leafy panicles.
   Fruits: ellipsoid, shortly beaked,
      furrowed.
   Site: gardens; M, W.
   Uses: culms-fishing rods; hedges,
      ornamental.

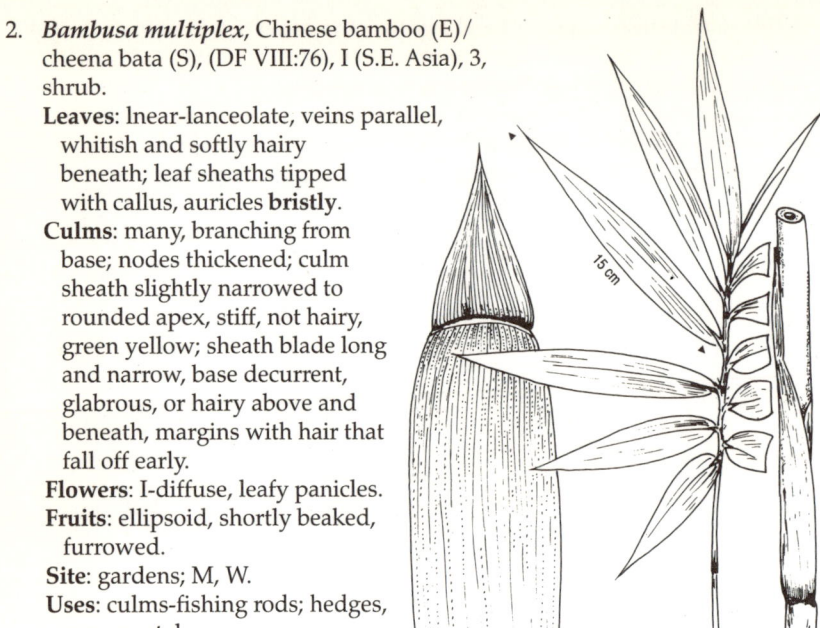

3. **Bambusa vulgaris**, common bamboo (E)/kaha una, kola una (S),
   (DF VIII:79), I (unknown), 20, tree-like clump.
   Leaves: linear-lanceolate, apex slender, twisted, veins parallel, margins
      and adjacent nerves rough; leaf sheaths slightly hairy, end in **smooth
      callus**, auricles **bristly** rounded.
   Culms: many, branching from base, green-striped
      yellow; nodes not prominent, with ring of hairs;
      culm sheath wide with rounded apex, retuse,
      hairy on both surfaces; sheath blade
      wide, subtriangular, pointed, hairy,
      base decurrent, ending in two rounded
      auricles fringed with bristles.
   Flowers: blooms rarely;
      I-straw-coloured panicles.
   Fruits: unknown.
   Site: paddy field
      bunds, water
      courses; DL, IN, W.
   Uses: culms-pipes,
      scaffolding, light
      construction,
      handicrafts; young
      shoots-edible.

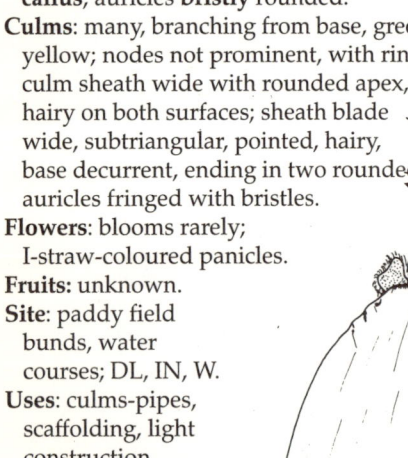
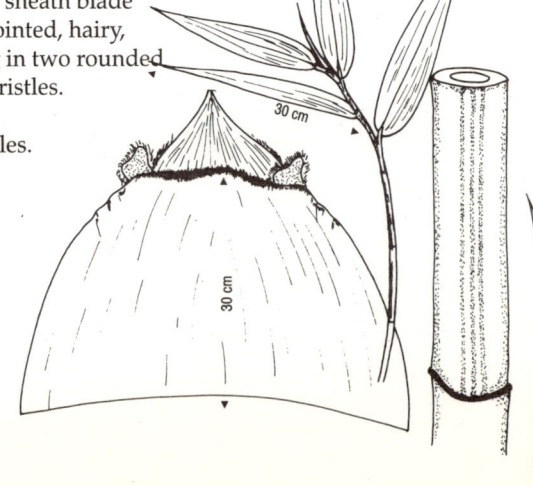

4. **Davidsea attenuata**, bamboo (E)/
   bata (S), (DF VIII:149), E, 7, shrub.
   Leaves: lanceolate, apex **twisted**,
   parallel veins, rather rough
   above, smooth and pale
   beneath, margins **ciliate**; leaf
   sheaths slightly hairy or
   without hairs or fringed with
   short hairs that fall early.
   Culms: many, branched from
   base, with whip-like curved tips
   bearing small leafy branchlets,
   culms dark green; culm sheath
   narrow, pale, soft white hairy;
   leaf sheaths recurved.
   Flowers: I-spike-like, leafy
   panicles.
   Fruits: elongate, beaked.
   Site: forest understory, gaps and
   fringes; M, LM.
   Uses: culms-basketry, for making
   boxes.

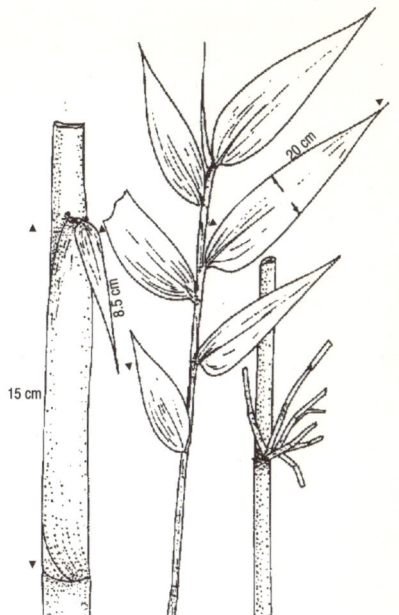

7

5. **Dendrocalamus giganteus**, giant bamboo (E)/yodha una (S), (DF VIII:154),
   I (Burma), 35, tree-like clump.
   Leaves: lanceolate, base rounded, hairy beneath when young.
   Culms: many, very large, basal nodes tufted, not branched, Leafy branches
   drooping at apex; culm sheaths very large, pale, glabrous within, hairy
   outside; sheath blades
   long, narrow,
   recurved with
   sides inrolled,
   with two narrow
   auricles.
   Flowers: I-panicle
   very large.
   Fruits: oblong,
   apex with a
   hairy crown.
   Site: home gardens,
   tea estates; LM, UW.
   Uses: culms-light
   construction;
   ornamental.

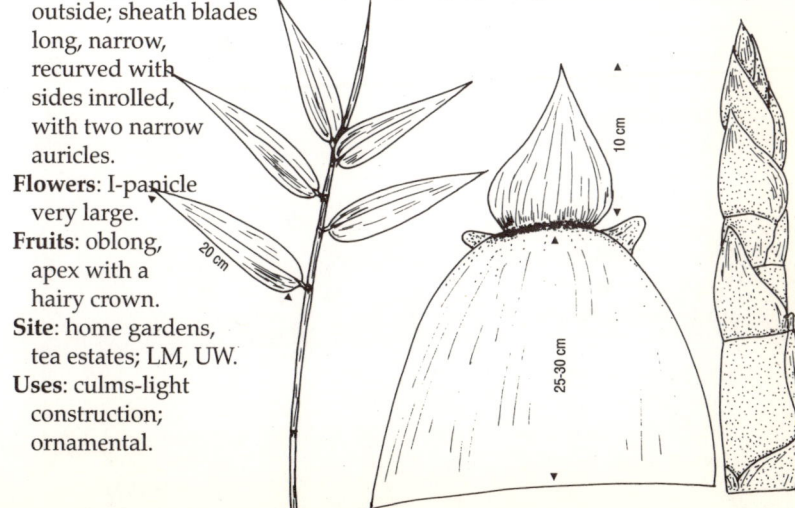

6. ***Ochlandra stridula***, bamboo (E)/bata li (S), (DF VIII:319), E, 3, shrub.

Leaves: oblong-lanceolate, apex long, rough, ending in hairs, margins stiff; leaf sheaths thickened at end, fringed with long, bristly, deciduous hairs.

Culms: many, branched from base, **green**, sometimes with purple blotches, nodes slightly zig-zag; culm sheaths glabrous, apex rounded, auricles bristly; sheath blade reflexed at right angles to culm.

Flowers: I-Spikes or spike-like leafy branches.

Fruits: ovoid, large, fleshy, beaked pericarp.

Site: understory, gaps and fringes of rain forest; UW.

Uses: culms-basketry, flutes, wattle; leaves-thatching.

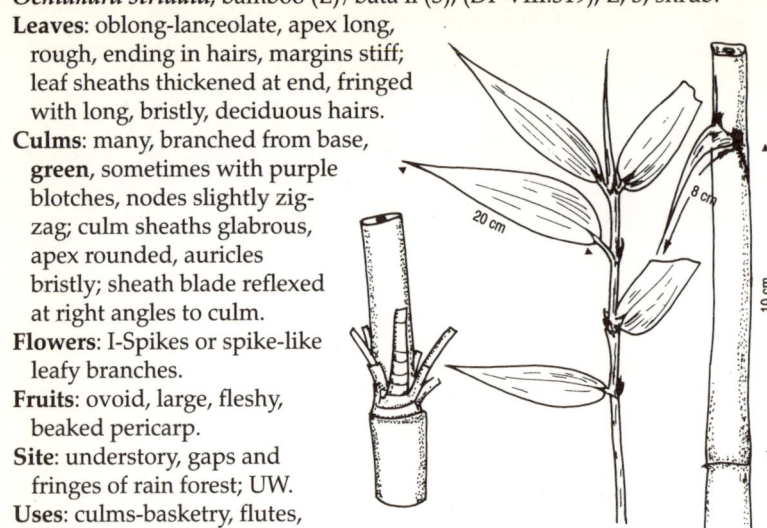

7. ***Pseudoxytenanthera monadelpha***, bamboo (E)/bata (S), (DF VIII:386), N, 4, shrub.

Leaves: lanceolate, acuminate, bristly, twisted apex, glabrous above, slightly hairy beneath, margins rough, midrib yellow.

Culms: many, branched from base, ends **whip-like, curved**; bearing small leafy branchlets, dull green, nodes prominent; culm sheath narrow, old sheaths light brown hairy, young sheaths shiny, glabrous; sheath blade with long narrow base, ending in large round auricles with bristly tips; bristles very long on young shoots.

Flowers: I-spikelets in large panicles.

Fruits: ovoid-oblong, glabrous.

Site: forest understory, gaps and fringes; M, LM.

Uses: culms-basketry.

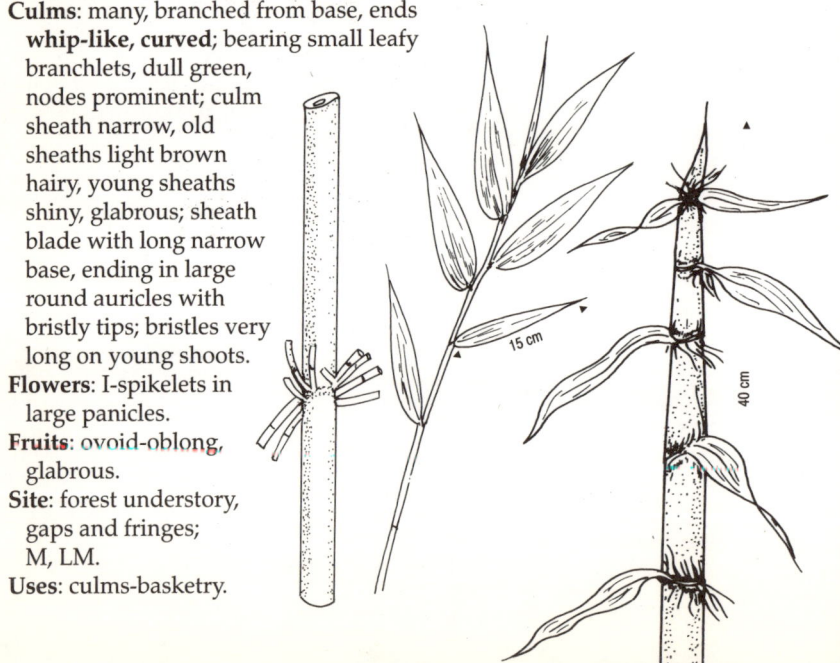

8. MUSACEAE

FAMILY DESCRIPTION - **Habit**: large herbs, tree-like, unbranched, pseudostems formed by leaf sheaths rolled round one another. **Leaves**: spiral, simple, with long petioles and expanded lamina. Midrib prominent, secondary veins pinnate and parallel. **Flowers**: bisexual or unisexual on monoecious plants. Inflorescence arises from the stem below ground and grows up through the pseudostem. Inflorescence bears an aggregation of spathaceous bracts, in whose axils are borne the flowers. The terminal bracts bear the male flowers. **Fruits**: fleshy berry, seedless or with few to numerous seeds.

FLOWER PARTS - PERIANTH: 6 petaloid segments, in two whorls. Segments free or variously united. ANDROECIUM: basically 6 stamens, one usually a staminode which is sometimes petaloid and the other 5 fertile. Anthers 2-celled, dehiscing lengthwise. Filaments free. GYNOECIUM: inferior, 3-carpellary, 3-locular. Ovules numerous, placentation axile. Style filiform. Stigmas usually 3, each sometimes branched.

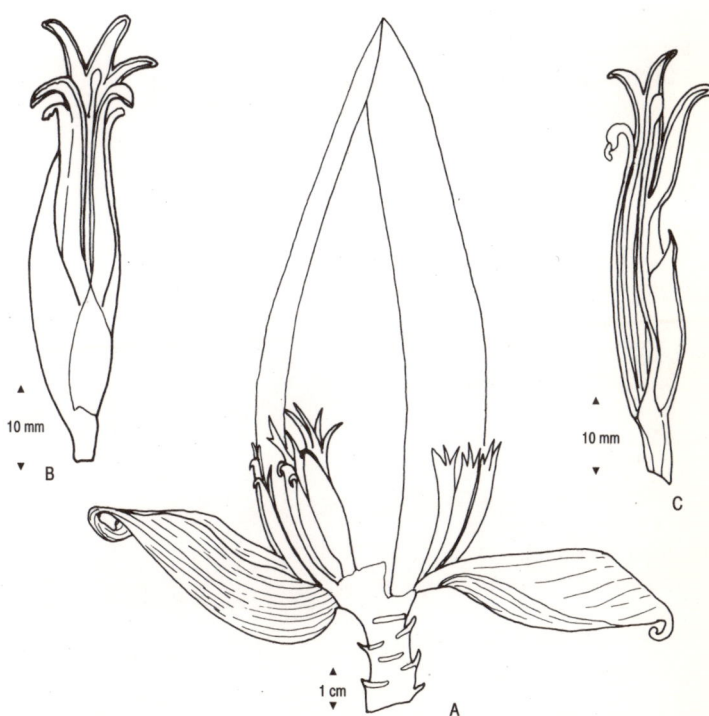

Key: male part of inflorescence (A), full flower (B) and half flower (C) of *Musa* x.

1. ***Musa* x**, banana, plantain (E)/kesel (S), (T IV:265),
   N, 4, large herb.
   Leaves: **bright green**, **very large**, oblong, **many
   parallel**, lateral veins, midrib very stout.
   Trunk: pseudostems, tending
   toward clumping from
   suckers; **mucilaginous**
   exudate.
   Flowers: I-terminal, stout,
   spike-like, lower part
   bisexual but functionally
   female, upper part
   functionally male; bracts
   large.
   Fruits: large, oblong or
   fusiform, fleshy.
   Site: home gardens;
   widespread.
   Uses: fruit-edible;
   inflorescence-edible;
   leaves-wrapping food;
   leaf sheath-fibre.

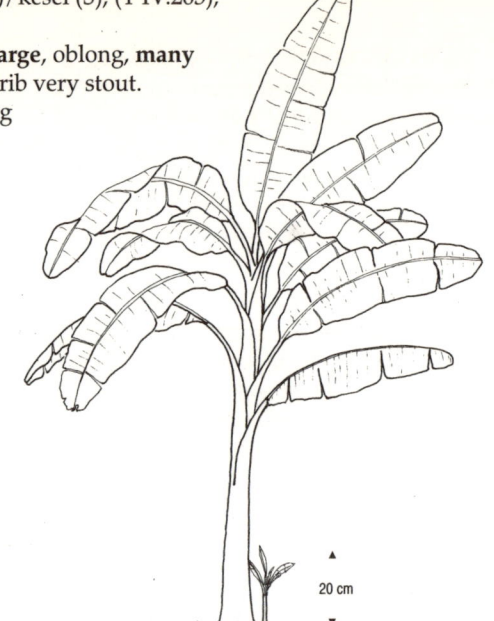

20 cm

2. *Ravenala madagascariensis*,
   traveller's palm (E), I (Madagascar),
   6, tree.
   Leaves: large, oblong,
   arranged fan-like; mid
   ribs stout.
   Trunk: unbranched,
   tendency toward
   clumping from suckers;
   mucilaginous exudate.
   Flowers: I-axillary,
   stout spike.
   Fruits: large, oblong,
   fleshy; seeds with blue aril.
   Site: home gardens, parks; W.
   Uses: ornamental.

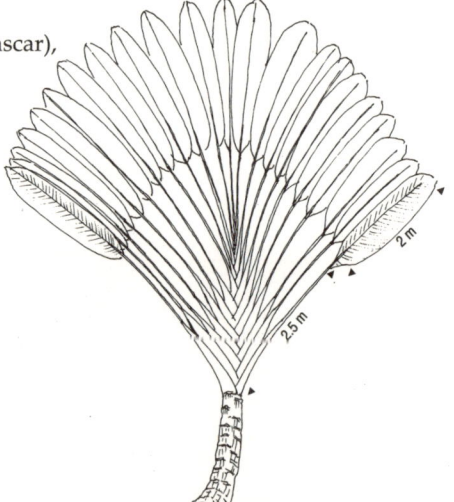

2 m

2.5 m

9. PALMAE (ARECACEAE*)

FAMILY DESCRIPTION - Habit: trees mostly erect and unbranched, or slender climbers. **Leaves**: spiral, usually terminally crowded, or along the length as in rattans, often very large, basal sheath tubular, splitting at maturity. Petiole smooth or dentate. Lamina simple (fan palms) or pinnate (feather palms). **Flowers**: unisexual, usually actinomorphic, monoecious or dioecious plants. Inflorescences mostly below or amongst the leaves, rarely above, often protected by large bracts or spathes when young. **Fruits**: usually a fleshy or fibrous drupe.

FLOWER PARTS - PERIANTH: 6 segments in 2 whorls of 3 each, free or united. Segments of outer whorl valvate in male and imbricate in female flowers. ANDROECIUM: stamens usually 6, occasionally more, in 2 whorls of 3 each. Anthers 2-locular, dehiscing by vertical slits. Filaments free and short. GYNOECIUM: superior, 1-3-carpellary and 1-3-locular. Ovules usually solitary in each loculus. Placentation basal, axile, erect or pendulous.

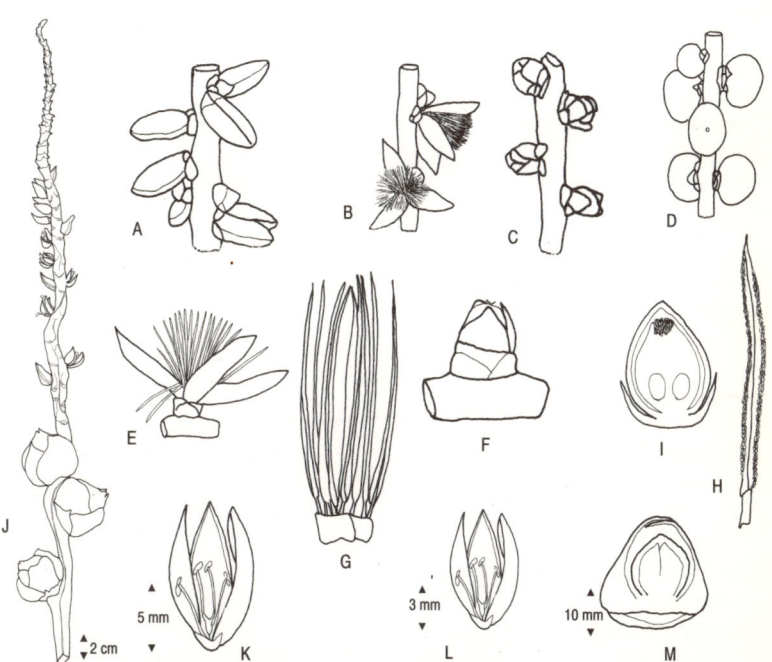

Key: part of the inflorescence, showing the arrangement of male and female flowers in triads (A), open male flowers (B), open female flowers after all the male flowers are shed (C), developing fruits (D); open male (E) and female flowers (F), androecium (G), a single stamen (H), and a female flower in longitudinal section (I) of *Caryota urens*. Part of inflorescence (J), a male flower in full (K), a male half-flower (L), and a female flower in vertical section (M) of *Cocos nucifera*.

* Alternative family name.

1. *Areca catechu*, betel nut palm (E)/puwak (S)/kamukai (T), (T IV:321),
   I (Indo-Malaya), 15, tree.
   Leaves: pinnate; leaflets
     numerous, linear,
     parallel-veined.
   Trunk: straight, **slender**,
     cylindrical, erect; B-grey,
     ringed.
   Flowers: pale
     yellowish;
     I-spadix arising
     below leaves,
     branched
     paniculately at
     base; spathes
     simple, compressed.
   Fruits: orange, ovoid
     drupes.
   Sites: home gardens; IN, W.
   Uses: flowers-temple offerings;
     seed-betel nut (masticatory);
     trunk-light construction;
     whole plant-medicinal.

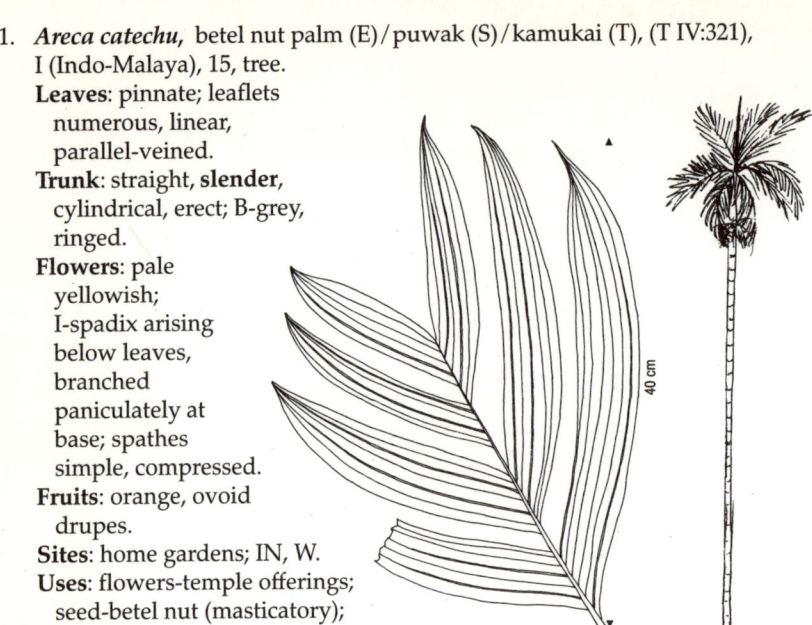

2. *Borassus flabellifer*, palmyra palm (E)/tal (S)/panai (T), (T IV:336),
   I (India), 20, tree.
   Leaves: **fan-shaped**, cleft into lanceolate to linear
     bifid lobes, margins of petiole **spiny.**
   Trunks: black, scarred above,
     swollen above middle and
     then narrowed again.
   Flowers: I-spadix among
     leaves, large, branches
     cylindical.
   Fruits: black, broadly
     ovoid drupes; seated on
     enlarged perianths;
     pulp yellow,
     fleshy, fibrous.
   Site: groves on sandy
     coastal areas; DC.
   Uses: fruit-edible;
     Inflorescence-beverage; leaves-
     thatching, basketry; stem-
     construction, salt resistant;
     whole plant-medicinal.

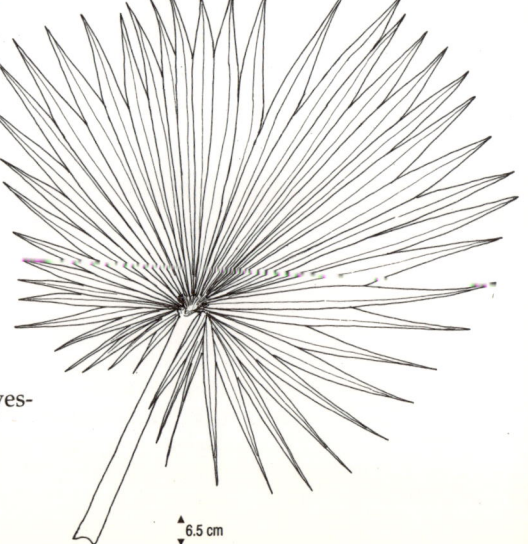

6.5 cm

3. *Caryota urens*, kitul palm (E)/kitul (S)/
    tippilipan (T), (T IV:324), N, 8, tree.
    Leaves: bipinnate, arched, **drooping,
    dark green**; fish tail-like leaflets in
    clusters or alternate, margins
    irregularly serrate;
    rachis stout; leaf
    sheath with fibrous
    netted
    margins
    Trunk: erect,
    cylindrical,
    ring-like
    leaf-scars.
    Flowers: spadix a dense
    drooping tassel.
    Fruits: red or yellow,
    globose drupes.
    Site: home gardens; rain
    forest subcanopy, gaps
    and fringes; LM, IN, W.
    Uses: leaves-fibre;
    I-beverage (toddy), jaggery
    (sugar); pith-edible starch.

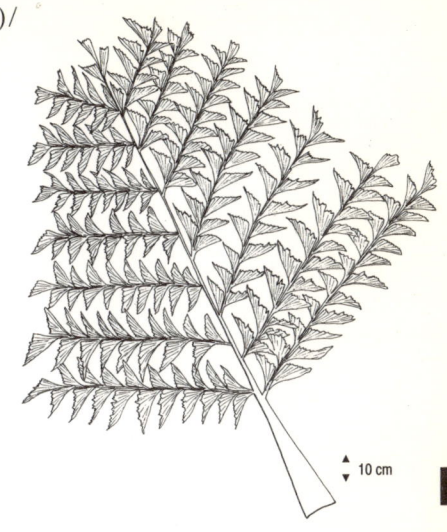

▲
▼ 10 cm

9

4. *Cocos nucifera*, coconut palm
    (E)/pol (S)/tennai (T), (T
    IV:337), I (unknown), 20, tree.
    Leaves: pinnate; leaflets **very
    narrow, pointed** apex.
    Trunk: **slender**, cylindrical,
    ring-like leaf-scars.
    Flowers: I-cream, axillary;
    spadix simply branched; spathe
    narrowly oblong, tapering.
    Fruits: brown, ovoid,
    large, fibrous husk.
    Site: plantations,
    home gardens; DL, IN, W.
    Uses: fruit-handicrafts; Inflorescence
    sap-beverage (toddy), jaggery,
    vinegar; leaves-thatching;
    husk-rope, coir; seed-oil, flesh
    edible; stem-construction,
    handicrafts.

PALMAE

5. ***Corypha umbraculifera***,
   talipot palm (E)/tala (S),
   (T IV:328), N, 15, tree.
   Leaves: **fan-like**, terminal,
   very large, deeply
   divided into pointed
   lobes, plicate; petiole very
   stout, **spiny**.
   Trunk: erect, straight,
   cylindrical, ring-like
   leaf-scars.
   Flowers: cream; I-spadix
   pyramidal, terminal, very
   large, branchlets forming
   pendulous spikes.
   Fruits: greyish olive, shortly stipitate,
   globose, rough drupe.
   Site: coastal groves, home gardens,
   roadsides; MC, W.
   Uses: fruit-stuns fish; leaves-mats,
   fans, umbrellas, for writing (ola
   leaves); pith-edible starch,
   medicinal; seed-beads, buttons.

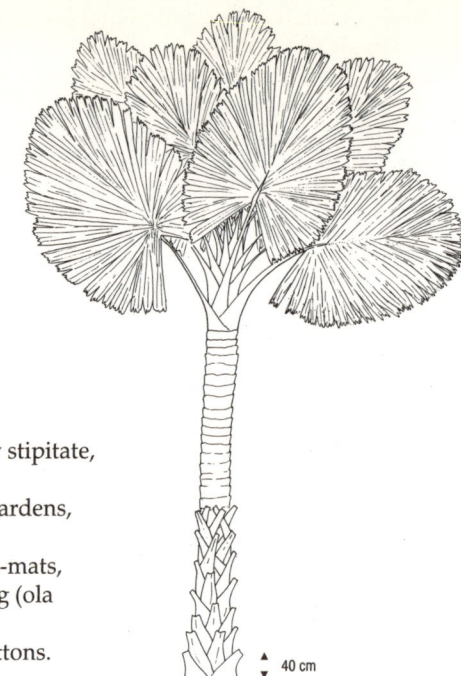

40 cm

6. ***Oncosperma fasciculatum***,
   katu kitul (S), (T IV:323), E, 10,
   tree.
   Leaves: pinnate; leaflets **12-18
   clustered**, lanceolate, pointed,
   drooping apex, scaly beneath;
   rachis **spiny** toward base.
   Trunk: base thickened,
   abundantly spined, black.
   Flowers: I-spadix without spines,
   very stout, drooping; spathes
   sparsely scaly, without spines.
   Fruits: black to purple, globose
   drupes.
   Site: exposed, steep, rocky areas;
   LM, UW, W.

50 cm

7. *Phoenix farinifera*, wild date palm (E)/ indi (S)/
   inchu (T), (T IV:327), N, 3, small tree.
   Leaves: pinnate; one or more leaflets
   reduced to **rigid spines**; rachis 4-sided.
   Trunk: mostly horizontal and
   underground, aerial stem very short,
   erect, covered with
   old leaf sheaths.
   Flowers: yellowish;
   I-spadix erect; spathe
   hard, compressed.
   Fruits: bright red to
   dull purple-black,
   ovoid to oblong,
   pointed, drupe.
   Site: thorn scrub; DL.
   Uses: fruit-edible, medicinal.

8. *Phoenix zeylanica*, wild date
   palm (E)/indi (S), (T
   IV:326), E, 6, small tree.
   Leaves: pinnate; leaflets
   many, long, linear to
   lanceolate, spiny ends.
   Trunk: aerial stem
   short, erect.
   Flowers: yellowish;
   I-spadix long, on
   stout peduncle.
   Fruits: red to violet-blue, obovoid
   to oblong, pointed drupe.
   Site: scrub; MC.
   Uses: fruit-edible, medicinal;
   leaves-hats, mats, boxes.

9. *Roystonea regia*, royal palm (E),
   I (Cuba), 10, tree.
   Leaves: pinnate; leaflets large,
   paired, long-pointed apex.
   Trunk: conspicuously swollen
   about the middle; B-grey.
   Flowers: whitish, small, borne
   below leaf bases.
   Fruits: light brown, elliptic drupes.
   Site: parks, avenues; W.
   Uses: ornamental.

24 cm

120 cm

10. PANDANACEAE

FAMILY DESCRIPTION - Habit: trees, shrubs or climbers with clasping aerial roots. Prop roots when present support the trunk. **Leaves**: in 3 ranks seemingly spiral, simple, glabrous. Lamina elongate with sheathing base, xeromorphic, margin and midrib beneath spiny. Veins parallel. **Flowers**: unisexual, in spadices, plants dioecious. **Fruits**: syncarp of drupes or berries.

FLOWER PARTS - PERIANTH: rudimentary or absent. Inflorescence enclosed by spathe or leaf- like bracts which sometimes form a 3-4-lobed cup. ANDROECIUM: stamens numerous, densely packed, separated or in clusters scattered over an axis. Anthers 2-loculed, basifixed, dehiscing by vertical slits. In male flowers rudimentary ovary present or absent. GYNOECIUM: superior with numerous carpels, coherent in bundles. Each carpel unilocular with 1 to many ovules. Placentation basal or parietal. Style short or absent.

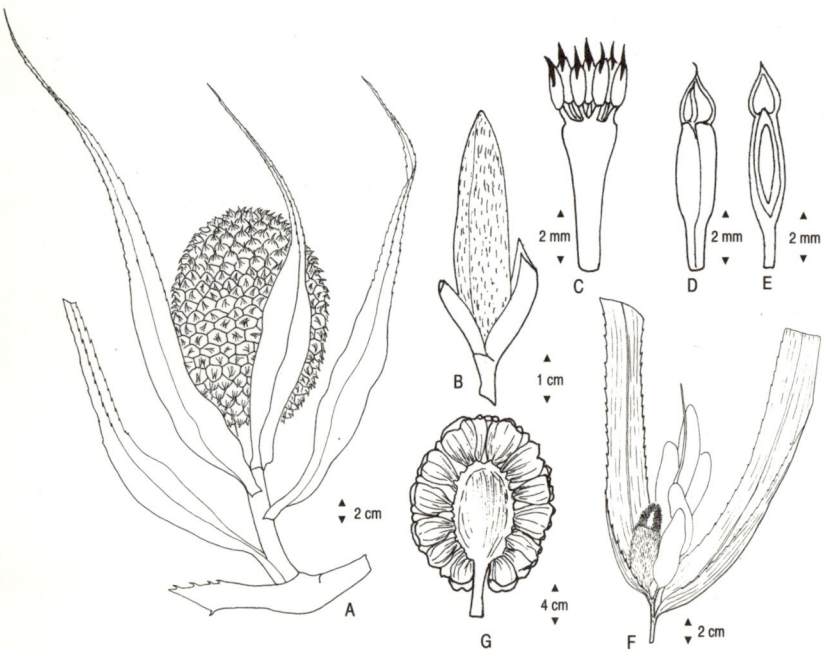

Key: fruiting head (A), male inflorescence (B), male flower with anthers (C), female flower in full (D), and female flower in half with a single seed (E) of *Pandanus ceylanicus*; male inflorescences (F) and female inflorescence in half (G) of *Pandanus odoratissimus*.

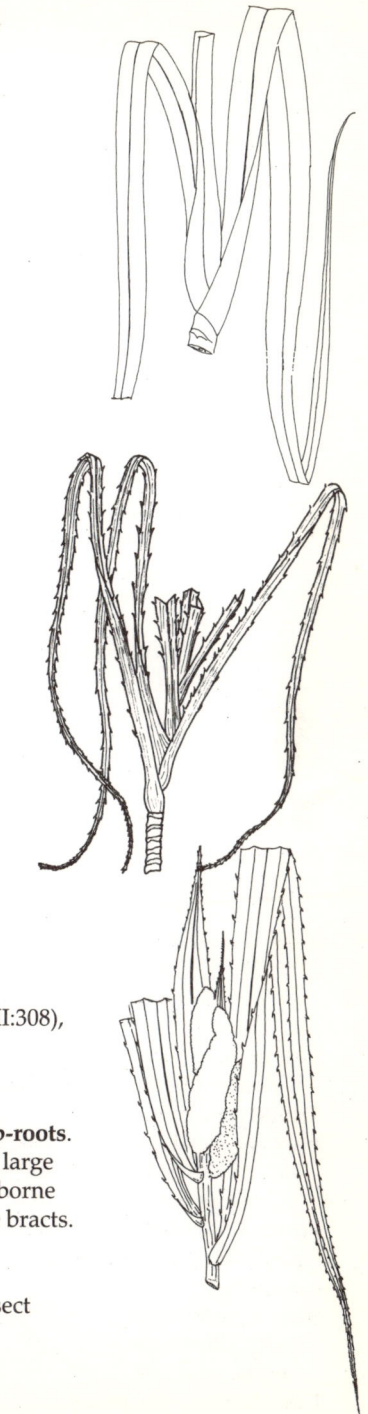

1. *Pandanus amaryllifolius*, rampeh (E,S),
   (DF II:319), I (Malesia), 3, shrub.
   Leaves: green, linear, margins at apex
      prickly; aromatic **musky smell**.
   Trunk: small, branched.
   Flowers: does not flower.
   Fruits: does not fruit.
   Site: home gardens; DL, IN, W.
   Uses: leaves-curry flavouring.

2. *Pandanus ceylanicus*, keyiya (S),
   (DF III:307), N, 3, prostrate shrub.
   Leaves: linear, leaf bases **reddish
      purple-bronze**, prickles
      greenish, dark-tipped.
   Trunk: branches erect or ascending,
      sparse, pale grey rings·of leaf
      scars; prop roots slender, few.
   Flowers: I-male terminal spikes,
      drooping; female terminal, a solitary
      head surrounded by a pale yellow
      spathe, small, subglobose.
   Fruits: subglobose syncarp of
      5-6- angled drupes.
   Site: along stream banks
      and paddy fields; W.
   Uses: leaves-fibre; roots-medicinal.

3. *Pandanus kaida*, weta keyiya (S), (DF III:308),
   N, 3, shrub.
   Leaves: linear; prickles **pale greenish**,
      stiffly erect or abruptly bent.
   Trunk: tree-like when not pruned; **prop-roots**.
   Flowers: I-terminal and bracteate; male large
      spicate raceme; female solitary or 2-3 borne
      together, each surrounded by leaf-like bracts.
   Fruits: syncarp of angled drupes.
   Site: paddy field bunds; IN, W.
   Uses: live fencing; medicinal; pollen-insect
      repellent; leaves-mats, boxes, hats.

10

PALMAE

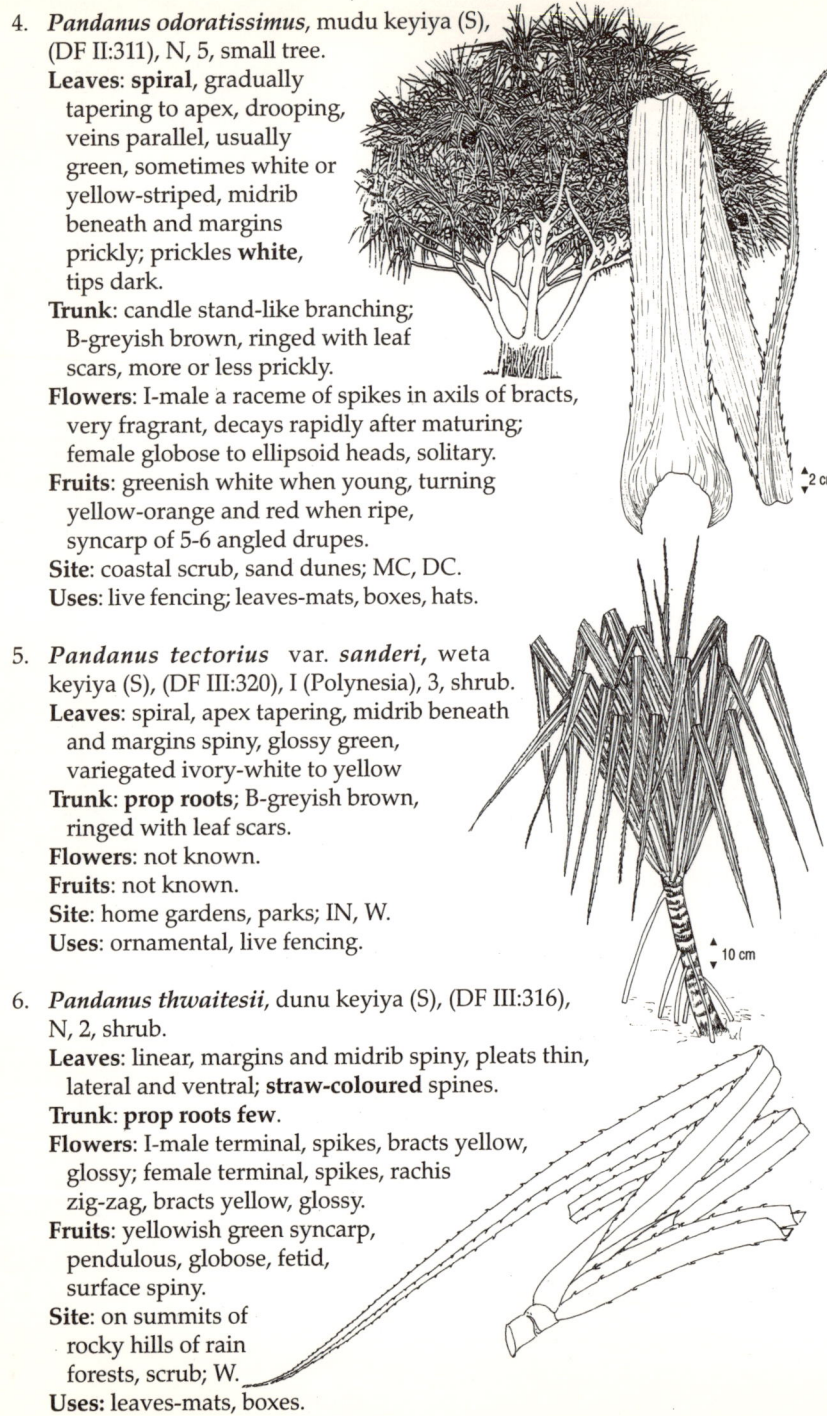

4. *Pandanus odoratissimus*, mudu keyiya (S),
   (DF II:311), N, 5, small tree.
   Leaves: **spiral**, gradually
      tapering to apex, drooping,
      veins parallel, usually
      green, sometimes white or
      yellow-striped, midrib
      beneath and margins
      prickly; prickles **white**,
      tips dark.
   Trunk: candle stand-like branching;
      B-greyish brown, ringed with leaf
      scars, more or less prickly.
   Flowers: I-male a raceme of spikes in axils of bracts,
      very fragrant, decays rapidly after maturing;
      female globose to ellipsoid heads, solitary.
   Fruits: greenish white when young, turning
      yellow-orange and red when ripe,
      syncarp of 5-6 angled drupes.
   Site: coastal scrub, sand dunes; MC, DC.
   Uses: live fencing; leaves-mats, boxes, hats.

   2 cm

5. *Pandanus tectorius* var. *sanderi*, weta
   keyiya (S), (DF III:320), I (Polynesia), 3, shrub.
   Leaves: spiral, apex tapering, midrib beneath
      and margins spiny, glossy green,
      variegated ivory-white to yellow
   Trunk: **prop roots**; B-greyish brown,
      ringed with leaf scars.
   Flowers: not known.
   Fruits: not known.
   Site: home gardens, parks; IN, W.
   Uses: ornamental, live fencing.

   10 cm

6. *Pandanus thwaitesii*, dunu keyiya (S), (DF III:316),
   N, 2, shrub.
   Leaves: linear, margins and midrib spiny, pleats thin,
      lateral and ventral; **straw-coloured** spines.
   Trunk: **prop roots few**.
   Flowers: I-male terminal, spikes, bracts yellow,
      glossy; female terminal, spikes, rachis
      zig-zag, bracts yellow, glossy.
   Fruits: yellowish green syncarp,
      pendulous, globose, fetid,
      surface spiny.
   Site: on summits of
      rocky hills of rain
      forests, scrub; W.
   Uses: leaves-mats, boxes.

11. ACANTHACEAE

FAMILY DESCRIPTION - Habit: herbs, shrubs or climbers. **Leaves**: opposite, simple. **Stipules**: absent. **Flowers**: bisexual, zygomorphic. Inflorescences often with conspicuous bracts. **Fruits**: loculicidal 2-valved capsule, often with a solid base. Seeds hard, usually compressed, with fine white hairs which become sticky on wetting.

FLOWER PARTS - CALYX: 4 or 5, united, imbricate or valvate, rarely reduced to a ring. COROLLA: 4 or 5, fused, 2-lipped or sometimes one-lipped, imbricate or contorted in bud. ANDROECIUM: stamens 4 or 2, epipetalous, didynamous when 4. Filaments free or partially united in pairs. Anther 2-locular, opening lengthwise. 1 or more staminodes often present. GYNOECIUM: superior, 2-locular, with 2 to numerous ovules in each loculus. Placentation axile. Style usually bifid.

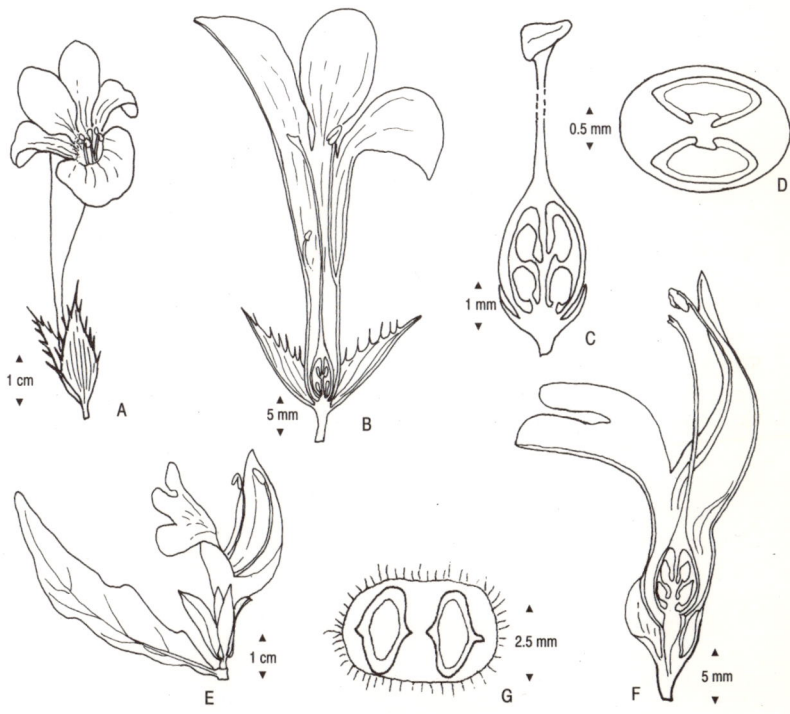

Key: full flower (A), half flower (B), longitudinal section of gynoecium (C), and transverse section of ovary (D) of *Barleria mysorensis*. Full flower (E), half flower (F), and transverse section of ovary (G) of *Justicia adhatoda*.

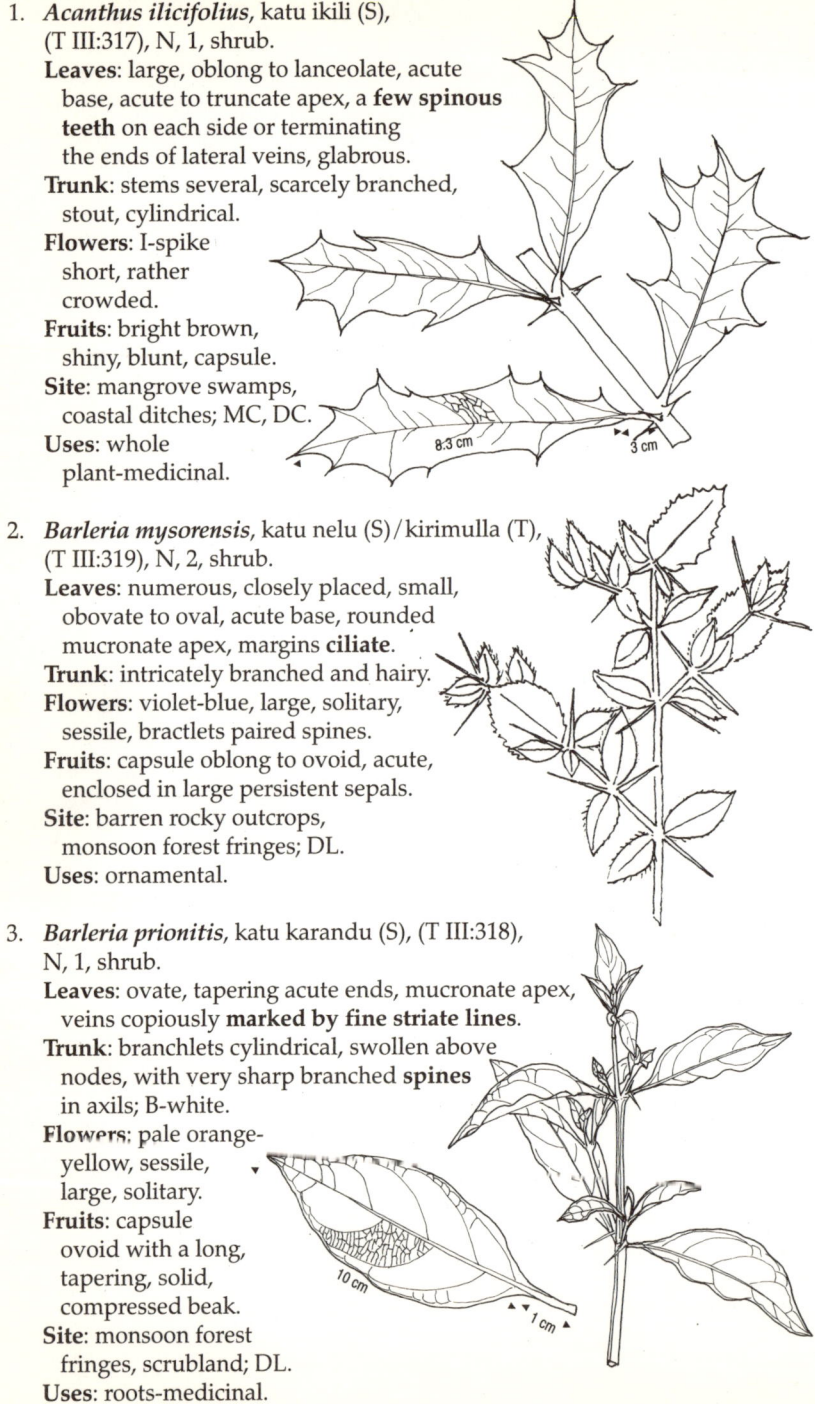

1. *Acanthus ilicifolius*, katu ikili (S),
   (T III:317), N, 1, shrub.
   Leaves: large, oblong to lanceolate, acute
   base, acute to truncate apex, a **few spinous
   teeth** on each side or terminating
   the ends of lateral veins, glabrous.
   Trunk: stems several, scarcely branched,
   stout, cylindrical.
   Flowers: I-spike
   short, rather
   crowded.
   Fruits: bright brown,
   shiny, blunt, capsule.
   Site: mangrove swamps,
   coastal ditches; MC, DC.
   Uses: whole
   plant-medicinal.

2. *Barleria mysorensis*, katu nelu (S)/kirimulla (T),
   (T III:319), N, 2, shrub.
   Leaves: numerous, closely placed, small,
   obovate to oval, acute base, rounded
   mucronate apex, margins **ciliate**.
   Trunk: intricately branched and hairy.
   Flowers: violet-blue, large, solitary,
   sessile, bractlets paired spines.
   Fruits: capsule oblong to ovoid, acute,
   enclosed in large persistent sepals.
   Site: barren rocky outcrops,
   monsoon forest fringes; DL.
   Uses: ornamental.

3. *Barleria prionitis*, katu karandu (S), (T III:318),
   N, 1, shrub.
   Leaves: ovate, tapering acute ends, mucronate apex,
   veins copiously **marked by fine striate lines**.
   Trunk: branchlets cylindrical, swollen above
   nodes, with very sharp branched **spines**
   in axils; B-white.
   Flowers: pale orange-
   yellow, sessile,
   large, solitary.
   Fruits: capsule
   ovoid with a long,
   tapering, solid,
   compressed beak.
   Site: monsoon forest
   fringes, scrubland; DL.
   Uses: roots-medicinal.

4. *Ecbolium viride*, (T III:341 & VI:229), N, 1, shrub.
   Leaves: large, oblong to oval to lanceolate,
   tapering base,
   acuminate to acute
   apex, **densely pubes-
   cent** beneath, margins
   entire to faintly crenate.
   Trunk: branches erect, cylindrical, thickened above nodes.
   Flowers: pale bluish green,
   large; I-spikes.
   Fruits: capsules pubescent.
   Site: scrub; DL.

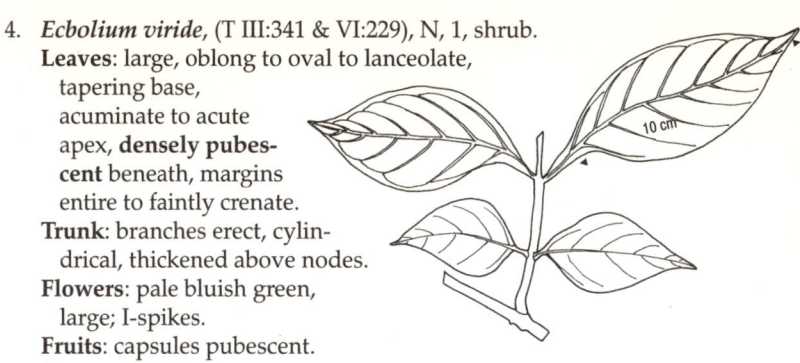

5. *Justicia adhatoda*, agaladara (S)/
   pavettai (T), (T III:338), N, 2, shrub.
   Leaves: large, lanceolate, tapering
   base, acuminate to subacute
   apex, margins **faintly crenate**,
   veins reticulate.
   Trunk: many, long,
   opposite,
   ascending
   branches;
   B-yellowish.
   Flowers: white;
   I-large, dense,
   axillary spikes.
   Fruits: pubescent,
   club-shaped capsules.
   Site: fences, home gardens; DL, W.
   Uses: live fencing; whole plant-medicinal;
   leaves-green manure.

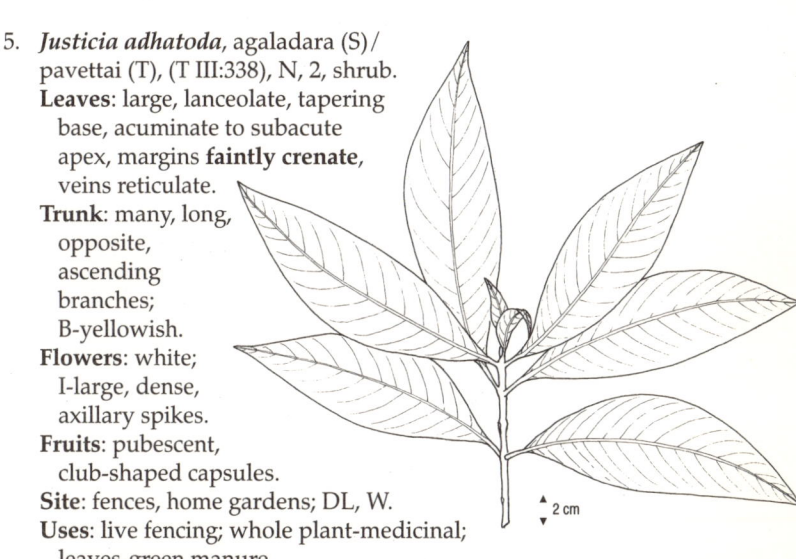

6. *Pseuderanthemum atropurpureum*,
   I (Polynesia), 2, shrub.
   Leaves: oblong oval to
   lanceolate, acuminate
   apex, variegated
   coppery-purple to
   greenish yellow
   and white.
   Trunk: erect; branches purplish,
   fleshy and twiggy.
   Flowers: purplish-rose with
   red markings, tubular.
   Fruits: capsules.
   Site: gardens; DL, IN, W.
   Uses: ornamental.

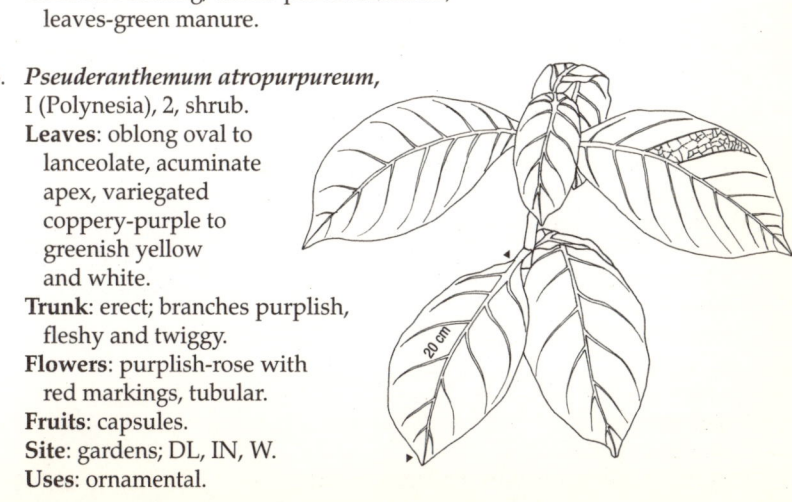

7. *Pseuderanthemum reticulatum*, golden
   eranthemum (E), I (Polynesia), 2, shrub.
   Leaves: yellow to greenish yellow, veins
      yellow on older foliage.
   Trunk: erect, **stout**, fleshy branches.
   Flowers: white and spotted
      purple,tubular; I-spikes terminal.
   Fruits: capsules.
   Site: gardens; DL, W.
   Uses: ornamental.

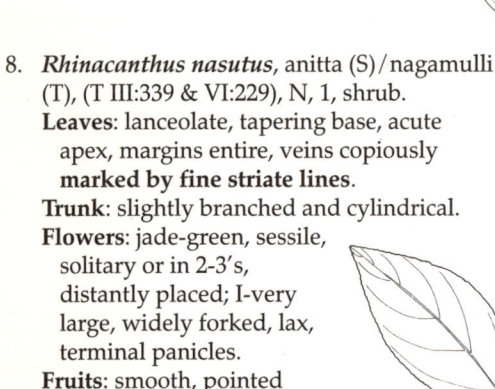

8. *Rhinacanthus nasutus*, anitta (S)/nagamulli
   (T), (T III:339 & VI:229), N, 1, shrub.
   Leaves: lanceolate, tapering base, acute
      apex, margins entire, veins copiously
      marked by fine striate lines.
   Trunk: slightly branched and cylindrical.
   Flowers: jade-green, sessile,
      solitary or in 2-3's,
      distantly placed; I-very
      large, widely forked, lax,
      terminal panicles.
   Fruits: smooth, pointed
      capsule, without hairs.
   Site: roadsides, monsoon forest fringes; DL.
   Uses: leaves, root, B-medicinal.

9. *Stenosiphonium cordifolium*, bunelu (S)/
   nelu (T), (T III:298 & VI:226), N, 1.5,
   shrub.
   Leaves: oval, tapering base, caudate-
      acuminate to obtuse apex, margins
      coarsely **serrate**, numerous **parallel**
      lateral veins; petiole length variable.
   Trunk: thickened stem nodes.
   Flowers: pale violet,
      darker spots on
      lobes; I-lax,
      axillary and
      terminal
      spikes.
   Fruits: capsules
      protrude beyond calyx.
   Site: scrub, monsoon forest understory; DL.

10. *Strobilanthes anceps*, nelu (S), (T III:307), E, 1, small shrub.
   Leaves: variable, oval, acute base, acuminate to acute apex,
      margins entire to slightly dentate, above **hairy**, below
      strongly **ciliate**; faintly aromatic.
   Trunk: stems swollen above nodes,
      subquadrangular to slightly
      winged and widely forked.
   Flowers: white, not very crowded;
      I-terminal and lateral spikes.
   Fruits: oblong capsule.
   Site: forest understory;
      M, LM.

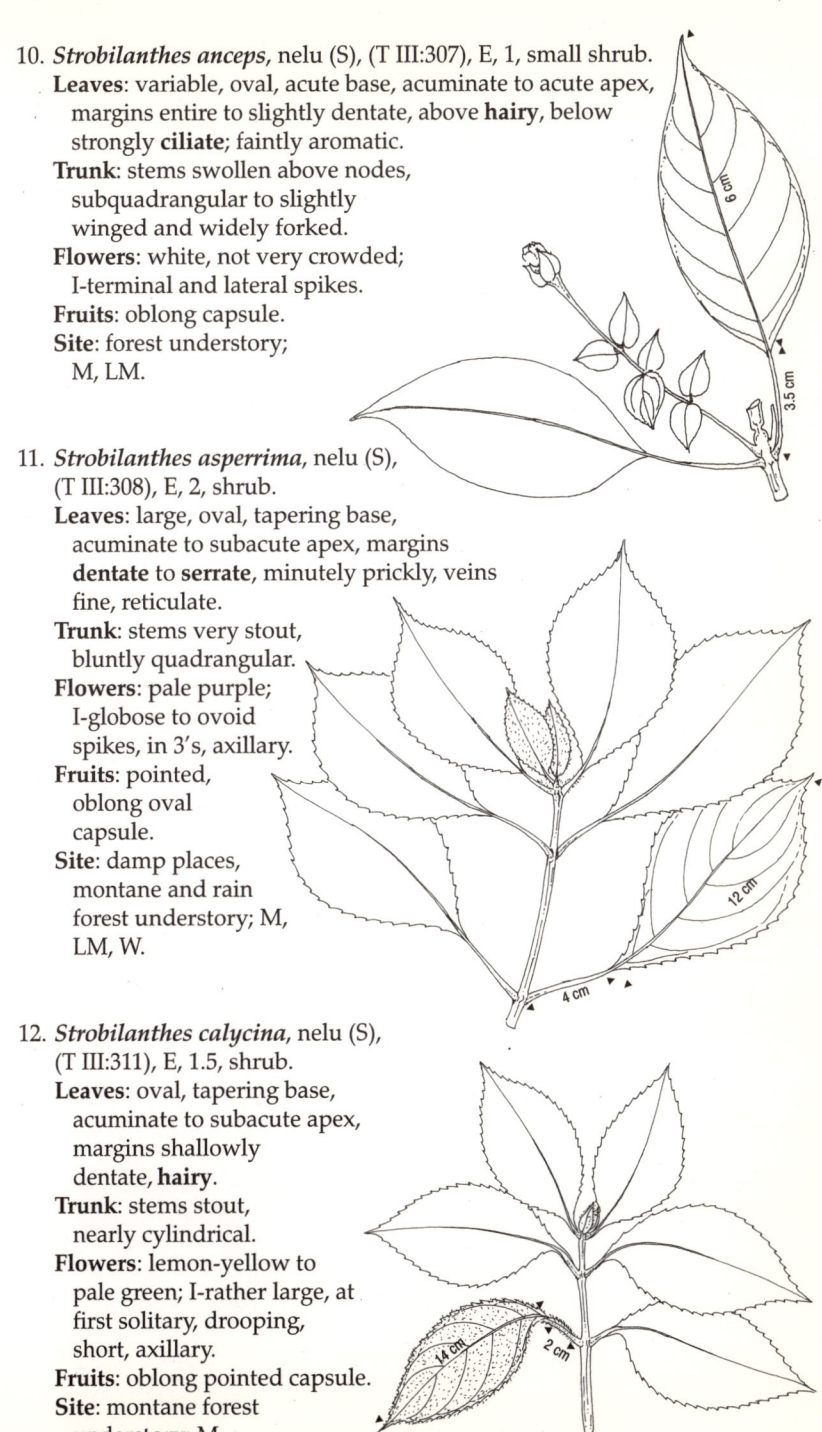

11. *Strobilanthes asperrima*, nelu (S),
   (T III:308), E, 2, shrub.
   Leaves: large, oval, tapering base,
      acuminate to subacute apex, margins
      dentate to **serrate**, minutely prickly, veins
      fine, reticulate.
   Trunk: stems very stout,
      bluntly quadrangular.
   Flowers: pale purple;
      I-globose to ovoid
      spikes, in 3's, axillary.
   Fruits: pointed,
      oblong oval
      capsule.
   Site: damp places,
      montane and rain
      forest understory; M,
      LM, W.

12. *Strobilanthes calycina*, nelu (S),
   (T III:311), E, 1.5, shrub.
   Leaves: oval, tapering base,
      acuminate to subacute apex,
      margins shallowly
      dentate, **hairy**.
   Trunk: stems stout,
      nearly cylindrical.
   Flowers: lemon-yellow to
      pale green; I-rather large, at
      first solitary, drooping,
      short, axillary.
   Fruits: oblong pointed capsule.
   Site: montane forest
      understory; M.

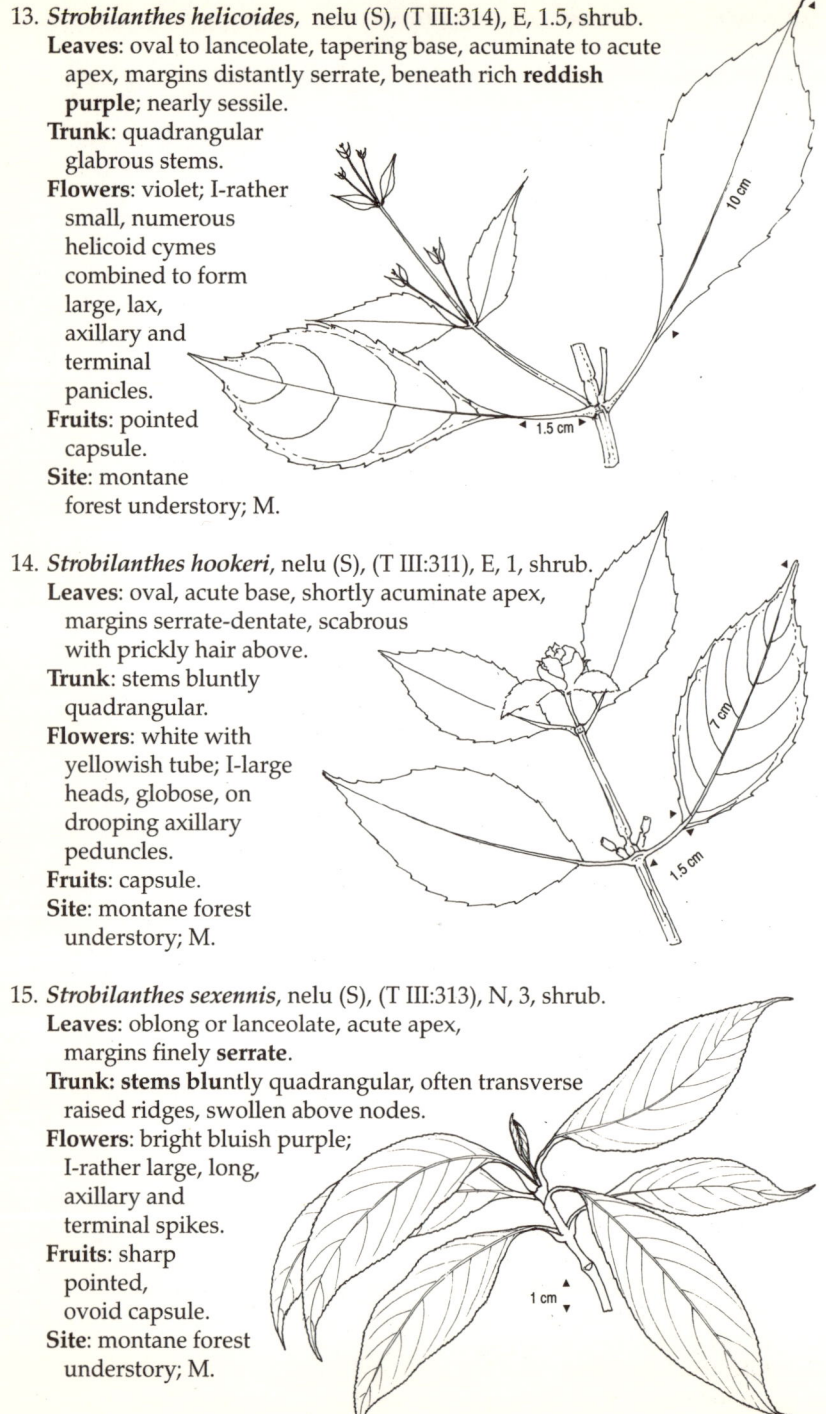

13. *Strobilanthes helicoides*, nelu (S), (T III:314), E, 1.5, shrub.
    Leaves: oval to lanceolate, tapering base, acuminate to acute
       apex, margins distantly serrate, beneath rich **reddish
       purple**; nearly sessile.
    Trunk: quadrangular
       glabrous stems.
    Flowers: violet; I-rather
       small, numerous
       helicoid cymes
       combined to form
       large, lax,
       axillary and
       terminal
       panicles.
    Fruits: pointed
       capsule.
    Site: montane
       forest understory; M.

14. *Strobilanthes hookeri*, nelu (S), (T III:311), E, 1, shrub.
    Leaves: oval, acute base, shortly acuminate apex,
       margins serrate-dentate, scabrous
       with prickly hair above.
    Trunk: stems bluntly
       quadrangular.
    Flowers: white with
       yellowish tube; I-large
       heads, globose, on
       drooping axillary
       peduncles.
    Fruits: capsule.
    Site: montane forest
       understory; M.

15. *Strobilanthes sexennis*, nelu (S), (T III:313), N, 3, shrub.
    Leaves: oblong or lanceolate, acute apex,
       margins finely **serrate**.
    Trunk: stems bluntly quadrangular, often transverse
       raised ridges, swollen above nodes.
    Flowers: bright bluish purple;
       I-rather large, long,
       axillary and
       terminal spikes.
    Fruits: sharp
       pointed,
       ovoid capsule.
    Site: montane forest
       understory; M.

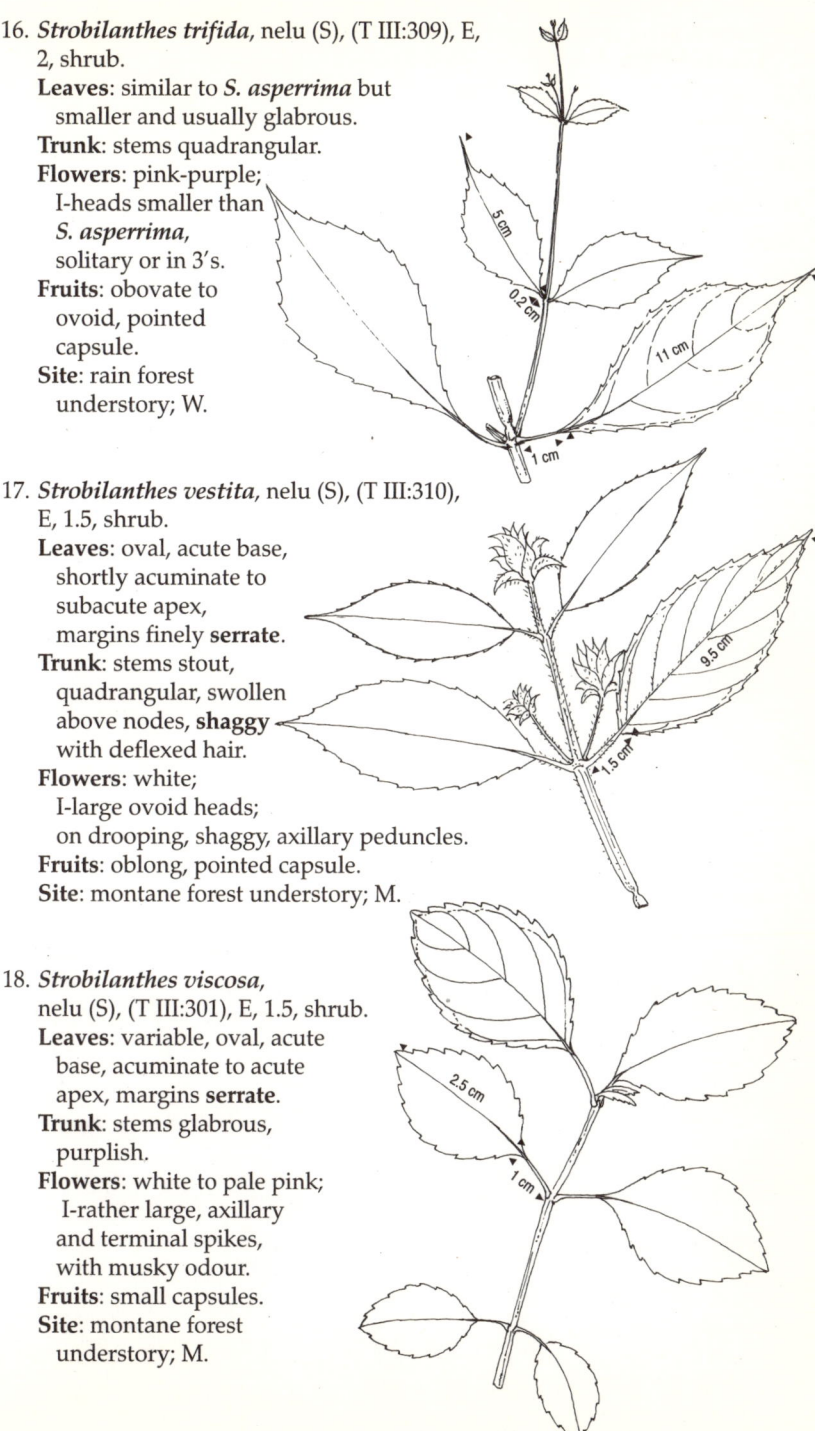

16. *Strobilanthes trifida*, nelu (S), (T III:309), E, 2, shrub.
   Leaves: similar to *S. asperrima* but smaller and usually glabrous.
   Trunk: stems quadrangular.
   Flowers: pink-purple; I-heads smaller than *S. asperrima*, solitary or in 3's.
   Fruits: obovate to ovoid, pointed capsule.
   Site: rain forest understory; W.

17. *Strobilanthes vestita*, nelu (S), (T III:310), E, 1.5, shrub.
   Leaves: oval, acute base, shortly acuminate to subacute apex, margins finely **serrate**.
   Trunk: stems stout, quadrangular, swollen above nodes, **shaggy** with deflexed hair.
   Flowers: white; I-large ovoid heads; on drooping, shaggy, axillary peduncles.
   Fruits: oblong, pointed capsule.
   Site: montane forest understory; M.

18. *Strobilanthes viscosa*, nelu (S), (T III:301), E, 1.5, shrub.
   Leaves: variable, oval, acute base, acuminate to acute apex, margins **serrate**.
   Trunk: stems glabrous, purplish.
   Flowers: white to pale pink; I-rather large, axillary and terminal spikes, with musky odour.
   Fruits: small capsules.
   Site: montane forest understory; M.

ACANTHACEAE

19. *Strobilanthes walkeri*, nelu (S), (T III:305), E, 1, small shrub.

Leaves: oval to lanceolate, tapering base, obtuse apex, margins **serrate**, **prickly hair** on both sides.

Trunk: bluntly quadrangular glabrous stems.

Flowers: white; I-solitary or 2-3 in axils, on stout, short quadrangular to slightly winged peduncles.

Fruits: large, oblong capsules.

Site: montane forest understory; M.

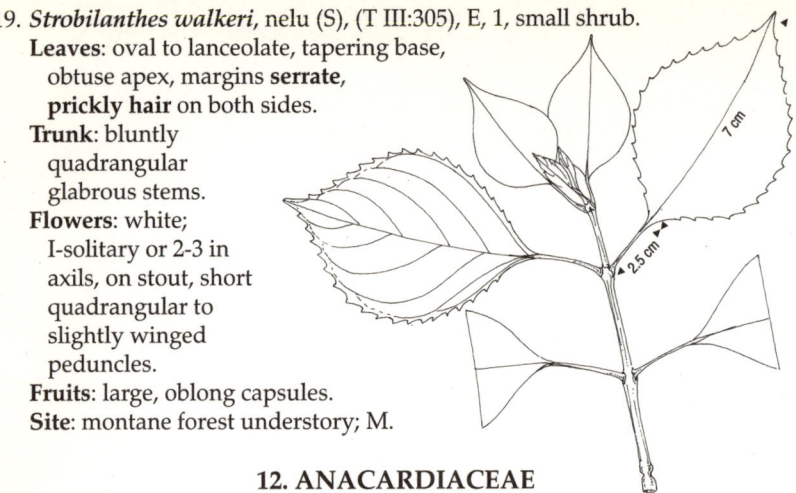

12. ANACARDIACEAE

FAMILY DESCRIPTION - Habit: trees or shrubs. Often with resinous bark. **Leaves**: spiral (rarely opposite or whorled). Simple or compound. **Stipules**: absent. **Flowers**: bisexual or unisexual, actinomorphic. **Fruits**: drupe.

FLOWER PARTS - CALYX: usually 5, free, valvate or imbricate, sometimes basally united. COROLLA: usually 5 (sometimes 4, rarely 3), free, valvate or imbricate, rarely united. ANDROECIUM: stamens 5-10, free, borne outside or on annular, conspicuous, often lobed nectary disk. Anthers 2-locular, opening lengthwise. GYNOECIUM: superior, 1-locular (rarely 2-5 locular), styles 1-3, usually separate; ovule solitary, pendulous.

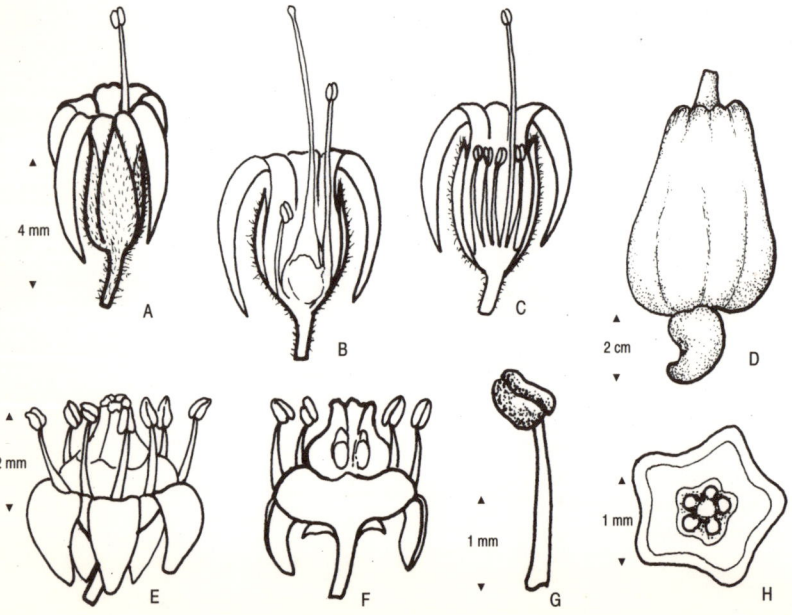

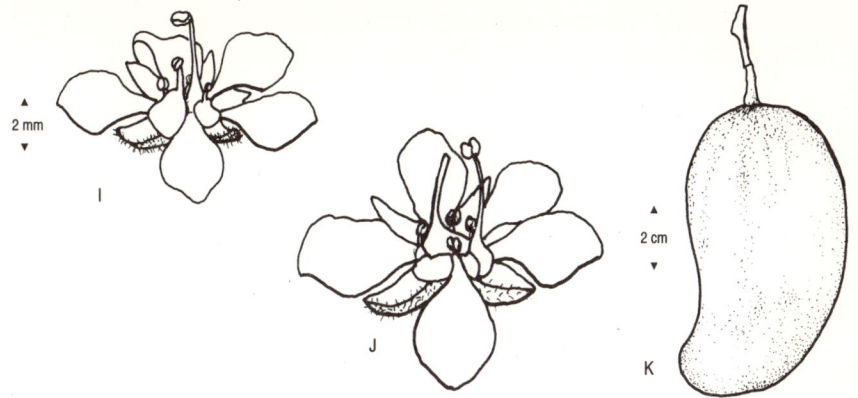

Key: full flower (A), half of a bisexual flower (B), half of a male flower (C), and fruit (D) of *Anacardium occidentale*. Full bisexual flower (E), half flower (F), single stamen (G), and transverse section of ovary (H) of *Spondias dulcis*. Full male flower (I), full bisexual flower (J), and fruit (K) of *Mangifera indica*.

1. *Anacardium occidentale*, cashew (E)/cadju (S)/montiri kai (T), (DF IV:8), I (Trop. America), 10, tree.
   Leaves: obovate, **rounded** apex, leathery.
   Trunk: B-smooth to fissured, light grey-brown; IB-red-brown, gummy; W-whitish to brown; exudate darkens on exposure.
   Flower: yellow, pink when old, fragrant, short-stalked; I-terminally crowded.
   Fruits: kidney-shaped nut on yellowish red, juicy stalk.
   Site: sandy coastal thickets, home gardens; DL, W.
   Uses: medicinal; fruit stalk, seed-edible; pericarp-cashew nutshell liquid used for brake linings and clutches; stem exudate-gum; bark exudate-ink for laundry marking; live fencing.

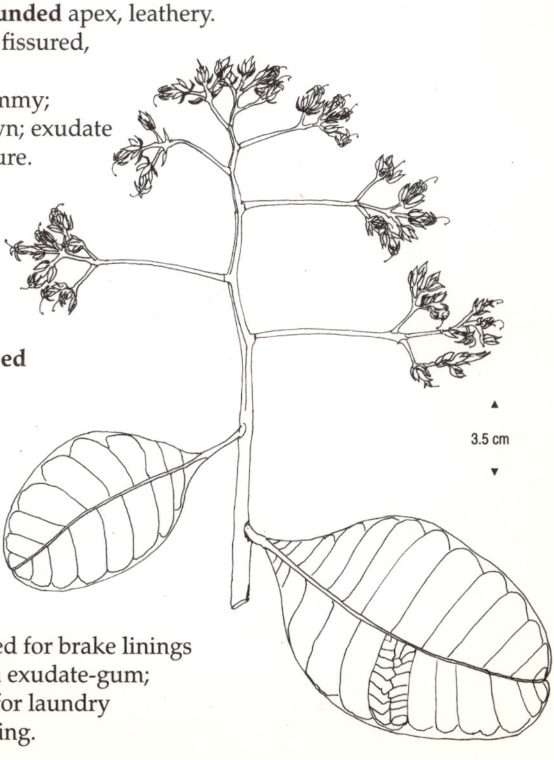

3.5 cm

2. *Campnosperma zeylanicum*, aridda (S), (DF IV:4), E, 30, tree.
   Leaves: terminally crowded, varying in size, elliptic to oblong, base tapered, apex rounded to retuse, lateral veins
   20-25 pairs nearly at right angles to midrib, beneath **rufous** scaly.
   Trunk: twigs conspicuously **leaf-scarred**; IB-red-brown layers.
   Flowers: I-axillary spikes.
   Fruits: purple, pulpy, apiculate drupe inside a low-rimmed disc.
   Site: rain forest canopy; W.
   Uses: ornamental; W-light construction.

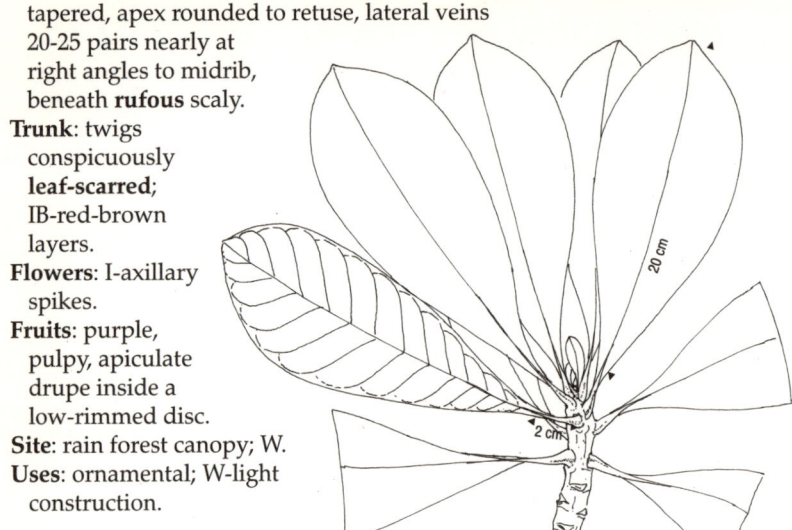

3. *Lannea coromandelica*, hik (S)/odi (T), (DF IV:21), N, 25, tree.
   Leaves: pinnate; leaflets paired with terminal, ovate, base rounded, slightly unequal, apex cordate to acuminate, entire or finely crenate, lateral veins 8 pairs, scattered stellate-hairy; deciduous.
   Trunk: B-scaly, fissured to smooth; IB-fibrous, striated red and white; sticky exudate.
   Flowers: I-axillary, sessile panicles.
   Fruits: flattened, kidney-shaped or rounded, in bunches on common stalk.
   Site: monsoon forest subcanopy, scrub; DL.
   Uses: young leaves and twigs-medicinal, edible; B-tanning; W-construction; gum-paint, lacquer.

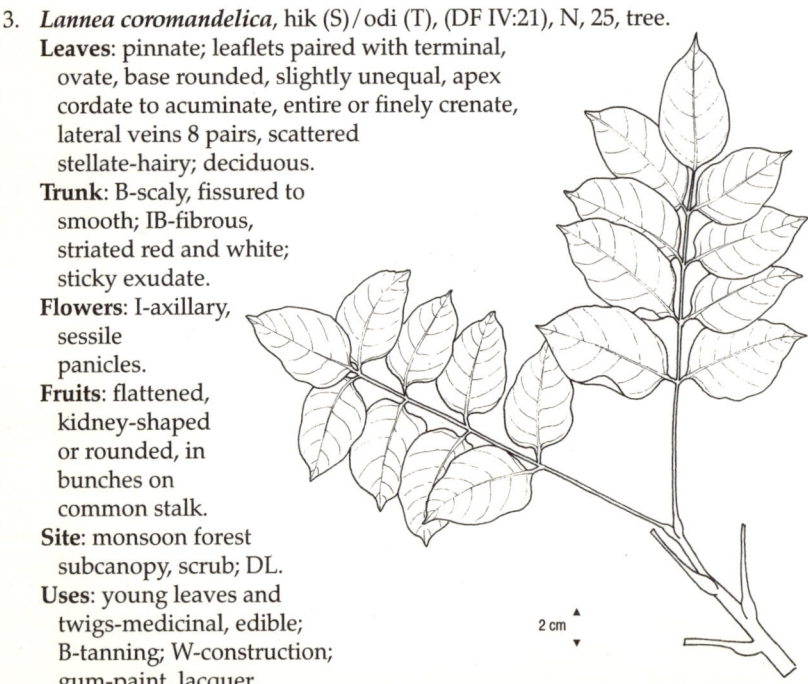

4. *Mangifera indica*, mango (E)/amba (s)/manga (T), (DF IV:6),
   I (India), 20, tree.
   Leaves: **lanceolate**, long pointed ends, **dark green**, red-brown flush,
   Trunk: B-smooth to
     furrowed, brown; IB-light
     brown; W,s-pale yellow;
     h-brown; exudate
     watery, astringent.
   Flowers: yellow-green
     to pink; I-showy,
     terminal clusters.
   Fruits: **large** elliptic,
     yellow-green
     drupe, fibrous pulp.
   Site: roadsides, home gardens; DL, IN, W.
   Uses: whole plant-medicinal; W-construction,
     furniture; fruit-edible.

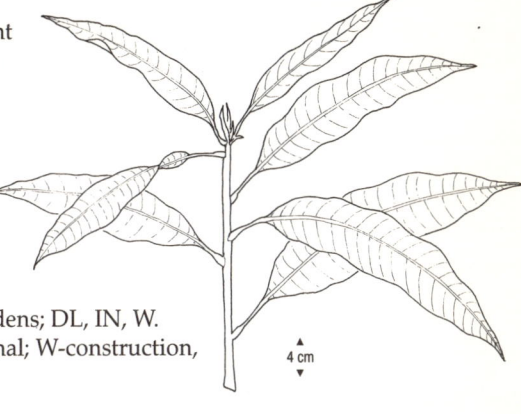

4 cm

12

5. *Mangifera zeylanica*, etamba (S), (DF IV:7), E, 30, tree.
   Leaves: lanceolate to ovate to oblong, base tapered,
     apex rounded or retuse (saplings acute), lateral
     veins about 7-10 pairs, coriaceous
     surface; **mango** smell.
   Trunk: B-rough, slightly
     fissured, corky lenticels,
     brownish grey;
     IB-**orange-yellow**;
     W,s-white.
   Flowers: I-large,
     terminal panicles.
   Fruits: small, ovoid drupes.
   Site: intermediate and rain
     forest canopy; IN, W.
   Uses: fruit-edible; W-light con-
     struction; whole plant-medicinal.

2 cm

ANACARDIACEAE

6. *Nothopegia beddomei*, bala (S), (DF IV:19), N, 10, tree.
   Leaves: oblong-lanceolate, base
     cuneate, acuminate apex, lateral
     veins up to 20 pairs disappearing
     into recurved margins.
   Trunk: B-closely fissured;
     IB-red-brown; exudate
     darkens on exposure,
     skin irritant.
   Flowers: I-in upper axils spicate,
     in lower axils paniculate, hairy.
   Fruits: purple, longitudinally striate drupe, copious pulp.
   Site: montane and rain forest sub canopy and understory; M, IN, W.

7. *Semecarpus coriacea*, badulla (S), (DF
IV:13), E, 10, tree.
   Leaves: oblong, base cuneate,
   rounded apex, lateral veins about
   6-10 pairs connected near margins,
   thick, coriaceous.
   Trunk: B-smooth,shining;
   exudate darkens on
   exposure, skin irritant.
   Flowers: I-terminal panicles.
   Fruits: small, compressed,
   apiculate drupe,
   scarcely cupped.
   Site: montane forest canopy; M.
   Uses: leaves, stem and flowers-medicinal.

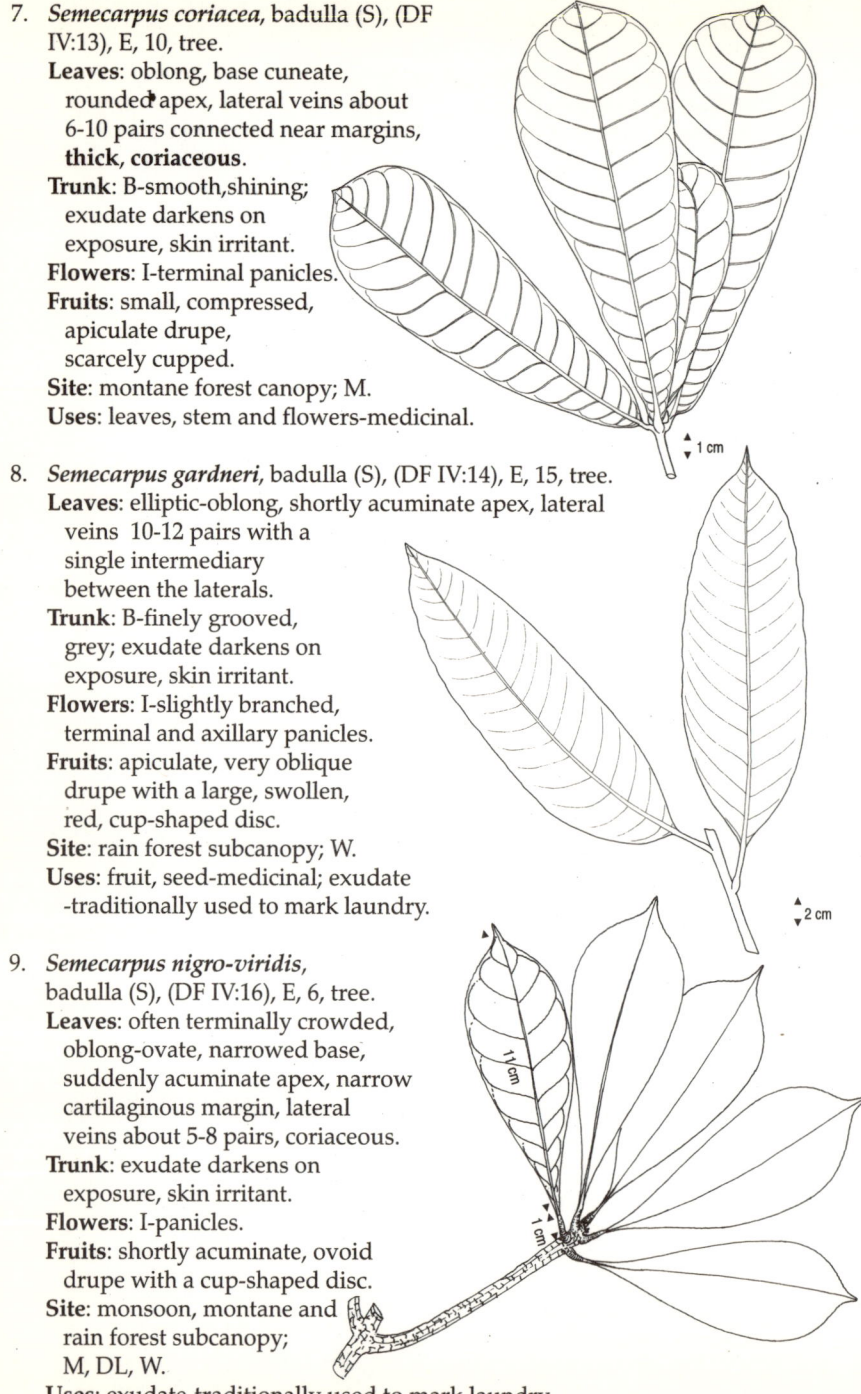

8. *Semecarpus gardneri*, badulla (S), (DF IV:14), E, 15, tree.
   Leaves: elliptic-oblong, shortly acuminate apex, lateral
   veins 10-12 pairs with a
   single intermediary
   between the laterals.
   Trunk: B-finely grooved,
   grey; exudate darkens on
   exposure, skin irritant.
   Flowers: I-slightly branched,
   terminal and axillary panicles.
   Fruits: apiculate, very oblique
   drupe with a large, swollen,
   red, cup-shaped disc.
   Site: rain forest subcanopy; W.
   Uses: fruit, seed-medicinal; exudate
   -traditionally used to mark laundry.

9. *Semecarpus nigro-viridis*,
   badulla (S), (DF IV:16), E, 6, tree.
   Leaves: often terminally crowded,
   oblong-ovate, narrowed base,
   suddenly acuminate apex, narrow
   cartilaginous margin, lateral
   veins about 5-8 pairs, coriaceous.
   Trunk: exudate darkens on
   exposure, skin irritant.
   Flowers: I-panicles.
   Fruits: shortly acuminate, ovoid
   drupe with a cup-shaped disc.
   Site: monsoon, montane and
   rain forest subcanopy;
   M, DL, W.
   Uses: exudate-traditionally used to mark laundry.

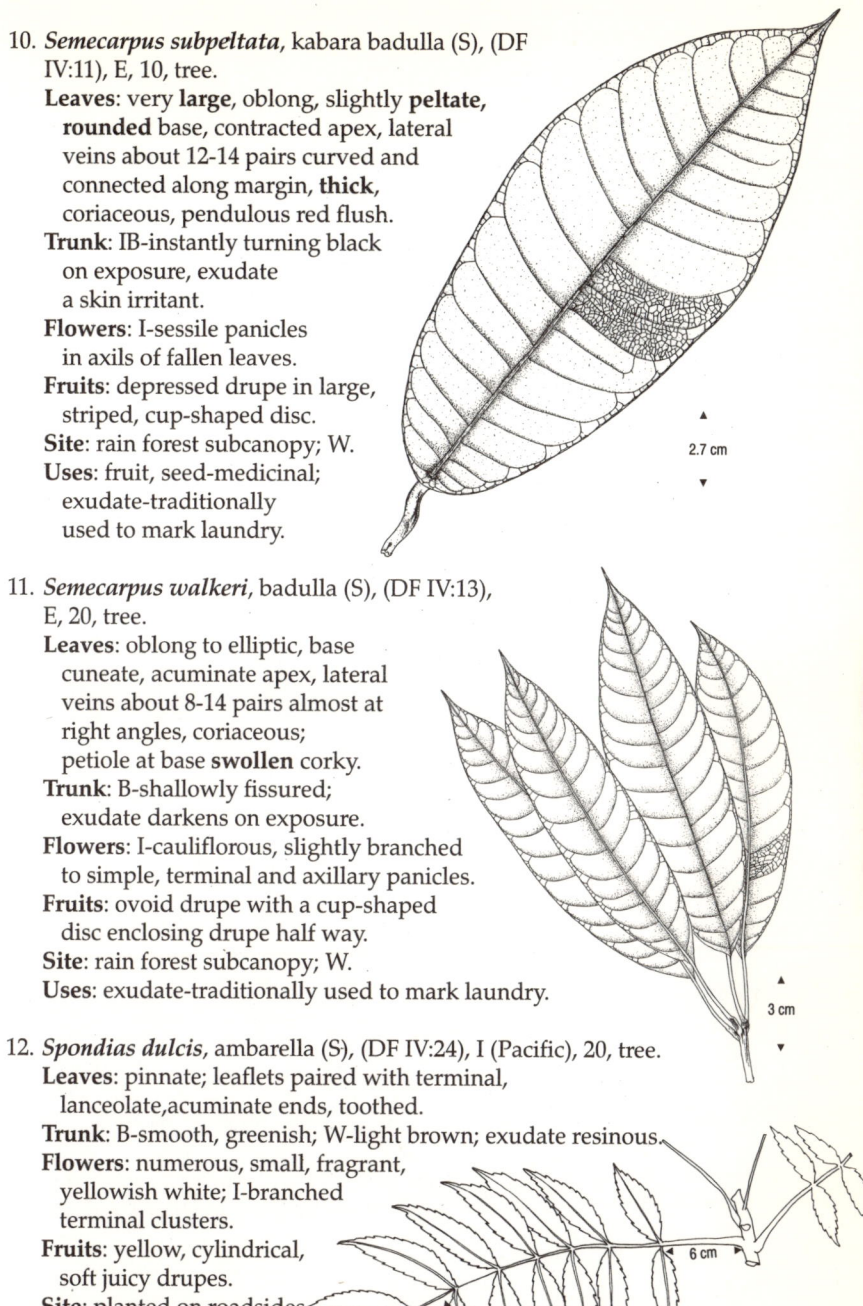

10. **_Semecarpus subpeltata_**, kabara badulla (S), (DF
   IV:11), E, 10, tree.
   Leaves: very **large**, oblong, slightly **peltate,
   rounded** base, contracted apex, lateral
   veins about 12-14 pairs curved and
   connected along margin, **thick**,
   coriaceous, pendulous red flush.
   Trunk: IB-instantly turning black
   on exposure, exudate
   a skin irritant.
   Flowers: I-sessile panicles
   in axils of fallen leaves.
   Fruits: depressed drupe in large,
   striped, cup-shaped disc.
   Site: rain forest subcanopy; W.
   Uses: fruit, seed-medicinal;
   exudate-traditionally
   used to mark laundry.

2.7 cm

11. **_Semecarpus walkeri_**, badulla (S), (DF IV:13),
   E, 20, tree.
   Leaves: oblong to elliptic, base
   cuneate, acuminate apex, lateral
   veins about 8-14 pairs almost at
   right angles, coriaceous;
   petiole at base **swollen** corky.
   Trunk: B-shallowly fissured;
   exudate darkens on exposure.
   Flowers: I-cauliflorous, slightly branched
   to simple, terminal and axillary panicles.
   Fruits: ovoid drupe with a cup-shaped
   disc enclosing drupe half way.
   Site: rain forest subcanopy; W.
   Uses: exudate-traditionally used to mark laundry.

3 cm

12. **_Spondias dulcis_**, ambarella (S), (DF IV:24), I (Pacific), 20, tree.
   Leaves: pinnate; leaflets paired with terminal,
   lanceolate,acuminate ends, toothed.
   Trunk: B-smooth, greenish; W-light brown; exudate resinous.
   Flowers: numerous, small, fragrant,
   yellowish white; I-branched
   terminal clusters.
   Fruits: yellow, cylindrical,
   soft juicy drupes.
   Site: planted on roadsides,
   home gardens; W.
   Uses: whole plant-
   medicinal;
   fruit-edible, sour.

6 cm

8 cm

22 cm

12

ANACARDIACEAE

13. ANNONACEAE

FAMILY DESCRIPTION - Habit: trees, shrubs or climbers. All woody. **Leaves**: spiral, simple, entire, distichous. **Stipules**: absent. **Flowers**: bisexual, actinomorphic. **Fruits**: when apocarpous with sessile or stalked, dehiscent or indehiscent, woody or succulent fruitlets. When syncarpous fleshy and many-seeded. Seeds large, ovoid or ellipsoid, glabrous, turgid or compressed.

FLOWER PARTS - CALYX: 3, free or partly united, slightly imbricate or valvate. COROLLA: 6, in two whorls, imbricate or valvate in each whorl. Free or partly united at base. ANDROECIUM: stamens numerous, free, spirally arranged or in trimerous whorls. Filaments very short. Anthers 2-locular opening lengthwise, often overtopped by truncate, enlarged connective. GYNOECIUM: superior, carpels numerous or few. Apocarpous or syncarpous. Styles separate. Ovules 1 to many. Placentation basal or parietal.

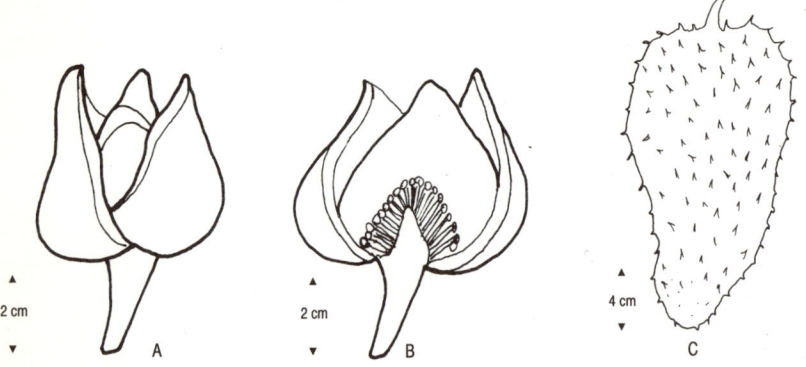

2 cm   2 cm   4 cm

A   B   C

Key: full flower (A), half flower (B) and fruit (C) of *Annona muricata*.

1. *Alphonsea sclerocarpa*, (DF V:30), N, 17, tree.
   Leaves: ovate to narrowly
      elliptical, obtuse to
      rarely acute
      base,tapering
      to rounded apex,
      lateral veins
      about 6-13 pairs.
   Trunk: young branches
      minutely pubescent.
   Flowers: I-pale yellow, sessile.
   Fruits: warty, hard, woody carpels,
      coarsely verrucose, finely
      yellowish tomentose.
   Site: scrub, monsoon forest; DL.

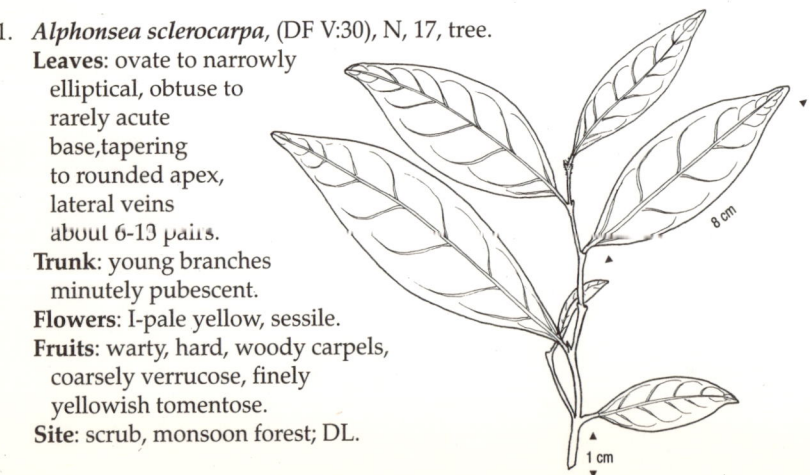

8 cm

1 cm

2. *Alphonsea zeylanica*, (DF V:29), E, 8, tree.
   Leaves: ovate to elliptic to lanceolate, ends acute,
   lateral veins about 7-12 pairs.
   Trunk: B-smooth, grey.
   Flowers: I-sessile,
   pubescent.
   Fruits: globose,
   densely fine tomentose,
   hard and woody.
   Site: intermediate
   and montane forest
   subcanopy; LM, IN.

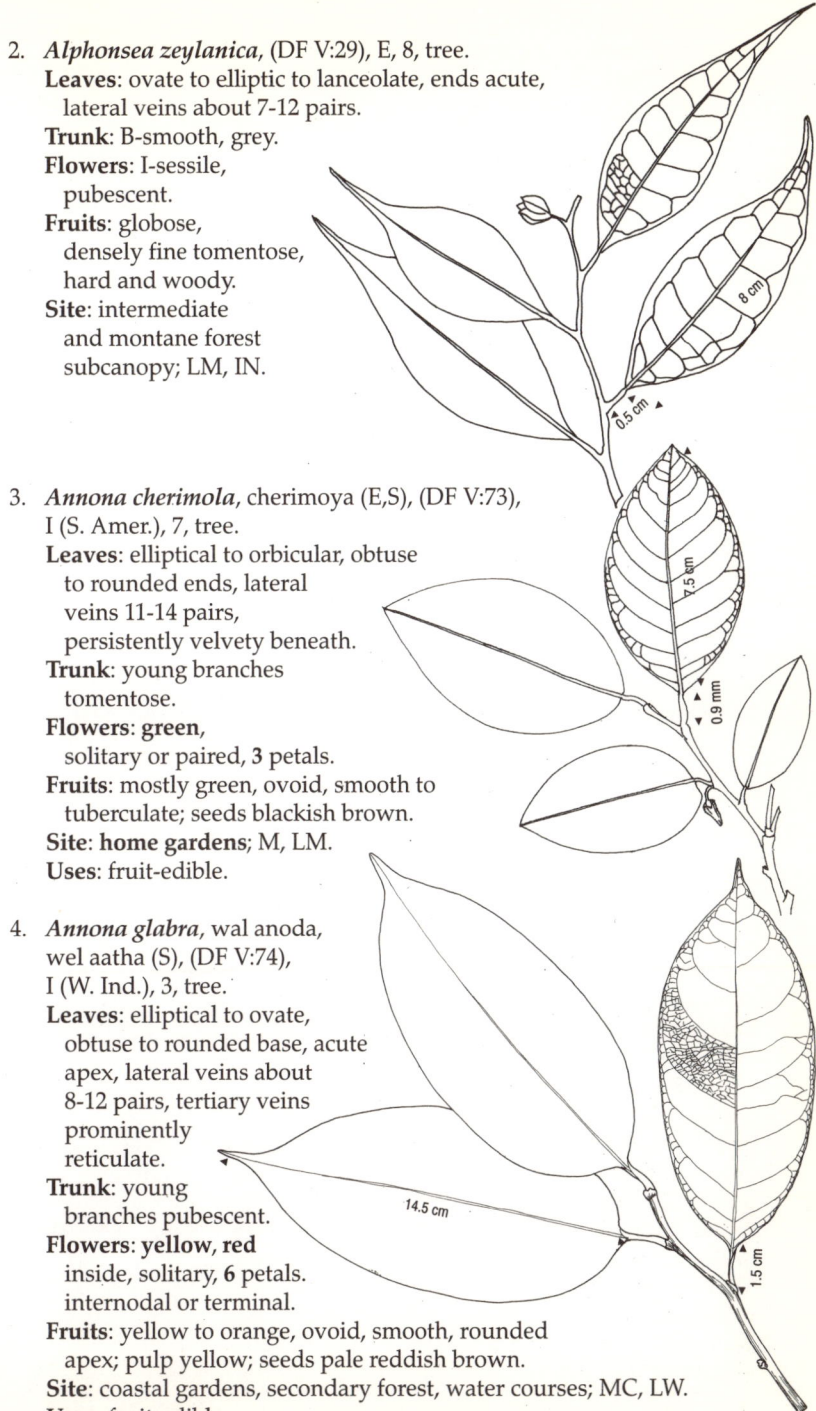

3. *Annona cherimola*, cherimoya (E,S), (DF V:73),
   I (S. Amer.), 7, tree.
   Leaves: elliptical to orbicular, obtuse
   to rounded ends, lateral
   veins 11-14 pairs,
   persistently velvety beneath.
   Trunk: young branches
   tomentose.
   Flowers: **green**,
   solitary or paired, **3** petals.
   Fruits: mostly green, ovoid, smooth to
   tuberculate; seeds blackish brown.
   Site: **home gardens**; M, LM.
   Uses: fruit-edible.

4. *Annona glabra*, wal anoda,
   wel aatha (S), (DF V:74),
   I (W. Ind.), 3, tree.
   Leaves: elliptical to ovate,
   obtuse to rounded base, acute
   apex, lateral veins about
   8-12 pairs, tertiary veins
   prominently
   reticulate.
   Trunk: young
   branches pubescent.
   Flowers: **yellow, red**
   inside, solitary, **6** petals.
   internodal or terminal.
   Fruits: yellow to orange, ovoid, smooth, rounded
   apex; pulp yellow; seeds pale reddish brown.
   Site: coastal gardens, secondary forest, water courses; MC, LW.
   Uses: fruit-edible.

ANNONACEAE

13

5. **Annona muricata**, soursop (E)/katu aatha (S),
   (DF V:72), I (W. Ind.), 8, tree.
   Leaves: oblong to ovate,
     acute ends; **aromatic**.
   Trunk: B-smooth, brown;
     IB-pinkish; W,s-whitish;
     h-brown, soft, light-weight.
   Flowers: **6 pale yellow** petals,
     broad green sepals, solitary.
   Fruits: green, fleshy, **spined**, egg-
     shaped to elliptic;pulp white;
     seeds blackish brown.
   Site: home gardens; widespread.
   Uses: leaves, fruit-medicinal,
     insecticidal (lice); fruit-edible.

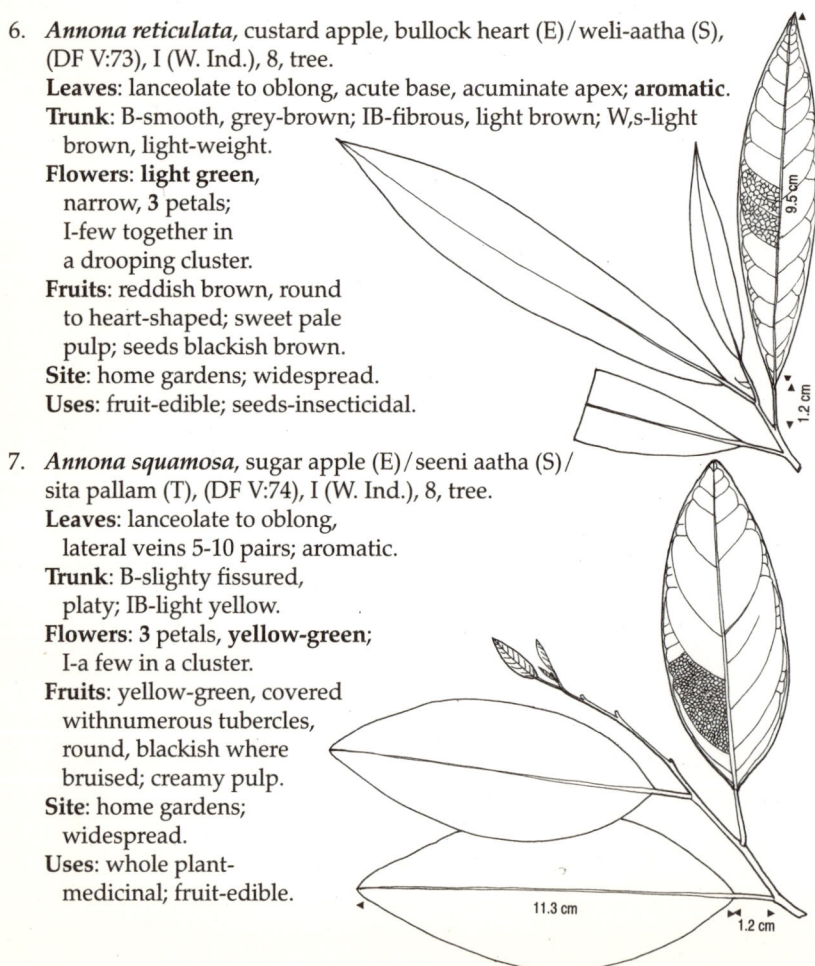

6. **Annona reticulata**, custard apple, bullock heart (E)/weli-aatha (S),
   (DF V:73), I (W. Ind.), 8, tree.
   Leaves: lanceolate to oblong, acute base, acuminate apex; **aromatic**.
   Trunk: B-smooth, grey-brown; IB-fibrous, light brown; W,s-light
     brown, light-weight.
   Flowers: **light green**,
     narrow, **3** petals;
     I-few together in
     a drooping cluster.
   Fruits: reddish brown, round
     to heart-shaped; sweet pale
     pulp; seeds blackish brown.
   Site: home gardens; widespread.
   Uses: fruit-edible; seeds-insecticidal.

7. **Annona squamosa**, sugar apple (E)/seeni aatha (S)/
   sita pallam (T), (DF V:74), I (W. Ind.), 8, tree.
   Leaves: lanceolate to oblong,
     lateral veins 5-10 pairs; aromatic.
   Trunk: B-slighty fissured,
     platy; IB-light yellow.
   Flowers: **3** petals, **yellow-green**;
     I-a few in a cluster.
   Fruits: yellow-green, covered
     withnumerous tubercles,
     round, blackish where
     bruised; creamy pulp.
   Site: home gardens;
     widespread.
   Uses: whole plant-
     medicinal; fruit-edible.

8. *Cananga odorata*, ylang ylang (E)/wana sapu (S),
    (DF V:69), I (S.E. Asia), 7, tree.
    Leaves: ovate to
       narrowly elliptic,
       obtuse to truncate
       base, acute to acuminate
       apex, glabrous, lateral
       veins about 8-11 pairs.
    Trunk: B-Smooth, grey.
    Flowers: petals green
       turning yellow
       with purple base,
       drooping, fragrant.
    Fruits: fruitlets several,
       ovoid to almost
       globose, smooth.
    Site: naturalized in
       secondary forest; W.

9. *Cyathocalyx zeylanica*, ipetta (S), (DF V:54),
    N, 25, tree.
    Leaves: large, elliptical to ovate to
       lanceolate, lateral veins 8-12 pairs.
    Trunk: branchlets minutely
       puberulous hairs;
       B-smooth, pale brown.
    Flowers: **pale apple-green**;
       sweet-scented; I-solitary or paired.
    Fruits: yellow to black, glabrous,
       fleshy, ovoid, rounded ends.
    Site: pioneers of
       disturbed rain forest; W.
    Uses: W-lacquered sticks;
       flowers-petals chewed
       with betel.

10. *Desmos zeylanica*, (DF V:10), N, 4, small tree.
    Leaves: narrowly elliptic to oblong, mostly
       obtuse to rounded base,acute apex, lateral
       veins 10-15 pairs,whitish beneath.
    Trunk: branches few, sparsely pubescent.
    Flowers: **reddish brown**; I-solitary
       or paired from axils
       of fallen leaves.
    Fruits: approximately
       20 carpels,
       2-3-jointed.
    Site: rain forest
       understory; W.

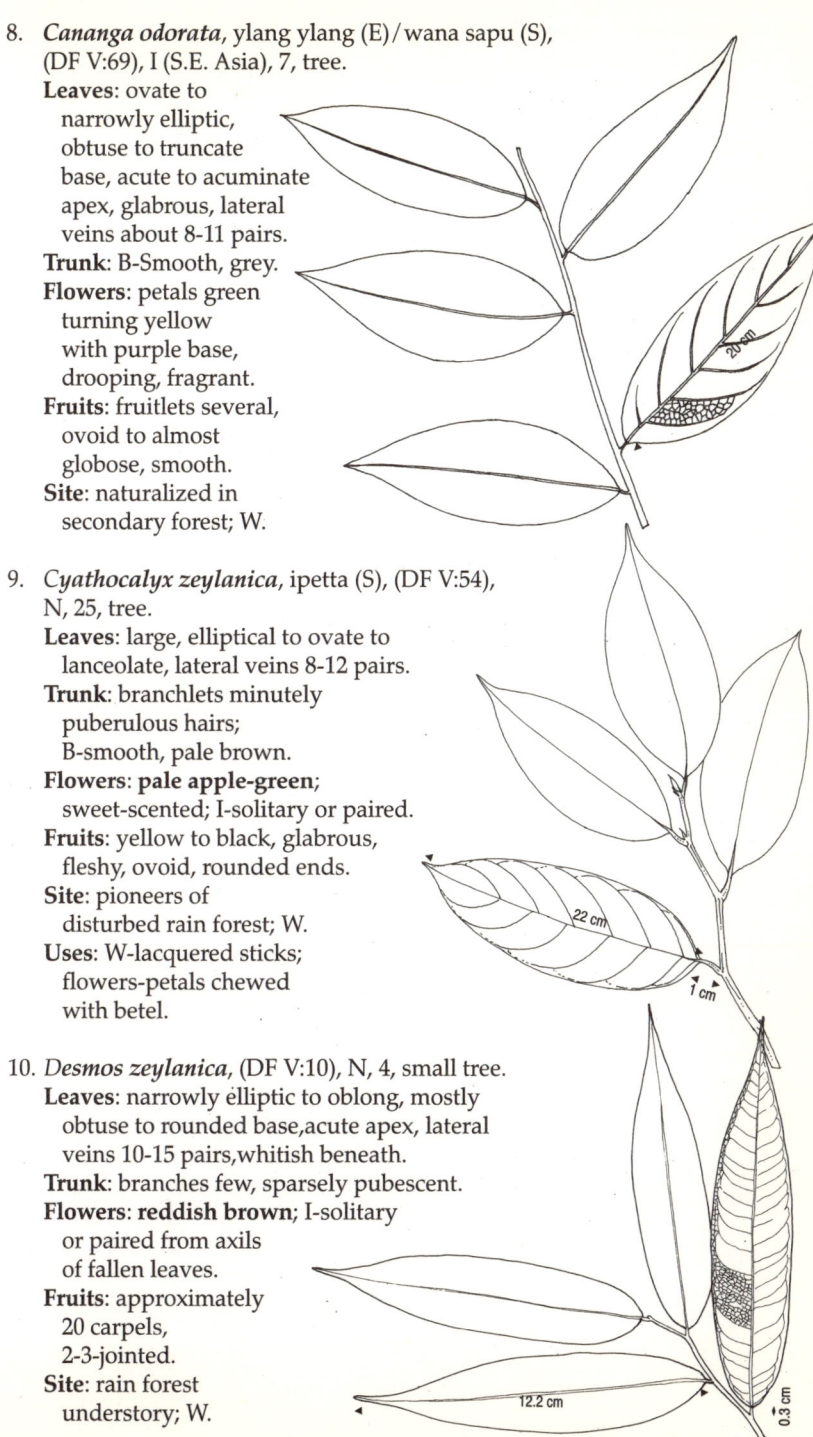

11. *Goniothalamus gardneri*, kalu kera (S),
    (DF V:63), E, 5, small tree.
    Leaves: oblong to narrowly
        elliptic, acute ends, lateral
        veins 14-20 pairs.
    Trunk: poorly branched,
        young branchlets
        sparsely pubescent.
    Flowers: **green to**
        yellow,
        solitary, axillary.
    Fruits: orange-yellow,
        up to 20 smooth fruitlets.
    Site: rain forest understory; W.

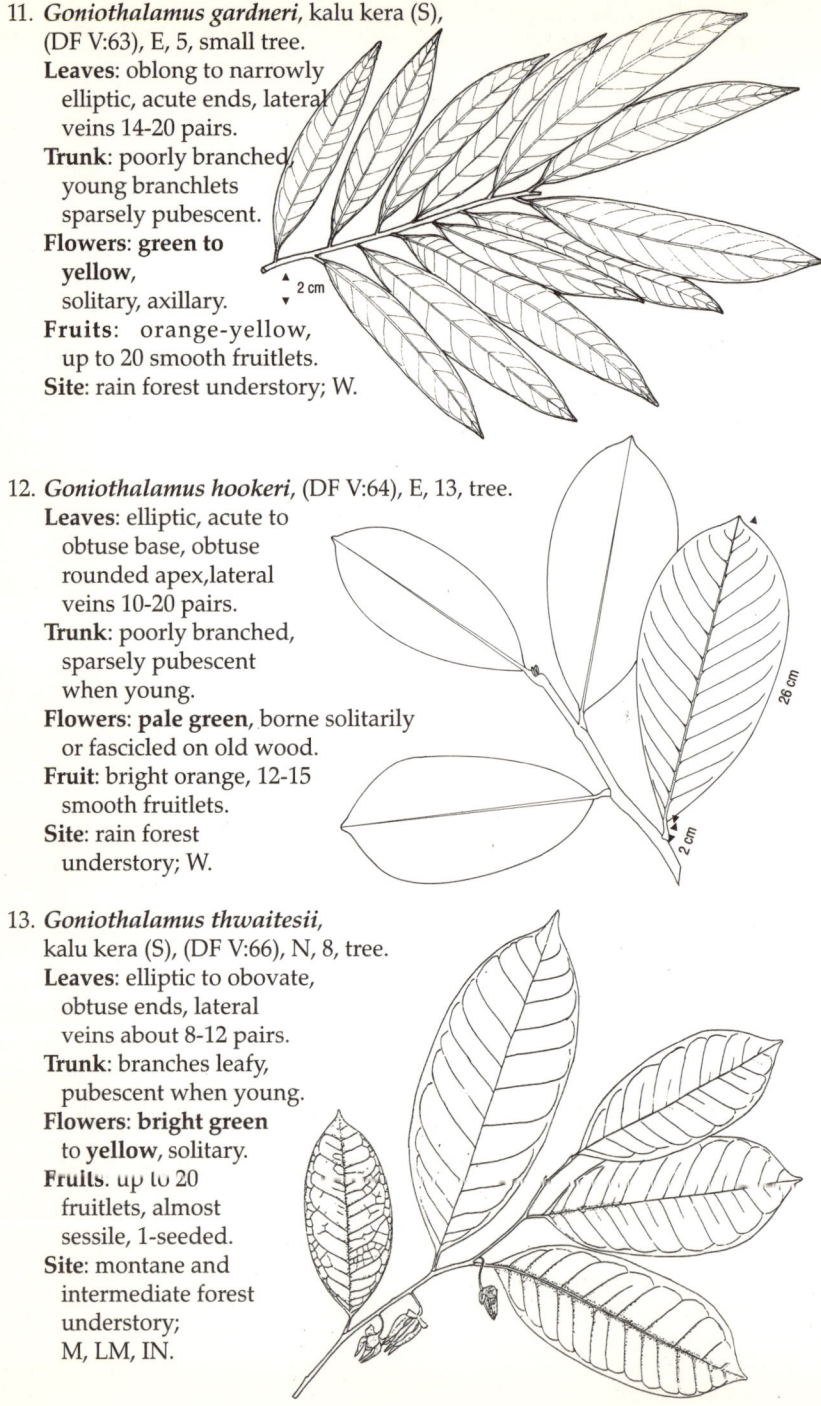

12. *Goniothalamus hookeri*, (DF V:64), E, 13, tree.
    Leaves: elliptic, acute to
        obtuse base, obtuse
        rounded apex,lateral
        veins 10-20 pairs.
    Trunk: poorly branched,
        sparsely pubescent
        when young.
    Flowers: **pale green**, borne solitarily
        or fascicled on old wood.
    Fruit: bright orange, 12-15
        smooth fruitlets.
    Site: rain forest
        understory; W.

13. *Goniothalamus thwaitesii*,
    kalu kera (S), (DF V:66), N, 8, tree.
    Leaves: elliptic to obovate,
        obtuse ends, lateral
        veins about 8-12 pairs.
    Trunk: branches leafy,
        pubescent when young.
    Flowers: **bright green**
        to **yellow**, solitary.
    Fruits. up to 20
        fruitlets, almost
        sessile, 1-seeded.
    Site: montane and
        intermediate forest
        understory;
        M, LM, IN.

14. *Miliusa indica*, kikili messa (S), (DF V:45), N, 6, tree.
    Leaves: rounded to subcordate
      base, obtuse apex,
      lateral veins 6-12
      pairs, sometimes
      pubescent beneath;
      petiole thick,
      flattened, swollen.
    Trunk: twigs densely
      fulvous hairy.
    Flowers: **yellow-purple**,
      drooping, solitary, axillary.
    Fruits: purple, ovoid, up to 10
      fruitlets; seeds yellow, smooth.
    Site: secondary forest, monsoon,
      intermediate and rain forest
      understory; DL, IN, W.

15. *Polyalthia cerasoides*, patta ul kenda (S), (DF V:35), N, 10, tree.
    Leaves: narrowly elliptic, obtuse to truncate base,
      long-acuminate to obtuse apex, lateral veins
      7-10 pairs, soft pubescence beneath.
    Trunk: branchlets yellow and
      densely pubescent.
    Flowers: green, solitary.
    Fruit: red, numerous roundish to
      ellipsoid, sparsely pubescent fruitlets.
    Site: monsoon forest, scrub; DL.

16. *Polyalthia coffeoides*,
    omara (S)/katilla (T),
    (DF V:34), N, 18, tree.
    Leaves: ovate to narrowly
      ovate, truncate to obtuse
      base, **tapering** to attenuate
      apex, lateral veins about
      10-14 pairs, margins **undulate**.
    Trunk: branchlets minutely
      densely puberulous.
    Flowers: **light-green**,
      solitary or a few
      from axillary peduncle.
    Fruits: dark purple, ovoid,
      bluntly pointed, smooth,
      shining fruitlets.
    Site: intermediate and
      monsoon forest
      understory; DL, IN.

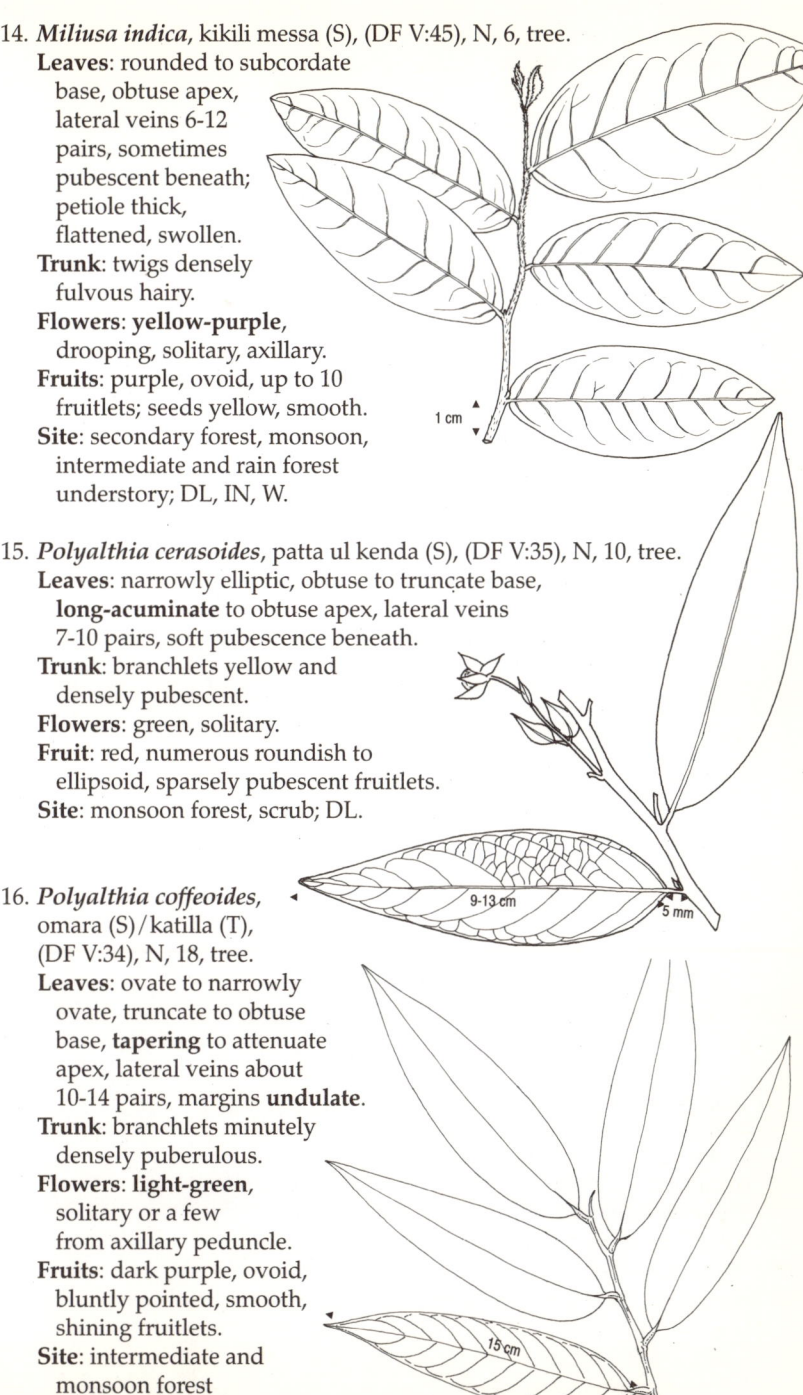

13

ANNONACEAE

17. *Polyalthia korinti*, ul kenda (S)/
    uluvintai (T), (DF V:36), N, 5, small tree.
    Leaves: ovate, elliptic or lanceolate,
       acute to obtuse base, acute apex,
       lateral veins 5-10 pairs.
    Trunk: young branchlets
       yellowish pubescent.
    Flowers: green,
       solitary or in 2's
       from leaf axils or
       woody tubercules
       of branches.
    Fruits: **bright crimson**,
       succulent, globose fruitlets.
    Site: riverine forest, mangroves,
       disturbed vegetation; DL, W.
    Uses: fruits-edible.

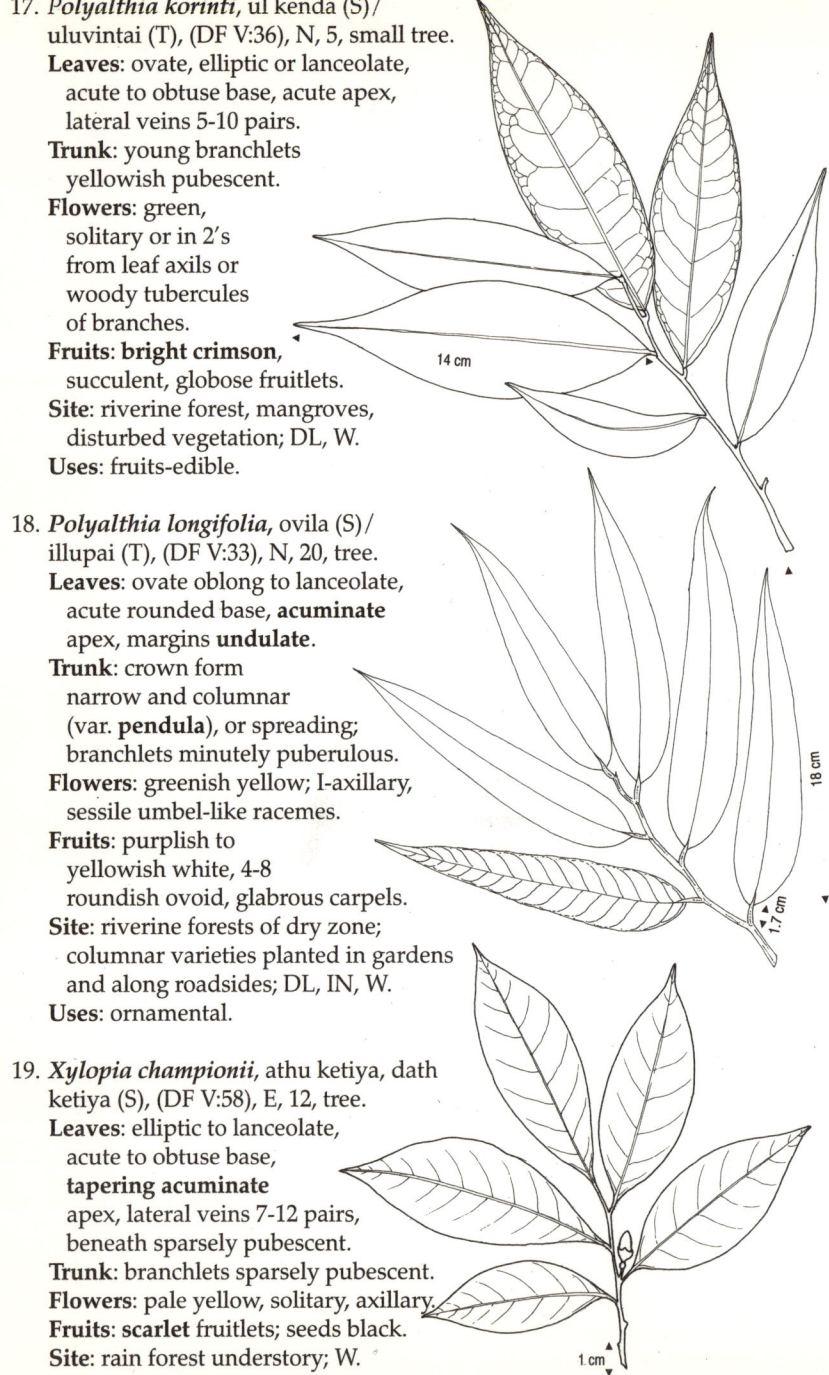

18. *Polyalthia longifolia*, ovila (S)/
    illupai (T), (DF V:33), N, 20, tree.
    Leaves: ovate oblong to lanceolate,
       acute rounded base, **acuminate**
       apex, margins **undulate**.
    Trunk: crown form
       narrow and columnar
       (var. **pendula**), or spreading;
       branchlets minutely puberulous.
    Flowers: greenish yellow; I-axillary,
       sessile umbel-like racemes.
    Fruits: purplish to
       yellowish white, 4-8
       roundish ovoid, glabrous carpels.
    Site: riverine forests of dry zone;
       columnar varieties planted in gardens
       and along roadsides; DL, IN, W.
    Uses: ornamental.

19. *Xylopia championii*, athu ketiya, dath
    ketiya (S), (DF V:58), E, 12, tree.
    Leaves: elliptic to lanceolate,
       acute to obtuse base,
       tapering acuminate
       apex, lateral veins 7-12 pairs,
       beneath sparsely pubescent.
    Trunk: branchlets sparsely pubescent.
    Flowers: pale yellow, solitary, axillary.
    Fruits: **scarlet** fruitlets; seeds black.
    Site: rain forest understory; W.

14. APOCYNACEAE

FAMILY DESCRIPTION - Habit: trees, shrubs or climbers, rarely perennial herbs. Latex present. **Leaves**: opposite or whorled (rarely spiral), simple, entire. **Stipules**: absent. **Flowers**: bisexual, actinomorphic, funnel-shaped. **Fruits**: usually 2 dehiscent follicles, more rarely indehiscent and fleshy. Seeds 1 to numerous, flattened, often winged or with long silky hairs.

FLOWER PARTS - Calyx: usually 5, rarely 4, free, often glandular inside, imbricate. Corolla: 5, united, usually funnel- or salver-shaped, contorted, rarely valvate or imbricate. Androecium: stamens 5, epipetalous, filaments free or rarely united, without coronal appendages. Anthers free or connivent around stigma, rarely adnate to stigma, 2-locular, opening lengthwise. 5 nectary glands near ovary base sometimes annular, reduced or absent. Gynoecium: superior, usually with two distinct carpels. Rarely 1-locular with two parietal placentas or 2-locular with placentas adnate to the septa.

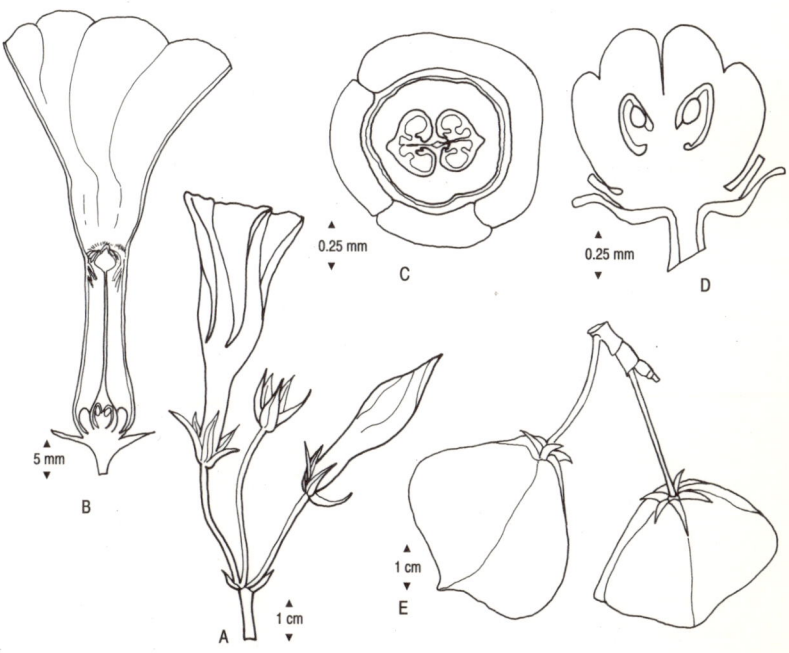

Key: inflorescence (A), half flower (B), transverse section of ovary surrounded by nectary (C), longitudinal section of ovary (D) and drupaceous fruits (E) of *Thevetia peruviana*.

1. *Allamanda cathartica*, yellow allamanda (E)/
   wal rukaththana (S), (DF IV:27),
   I (S. Amer.), 1, shrub.
   Leaves: whorled in **4's**,
      light green, pointed ends,
      entire margins, thick, smooth.
   Trunk: sprawling habit; milky latex.
   Flowers: **large**, **bright yellow**, tubular,
      in conspicuous terminal clusters.
   Fruits: spiny capsules.
   Site: home gardens, parks; widespread.
   Uses: ornamental; flowers-temple offerings.

   2 cm

2. *Alstonia macrophylla*, hawari nuga (S)/
   velai maram (T), (DF IV:41), I (Malesia), 20, tree.
   Leaves: in **whorls** of **3-4**, cuneate base, abruptly
      short acuminate apex, lateral veins distant,
      entire margin; young leaves
      densely pubescent beneath.
   Trunk: B-smooth, pale brown-
      grey; W-pale yellow-white,
      light-weight; milky latex.
   Flowers: white; I-cymes
      lax, in umbels.
   Fruits: **pendulous**
      follicles, glabrous
      when ripe.
   Site: secondary forest,
      reforested areas; IN, W.
   Uses: W-construction timber.

   23 cm

3. *Alstonia scholaris*, rukaththana (S)/ elilaippalai (T),
   (DF IV:42), 20, N, tree.
   Leaves: in **whorls** of **5-10**, cuneate base, rounded
      apex, entire margin, lateral veins numerous.
   Trunk: B-astringent; W-light, soft; milky latex.
   Flowers: white; I-dense cymes,
      clustered in umbels, fragrant.
   Fruits: **pendulous** follicles,
      glabrous when ripe.
   Site: particularly along
      stream ways in monsoon
      forest, rain forest; DL, W.
   Uses: live fencing;
      whole plant-
      medicinal;
      W-coffins,
      wood carving.

   2 cm

4. **Carissa carandas**, maha karamba (S)/kalaka (T), (DF IV:37), N, 3, shrub.
   Leaves: opposite, obovate to oblong, cuneate base, obtuse apex, entire margin.
   Trunk: simple **spines** at alternate nodes; milky latex.
   Flowers: white or pink, in 3's, in axillary and terminal clusters.
   Fruits: reddish purple berries.
   Site: home gardens; DL.
   Uses: leaves, fruit, root-medicinal; fruit-edible.

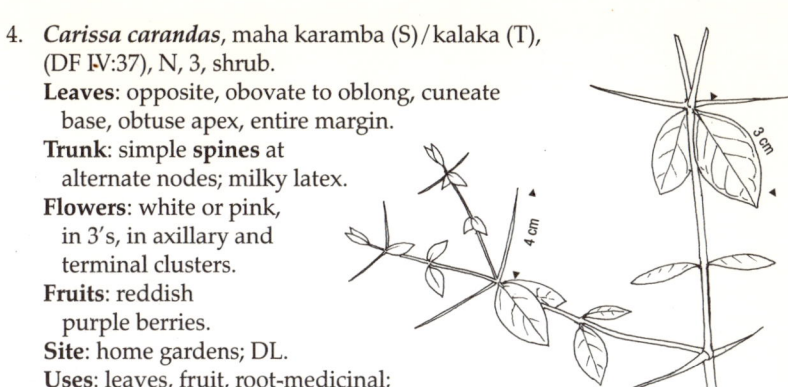

5. **Carissa grandiflora**, Natal plum (E)/damson (S), I (S. Africa), 3, spiny shrub.
   Leaves: ovate, rounded base, acute apex, lateral veins inconspicuous, thickly coriaceous.
   Trunk: many branched, spiny twigs; latex milky.
   Flowers: white, large, fragrant, solitary and terminal.
   Fruits: purplish-red, oblong to rotund, milky juice.
   Site: home gardens; DL, IN, W.
   Uses: hedges; fruit-edible (jam).

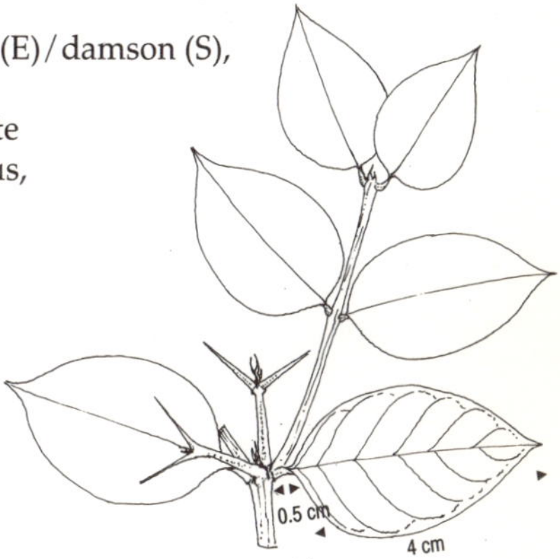

6. **Carissa spinarum**, heen karamba (S)/kilatti (T), (DF IV:35), N, 3, tree.
   Leaves: opposite, **rhomboid** to broadly ovate, acute to rounded base, acute and apiculate apex, entire margins, coriaceous.
   Trunk: sometimes slightly scandent; **spines** simple or forked;milky latex.
   Flowers: white or pink.
   Fruits: black, shiny berries.
   Site: disturbed areas, roadsides; DL.
   Uses: flowers, fruits-edible.

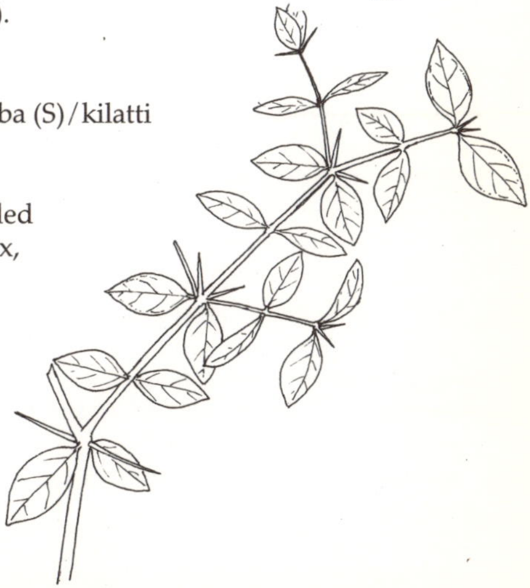

7. *Cerbera odollam*, gon kaduru (S)/nangi ma (T),
    (DF IV:53), N, 7, tree.
    Leaves: terminally **crowded**, tapering
      base, acuminate apex, entire margin.
    Trunk: branchlets **whorled**.
    Flowers: white with a
      yellow throat; I-terminal.
    Fruits: green, glabrous
      drupes, tinged
      pink when ripe.
    Site: coastal forest
      behind mangroves,
      hedges; DL, W.
    Uses: living fences, paddy
      field dividers, green manure.

8. *Nerium oleander*, oleander (E)/kaneru (S)/alari (T),
    (DF IV:28), I (Mediterranean), 5, shrub.
    Leaves: whorled, **narrow**, lanceolate, acuminate
      ends, lateral veins **parallel**, **leathery**.
    Trunk: B-dark; latex watery, poisonous.
    Flowers: red, pink to white, **large**,
      in terminal showy clusters.
    Fruits: long pod-like follicles.
    Site: home gardens;
      widespread.
    Uses: ornamental, shade
      tree, poisonous; stem,
      leaves, roots-medicinal.

9. *Pagiantha dichotoma*, divi
    kaduru (S)/nandi battai (T),
    (DF IV:39), N, 5, small tree.
    Leaves: oblong to elliptic,
      base tapering, rounded to
      very short acuminate apex,
      lateral veins obscure; petiole base
      with rudimentary rim of semi-circular
      stipules; young parts shiny, resinous.
    Trunk: latex, milky and poisonous.
    Flowers: corolla white, throat and
      tube yellow; I-dichasially branched,
      cymes from axils of terminal pair of leaves.
    Fruits: orange when ripe, pendulous capsules
      obliquely ovoid, fleshy, dehiscent,
      glabrous; crimson pulp.
    Site: secondary forest; IN, W.
    Uses: W-mask carving.

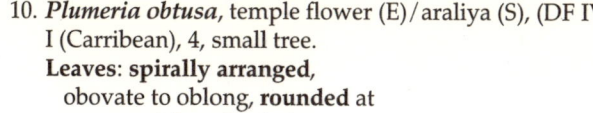

10. **Plumeria obtusa**, temple flower (E)/araliya (S), (DF IV:29), I (Carribean), 4, small tree.
    Leaves: spirally arranged, obovate to oblong, **rounded** at apex; petiole softly hairy.
    Trunk: branches **stout** and fleshy; latex milky, poisonous.
    Flowers: white with yellow throat, not tinged with red; I-terminal flat-topped cyme.
    Fruits: paired pod-like follicles.
    Site: home gardens; widespread.
    Uses: ornamental; flowers-temple offering.

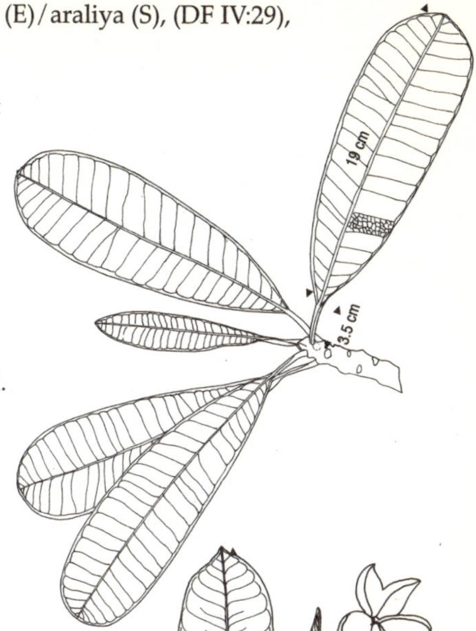

11. **Plumeria rubra**, temple flower (E)/araliya (S), (DF IV:29), I (C. Amer.), 4, small tree.
    Leaves: spirally arranged, large, oblong to obovate, acute or acuminate apex; petiole without hairs.
    Trunk: branches **stout**, fleshy; milky, poisonous latex.
    Flowers: red-yellow or white, tinged with pink or purple on at least outside of tube, **large**, long-stalked; I-terminal flat-topped cymes.
    Fruits: paired follicles.
    Site: home gardens; widespread.
    Uses: ornamental; flowers-temple offering.

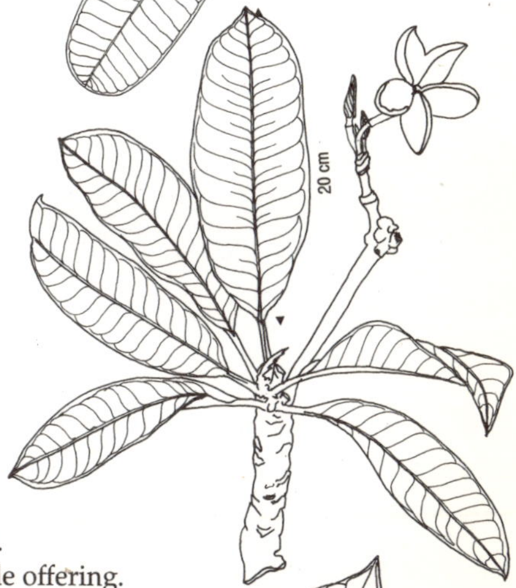

12. **Rauvolfia densiflora**, (DF IV:47), N, 3, shrub.
    Leaves: whorls of 3, lanceolate to obovate, tapering base, acuminate apex, lateral veins strongly arched.
    Trunk: sparse milky latex.
    Flowers: white tinged with violet; I-laxly branched cymes.
    Fruits: bluish-grey when ripe, fleshy, indehiscent.
    Site: disturbed and secondary forests; M.

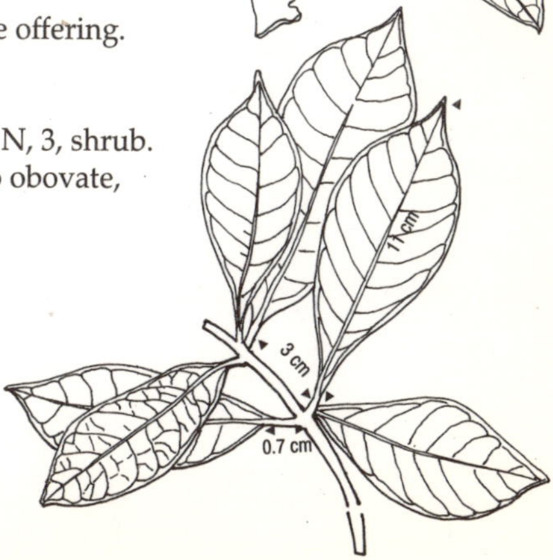

14

APOCYNACEAE

13. *Rauvolfia serpentina*, ekaweriya (S)/chiran
ampelpodi (T), (DF IV:49), N, 1, shrub.
Leaves: **whorls of 3**, lanceolate,
tapering base, acute apex.
Trunk: milky latex.
Flowers: petals white tinged
with violet, calyx red; I-dense cymes.
Fruits: blackish-purple, fleshy, indehiscent.
Site: secondary forest; IN, W.
Uses: leaves-medicinal.

3 cm

14. *Tabernaemontana divaricata*, crepe jasmine (E)/
wathu sudda (S)/nandi battai (T),
(DF IV:27), I (India), 2, shrub.
Leaves: elliptic, acuminate apex,
distant lateral veins.
Trunk: sometimes scandent; milky latex.
Flowers: white; I-cymes penduncled,
dichasially branched, terminal
but often overtopped by
axillary branches.
Fruits: spindle-shaped
capsules with acute beak,
fleshy, dehiscent,
orange inside.
Site: home gardens,
parks; widespread.
Uses: root, sap-medicinal;
ornamental.

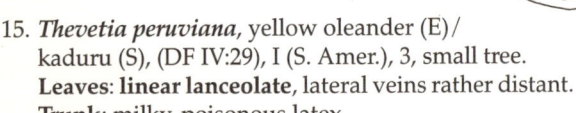

15. *Thevetia peruviana*, yellow oleander (E)/
kaduru (S), (DF IV:29), I (S. Amer.), 3, small tree.
Leaves: **linear lanceolate**, lateral veins rather distant.
Trunk: milky, poisonous latex.
Flowers: **yellow**; I-cymes
monochasially branched,
terminal but often
overtopped by
axillary branches.
Fruits: broadly
turbinate drupe.
Site: home gardens,
roadsides; DL, IN, W.
Uses: ornamental,
live fences.

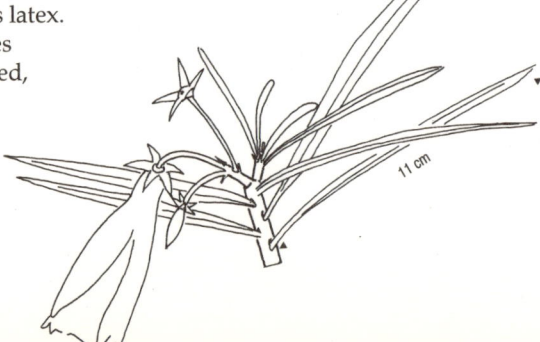

11 cm

16. *Walidda antidysenterica*, wal idda (S), (DF IV:61), E, 2, shrub.
   Leaves: oval to lanceolate, cuneate base, strongly
      acuminate apex, veins reticulate.
   Trunk: few erect branches;
      milky latex.
   Flowers: white, rather large,
      few on longish pedicels,
      mouth of corolla
      tube with a series
      of erect scales
      (corona); I-erect
      cymes.
   Fruits: follicles
      linear, cylindrical
      and glabrous.
   Site: monsoon,
      intermediate, rain
      forest, scrub; DL, IN, W.
   Uses: whole plant-medicinal;
      ornamental.

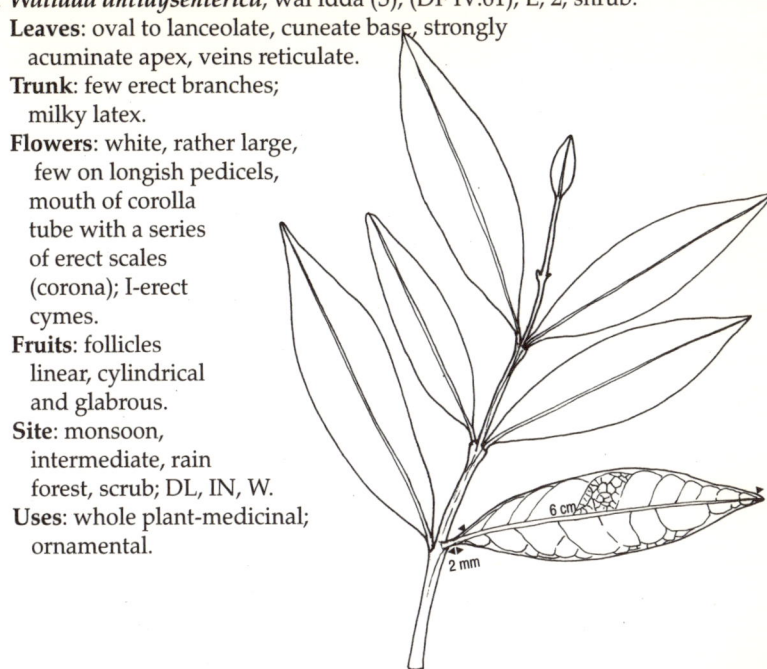

14

17. *Wrightia angustifolia*, velai pal madankai (T), (DF IV:64), E, 15, tree.
   Leaves: opposite, narrowly lanceolate, tapering base,
      long acuminate but obtuse apex, glabrous,
      yellowish-green.
   Trunk: **willow-like**;
      branchlets drooping;
      milky latex.
   Flowers: creamy white,
      malodorous, mouth of corolla
      with series of erect scales
      (corona); I-lax cymes.
   Fruits: cylindrical follicles,
      two longitudinal grooves,
      dehiscent when dry;
      seeds numerous,
      compressed,
      tuft of hair
      at one
      end.
   Site: scrub;
      DL, W.

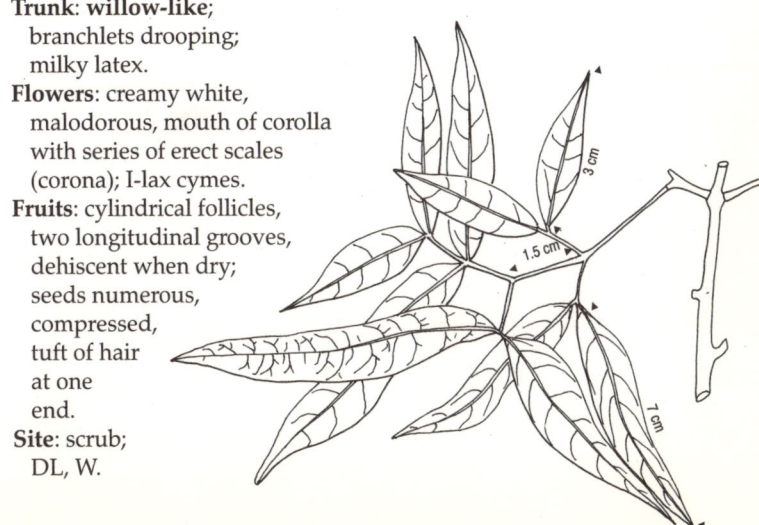

APOCYNACEAE

15. AQUIFOLIACEAE

FAMILY DESCRIPTION - Habit: trees or shrubs. **Leaves**: almost always spiral, rarely in pseudo-whorls or opposite. Simple, often with resiniferous or lactiferous cells in mesophyll. **Stipules**: small or absent. **Flowers**: usually unisexual (plants dioecious), actinomorphic, in axillary, supra-axillary or terminal fascicles, cymes or racemes, rarely solitary. **Fruits**: drupe with as many pyrenes as carpels.

FLOWER PARTS - CALYX: 4-5, united, imbricate. COROLLA: 4-5, free or united at base, imbricate or valvate. ANDROECIUM: stamens 4-5, often epipetalous. Anthers 2-locular, opening lengthwise. Nectary disk absent. GYNOECIUM: superior, 3 or more- locular; style terminal or absent. Ovules 1-2 in each loculus and pendulous from apex.

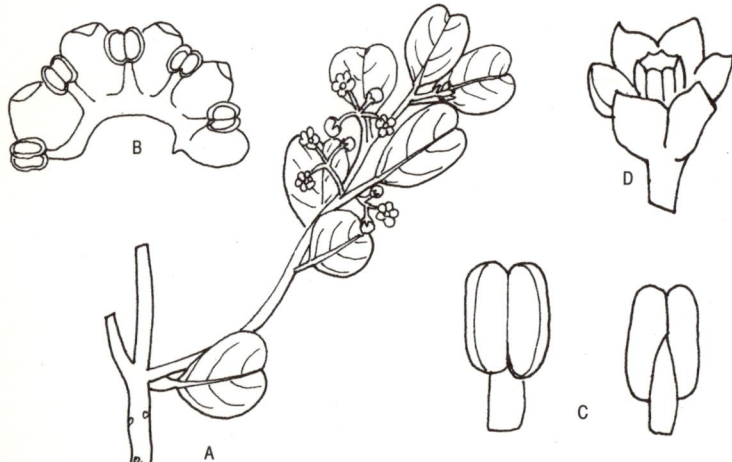

Key: inflorescence (A), corolla opened out to show the epipetalous stamens (B), stamen, front and back (C), and gynoecium surrounded by calyx (D) of *Ilex walkeri*

1. *Ilex walkeri*, (T I:264), N, 5, small tree.
   Leaves: numerous, closely spaced, variable, rotund, oblong or oval, apex variable, margins entire or serrate at apex, revolute; midrib and petiole dark purple.
   Trunk: B-longitudinally furrowed, grey.
   Flowers: white, very small; I-umbellate fascicles.
   Fruits: nearly globular drupe.
   Site: montane forest canopy; M.

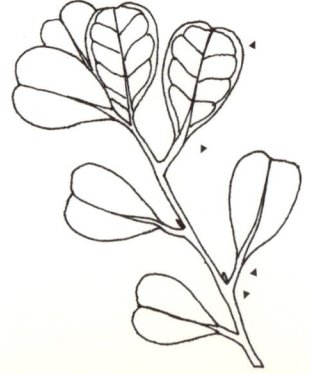

16. ARALIACEAE

FAMILY DESCRIPTION - Habit: trees, shrubs, woody epiphytes or rarely herbs. **Leaves**: spiral, rarely opposite or whorled. Palmately or pinnately compound. **Stipules**: adnate to petiole. **Flowers**: bisexual, actinomorphic, usually terminal umbels or heads, rarely solitary. **Fruits**: drupe with as many pyrenes as carpels, or a berry.

FLOWER PARTS - CALYX: 5, small, entire or toothed. Sometimes reduced. COROLLA: usually 5, free or united, valvate or slightly imbricate. ANDROECIUM: stamens usually 5, twice the number of petals or numerous, free. Anthers 2-locular, opening lengthwise. GYNOECIUM: Inferior, 1 or more locules with as many free or united styles, which are basally swollen and confluent with epigynous nectary disk. Ovule solitary in each loculus, pendulous, anatropous.

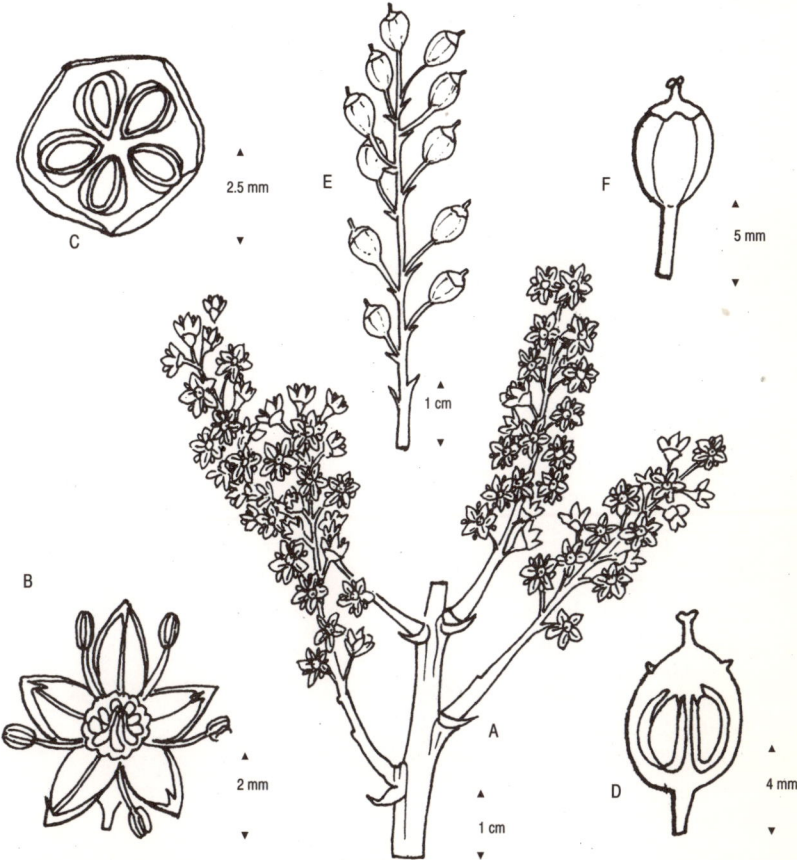

Key: inflorescence (A), flower in full (B), ovary in transverse section (C), gynoecium in longitudinal section (D), part of infructescence (E) and single fruit (F) of *Schefflera racemosa*.

1. *Schefflera racemosa*, (T II:283 & VI:139), N, 5, small tree.
    Leaves: palmately compound; 5-9 stalked leaflets, oblong-ovate,
       rounded base, acuminate
       twisted apex, veins
       reticulate; young parts
       with dense **orange scales**.
    Trunk: branchlets **stout,
       leaf-scarred**.
    Flowers: pale green,
       numerous; I-stalked
       umbellate racemes
       spreading
       divaricately in a
       terminal panicle.
    Fruits: ovoid, tipped
       with beak, 5-9 blunt ribs.
    Site: montane forest canopy; M.

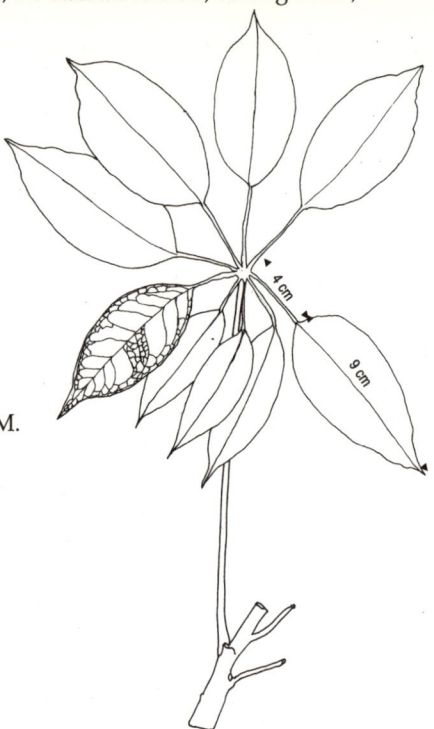

2. *Schefflera stellata*, itta (S), (T II:283 & VI:139)
    N, 10, epiphytic tree.
    Leaves: palmately compound; 4-7
       leaflets, acute ends.
    Trunk: sometimes scandent;
       branchlets **stout,
       leaf-scarred**.
    Flowers: yellow, bracts
       membranous, deciduous;
       I-stalked umbels on large
       terminal panicles.
    Fruits: yellow, club-shaped,
       faintly ribbed.
    Site: rain forest understory; W.
    Uses: leaves-medicinal.

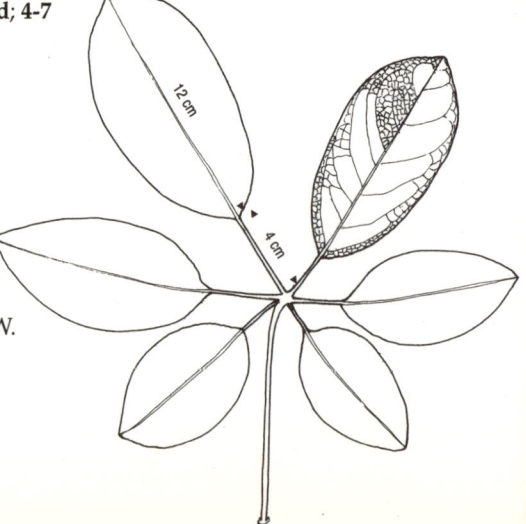

17. ARISTOLOCHIACEAE

FAMILY DESCRIPTION - **Habit**: perennial shrubs or herbs, sometimes twining. **Leaves**: spiral, simple. **Stipules**: Absent. **Flowers**: bisexual, zygomorphic, or rarely actinomorphic. Solitary or in terminal or lateral racemes or cymes. Often smelling of rotting meat. **Fruits**: capsular, sometimes with fleshy endocarp, many-seeded, rarely follicular or indehiscent and 1-seeded, sometimes dehiscing from the base upwards and hanging like an inverted parachute. Seeds numerous, 3-sided or flattened, sometimes winged.

FLOWER PARTS - CALYX: 3, united, enlarged and petaloid, usually with s-shaped tube. COROLLA: absent or small. 3, united. ANDROECIUM 6-40 in 1 or 2 whorls, free or united with style to form a gynostegium. Anthers free or adnate, extrorse, opening longitudinally. GYNOECIUM: inferior, 4-6-locular, sometimes with incomplete partitions. Ovules numerous in each loculus. Style thick, short, 3 to many stigmatic lobes.

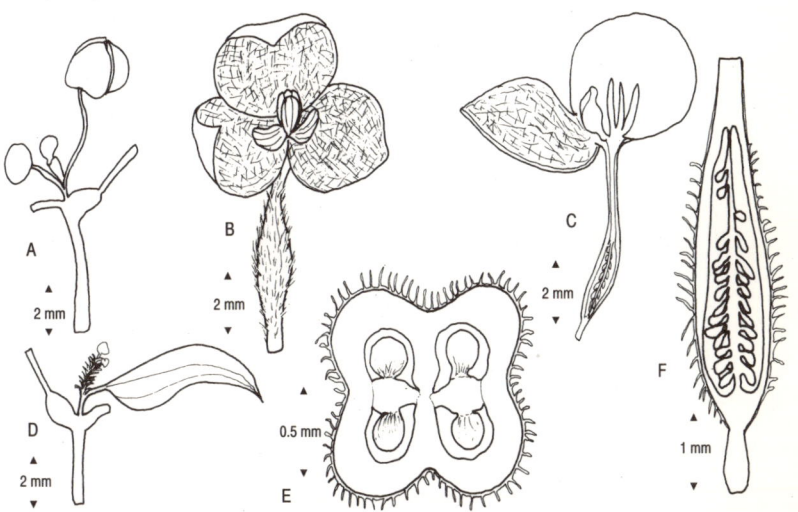

Key: Inflorescence (A), full flower (B), half flower (C), fruit (D), and transverse (E) and longitudinal (F) sections of ovary of *Thottea siliquosa*.

1. ***Thottea siliquosa***, tapasara bulath (S), (T III:421 & VI:245), N, 3, shrub.
   Leaves: linear-lanceolate, attenuate, acute ends, 3 veins at base, minutely pubescent beneath; **aromatic**.
   Trunk: erect, slender; B-smooth, yellowish; twigs **swollen** above nodes, young parts finely pubescent.
   Flowers: purple to greenish; I-irregularly umbellate cymes.
   Fruits: obtuse, 4-sided capsule.
   Site: rain forest understory; LM, W.
   Uses: leaves chewed with betel.

18. ASCLEPIADACEAE

FAMILY DESCRIPTION - Habit: twining or erect shrubs, herbs or rarely trees. Latex present. **Leaves**: opposite, whorled, rarely spiral, simple. **Stipules**: absent. **Flowers**: usually bisexual, actinomorphic. Inflorescence cymose. **Fruits**: 2 distinct follicles, often only 1 developing. Seeds usually flattened, crowned with long silky hairs.

FLOWER PARTS - CALYX: 5, free or basally united, imbricate or valvate. COROLLA: 5, united usually into short tube, lobes contorted or valvate. ANDROECIUM: 5, epipetalous. Filaments connate into a fleshy tube usually with fleshy scales or processes on the outside. Anthers united or distinct, adnate by the broad connective to the stigma, usually prolonged into dilated spreading appendages. Pollen grains combined into granular or waxy masses (pollinia). GYNOECIUM: superior, 2-locular, free, united only by the stigma. Styles free. Stigma 1, peltately dilated and disc-like, convex, conical or beaked. Ovules numerous with marginal placentation.

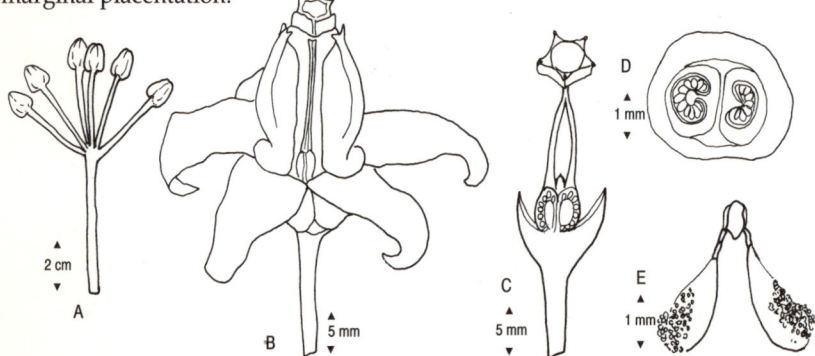

Key: inflorescence (A), full flower (B), longitudinal section of gynoecium (C), transverse section of ovary (D) and pollinia (E) of *Calotropis gigantea*.

1. *Calotropis gigantea*, wara (S)/manakkovi (T), (T III:148), N, 3, shrub.
   Leaves: **large**, oblong-ovate, cordate base, acute apex, **cottony tomentose, fetid odour** when crushed; nearly sessile.
   Trunk: B-furrowed, yellowish white; branches stout, fine cottony pubescent; milky latex.
   Flowers: pale violet or white; I-large, cymes 2-pronged, irregularly umbellate.
   Fruits: thick fleshy follicles.
   Site: roadsides, scrub; DL, W.
   Uses: whole plant-medicinal; stem fibre-fishing lines.

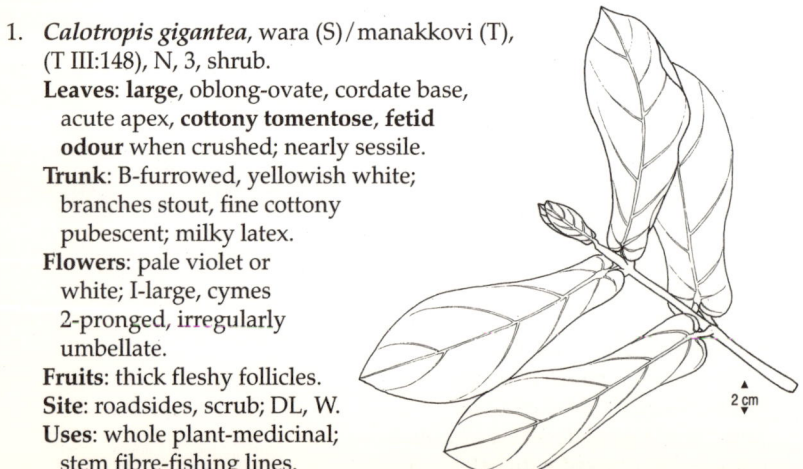

19. AVICENNIACEAE

FAMILY DESCRIPTION - Habit: shrubs or trees of maritime or saline regions. Pneumatophores present. **Leaves**: decussate, simple, entire. **Stipules**: absent. **Flowers**: bisexual, actinomorphic. Inflorescence cymose or racemose. **Fruits**: capsule, with fleshy, tomentose exocarp. Dehiscent by 2 valves. Embryo viviparous or semiviviparous.

FLOWER PARTS - CALYX: 5, basally united, persistent in fruit. COROLLA: 5, united. ANDROECIUM: stamens 4, epipetalous, equal or didynamous. GYNOECIUM: superior. Carpels 2, united, with a free central, often 4-winged placenta. Ovules 4, pendant, hanging from central placenta.

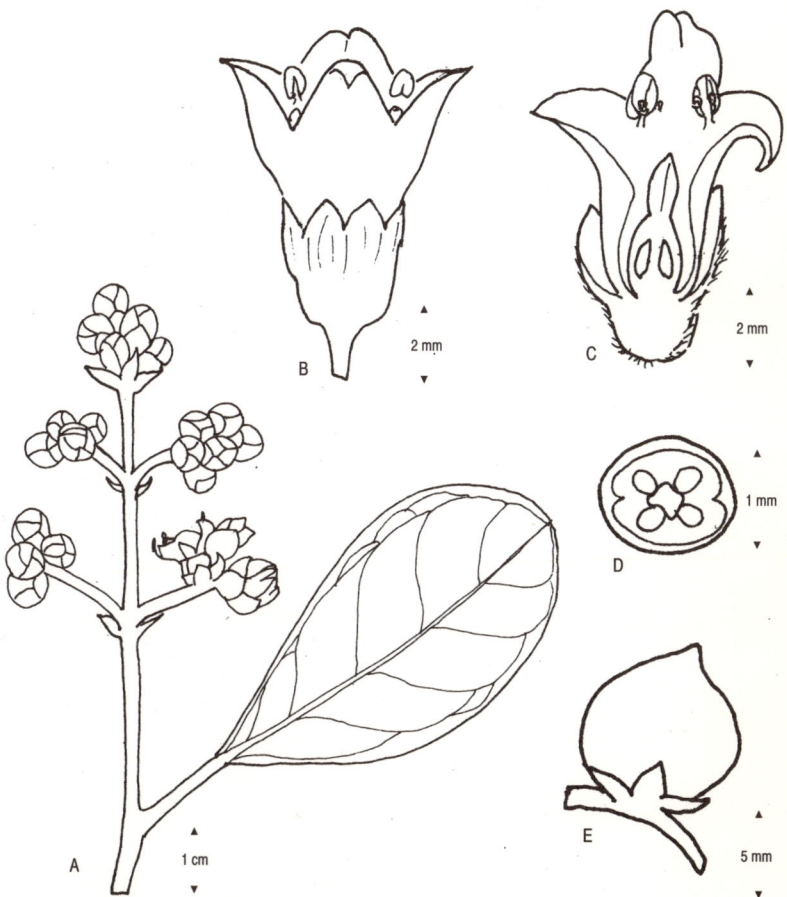

Key: inflorescence (A), full flower (B), and half flower (C), ovary in transverse section (D) and fruit (E) of *Avicennia officinalis*.

1. *Avicennia marina*, kanna (T), (DF IV:127),
   N, 10, shrub to tree.
   Leaves: very **dark green** above, silvery
     beneath, lanceolate to elliptic, abruptly
     acute apex, entire subrevolute margin.
   Trunk: B-smooth, greyish white; IB-green;
     W-pale, heavy, hard, **turnip-like** odour;
     twigs densely buff puberulent;
     Branched root pneumatophores.
   Flowers: white to orange-yellow;
     I-axillary or terminal pani-
     cles, 3-5 in a cluster.
   Fruits: yellowish green to
     greyish green, ovoid,
     dehiscent on tree.
   Site: saltwater tidal
     mudflats; MC, DC.
   Uses: branches for brush-pile fishing;
     W-firewood, boat construction; roots-medicinal.

2. *Avicennia officinalis*, kanna (T), (DF IV:132), N, 25, tree.
   Leaves: obovate to oblong, rounded ends, shiny above, adpressed
     tomentose and **resinous dots** beneath; **crown scraggly**.
   Trunk: B-yellowish green-black, rough; IB-white;
     W,s-grey; h-darker, hard, heavy;
     branched root pneumatophores.
   Flowers: yellow to brown,
     unpleasantly scented, solitary
     or paired; I-1 to 3 branched
     cymes in terminal panicles.
   Fruits: greenish-purple, broadly
     ovate, subcordate base.
   Site: saltwater tidal
     mudflats; MC, DC.
   Uses: branches for brush-pile
     fishing; W-firewood,
     pilings, boat construction;
     B-astringent, tanning
     agent; roots-medicinal.

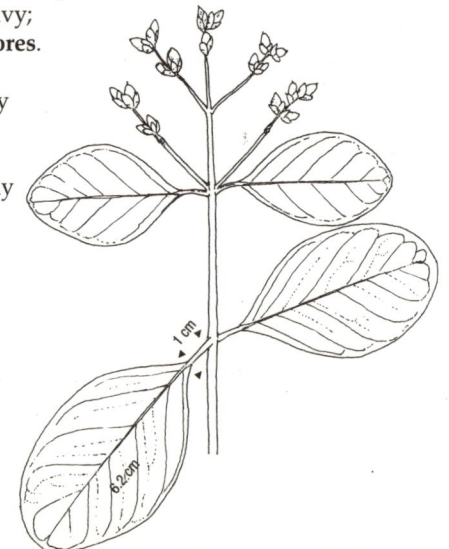

20. BIGNONIACEAE

FAMILY DESCRIPTION - **Habit**: trees, lianas, shrubs, rarely herbs. **Leaves**: opposite, sometimes whorled, rarely spiral. Palmately or pinnately compound, rarely simple. **Stipules**: absent. **Flowers**: bisexual, often funnel-shaped, zygomorphic in cymes or racemes. Sometimes solitary. Often showy. **Fruits**: bivalved capsule with a septum between that separates from the valves. Rarely fleshy and indehiscent. Seeds numerous, flat, winged.

FLOWER PARTS - CALYX: 5, united, sometimes bilobed. COROLLA: 5, united, often 2-lipped, imbricate or rarely valvate. ANDROECIUM: stamens 4 in 2 pairs. Staminode present or absent. Anthers 2-locular, opening lengthwise. Annular or cupular nectary disc usually around ovary. GYNOECIUM: Superior, 2 fused carpels, either bilocular with 2 axile placentas or unilocular with 2-4 parietal placentas. Ovules numerous.

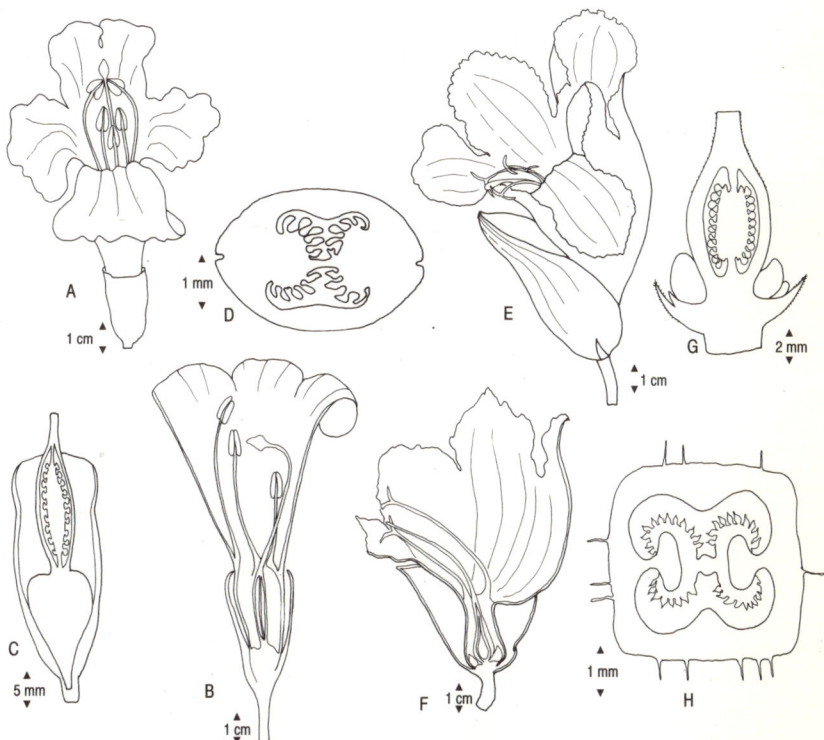

Key: full flower (A), half flower (B), and longitudinal (C) and transverse (D) sections of ovary of ***Oroxylum indicum***. Full flower (E), half flower (F), and longitudinal (G) and transverse (H) sections of ovary of ***Spathodea campanulata***.

1. *Crescentia cujete*, rum tree (E,S),
   I (C. & S. Amer.), 5, small tree.
   Leaves: whorled 3-5 together on
      short side shoots, spatulate,
      narrowed and acute base,
      acute apex, wavy margins.
   Trunk: B-smooth to fissured,
      light grey; W,s-pinkish
      brown;h-light brown.
   Flowers: light green, bell-shaped,
      borne singly on trunk.
   Fruits: green to brown, very
      large, hard, gourd-like.
   Site: home gardens;
      DL, IN, W.
   Uses: ornamental.

2. *Dolichandrone spathacea*, diya danga (S),
   (DF II:392), N, 15, tree.
   Leaves: pinnate; leaflets paired with terminal,
      ovate to lanceolate, acute to rounded base,
      acuminate to caudate apex, entire;
      rachis joints swollen.
   Trunk: often branching near base.
   Flowers: petals white; I-2 to 8 flowers.
   Fruits: smooth to obscurely
      ribbed, very large capsules;
      seeds corky, winged.
   Site: **mangrove swamps**,
      tidal marshes; MC.

3. *Jacaranda mimosifolia*, fern tree (E),
   (DF II:388), I (C. Amer.), 10, tree.
   Leaves: bipinnately compound;
      mimosa-like; leaflets paired
      or alternate with terminal.
   Trunk: B-rough, pale brown;
      twigs green, lenticellular.
   Flowers: **mauve**;
      I-conspicuous,
      compound cymes
      in axillary and
      terminal
      panicles.
   Fruits: oblong,
      flat capsule.
   Site: gardens,
      roadsides; M, IN, W.
   Uses: ornamental.

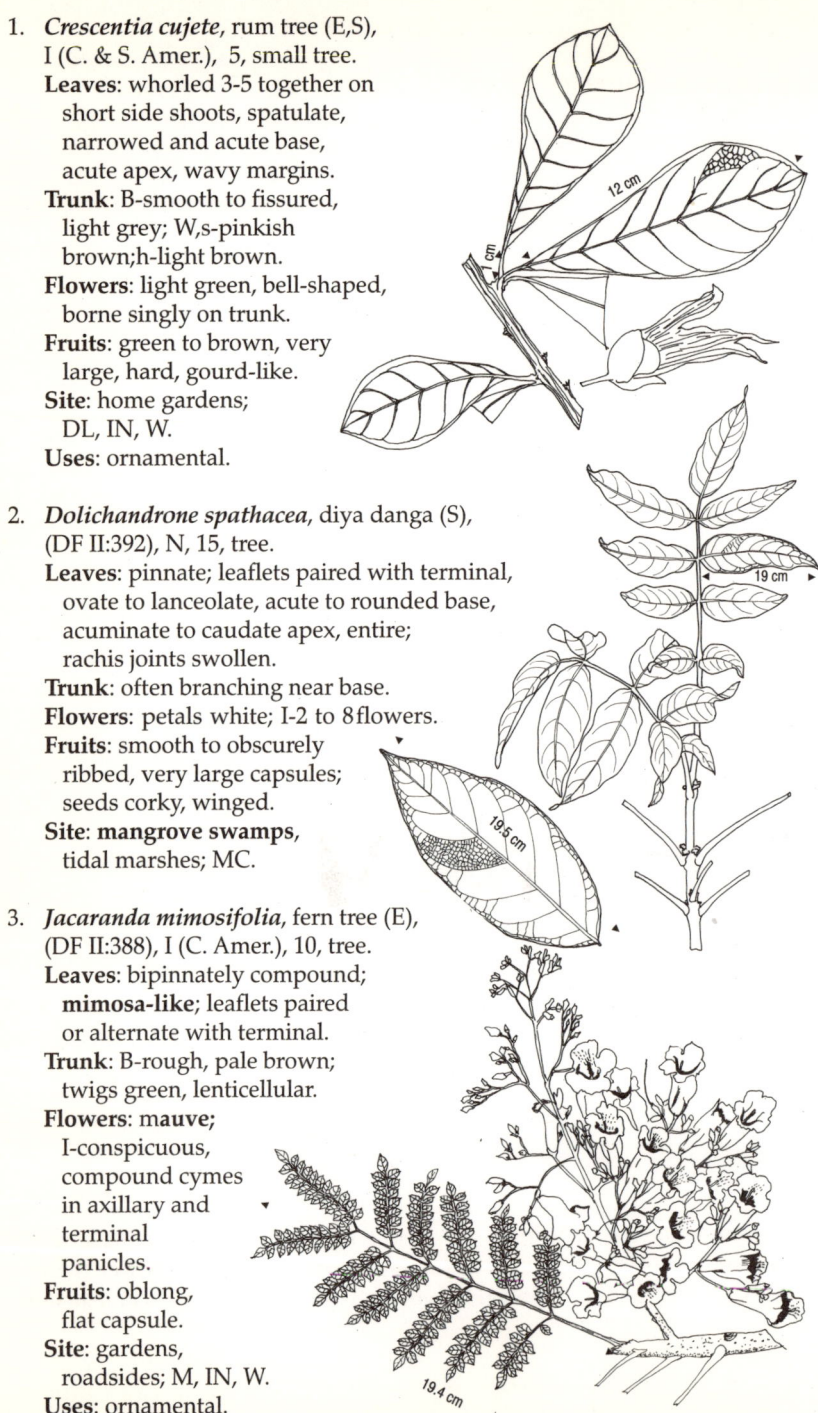

4. *Oroxylum indicum*, totila (S), (DF II:389), N, 8, tree.
   Leaves: bipinnate; leaflets paired with terminal,
      deltoid-ovate; joints of rachis
      swollen, with corky lenticels.
   Trunk: B-numerous large corky
      lenticels, yellowish grey.
   Flowers: petals deep maroon-
      reddish, creamy yellow inside;
      I-large terminal racemes on
      stout branch-like peduncles
      extending **prominently**
      above foliage.
   Fruits: persistent, long, flat
      capsule, tapering ends;
      seeds with winged margin.
   Site: disturbed forest areas; W.
   Uses: live fencing; roots, seeds-medicinal.

5. *Spathodea campanulata*, African tulip tree (E)/kudella
   gaha (S), (DF II:393), I (W. Africa), 30, tree.
   Leaves: pinnate; leaflets paired
      with terminal, elliptic
      to obovate.
   Trunk: often buttressed;
      B-smooth, light brown;
      IB-white;W-whitish, soft.
   Flowers: **bright orange-red**;
      I-terminal clusters.
   Fruits: green to brown, large,
      erect capsules.
   Site: home gardens, roadsides; widespread.
   Uses: ornamental, shade-tree.

6. *Stereospermum colais*, lunu madala (S)/padri (T), (DF II:395),
   N, 25, tree.
   Leaves: pinnate; leaflets paired with terminal,
      elliptic-oblong, acute base, caudate apex.
   Trunk: B-thick, rough, greyish yellow.
   Flowers: throat pink to yellow,
      purple flecking, calyx
      yellowish purple;
      I-short peduncles.
   Fruits: curved to spirally
      twisted, ridged
      capsules of varying
      length; seeds winged.
   Site: rain forest and moist
      coastal forest; MC, W.
   Uses: ornamental.

7. *Tabebuia rosea*, pink tabebuia (S), (DF II:387), I (S. Amer.), 25, tree.
   Leaves: **palmately** compound;
     leaflets elliptic, pointed ends.
   Trunk: B-rough, furrowed, grey.
   Flowers: **pink, showy**;
     I-terminal and
     axillary corymbs.
   Fruits: dark brown,
     cigar-like capsules.
   Site: parks, home
     gardens, forestry
     plantations. LM, IN, W.
   Uses: W-heavy construc-
     tion, furniture, cabinetwork;
     ornamental, shade tree.

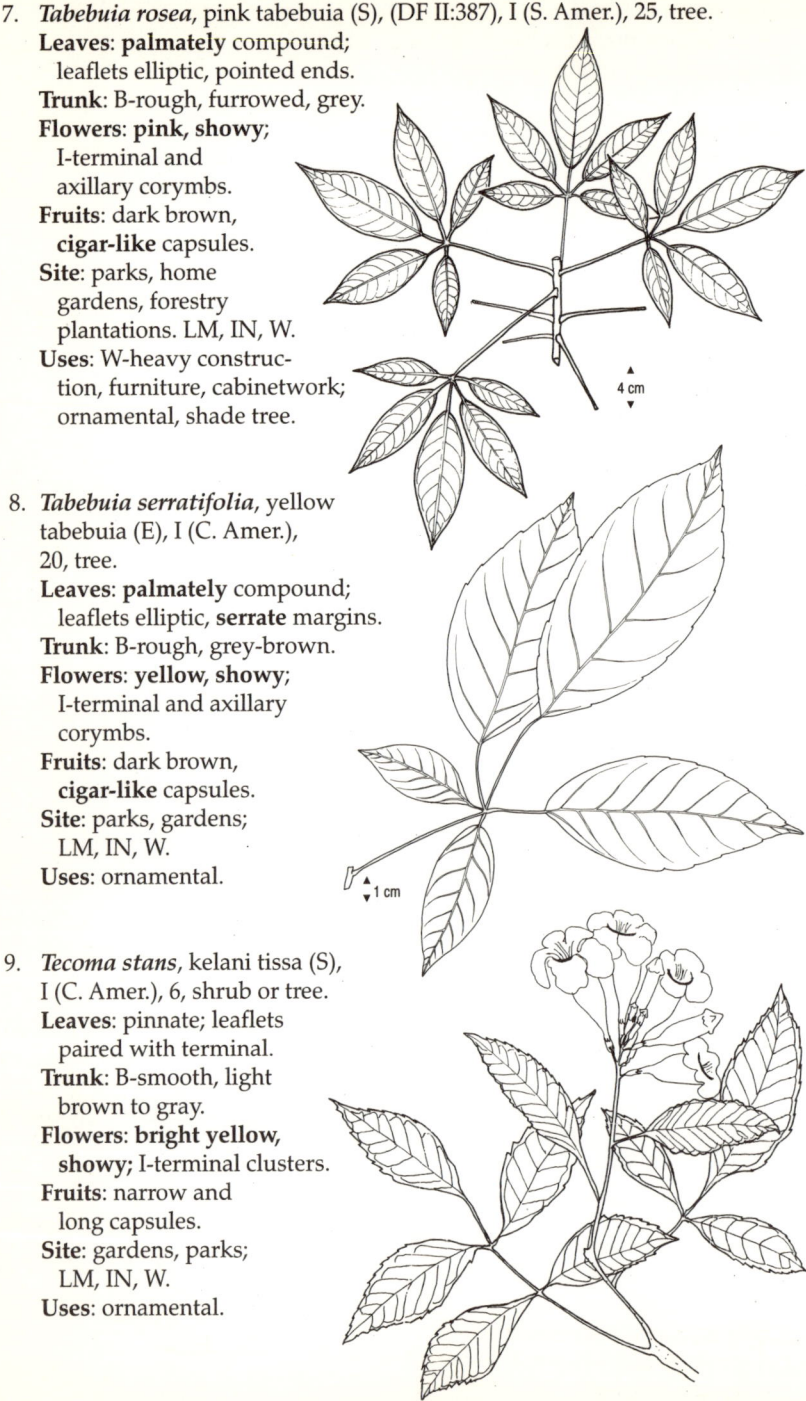

4 cm

8. *Tabebuia serratifolia*, yellow
   tabebuia (E), I (C. Amer.),
   20, tree.
   Leaves: **palmately** compound;
     leaflets elliptic, **serrate** margins.
   Trunk: B-rough, grey-brown.
   Flowers: **yellow, showy**;
     I-terminal and axillary
     corymbs.
   Fruits: dark brown,
     cigar-like capsules.
   Site: parks, gardens;
     LM, IN, W.
   Uses: ornamental.

1 cm

9. *Tecoma stans*, kelani tissa (S),
   I (C. Amer.), 6, shrub or tree.
   Leaves: pinnate; leaflets
     paired with terminal.
   Trunk: B-smooth, light
     brown to gray.
   Flowers: **bright yellow,
     showy;** I-terminal clusters.
   Fruits: narrow and
     long capsules.
   Site: gardens, parks;
     LM, IN, W.
   Uses: ornamental.

21. BIXACEAE

FAMILY DESCRIPTION - Habit: trees, shrubs or rhizomatous herbs, red or orange exudate. **Leaves**: spiral, palmately lobed or veined, simple. **Stipules**: present. **Flowers**: bisexual, actinomorphic, in racemes or panicles. **Fruits**: densely hairy or smooth, loculicidal capsule. Seeds glabrous or woolly.

FLOWER PARTS - CALYX: 5, free, imbricate, deciduous. COROLLA: 5, free, imbricate or convolute. ANDROECIUM: stamens numerous. Filaments free. Anthers horseshoe-shaped, opening by short slits at the top. Nectary disc within androecium or androecium on it. GYNOECIUM: superior, 1 locule with partitions; placentation partly axile partly parietal or wholly parietal. Style slender. Stigma 2-lobed. Ovules numerous.

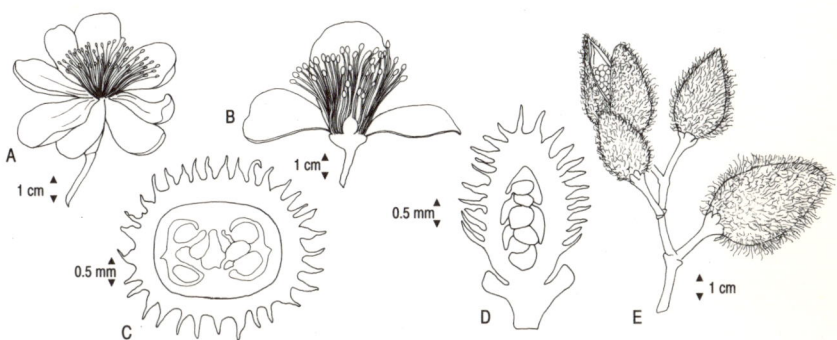

Key: full flower (A), half flower (B), and transverse (C) and longitudinal (D) sections of ovary, and fruits (E) of *Bixa orellana*.

1. *Bixa orellana*, anatto (E)/rata kaha (S), (DF V:121), I (S. Amer.), 5, shrub to small tree.
   Leaves: spiral, ovate, **heart**-shaped base, acuminate apex, palmately veined; petiole long.
   Trunk: B-smooth to warty lenticels, light brown; IB-pinkish orange; exudate **orange**.
   Flowers: pinkish to purplish white, large, **showy**.
   Fruits: reddish brown, rounded, capsules with dense prickles; seeds orange.
   Site: home gardens; widespread.
   Uses: ornamental; B-twine/rope; leaves, roots- medicinal; seeds-food colouring, dye.

22. BOMBACACEAE

FAMILY DESCRIPTION - Habit: trees. **Leaves**: spiral, simple or palmately compound. **Stipules**: present. **Flowers**: bisexual, actinomorphic, large, showy. Solitary or in short cymes. Often with epicalyx. **Fruits**: loculicidal capsules, rarely fleshy and indehiscent. Seeds often embedded in hairs from the wall of the fruit.

FLOWER PARTS - CALYX: 5, free or basally united, valvate, glandular hairs (nectaries) at base. COROLLA: 5, free, convolute, sometimes absent. ANDROECIUM: stamens 5 to numerous, free or united into 5-15 bundles or into a tube. Anthers 1-locular, opening by longitudinal slit. GYNOECIUM: superior, 2-5-locular. Style simple or lobed. Ovules 2 or more per loculus. Placentation axile.

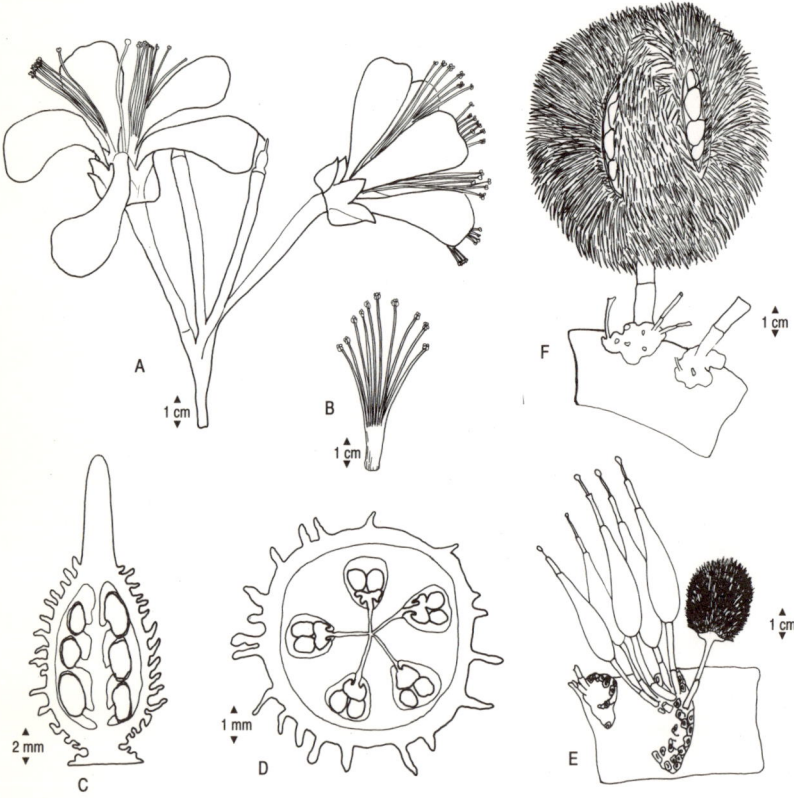

Key: inflorescence with full flowers (A), bundle of fused stamens (B), and longitudinal (C) and transverse (D) sections of ovary of *Durio zibethinus*. Infructescence bearing a young fruit and mature flowers (E), and mature fruit (F) of *Cullenia rosayroana*.

1. ***Adansonia digitata***, baobab (E)/
   aliya gaha (S)/papparappuli (T),
   (DF I:67), I (Africa), 20, tree.
   Leaves: palmately compound;
   leaflets oblong ovate, cuneate
   base, acuminate apex,
   densely pubescent.
   Trunk: B-smooth;
   W-light, spongy.
   Flowers: white,
   showy, pendulous,
   axillary, solitary.
   Fruits: pointed to
   rounded ends,
   dense velvety
   hairs, oblong capsule.
   Site: arid north-western
   dry coastal thorn scrub; DC.
   Uses: B-rope; young leaves-vegetable;
   capsules-fishing floats, beverage;
   stem, leaves, fruit-medicinal.

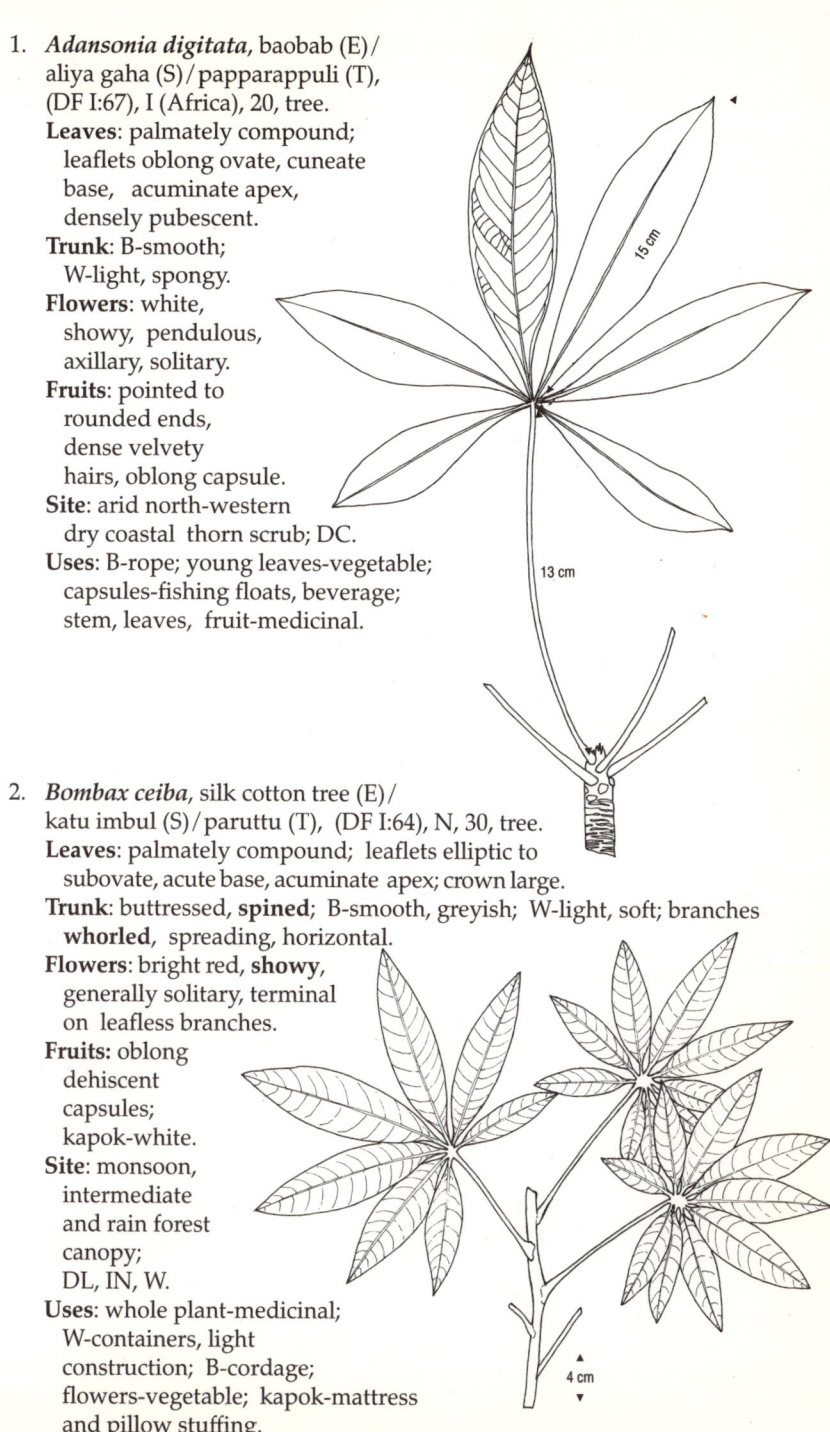

2. ***Bombax ceiba***, silk cotton tree (E)/
   katu imbul (S)/paruttu (T), (DF I:64), N, 30, tree.
   Leaves: palmately compound; leaflets elliptic to
   subovate, acute base, acuminate apex; crown large.
   Trunk: buttressed, **spined**; B-smooth, greyish; W-light, soft; branches
   whorled, spreading, horizontal.
   Flowers: bright red, **showy**,
   generally solitary, terminal
   on leafless branches.
   Fruits: oblong
   dehiscent
   capsules;
   kapok-white.
   Site: monsoon,
   intermediate
   and rain forest
   canopy;
   DL, IN, W.
   Uses: whole plant-medicinal;
   W-containers, light
   construction; B-cordage;
   flowers-vegetable; kapok-mattress
   and pillow stuffing.

22

BOMBACACEAE

3. ***Ceiba pentandra***, kapok tree (E)/pulun imbul (S), (DF I:70), I (S. Amer.), 20, tree.
   Leaves: palmately compound; leaflets with acute base, acuminate apex, entire.
   Trunk: B-smooth, **greenish**; branches whorled.
   Flowers: petals yellowish white; I-terminal clusters.
   Fruits: green to brown, ellipsoid capsule; kapok white to greyish.
   Site: home gardens, cultivated; DL, W.
   Uses: live fencing, whole plant-medicinal; kapok-mattress and pillow stuffing, sound and heat insulation.

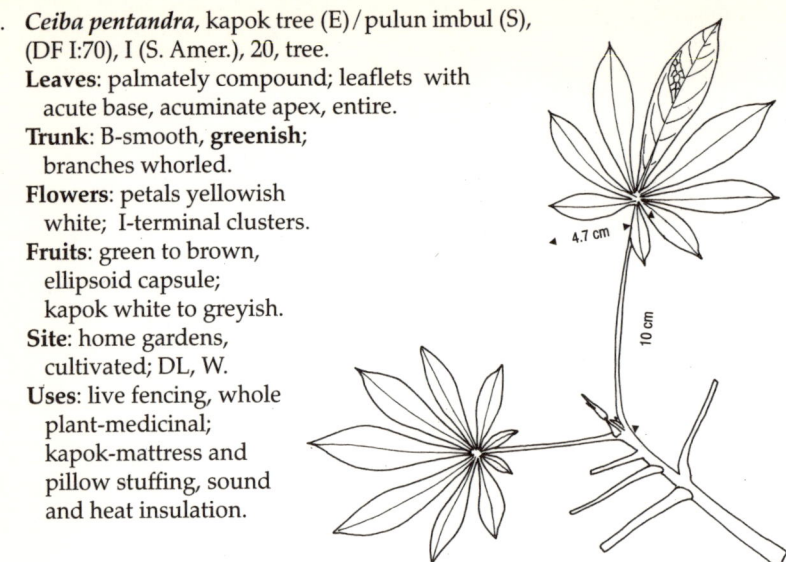

4. ***Cullenia ceylanica***, kata boda (S), (DF I:60), E, 36, tree.
   Leaves: deep green above, beneath **golden** and **scaly**, narrowly elliptic, rounded base, acuminate apex, main veins prominent, other veins inconspicuous; petiole swollen at apex.
   Trunk: B-smooth, grey; W-pale red, fine grained, light.
   Flowers ramiflorous, covered by densely tawny scales, conspicuous.
   Fruits: **globular**, **spiny**, dehiscent capsule, white inside; brown seeds with orange-brown aril.
   Site: rain forest canopy; W.
   Uses: W-furniture, panelling, decorative work.

5. *Cullenia rosayroana*, kata boda (S), (DF I:62), E, 30, tree.
   Leaves: dark green above, **golden** and **scaly**
   beneath; narrowly oblong-elliptic,
   rounded base, acuminate apex, main
   vein prominent, other veins
   inconspicuous; petiole not
   swollen at apex.
   Trunk: B-smooth.
   Flowers: **ramiflorous**,
   covered by densely
   tawny scales,
   conspicuous.
   Fruits: **globular**, **spiny**,
   dehiscent capsule,
   white inside; dark
   brown seeds;
   whitish aril.
   Site: rain forest canopy; W.
   Uses: W-furniture, panelling,
   decorative work.

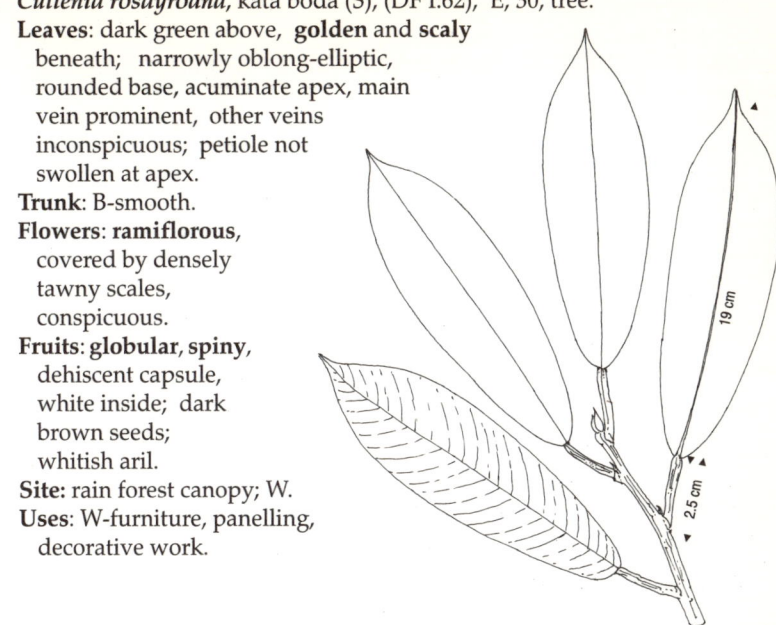

6. *Durio zibethinus*, durian (E,S), (DF I:59), I (Malesia), 30, tree.
   Leaves: **2-ranked**; narrowly oblong to elliptic, acuminate apex.
   Trunk: columnar; B-vertically fissured, greyish brown.
   Flowers: **ramiflorous**, covered by densely tawny scales, conspicuous.
   Fruits: large **spiny** capsule, **malodorous**; aril
   tasting of a caramel, banana and vanilla mix.
   Site: home gardens. W.
   Uses: W-light construction;
   fruit-edible and much desired!

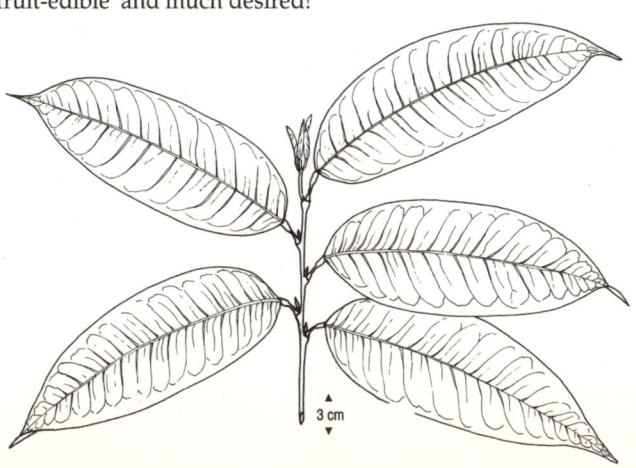

22

BOMBACACEAE

23. BORAGINACEAE

FAMILY DESCRIPTION - Habit: trees, shrubs, frequently herbs, rarely lianas. **Leaves**: spiral, rarely opposite, simple, usually entire. **Stipules**: absent. **Flowers**: usually bisexual, actinomorphic, usually in cymes, rarely solitary and axillary. **Fruits**: achenes or drupes with 1-4 seeds.

FLOWER PARTS - CALYX: usually 5, rarely 4 or 6, free or united, imbricate, rarely valvate.   COROLLA: usually 5, rarely 4 or 6, united, imbricate or convolute. Sometimes tube with hairy appendages between lobes. ANDROECIUM: stamens usually 5, rarely 4 or 6, epipetalous. Anthers 2-locular, opening lengthwise. Nectary disc absent or present. GYNOECIUM: superior, 2-locular with 2 ovules per loculus or 4-locular with 1 ovule in each, entire or deeply 4-lobed, with 1 style.

Key: part of inflorescence (A), full flower (B), gynoecium surrounded by calyx (C), and gynoecium in longitudinal (D) and transverse (E) sections of *Ehretia laevis*. Fruit (F) of *Ehretia buxifolia*. Inflorescence (G), full bisexual flower (H), opened out male and female flowers (I, J), and fruits (K) of *Cordia dichotoma*.

1. ***Carmona retusa***, heen tambala (S)/pakkuvetti (T),
    (DF VII:5), N, 2, shrub.
    Leaves: **small**, very **numerous**, clustered,
        spoon-shaped, base tapering, apex
        toothed; petiole nearly absent.
    Trunk: B-cracked, reddish brown;
        branches numerous, divaricate.
    Flowers: white, small; I-solitary
        or paired, axillary.
    Fruits: small, globose, pointed,
        shiny drupe.
    Site: monsoon forest understory
        and openings, scrub; DL, IN.
    Uses: ornamental;
        root-medicinal.

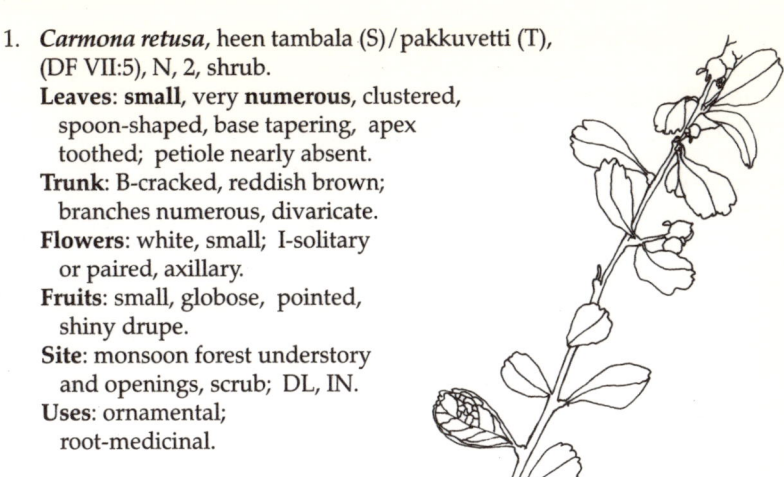

2 cm

2. ***Cordia curassavica***, lolu (S), (DF VII:10), I (S. & C. Amer.),
    3, shrub.
    Leaves: lanceolate to narrowly elliptic or ovate,
        wedge-shaped to acute base that tapers along
        petiole, acute apex, margins **serrate**,
        occasionally undulate, hairy on main
        veins beneath; petiole nearly absent.
    Trunk: B-dark brown, scatttered
        with lenticels; twigs with small,
        globose, waxy particles.
    Flowers: bisexual; white;
        I-dense spike.
    Fruits: bright
        red, small,
        enclosed in
        calyx; seed
        ovoid.
    Site: scrub,
        roadsides; DL.

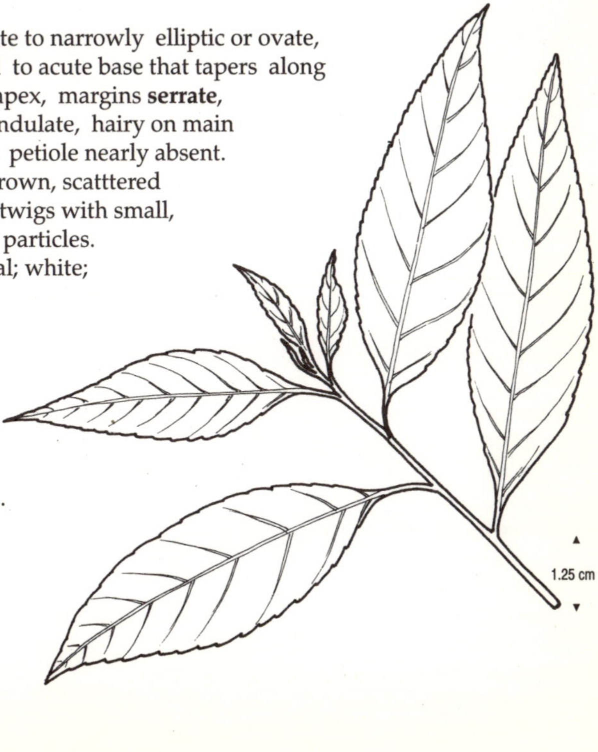

1.25 cm

3. **Cordia dichotoma**, lolu (S), (DF VII:17), N, 5, tree.
    Leaves: elliptic, occasionally ovate, base obtuse to rounded, apex obtuse
        to acute, margins entire, occasionally undulate to crenate near apex,
        upper surface with numerous calcified spots,
        lower surface with small tufts of hair
        in the axils of the main veins.
    Trunk: twigs without hairs.
    Flowers: bisexual
        and male; white,
        tubular; I-cymose
        panicle.
    Fruits: yellow to orange,
        ovoid to globose drupe,
        mucilaginous, borne on a
        persistent saucer-shaped calyx.
    Site: monsoon forest understory; DL.
    Uses: fruit-medicinal.

    3.3 cm

4. **Cordia monoica**, naruvilli (T), (DF VII:12),
    N, 3, small tree.
    Leaves: ovate to oval, rounded to acute base,
        acute to obtuse apex, margins
        undulate to repand, **rough
        tomentose** above.
    Trunk: B-smooth, grey,
        more or less densely
        tomentose.
    Flowers: bisexual and male;
        white, subsessile;
        I-dense corymbs.
    Fruits: **bright yellow**, ovate to
        ovoid, pointed, smooth, drupe;
        enlarged persistent calyx.
    Site: monsoon forest; DL.

    2.3 cm

5. **Cordia sinensis**, lolu (S), (DF VII:12), N, 4,
    shrub to small tree.
    Leaves: oblong to elliptic-oblong, obtuse base, margins
        entire, surface rough, with short hairs and dense tufts
        of hairs in axils of secondary veins beneath.
    Trunk: twigs are without hairs.
    Flowers: bisexual; white; I-small,
        paniculate cymes.
    Fruits: orange at maturity,
        mucilaginous drupes;
        borne on
        saucer-shaped calyx.
    Site: monsoon forest
        understory; DL.

    11.5 cm

    3.5 cm

6. **Cordia subcordata,** lolu (S), (DF VII:9), N, 8, tree.
   Leaves: widely ovate, base obtuse to rounded, apex
   acute to obtuse, margins unevenly wavy, slightly
   hairy above, with dense tufts of hair in the axils
   of veins and along main veins beneath.
   Trunk: twigs often with
   widely scattered hairs.
   Flowers: bisexual; orange-red,
   small; I-loose cyme
   sparsely branched.
   Fruits: smooth, lustrous
   drupe, inside corky,
   completely enclosed
   in calyx.
   Site: scrub on sandy
   seashore; DC.

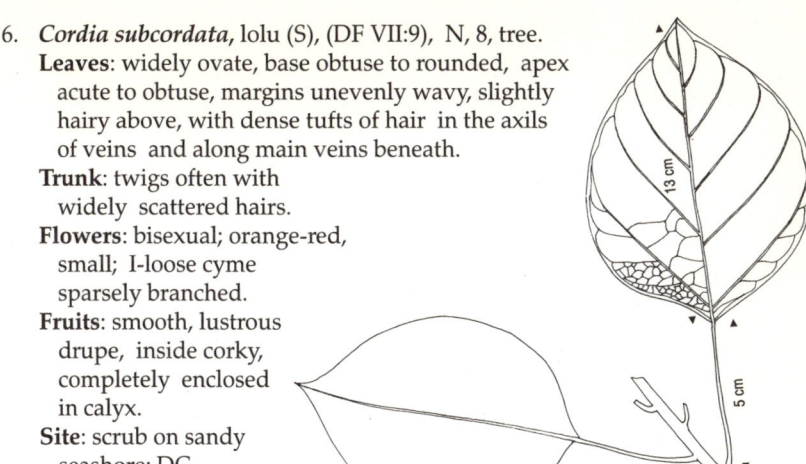

7. **Ehretia laevis**, addula (T), (DF VII:21), N, 8,
   shrub to tree.
   Leaves: variable in shape, elliptic to narrowly
   ovate or obovate, base acute to wedge-shaped,
   apex acute, rarely obtuse, margins entire,
   shiny above with scattered hairs
   beneath; buds black, resinous.
   Trunk: B-pale grey.
   Flowers: white, numerous;
   I-terminal, small, scorpioid cymes.
   Fruits: bright **orange-red**, nearly
   globose drupe, tipped with style.
   Site: scrub, monsoon forest; DL.

8. **Tournefortia argentea**, Karan (T),
   (DF VII:19), N, 5, shrub or small tree.
   Leaves: oblanceolate, spiral, terminally
   crowded at ends of branches. Base long
   and tapering, apex acute, nearly sessile.
   Trunk: young stems densely silky hairy.
   Flowers: white, small, softly hairy
   outside; I-conspicuous cymes.
   Fruits: globose drupe
   with spongy tissue.
   Site: scrub along sandy
   seashores; DC.

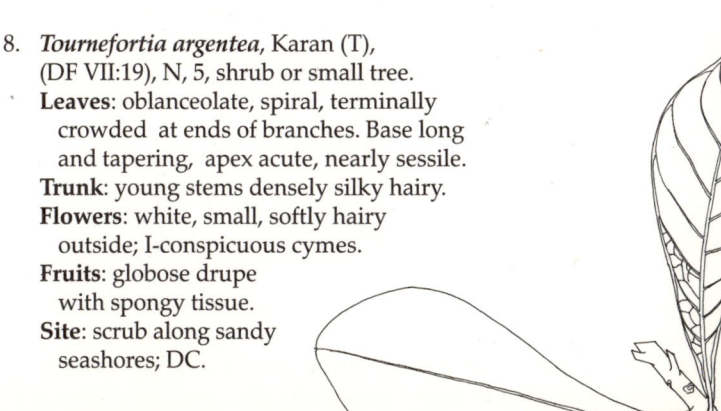

24. BURSERACEAE

FAMILY DESCRIPTION - **Habit**: trees or shrubs with resin ducts in bark. **Leaves**: spiral, rarely opposite, pinnate or trifoliolate, rarely unifoliolate. **Stipules**: absent or present. **Flowers**: bisexual or unisexual, actinomorphic, in dense panicles, rarely racemes or heads. **Fruits**: drupe, rarely a capsule.

FLOWER PARTS - CALYX: 3-5, free or united at base, valvate or imbricate. COROLLA: 3-5, rarely absent, free or variously united, imbricate, rarely valvate. ANDROECIUM: stamens 3-5 or double the number of the petals, borne outside or on a nectary disc. Filaments free, rarely united. Anthers 2-locular with longitudinal slits. Staminodes present in female flowers. GYNOECIUM: superior, 2-5-locular. Ovules 2 or 1 per loculus, pendulous on axile placenta.

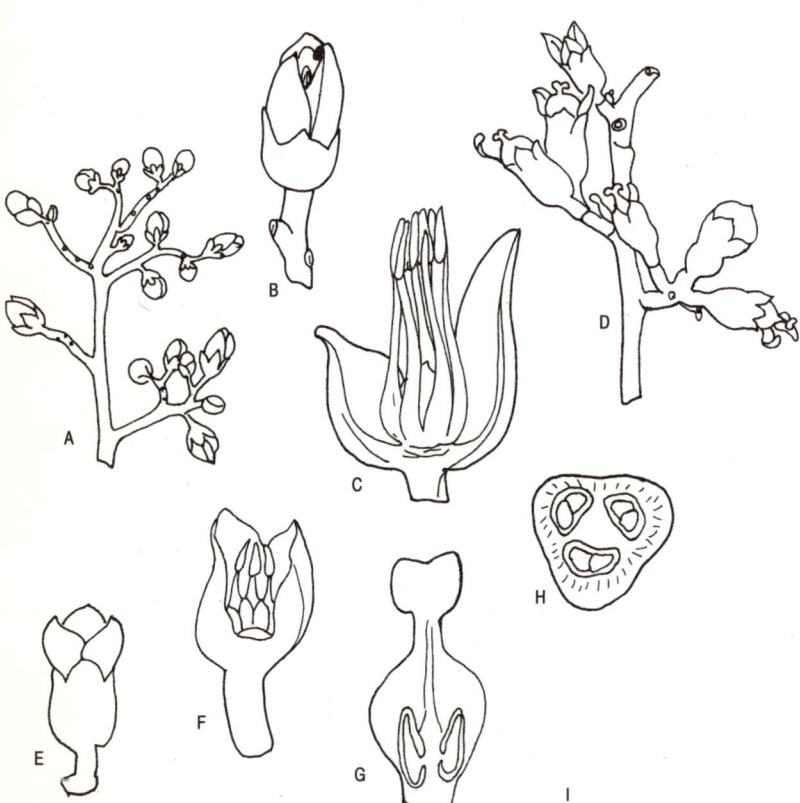

Key: male inflorescence (A), full male flower (B), half male flower (C), inflorescence bearing bisexual flowers (D), bisexual flower in full (E), half bisexual flower with the gynoecium removed (F), gynoecium in longitudinal (G), and transverse (H) sections, and mature fruit (I) of *Canarium zeylanicum*.

1. *Canarium zeylanicum*, kekuna (S)/ pakkilipal (T), (T I:239), E, 30, tree.
    Leaves: **pinnate**, rachis woody, brown; leaflets paired
    with terminal, broadly oblong-oval,
    subcordate base, shortly
    acuminate apex, margin entire.
    Trunk: B-smooth, pale,
    thin; W-white, soft, light;
    resinous and **aromatic**;
    young branchlets
    fulvous pubescent.
    Flowers: pale greenish yellow,
    on short pedicels; I-narrow,
    terminal, rufous-tomentose
    panicles.
    Fruits: brownish, ovoid,
    faintly trigonous, drupe.
    Site: rain forest canopy; W.
    Uses: W-light construction;
    B-medicinal; resin-incense;
    seeds-edible.

4.8 cm

24

BURSERACEAE

2. *Commiphora caudata*, siviya gas (S) /kilivai (T),
    (T I:238), N, 10, tree.
    Leaves: pinnate; leaflets paired
    with terminal, ovate, unequal
    base, attenuate-caudate apex.
    Trunk: B-smooth to flaky, pinkish;
    strongly resinous.
    Flowers: pinkish or red,
    numerous, small;
    I-cymes from the ends
    of branches behind
    the new leaves.
    Fruits: pendulous,
    broadly ovoid
    drupe, supported
    on persistent
    calyx.
    Site: monsoon forest
    rocky hills;
    DL.
    Uses: live
    fencing.

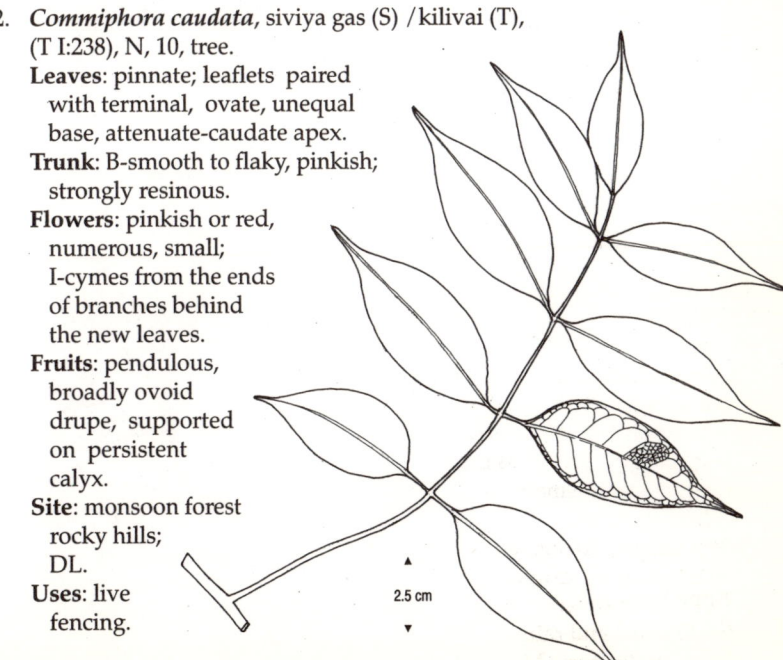

2.5 cm

25. BUXACEAE

FAMILY DESCRIPTION - Habit: trees, shrubs or rarely herbs. **Leaves**: opposite, rarely spiral, simple. **Stipules**: absent. **Flowers**: unisexual, (plants monoecious or rarely dioecious), actinomorphic, in heads or spikes. Female flowers larger than male, and fewer or solitary. **Fruits**: capsule, dehiscent loculicidally. Less often a drupe. Seeds black, shiny.

FLOWER PARTS - Calyx: usually 4, imbricate. Corolla: absent. Androecium: stamens 4-6, rarely more. Anthers large, sessile or borne on fairly long filaments. Gynoecium: superior, 3-locular, styles widely separated or contiguous. Ovules 1-2 per loculus, pendulous.

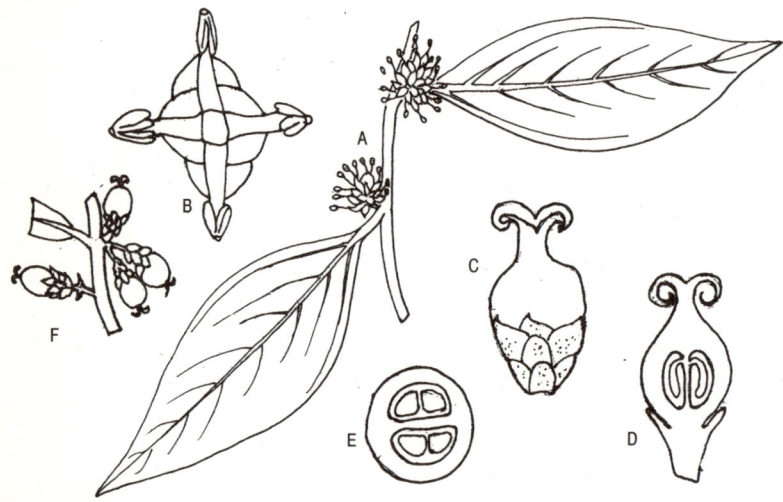

Key: inflorescences (A), flower from above (B), gynoecium in full (C), in longitudinal section (D), and in transverse section (E) and fruits (F) of *Sarcococca zeylanica.*

1. *Sarcococca zeylanica*, (T IV:9), E, 4, shrub.

   Leaves: ovate to lanceolate, variable tapering base, often sickle-shaped, base appears distinctly 3-veined.

   Trunk: much branched; branchlets **long, twiggy, glabrous green.**

   Flowers: green; I-small axillary racemes.

   Fruits: purple, oblong to ovoid, smooth, drupe, tipped with style.

   Site: montane and rain forest understory; M.

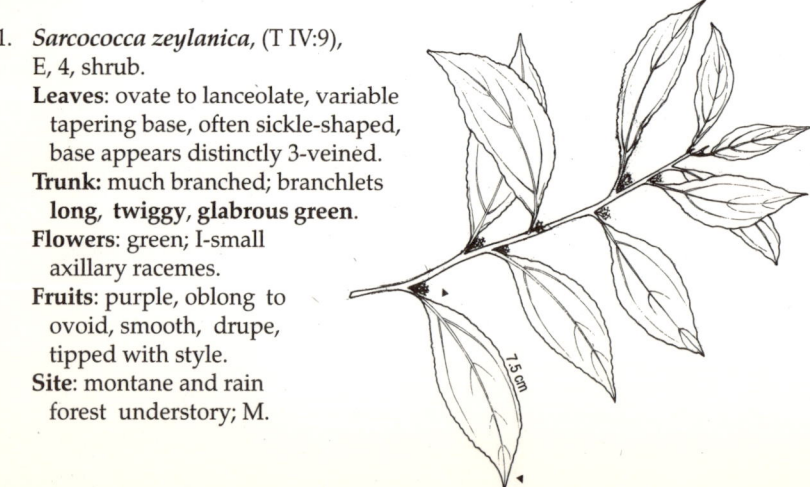

26. CAPPARIDACEAE

FAMILY DESCRIPTION - **Habit**: shrubs, herbs or rarely trees. **Leaves**: spiral, rarely opposite, simple, trifoliolate or palmately compound. **Stipules**: absent or small and spiny. **Flowers**: bisexual, actinomorphic or rarely zygomorphic, in axillary or terminal racemes. **Fruits**: capsule or berry. Seeds usually kidney shaped.

FLOWER PARTS - CALYX: usually 4, free or partially united, imbricate or valvate. COROLLA: usually 4, rarely 2 or 6, or absent, united, often clawed. ANDROECIUM: stamens few to many, some staminodal. Anthers 2-locular, dehiscing longitudinally. GYNOECIUM: receptacle prolonged into gynophore or androgynophore. Carpels 2-12, ovary 1-locular; placentation parietal.

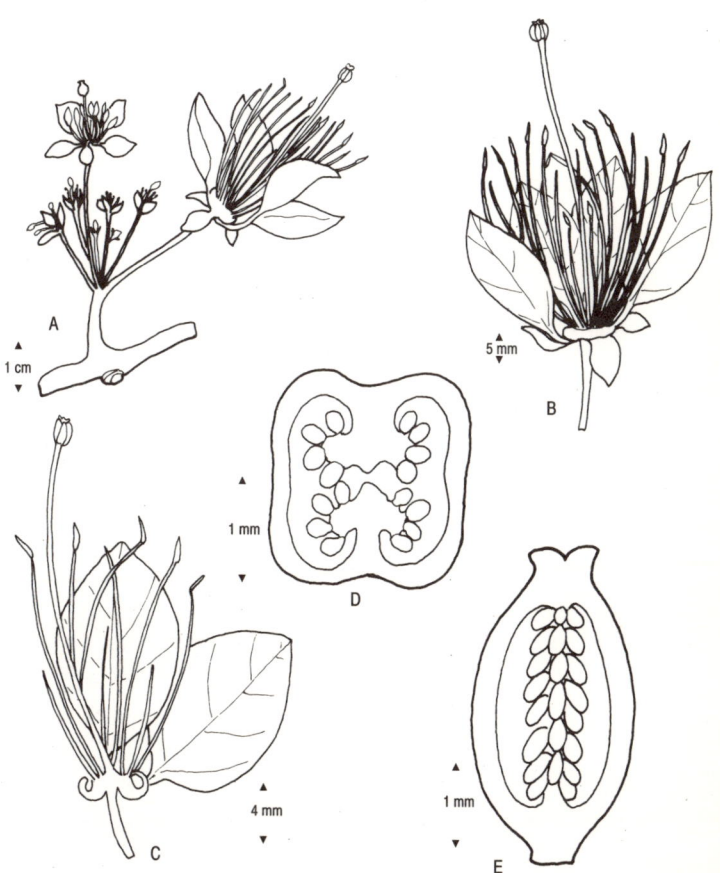

Key: inflorescence (A), full flower with petal removed (B), half flower (C), and transverse (D) and longitudinal (E) sections of ovary of *Crateva adansonii*.

1. *Capparis zeylanica*, kattodi (T), (T I:61), N, 2, shrub.

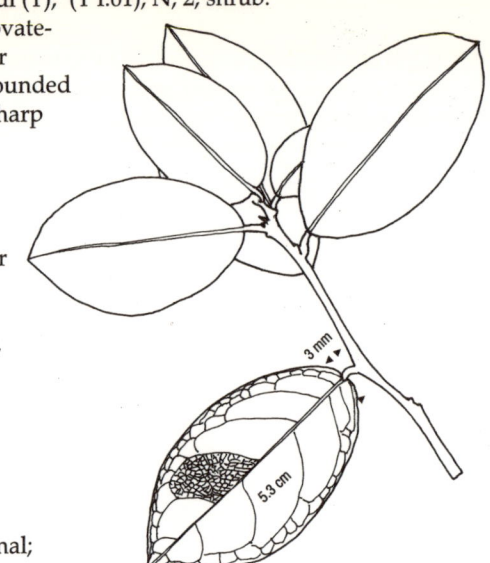

Leaves: ovate-elliptic or ovate-lanceolate, apex acute or rounded-obtuse, base rounded to subcordate, 2 small sharp stipular spines.

Trunk: much-branched, young stems greyish or brownish tomentose.

Flowers: white, upper pair of petals yellow at the base, becoming reddish violet with age; axillary, solitary or in pairs.

Fruits: subglobose or irregularly ovoid, many seeded.

Site: scrub, monsoon forest; DL.

Uses: whole plant-medicinal; fruit-edible.

2. *Crataeva adansonii*, lunu warana (S)/ navala (T), (T I:59), N, 8, small tree.

Leaves: pinnate; leaflets broadly oval, base tapering, acuminate apex, margin entire; deciduous.

Trunk: W-yellowish white, hard; young twigs prominently **leaf-scarred**.

Flowers: greenish-white; I-large, clustered corymbs, axillary.

Fruits: globose berry on long thickened woody stalk.

Site: scrub, monsoon forest; DL.

Uses: whole plant-medicinal; W-sandals.

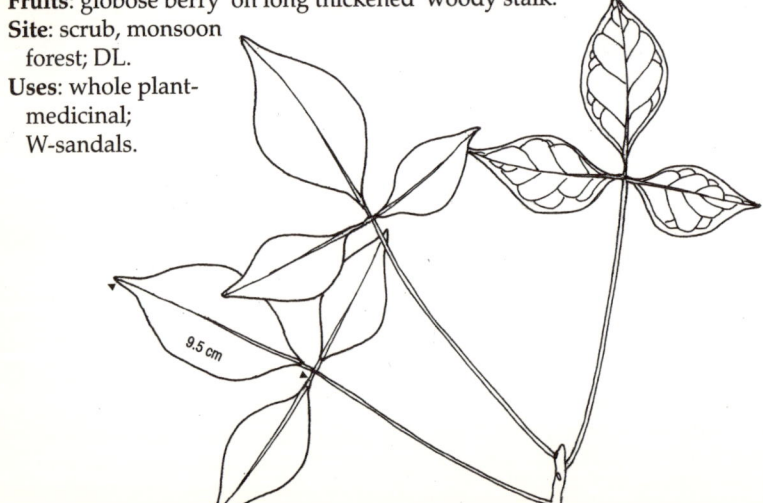

27. CARICACEAE

FAMILY DESCRIPTION - **Habit**: trees or shrubs, rarely herbs. Milky latex. **Leaves**: spiral, simple, lobed to palmate. **Stipules**: absent or spine-like. **Flowers**: unisexual, rarely bisexual. **Fruits**: large berry with numerous seeds.

FLOWER PARTS - CALYX: both male and female flowers 5-lobed or toothed, small. COROLLA: male flowers, 5 petals, united basally into a slender tube, free ends contorted or valvate. Female flowers, 5 petals, usually free. ANDROECIUM: stamens 10, epipetalous. Filaments free or united at base. Anthers dehiscing longitudinally. GYNOECIUM: superior, with 5 fused carpels. One loculus. Placentation parietal. Style short or absent.

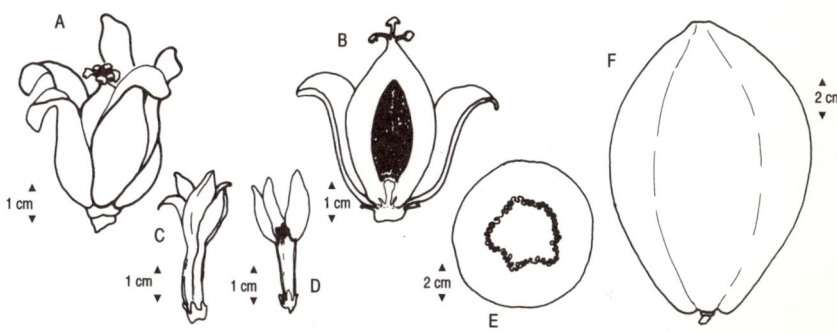

Key: full female flower (A) half female flower (B), full male flower (C), half male flower (D), transverse section of ovary (E) and fruit (F) of *Carica papaya*.

1. *Carica papaya*, papaya, papaw (E)/ gas labu, papol (S), I (S. Amer.), 7, small tree.
   Leaves: large, **palmate, pointed lobes**; petioles long.
   Trunk: **unbranched**; B-greenish brown; IB-yellow-green; W-soft, succulent; **milky** latex.
   Flowers: whitish to pale yellow; I-branched clusters or flowers solitary.
   Fruits: **orange, large** berry, soft orange flesh.
   Site: home gardens; widespread.
   Uses: whole plant-medicinal; fruit-edible; latex-source of papain.

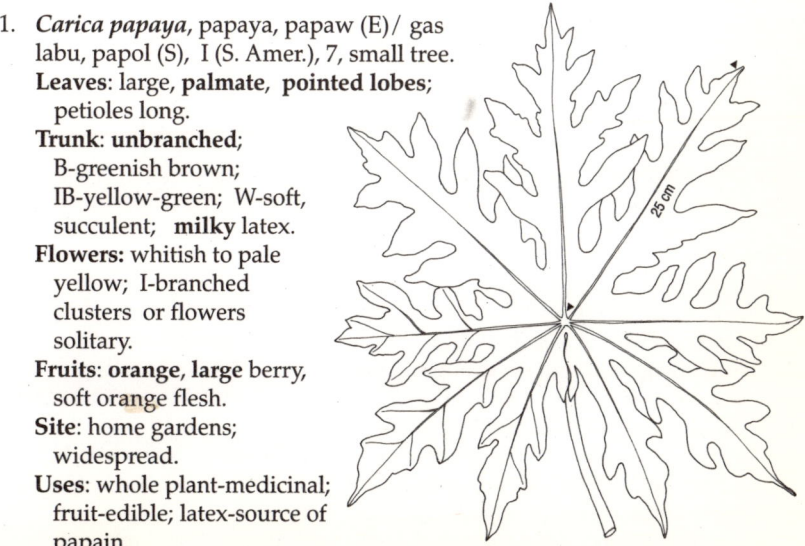

28. CASUARINACEAE

FAMILY DESCRIPTION - **Habit**: trees or shrubs with jointed twigs. Roots with nitrogen fixing bacterial nodules. **Leaves**: reduced to many-toothed sheaths around nodes. **Stipules**: absent. **Flowers**: unisexual, (plants monoecious or dioecious). Male flowers in spikes, female in heads. **Fruits**: one-seeded, winged nuts, crowded into a cone.

FLOWER PARTS - CALYX: absent in both male and female flowers. COROLLA: absent in both male and female flowers. ANDROECIUM: central solitary stamen. Filament lengthens during flowering. Anther 2-locular, opening lengthwise. GYNOECIUM: superior, 1-locular, style short.

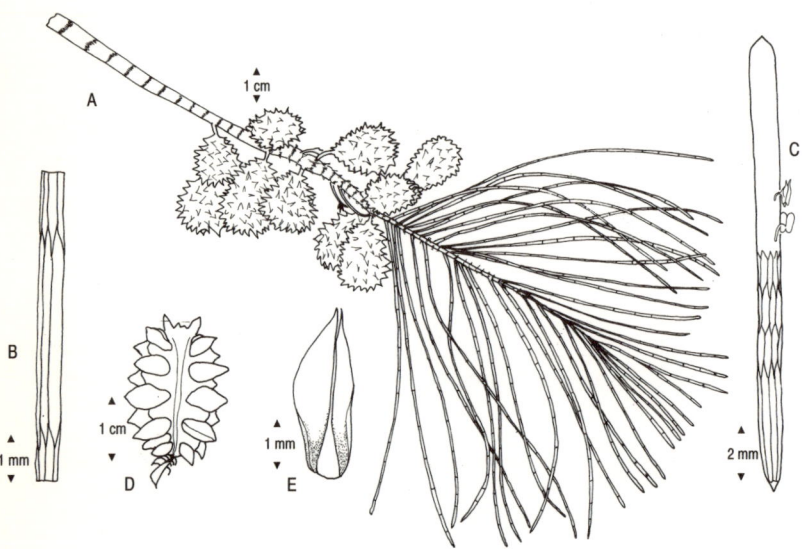

Key: twig bearing cones (A), portion of twig showing jointed segments and scale leaves (B), male inflorescence (C), longitudinal section of a cone (D), and winged nut (E) of *Casuarina equisetifolia*.

1. *Casuarina equisetifolia*, casuarina (E)/kassa (S), I (Australia), 35, tree.
   Leaves: scale-like, basally united, forming a sheath around the twig.
   Trunk: twigs dark green, **drooping, needle-like**. B-smooth to deeply furrowed, light grey-brown; B-reddish; W,s-pink to light brown; h-dark brown, hard, heavy, termite-susceptible.
   Flowers: numerous, small, inconspicuous; I-light brown spikes and heads.
   Fruits: multiple, grey-green, **cone-like**.
   Site: roadsides, coastal sands, plantations; DL, W.
   Uses: windbreaks; erosion control; W-general construction, fuelwood.

29. CELASTRACEAE

FAMILY DESCRIPTION - Habit: trees, shrubs or lianas. **Leaves**: spiral or opposite, simple. **Stipules**: small or absent. **Flowers**: bisexual, rarely unisexual, actinomorphic, small, in cymes or racemes. **Fruits**: berry, capsule, drupe or samara. Seeds with arils, wings, angular or compressed.

FLOWER PARTS - CALYX: 4-5 lobed, imbricate, very rarely valvate, sometimes with a basal tube. COROLLA: 5, rarely absent, imbricate or rarely valvate. ANDROECIUM: stamens 4 or 5, rarely more, on, without or within an often flat, fleshy nectary disc. Anthers 2-locular, opening lengthwise. GYNOECIUM: superior, rarely half-inferior, free or adherent to nectary disc. 1-5-locular. Style short, 3-lobed. Ovules 2 per loculus.

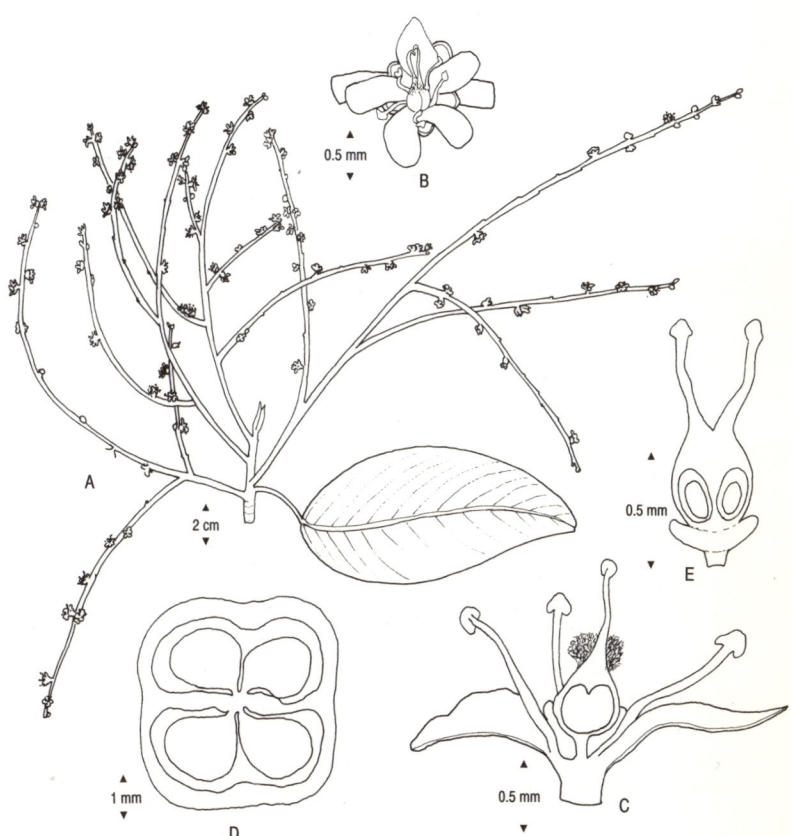

Key: inflorescence (A), full flower (B), half flower (C), transverse section (D), and longitudinal section of gynoecium (E) of *Bhesa ceylanica*.

1. *Bhesa ceylanica*, pelan (S)/konnai (T), (T I:274), E, 25, tree.
    Leaves: spiral, oval to lanceolate, base rounded,
        acute, **often twisted** apex, lateral
        veins 12-15, **parallel, curved**.
    Trunk: B-smooth, thick, dark grey;
        W-pale yellowish brown, heavy.
    Flowers: green, sessile; laxly
        arranged on branches of
        terminal panicle.
    Fruits: irregularly ovoid, blunt,
        red, glabrous capsule.
    Site: rain forest canopy; LM, W.
    Uses: W-heavy construction,
        furniture.

2. *Cassine glauca*, neralu (S)/ piyan (T),
        (T I:271), E, 10, tree.
    Leaves: opposite, oblong, acute base, acute
        apex, margin **shallowly serrate to entire**.
    Trunk: **dichotomously** branched; b-warted, thick,
        brownish grey; W-heavy, hard, reddish brown.
    Flowers: pale yellowish green; I-numerous,
        axillary, spreading, paniculate cymes.
    Fruits: ovoid, apiculate, glabrous drupe.
    Site: coastal scrub, monsoon forest; DL,DC.
    Uses: leaf, roots-medicinal;
        W-turnery, utensils, cart axels.

3 *Euonymus walkeri*, (T I:267),
        E, 3, small tree.
    Leaves: opposite, shape variable,
        acute base, apex variable,
        margin **shallowly scalloped**.
    Trunk: ib-**brilliant orange-red**; branches
        dichotomous; twigs somewhat quadrangular.
    Flowers: greenish-crimson, solitary or in 3's,
        sessile, axillary.
    Fruits: pear-shaped
        capsule; very
        large yellow aril.
    Site: rain forest
        understory; W.

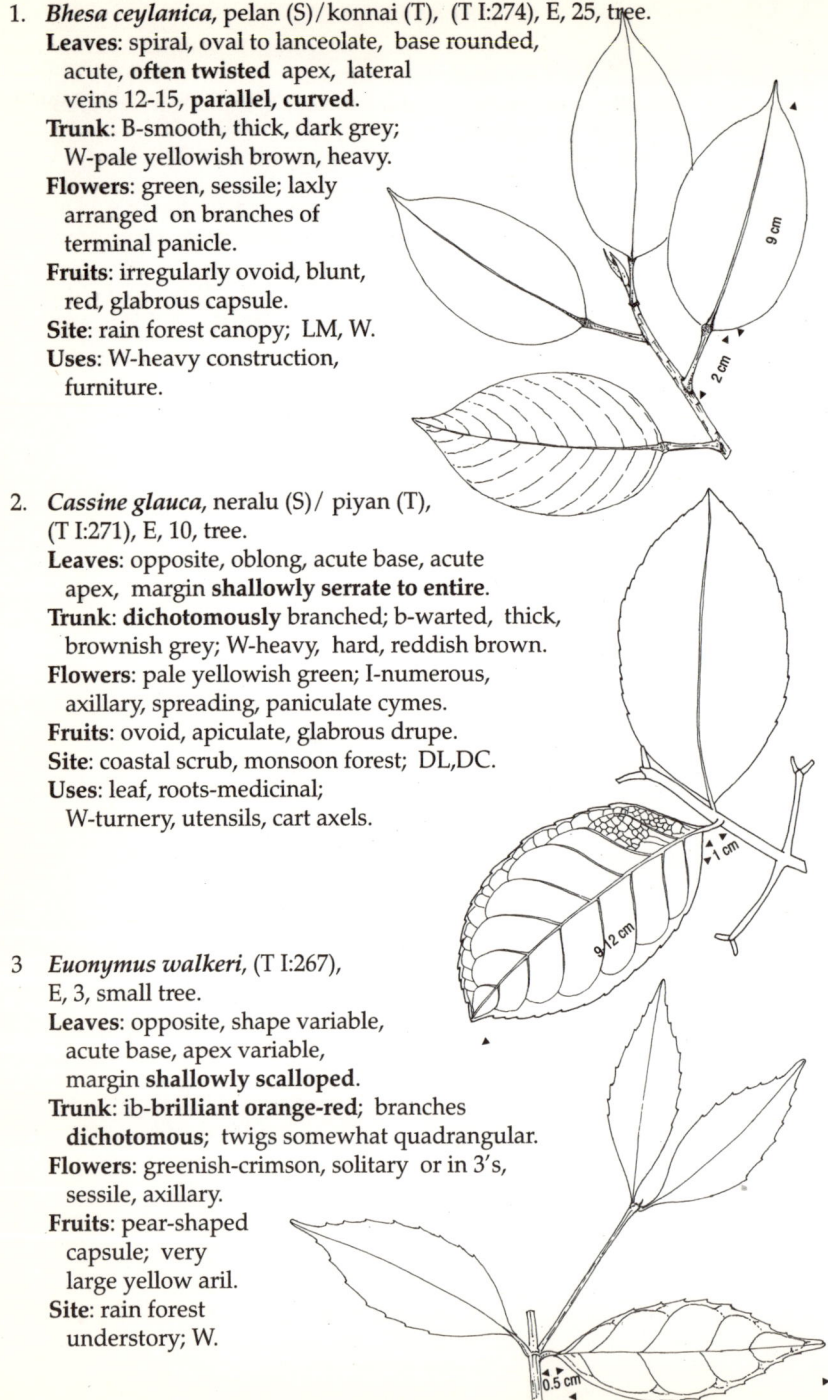

4. *Glyptopetalum zeylanicum*, (T I:268),
   N, 4, small tree.
   Leaves: opposite, oval to
   lanceolate, tapering ends,
   margins **serrate**.
   Trunk: B-smooth, grey.
   Flowers: pale green;
   I-dichotomously paniculate,
   extra-axillary cymes.
   Fruits: capsule; seed
   with crimson aril.
   Site: rain forest understory; W.

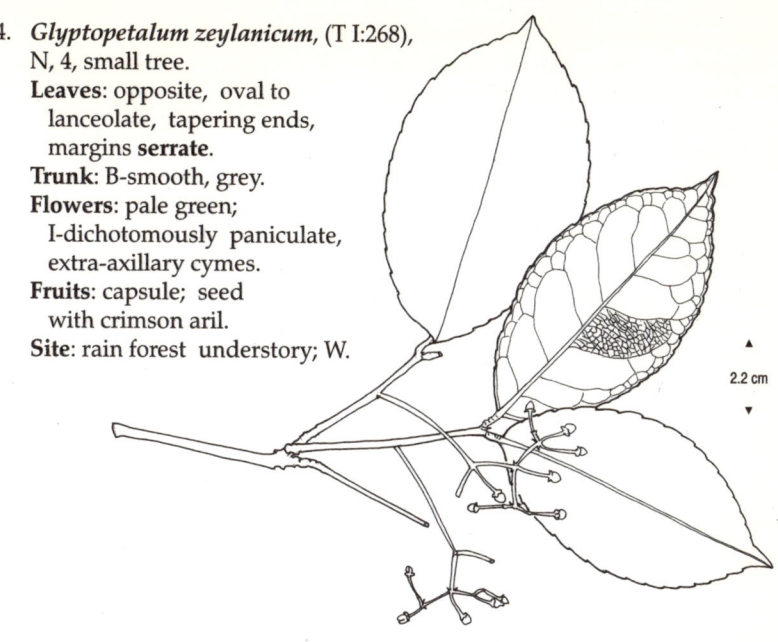

2.2 cm

29

5. *Kokoona zeylanica*, kokum (S), (T I:270), E, 30, tree.
   Leaves: opposite, obovate, cuneate base, rounded to retuse apex,
   margins entire to faintly serrate, beneath **glandular dots**.
   Trunk: B-rough, corky, grey; IB-**bright yellow**; W-pale yellowish brown.
   Flowers: dull yellowish brown;
   I-axillary panicles.
   Fruits: oblong ovoid, bluntly
   3-sided capsule; seeds winged.
   Site: rainforest canopy; W.
   Uses: leaves, bark-medicinal,
   powdered form-cosmetic;
   IB-polishing gold, snuff;
   W-light construction.

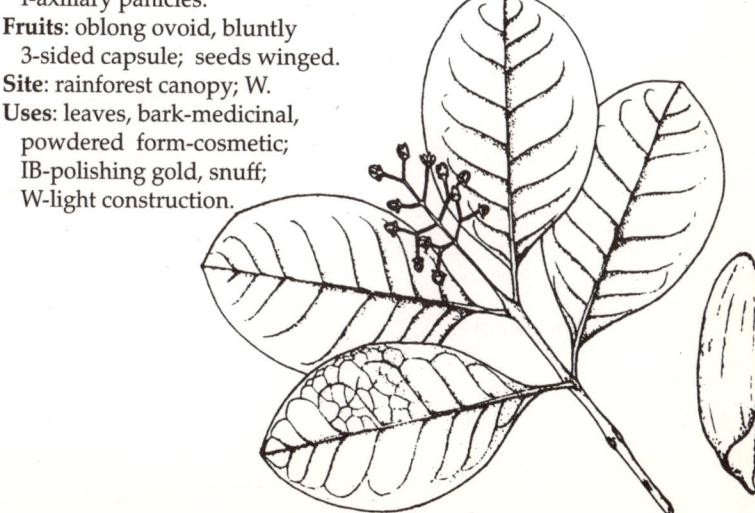

CELASTRACEAE

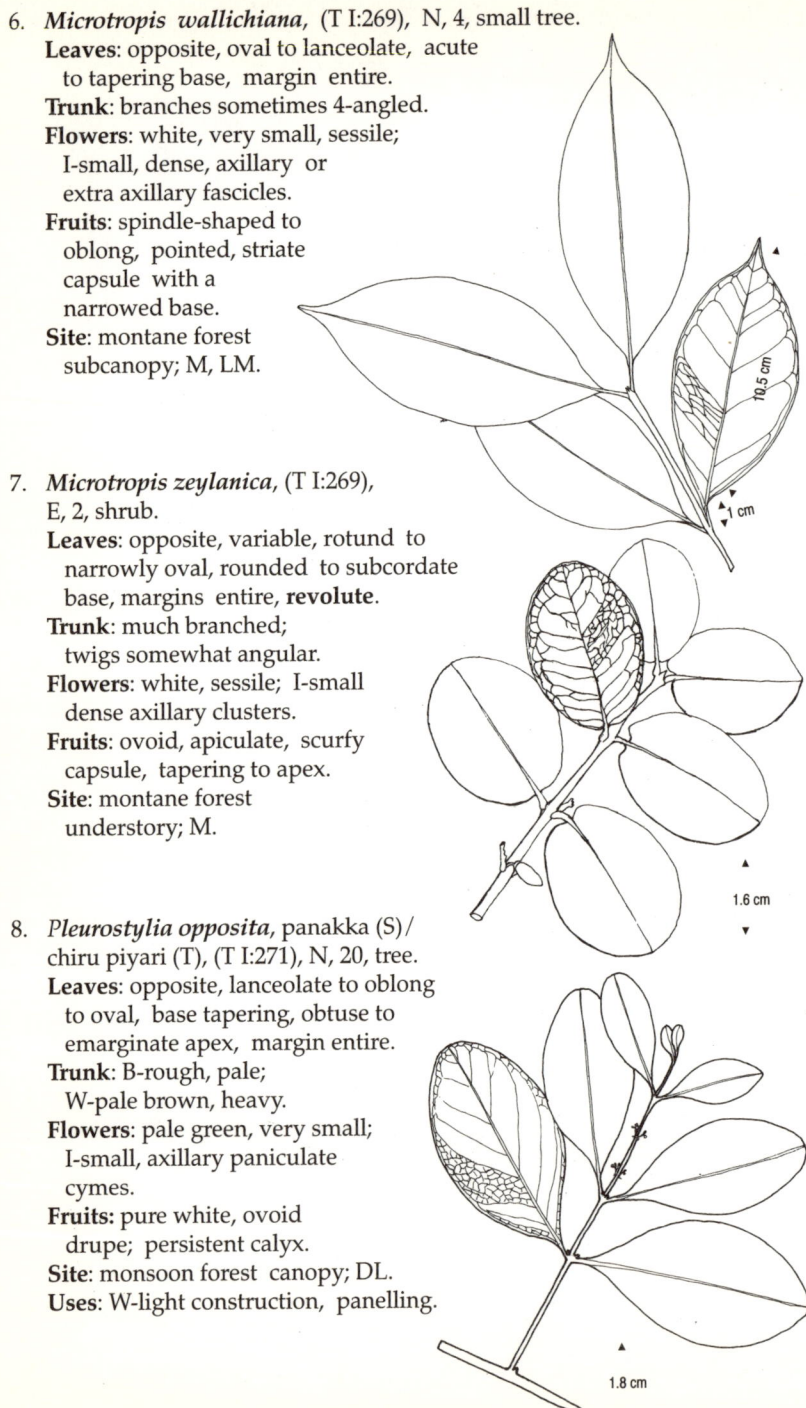

6. *Microtropis wallichiana*, (T I:269), N, 4, small tree.
   Leaves: opposite, oval to lanceolate, acute
       to tapering base, margin entire.
   Trunk: branches sometimes 4-angled.
   Flowers: white, very small, sessile;
       I-small, dense, axillary or
       extra axillary fascicles.
   Fruits: spindle-shaped to
       oblong, pointed, striate
       capsule with a
       narrowed base.
   Site: montane forest
       subcanopy; M, LM.

7. *Microtropis zeylanica*, (T I:269),
   E, 2, shrub.
   Leaves: opposite, variable, rotund to
       narrowly oval, rounded to subcordate
       base, margins entire, **revolute**.
   Trunk: much branched;
       twigs somewhat angular.
   Flowers: white, sessile; I-small
       dense axillary clusters.
   Fruits: ovoid, apiculate, scurfy
       capsule, tapering to apex.
   Site: montane forest
       understory; M.

8. *Pleurostylia opposita*, panakka (S) /
   chiru piyari (T), (T I:271), N, 20, tree.
   Leaves: opposite, lanceolate to oblong
       to oval, base tapering, obtuse to
       emarginate apex, margin entire.
   Trunk: B-rough, pale;
       W-pale brown, heavy.
   Flowers: pale green, very small;
       I-small, axillary paniculate
       cymes.
   Fruits: pure white, ovoid
       drupe; persistent calyx.
   Site: monsoon forest canopy; DL.
   Uses: W-light construction, panelling.

30. CHLORANTHACEAE

FAMILY DESCRIPTION - **Habit**: trees, shrubs or herbs. **Leaves:** opposite, simple. Petioles more or less united at base. **Stipules**: small, interpetiolar. **Flowers**: bisexual or unisexual, in spikes, panicles or heads. **Fruits**: drupe, small, ovoid or globose.

FLOWER PARTS - CALYX: absent in male flowers. Present in female flowers where it is adnate   to ovary and often minutely 3-dentate at apex. COROLLA: absent in both male and female flowers. ANDROECIUM: stamens 1 to 5, united into a mass. GYNOECIUM: inferior or half-inferior, 1-locular. Stigma sessile or on a short   style. Ovule solitary and pendulous.

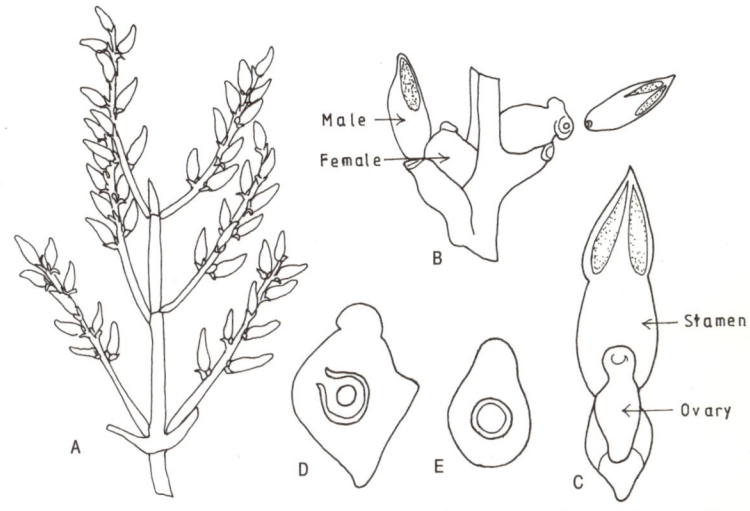

Key: inflorescence (A), part of inflorescence showing male and female flowers (B, C), the gynoecium in longitudinal (D) and transverse (E) section showing single pendulous ovule of *Sarcandra chloranthoides*.

1. *Sarcandra chloranthoides*, (T III:433), N, 2, shrub.
   Leaves: rather large, lanceolate, tapering ends, very acute apex, coarsely **spinous-serrate**.
   Trunk: stems cylindrical, glabrous, dark green.
   Flowers: sessile, stamens pinkish yellow; I-short, lax, spikes forming terminal panicles.
   Fruits: purplish black, fleshy drupes, broadly ovoid, pointed.
   Site: rain forest understory and openings; LM, W.

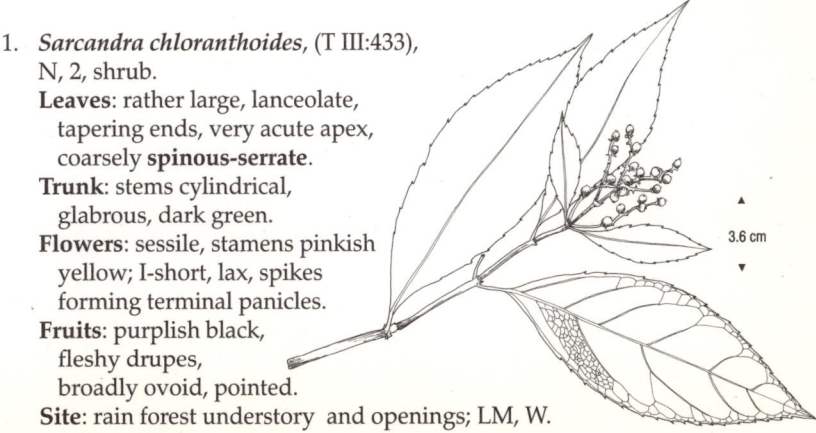

3.6 cm

31. CLUSIACEAE (GUTTIFERAE*)

FAMILY DESCRIPTION - **Habit**: trees, shrubs, lianas or herbs. Milky, yellow or pink latex. **Leaves**: opposite or whorled, simple, usually entire. **Stipules**: absent. **Flowers**: bisexual or unisexual, actinomorphic, in terminal cymes. **Fruits**: drupe, berry or septicidal capsule. Seeds often arillate.

FLOWER PARTS - CALYX: 2-6 or rarely more, free, imbricate. COROLLA: 2-6 or rarely more, free or sometimes basally united. ANDROECIUM: stamens numerous, free or united into 2-5 bundles. Anthers 2-locular, opening lengthwise. Staminodes often present in female flowers. GYNOECIUM: superior, 1 to many-locular. Ovules 1-many per loculus. Styles as many as locules, basally united sometimes. Stigma lobed or peltate. Placentation axile, rarely parietal.

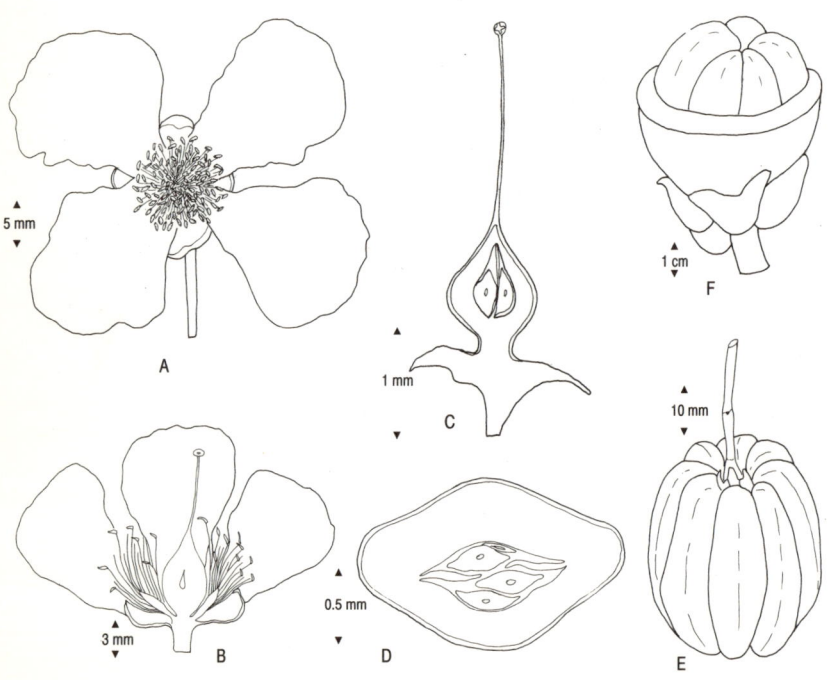

Key: full flower (A), half flower (B), longitudinal section of gynoecium (C), and transverse section of ovary (D) of *Mesua nagassarium*. Mature berries of *Garcinia quaesita* (E) and *Garcinia mangostana* (F) cut open to show arillate seeds.

* Alternative family name.

1. *Calophyllum bracteatum*, walu kina (S), (DF I:97), E, 20, tree.
   Leaves: **dimorphic**; ordinary leaves oblong, long pointed apex, veins
   parallel, prominulous; other leaves are in 2-8 pairs in between ordinary
   leaves, small, overlapping; flush **whitish**, **limp**, **pendulous**.

   Trunk: B-deeply fissured,
   yellowish stripes;
   IB-beefy red; branches
   pendulous;
   branchlets
   quadrangular,
   densely rusty
   tomentose; latex
   resinous, whitish.
   Flowers: white; I-axillary,
   rusty tomentose,
   pendulous racemes.
   Fruits: pendulous,
   smooth, globose drupes.
   Site: rain forest subcanopy along
   stream ways and river courses; W.
   Uses: ornamental; W-heavy construction,
   furniture.

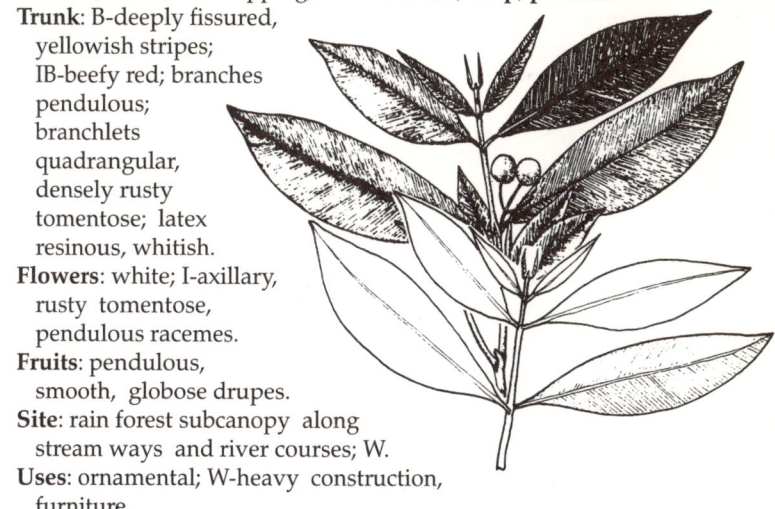

2. *Calophyllum calaba*, guru kina (S), (DF I:95), E, 10, tree.
   Leaves: elliptic to ovate, base shortly cuneate, acute to
   acuminate apex, lateral veins parallel, prominent.

   Trunk: B-young smooth, mature rough, deeply
   fissured, lenticellate, **bright
   yellow-orange**;
   IB-beefy red;
   clear exudate.
   Flowers: I-racemes
   in axils of upper
   leaves.
   Fruits: bright
   orange, ellipsoid
   to subglobose
   drupe.
   Site: monsoon and
   intermediate forest
   subcanopy;
   DC, DL, IN.
   Uses: fruit-edible.

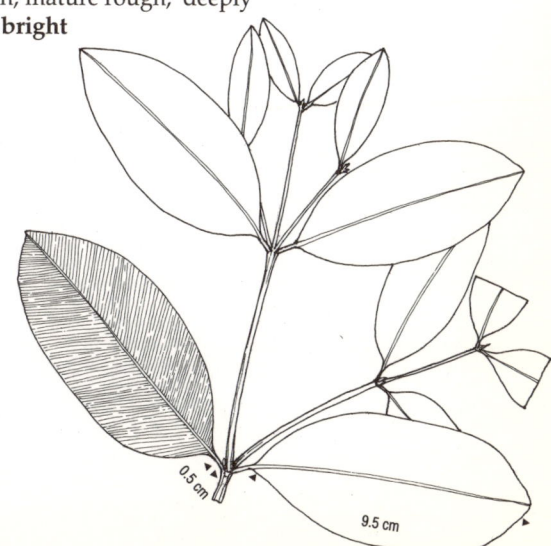

0.5 cm

9.5 cm

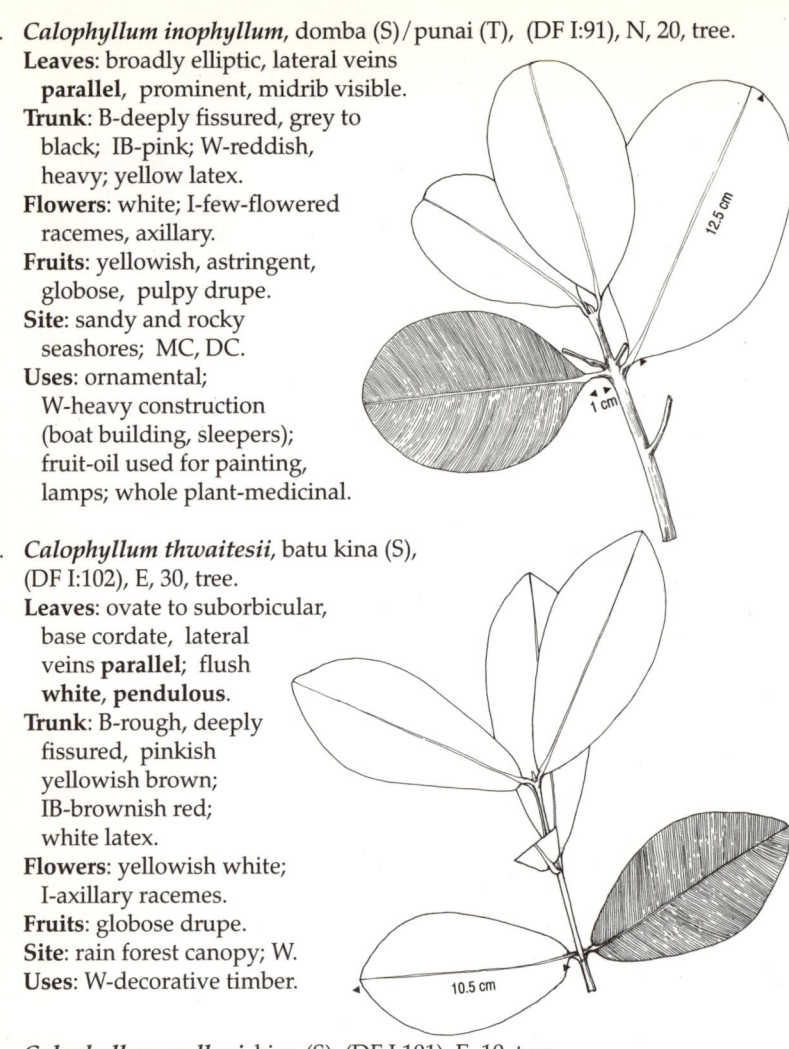

3. *Calophyllum inophyllum,* domba (S)/punai (T), (DF I:91), N, 20, tree.
   Leaves: broadly elliptic, lateral veins
      parallel, prominent, midrib visible.
   Trunk: B-deeply fissured, grey to
      black; IB-pink; W-reddish,
      heavy; yellow latex.
   Flowers: white; I-few-flowered
      racemes, axillary.
   Fruits: yellowish, astringent,
      globose, pulpy drupe.
   Site: sandy and rocky
      seashores; MC, DC.
   Uses: ornamental;
      W-heavy construction
      (boat building, sleepers);
      fruit-oil used for painting,
      lamps; whole plant-medicinal.

4. *Calophyllum thwaitesii,* batu kina (S),
   (DF I:102), E, 30, tree.
   Leaves: ovate to suborbicular,
      base cordate, lateral
      veins **parallel**; flush
      white, pendulous.
   Trunk: B-rough, deeply
      fissured, pinkish
      yellowish brown;
      IB-brownish red;
      white latex.
   Flowers: yellowish white;
      I-axillary racemes.
   Fruits: globose drupe.
   Site: rain forest canopy; W.
   Uses: W-decorative timber.

5. *Calophyllum walkeri,* kina (S), (DF I:101), E, 10, tree.
   Leaves: **obovate** to **rotund**, lateral veins parallel but hardly
      visible, midrib prominent; young flush **bright red**.
   Trunk: B-fissured, cracked, yellowish;
      IB-brownish to light red.
   Flowers: pinkish white; I-axillary racemes.
   Fruits: pale yellow with brown spots,
      erect, globose, smooth drupe.
   Site: **montane** forest canopy; M, LM.
   Uses: W-construction timber;
      fruit-edible (jelly);
      seed-medicinal oil.

6. *Clusea major*, (DF I:110), I (C. Amer.), 7, tree.
    Leaves: thick, **round slightly notched** apex; **horizontally branched** crown.

    Trunk: B-fissured and warty, grey; IB-pink
    brown; W-reddish brown, hard,
    heavy; **prop and aerial
    roots**; **yellow** latex.
    Flowers: showy, white.
    Fruits: yellow-green to brown,
    rounded, fleshy, septicidal
    capsule splitting into
    parts; orange pulp.
    Site: 'strangling tree'
    of disturbed forest,
    often prostrate on
    rocks or cliffs; IN, W.
    Uses: latex-caulking;
    W-posts, fuelwood.

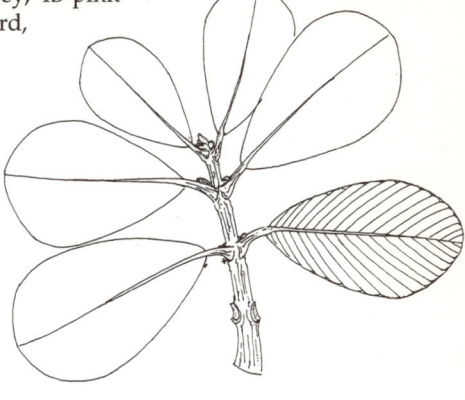

7. *Garcinia echinocarpa*, madol (S),
    (DF I:79), N, 10, tree.
    Leaves: obovate to
    oblanceolate, tapered base,
    pointed apex,   lateral veins
    approximately **25** pairs, with
    intermediaries between,
    midrib prominent.
    Trunk: **stilt roots**, B-rough,
    peeling, lenticellate, red
    brown; IB-red brown;
    branchlets angular; white latex.
    Flowers: female flowers pale green, solitary,
    almost sessile, axillary at base of 2 apical leaves.
    Fruits: green to yellowish green, spiny berry; pulp white.
    Site: rain forest understory; M, LM, W.
    Uses: W-roof shingles.

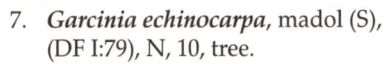

7 cm

1.3 cm

8. *Garcinia hermonii*, madol (S), (DF I:81),
    E, 20, tree.
    Leaves: elliptic to lanceolate, acute base,
    acuminate apex, lateral veins **50-60**
    pairs, with a few intermediaries.
    Trunk: B-smooth to rough, peeling, dark
    rusty; inside beefy red; white latex.
    Flowers: pale green.
    Fruits: pale green, subglobose berry
    with soft spines; orange mealy pulp.
    Site: rain forest understory; W.

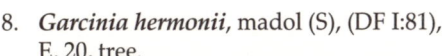

31

CLUSIACEAE

9. *Garcinia mangostana*, mangosteen (E,S), (DF I:88), I (Malesia), 20, tree.
    Leaves: elliptic to oblong, base obtuse, short
        pointed apex, lateral veins parallel,
        double intra-marginal vein.
    Trunk: B-smooth, dark;
        IB-yellowish; W-red, hard,
        heavy; **yellowish** latex.
    Flowers: 4 fleshy petals;
        I-axillary (pseudo-
        terminal).
    Fruits: purplish black berry,
        globose; enlarged sepals;
        spongy pulp.
    Site: **home gardens**; IN, W.
    Uses: fruit-edible; whole
        plant-medicinal.

10. *Garcinia morella*, gokatu (S), (DF I:77), N, 20, tree.
    Leaves: obovate to oblanceolate, tapered
        base, short pointed apex.
    Trunk: B-becoming deeply fissured, dark
        brown; IB-white; W-yellowish, hard,
        mottled; **yellow** latex.
    Flowers: white to pink, sessile in axils;
        male 2-3 together; female solitary.
    Fruits: yellowish, globose, smooth
        berry; spreading sepals.
    Site: rain forest canopy,
        subcanopy; LM, W.
    Uses: whole plant-medicinal; latex-dye, painting,
        mordant; fruits-spice, preservative.

11. *Garcinia quaesita*, kana goraka, honda
    goraka (S), (DF I:74), E, 15, tree.
    Leaves: oblanceolate to subovate, base
        tapered, lateral veins **8-10** pairs
        with intermediaries between.
    Trunk: B-rough, cracked, peeling,
        blackish; IB-brown red to
        yellow; **dark yellow** latex.
    Flowers: white; axils of upper leaves;
        male clustered; female solitary.
    Fruit: red to orange berry, globose,
        6-9-grooved; pulp very acid.
    Site: intermediate and rain
        forest understory,
        home gardens; IN, W.
    Uses: fruit-spice, preservative.

12. *Garcinia spicata*, ela gokatu (S), (DF I:84), N, 12, tree.
    Leaves: elliptic to ovate, base obtuse,
       lateral veins about **15-18** pairs,
       intermediaries and midrib
       prominent.
    Trunk: B-smooth, pale brown;
       IB-white; W-yellowish
       white, heavy, hard; twigs
       densely **pubescent**;
       white to yellow latex.
    Flowers: white; I-axillary,
       raceme-like clusters.
    Fruits: yellow, subglobose
       to oblong berry; bad smell.
    Site: riverine and monsoon forest; DL.
    Uses: W-construction.

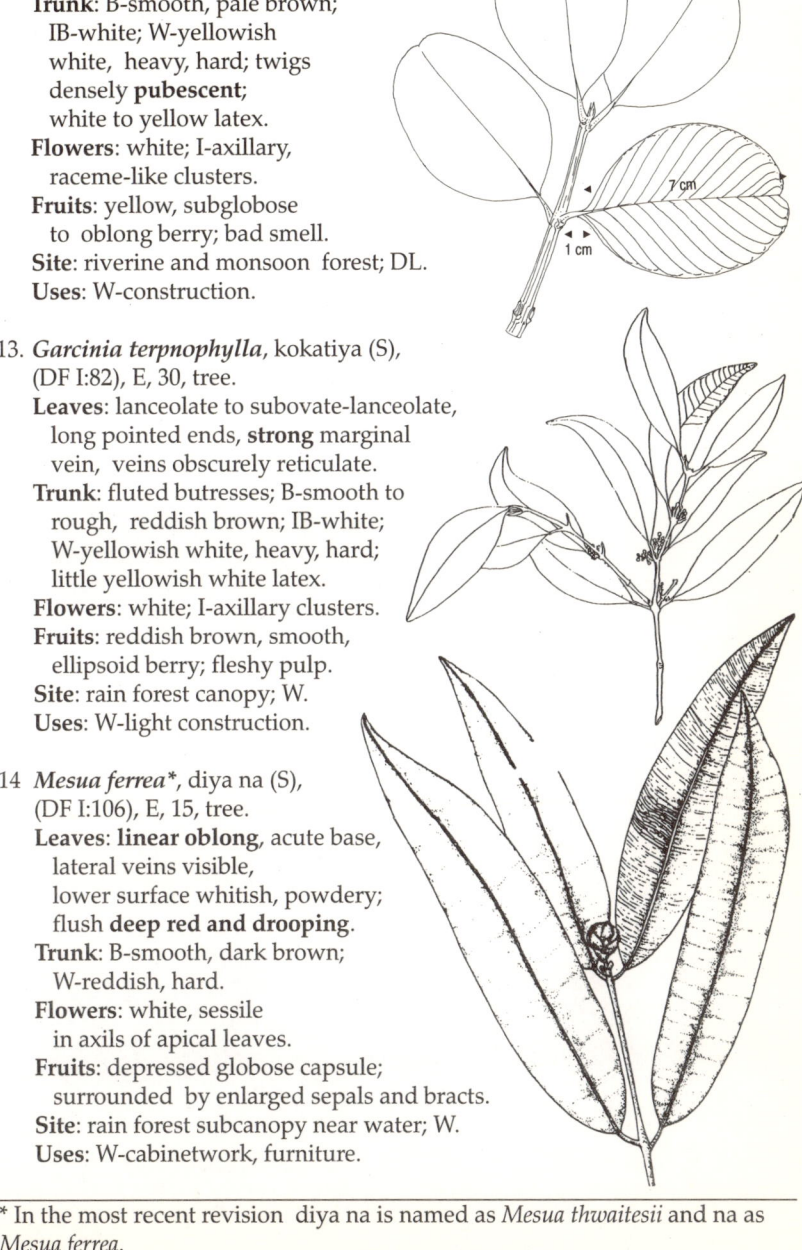

31

CLUSIACEAE

13. *Garcinia terpnophylla*, kokatiya (S),
    (DF I:82), E, 30, tree.
    Leaves: lanceolate to subovate-lanceolate,
       long pointed ends, **strong** marginal
       vein, veins obscurely reticulate.
    Trunk: fluted butresses; B-smooth to
       rough, reddish brown; IB-white;
       W-yellowish white, heavy, hard;
       little yellowish white latex.
    Flowers: white; I-axillary clusters.
    Fruits: reddish brown, smooth,
       ellipsoid berry; fleshy pulp.
    Site: rain forest canopy; W.
    Uses: W-light construction.

14 *Mesua ferrea**, diya na (S),
    (DF I:106), E, 15, tree.
    Leaves: **linear oblong**, acute base,
       lateral veins visible,
       lower surface whitish, powdery;
       flush **deep red and drooping**.
    Trunk: B-smooth, dark brown;
       W-reddish, hard.
    Flowers: white, sessile
       in axils of apical leaves.
    Fruits: depressed globose capsule;
       surrounded by enlarged sepals and bracts.
    Site: rain forest subcanopy near water; W.
    Uses: W-cabinetwork, furniture.

* In the most recent revision diya na is named as *Mesua thwaitesii* and na as
Mesua ferrea.

15. *Mesua nagassarium*, na (S)/naka (T), (DF I:106), N, 30, tree.

Leaves: very variable, **linear-lanceolate**, prominent midrib, fine lateral **parallel** veins, lower surface whitish; flush **deep red** and **drooping**.

Trunk: fluted butresses; B-peeling flakes, ash grey to brown; IB-fibrous, reddish brown; W,s-creamy white to pinkish; h-dark red; resin aromatic.

Flowers: white, solitary, fragrant, in axils of upper leaves.

Fruits: ovoid to globose capsule.

Site: intermediate and rain forest canopy; home gardens; IN, W.

Uses: flowers-temple offering; national tree; ornamental; sacred; W-heavy construction, perfumery, cosmetics, cabinetwork; whole plant-medicinal.

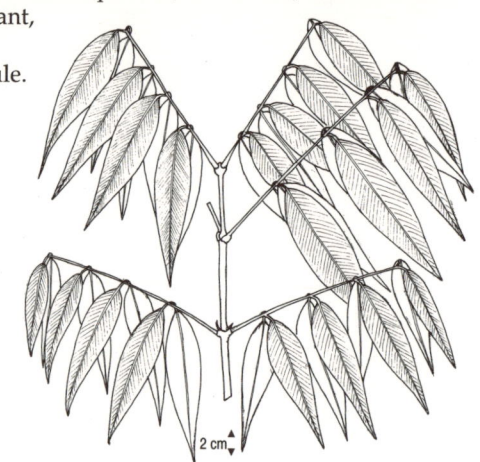

32. COCHLOSPERMACEAE

FAMILY DESCRIPTION - **Habit**: trees or shrubs. Coloured exudate. **Leaves**: spiral, simple. Palmately lobed. **Stipules**: present. **Flowers**: bisexual, actinomorphic, showy, in racemes or panicles. **Fruits**: 3-5-valved capsule. Seeds glabrous or covered with woolly hairs.

FLOWER PARTS - CALYX: 5, free, imbricate, deciduous. COROLLA: 5, free, imbricate or convolute. ANDROECIUM: stamens numerous. Filaments free, equal or some longer than others. Anthers 2-locular with terminal pore-like slits. GYNOECIUM: superior, 1-3-locular with parietal placentation. Style simple with minutely denticulate stigma.

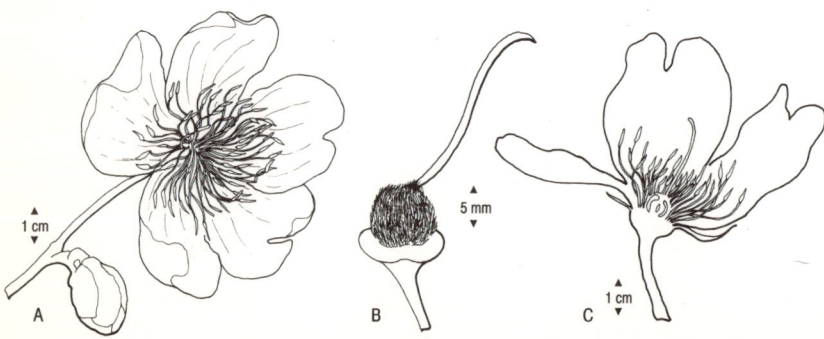

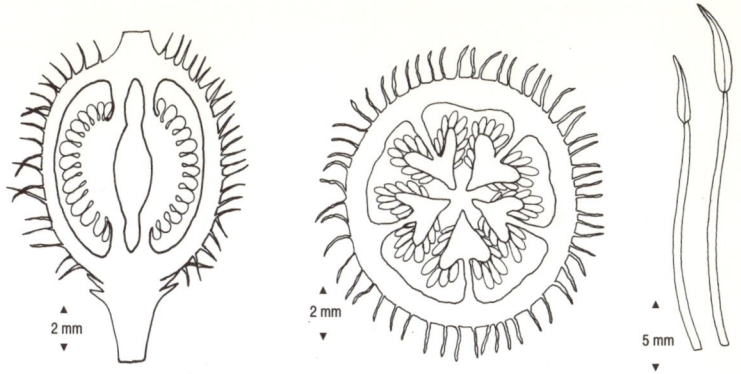

Key: full flower (A), half flower (B), gynoecium (C), longitudinal (D) and transverse (E) sections of ovary, and stamens (F) of *Cochlospermum religiosum*.

32

1. *Cochlospermum religiosum*, yellow silk cotton tree (E)/kinihiriya (S)/
    konga (T), (DF V:123), I (India), 8, tree.
    Leaves: **palmately lobed**, heart-shaped, acute, 5 - 7 main veins,
      sparsely hairy above, below densely hairy on veins; deciduous.
    Trunk: B-smooth, grey; exudate a skin irritant.
    Flowers: **bright yellow**, conspicuous; I-terminal,
      simple raceme or loosely branched panicle,
      blooming after leaf-fall.
    Fruits: obovoid, ribbed capsule;
      seeds covered with woolly hair:
      stalk purplish, jointed at base
    Site: temples, gardens; DL, IN
    Uses: exudate-medicinal;
      flowers-temple
      offerings; fruit
      hairs-stuffed in a
      pillow induces sleep.

COCHLOSPERMACEAE

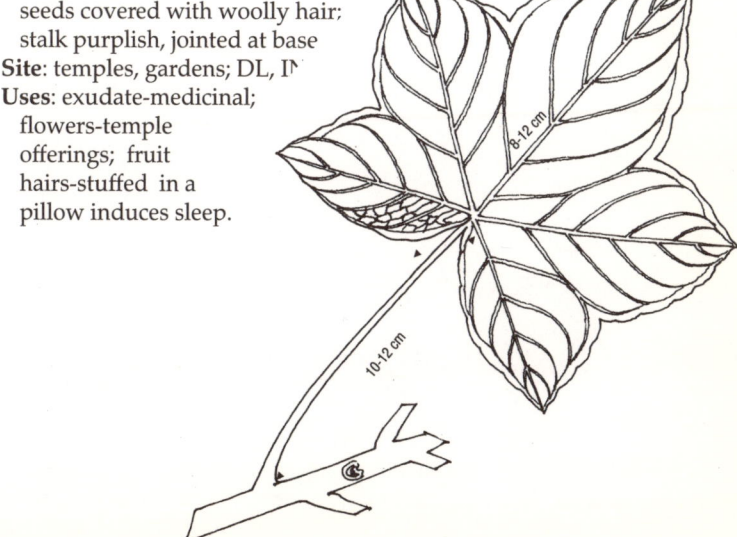

141

33. COMBRETACEAE

FAMILY DESCRIPTION - Habit: trees or shrubs, often scandent. **Leaves**: spiral, opposite or whorled, simple and entire. Leaf base often with 2 gland-containing cavities. **Stipules**: minute or absent. **Flowers**: bisexual, actinomorphic, in spikes, racemes or heads. **Fruit**: one-seeded drupe, usually indehiscent, water -dispersed, generally ribbed, the ribs wing-like. Rarely dry and dehiscent.

FLOWER PARTS - CALYX: 4-8, basally united in a tube adnate to the ovary, valvate or imbricate. COROLLA: 4-8, or absent, free, small, imbricate or valvate. ANDROECIUM: stamens 4-10, rarely more. Anthers versatile, opening lengthwise by slits. GYNOECIUM: inferior, 1-locular. Style simple. Ovules 2-6, pendulous.

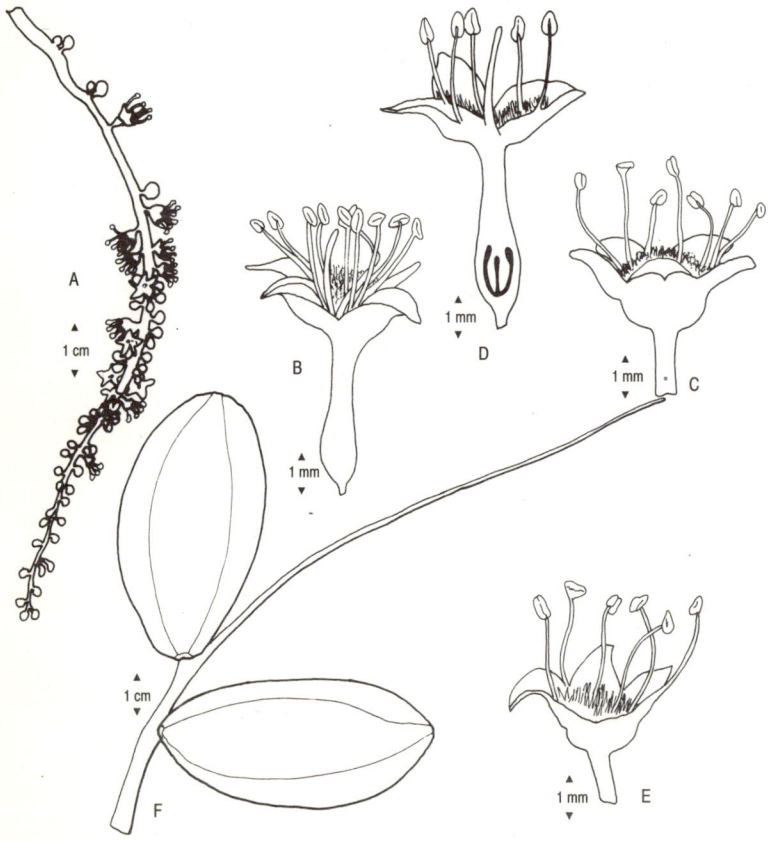

Key: inflorescence (A), full bisexual flower (B), full male flower (C), half bisexual flower (D), half male flower (E) and mature fruits (F) of *Terminalia catappa*.

1. *Anogeissus latifolius*, dawu (S)/vekkali (T),
   (T II:162), N, 20, tree.
   Leaves: oblong-oval, rounded base, obtuse to
       pointed apex, margins slightly undulate;
       petiole **pink**; flush pink; deciduous.
   Trunk: B-smooth, whitish grey;
       W, h-extremely hard.
   Flowers: pale greenish yellow;
       I-sessile, small dense heads.
   Fruits: very small drupes, broadly
       winged along edges, beaked, brown,
       crowded in small globular heads.
   Site: monsoon and intermediate
       forest canopy, savannah; DL, IN.
   Uses: W-turnery; whole plant-medicinal.

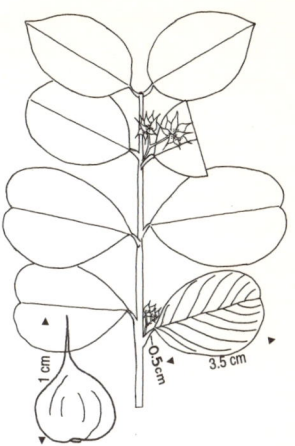

2. *Lumnitzera racemosa*, beriya (S)/tipparathai (T),
   (T II:162), N, 3, shrub to small tree.
   Leaves: **spathulate** to oblong, tapering base, rounded
       apex, margins **shallowly crenate**.
   Trunk: B-purplish, smooth, fleshy;
       W-pale, reddish-grey, heavy.
   Flowers: white, small, sessile; I-spike.
   Fruits: oblong to ovoid capsule.
   Site: **mangroves**; MC, DL.

3. *Terminalia arjuna*, kumbuk (S)/
   marutu (T), (T II:160 & VI:109), N, 30, large tree.
   Leaves: oblong-oval, rounded tapering ends,
       margins **shallowly crenate to serrate**,
       2 prominent **glands** at petiole apex.
   Trunk: B-thick, smooth to flaky, pinkish to
       greenish white; W-greyish brown with
       dark streaks, heavy, strong;
       horizontally spreading branches.
   Flowers: greenish white, honey-scented; I-sessile
       spikes rather lax, in axillary or terminal panicles.
   Fruits: obovate to ovoid,
       base narrowed, apex
       bluntly pointed, hard, fibrous
       drupe with 5 stiff wings.
   Site: **along water courses** of
       monsoon forest; DL, IN.
   Uses: leaves, B-medicinal,
       burnt for lime (betel);
           W-heavy construction timber.

4. *Terminalia bellirica*, bulu (S)/tanti (T), (T II:159), N, 25, large tree.

Leaves: terminally **clustered**, whorled, obovate to oval, tapering base, rounded to shortly acuminate apex, main veins prominent beneath.

Trunk: buttressed; B-brown, vertically furrowed; W-greyish yellow, hard; **horizontally** branching.

Flowers: pale greenish yellow, scented, small, numerous; I-axillary spikes, crowded terminally.

Fruits: pear-shaped drupe, narrowed base, irregular surface, with fine brown tomentum.

Site: intermediate and rain forest canopy and savannah; IN, W.

Uses: shade tree; W-heavy construction; fruit, seed, stem-medicinal.

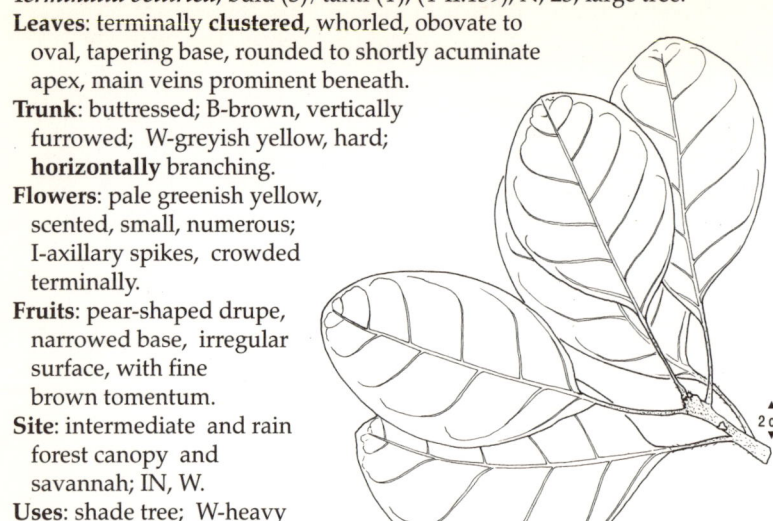

5. *Terminalia catappa*, Indian almond (E)/ kottamba (S), (W:234), I (Malesia), 20, tree.

Leaves: terminally **clustered**; large, leathery, obovate, **reddish** before falling.

Trunk: B-smoothish to fissured, grey; IB-pink-brown; W,s-light brown; h-reddish, hard, heavy, termite-susceptible; branches horizontally spreading, layered **pagoda-like**.

Flowers: greenish white; I-narrow, lateral or terminal racemes.

Fruits: greenish, large, elliptic, slightly compressed, fibrous drupe.

Site: roadsides, home gardens; salt-tolerant; sandy soils; widespread.

Uses: shade tree, ornamental; leaves-medicinal; W-fuelwood; seed-edible.

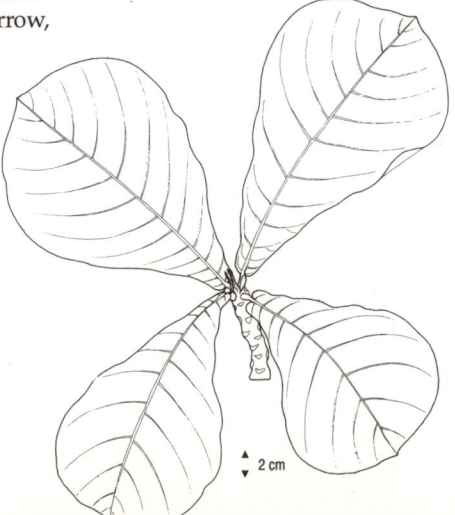

6. *Terminalia chebula*, aralu (S)/kadukkay (T), (T II:159), N, 25, tree.
   Leaves: oblong-oval, rounded to cordate base,
     obtuse apex, pubescent; petiole with
     2 prominent **glands** at apex.
   Trunk: crooked; B-grey-brown; W-heavy,
     very hard, dark brown with
     purplish tinge; young parts
     of branchlets pubescent.
   Flowers: greenish white;
     I-terminal spikes.
   Fruits: yellowish
     green, pendulous,
     ovoid drupe.
   Site: monsoon forest
     canopy, savannah;
     DL, IN.
   Uses: fruit, B-medicinal;
     fruit extract-traditionally
     used for marking laundry.

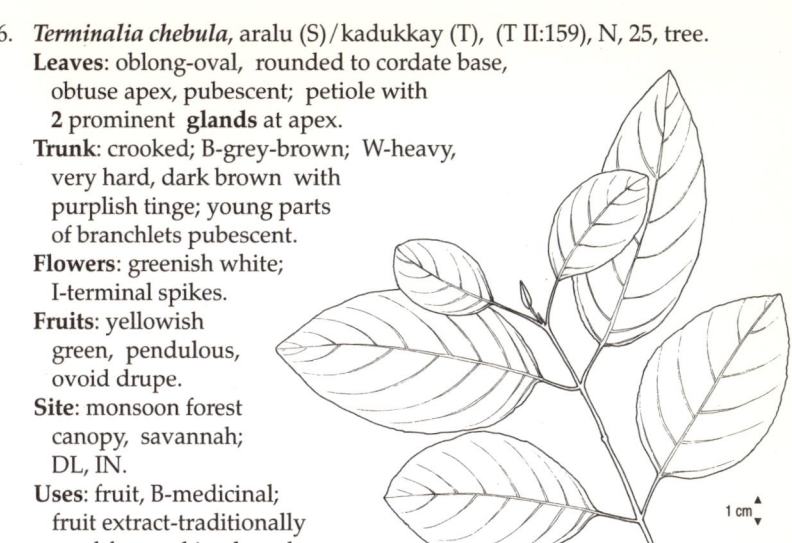

1 cm

33

7. *Terminalia zeylanica*, hampalanda (S), (T II:160), E, 25, tree.
   Leaves: ovate-oval, rounded **often**
     unequal-sided base, tapering acuminate
     apex, margins somewhat undulate.
   Trunk: B-smooth, pale brown;
     W-orange-brown, heavy.
   Flowers: pale sulphur-yellow,
     very small sessile; I-spikes
     in terminal and axillary panicles.
   Fruits: olive green with white dots,
     oblong-ovoid, blunt, cylindrical,
     smooth drupe.
   Site: rain forest canopy;
     LM, W.
   Uses: W-construction
     timber, furniture.

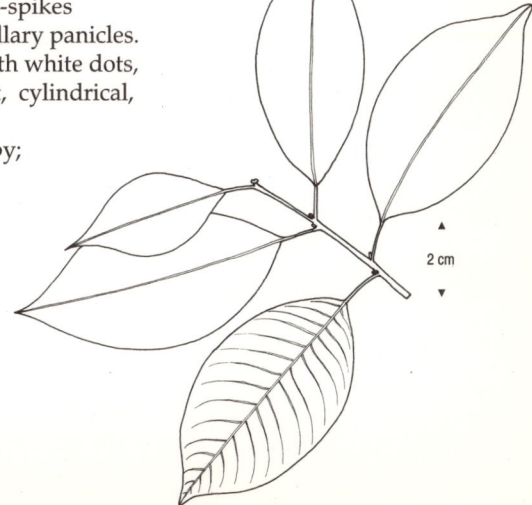

2 cm

COMBRETACEAE

34. COMPOSITAE

FAMILY DESCRIPTION - Habit: herbs, shrubs or climbers, rarely trees. **Leaves**: spiral, rarely opposite or whorled, simple or compound. **Stipules**: absent. **Flowers**: bisexual, unisexual or sterile, (plants rarely dioecious), actinomorphic or zygomorphic, in heads, surrounded by an involucre. Outer flowers on head female or sterile, often ligulate (ray florets). Inner ones tubular, bisexual or functionally male (disc florets). In some, tubular flowers only, in others ligulate only. **Fruits**: cypsela with persistent pappus formed by calyx.

FLOWER PARTS - CALYX: modified and thread-like, persistent in fruit. COROLLA: 5, united   basally, valvate. ANDROECIUM: stamens 5, rarely 4, epipetalous. Filaments free from each other. Anthers united into a tube (syngenesious), very rarely free, 2-locular and opening lengthwise. GYNOECIUM: inferior, 2 carpels, loculus, 1 basal ovule, 1 style, usually with 2 branches.

Key: head (A), bisexual disc floret (B), and sterile ray floret (C) of *Tithonia diversifolia.* Inflorescences of *Vernonia arborea* (D).

1. ***Tithonia diversifolia***, wild sunflower (E)/
   natta suriya, wal suriya kantha (S),
   (DF I:218), I (S. Amer.), 3, shrub.
   Leaves: alternate, **lower ones
   3-5-lobed**, margins **scalloped
   to serrate**, three main veins
   arising at base; upper
   ones oblanceolate to
   ovate-deltoid; petiole
   winged, pubescent.
   Trunk: stem striate,
   sparsely pubescent.
   Flowers: yellow,
   conspicuous.
   Fruits: dark brown
   achenes.
   Site: roadsides, waste
   ground; LM, UW.
   Uses: green manure,
   cattle fodder.

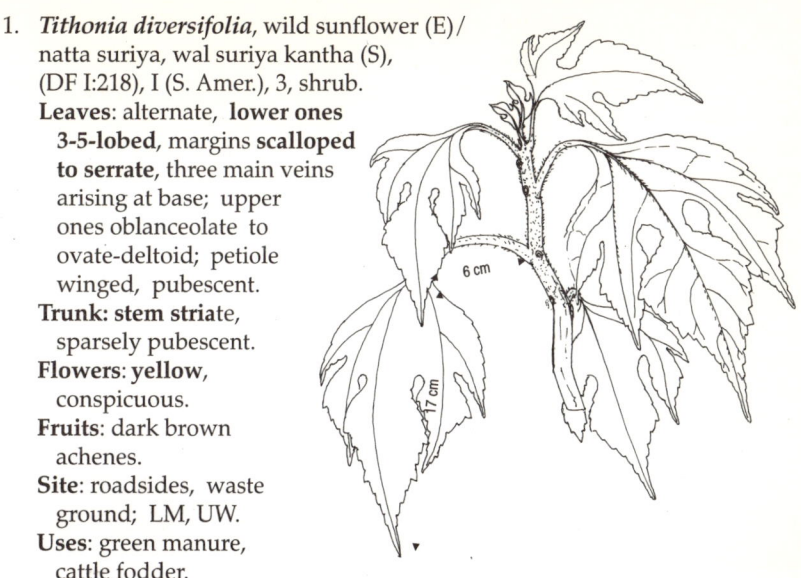

34

2. *Vernonia arborea*, kobo mella (S), (DF I:122), N, 12, small tree.
   Leaves: ovate to oblong-elliptic, rounded base, long pointed apex,
   margins entire, underneath **pubescent**.
   Trunk: B-whitish; twigs **yellowish
   tomentum**.
   Flowers: white or violet;
   I-terminal panicles.
   Fruits: angular achenes
   with about 10 ribs;
   pappus dull white,
   slightly hairy.
   Site: rain forest understory,
   openings and fringes,
   scrub; W.

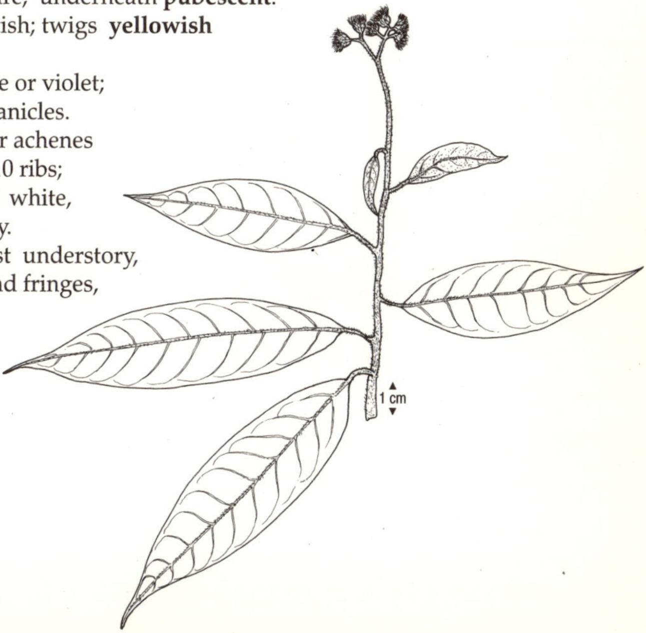

COMPOSITAE

35. CONNARACEAE

FAMILY DESCRIPTION - **Habit**: trees, shrubs or lianas. Bark, fruit and seed often poisonous. **Leaves**: spiral, imparipinnate. **Stipules**: absent. **Flowers**: bisexual, rarely unisexual. Actinomorphic or slightly zygomorphic. **Fruits**: dry, dehiscent, opening along ventral or both sutures, or a nut. Seeds often arillate.

FLOWER PARTS - CALYX: usually 5, sometimes basally united, imbricate or valvate. COROLLA: usually 5, free or basally united, imbricate or rarely valvate. ANDROECIUM: stamens 5 or 10, when 10 five of them sometimes staminodes. Filaments often united at base. Anthers 2-locular, opening lengthwise. Nectary disc absent or present. GYNOECIUM: superior, 1-5 carpels, free or basally united. Each carpel with terminal style and capitate stigma. Ovules 2 per carpel and marginal.

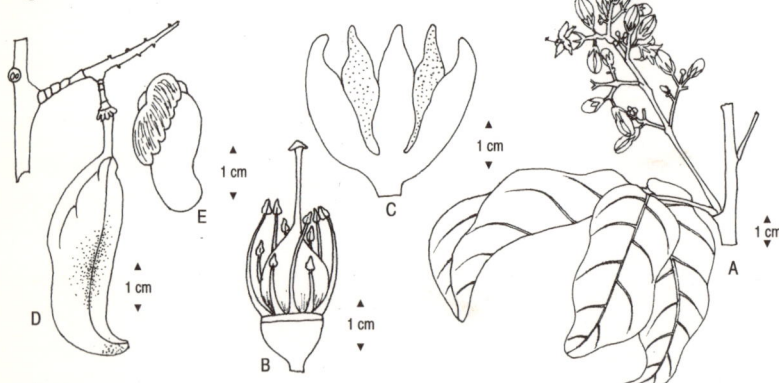

Key: inflorescence (A), flower with calyx and corolla removed (B), corolla (C), developing fruit (D) and seed with aril (E) of *Connarus monocarpus*.

1. *Connarus monocarpus*, radaliya (S)/chettupulukodi (T), (DF I:283), N, 5, scandent shrub.
    Leaves: pinnate; leaflets paired with terminal, rounded base, twisted, acuminate apex, shiny dark green above, paler with minute scales beneath, venation reticulate, young parts **tomentose**.
    Trunk: B-reticulate, striate, many lenticels; IB-beefy red; W,s-mottled pale pink; exudate watery.
    Flowers: I-terminal, erect, pyramidal, paniculate cymes.
    Fruits: green to scarlet, irregularly obovoid follicle, persistent calyx at base; seed single, shiny, black, with pale cream aril.
    Site: sandy soils, dunes, secondary forest; DL, DC.
    Uses: medicinal.

36. CORNACEAE

FAMILY DESCRIPTION - Habit: trees, shrubs or rarely rhizomatous herbs. **Leaves**: opposite or rarely spiral. **Stipules**: absent. **Flowers**: bisexual or unisexual, the plants then usually monoecious, actinomorphic, in cymes or cymose heads. **Fruits**: drupe with endocarp bearing 1-5 longitudinal grooves. Rarely a berry.

FLOWER PARTS - CALYX: 4-5, united basally to form a tube adnate to ovary. COROLLA: 4 to 5   or sometimes absent, free, valvate or imbricate. ANDROECIUM: stamens 4-5, free. Anthers 2-locular, opening lengthwise. Disk present. GYNOECIUM: inferior, 1 to 5-locular. Style simple or lobed. 1 Ovule per loculus, pendulous.

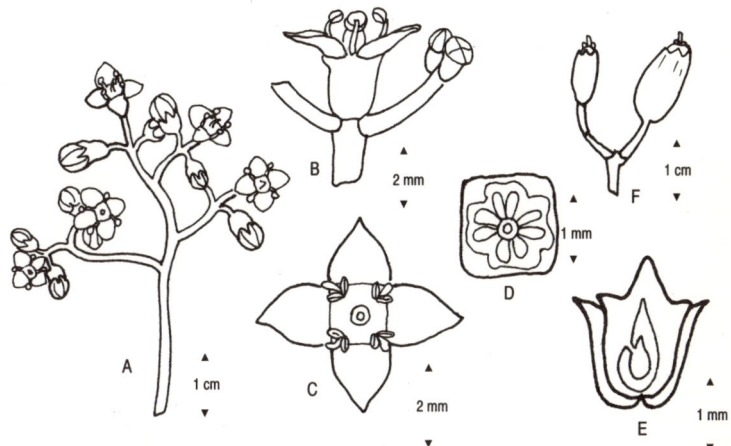

Key: inflorescence (A), side view of full flower (B), view of full flower from above (C), ovary in transverse (D), and longitudinal (E) section and fruits (F) of *Mastixia tetrandra*.

1. *Mastixia arborea*, (T II:287), N, 10, tree.
   Leaves: obovate to oval, tapering base,
       shortly acuminate apex,
       margins revolute.
   Trunk: no description.
   Flowers: green, in 3's;
       I-terminal, corymbose
       panicles.
   Fruits: purplish green,
       oblong to ovoid,
       smooth, drupe.
   Site: montane forest
       subcanopy; M, LM.

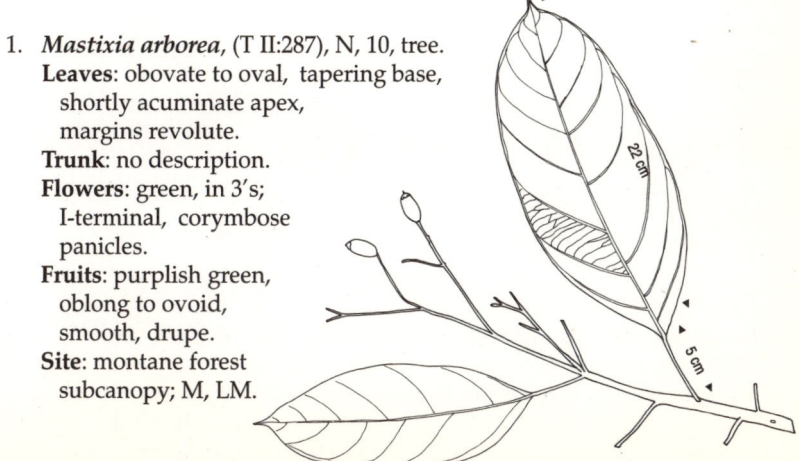

2. *Mastixia tetrandra*, diyataliya (S), (T II:287), E, 20, tree.
    Leaves: oval to lanceolate, acute base, caudate,
    acuminate to obtuse apex, **resinous** odour.
    Trunk: B-smooth, pale, **resinous**; W-red, heavy,
    splitting; twigs numerous, stout, puberulous.
    Flowers: pale green; I-small, lax, terminal,
    corymbose panicles, sometimes large.
    Fruits: ovoid-oblong drupe,
    crowned with small calyx.
    Site: rain forest subcanopy; at higher
    elevations, montane variety
    occupies forest canopy; M, W.
    Uses: resin-incense.

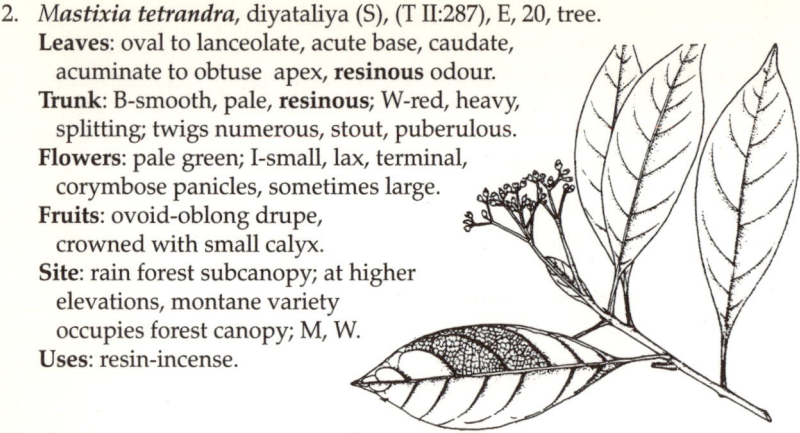

37. DAPHNIPHYLLACEAE

FAMILY DESCRIPTION - Habit: trees or shrubs. **Leaves**: spiral, simple, entire.
Stipules: absent. **Flowers**: unisexual, (plants dioecious), actinomorphic, in
axillary racemes. **Fruits**: drupe.

FLOWER PARTS - CALYX: 2-4 or absent, imbricate. COROLLA: absent. ANDROECIUM:
in male flower stamens 5-12, free. Anthers 2-locular, opening lengthwise. No
rudimentary ovary. GYNOECIUM: superior, 2-4-locular. Styles 1-2, basally united,
very short, curved to circinate. Ovules 2 per loculus, pendulous.

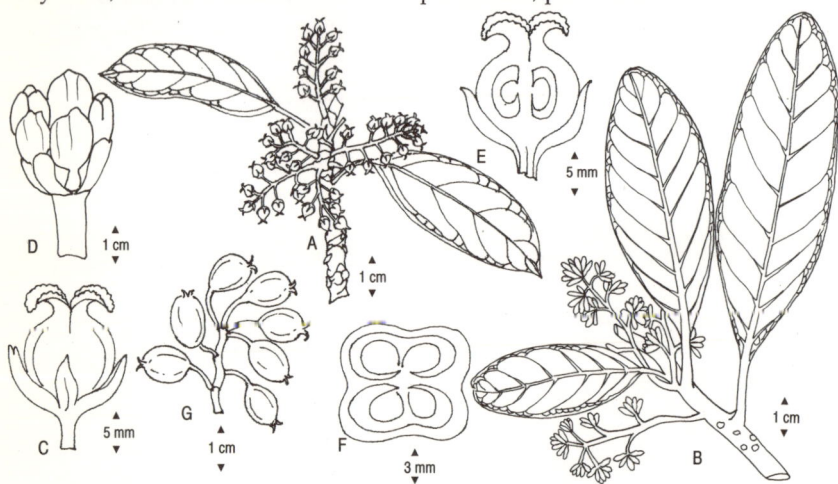

Key: inflorescence bearing twig with female (A) and male flowers (B), full
female flower (C), full male flower (D), longitudinal (E) and transverse (F)
sections of ovary, and fruits (G) of *Daphniphyllum neilgherrensis*.

1. *Daphniphyllum neilgherrensis*, (T IV:43 & VI:262), N, 7, tree.
    Leaves: numerous, **terminally crowded**, obovate, tapering base, rounded
       or pointed apex, veins **reticulate**, whitish beneath; flush **pale pink**.
    Trunk: branchlets **stout**, **leaf scars** prominent; young parts glabrous.
    Flowers: yellowish; I-short, axillary racemes at lower leaves.
    Fruits: ovoid, glabrous, tipped with style.
    Site: montane forest subcanopy; M.

38. DATISCACEAE

FAMILY DESCRIPTION - Habit: trees or perennial herbs. **Leaves**: spiral,
simple or pinnate. **Stipules**: absent. **Flowers**: unisexual, (plants dioecious),
polygamous or rarely bisexual, actinomorphic, in axillary spikes or panicles.
Fruits: capsule dehiscing apically between the styles, many seeded.

FLOWER PARTS - CALYX: in male flowers, sepals 4-8, sometimes united. In
female and bisexual   flowers, sepals 3-8 on summit of ovary or adnate to it.
COROLLA: in male   flowers, petals 6-8 or absent. In female and bisexual
flowers,petals absent. ANDROECIUM: in male flowers, stamens 4-25, free; anthers
2-locular, opening lengthwise; rudimentary ovary small or absent. In female or
bisexual flowers stamens same as in male or reduced to staminodes. GYNOECIUM:
inferior, 3-8 carpels, 1-locular with parietal placentation. Ovules numerous.

38

DATISCACEAE

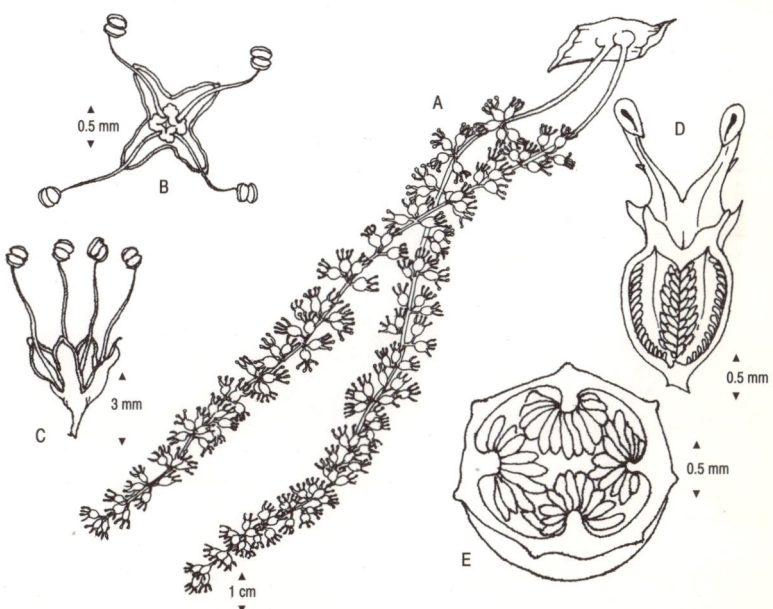

Key: part of inflorescence (A), male flower from above (B), from the side (C),
longitudinal section of bisexual flower, (D) and transverse section of ovary (E)
of *Tetrameles nudiflora*.

1. *Tetrameles nudiflora*, nigunu (S)/chini (T), (DF 6:108), N, 30, tree.
   Leaves: **large**, ovate, cordate base, acuminate to acute
     apex, irregularly **dentate**, veins reticulate,
     below **yellow hairy**; deciduous.
   Trunk: buttressed; B-smooth, fibrous, grey;
     W-soft, light pale yellow; twigs
     leaf-scarred; young parts puberulous.
   Flowers: yellowish green;
     I-numerous panicles
     from base of new leaf growth.
   Fruits: truncate, 8-ribbed,
     slightly rough.
   Site: roadsides and
     secondary forest; IN.

39. DICHAPETALACEAE

FAMILY DESCRIPTION - **Habit**: trees, shrubs or lianas. Usually very poisonous and with characteristic unicellular hairs bearing warty papillae. **Leaves**: spiral, simple, entire. **Stipules**: present. **Flowers**: bisexual, rarely unisexual, actinomorphic or slightly zygomorphic, in axillary cymes. **Fruits**: drupe, small, ovoid or globose.

FLOWER PARTS - CALYX: 5 free or basally united, imbricate. COROLLA: petals 5, usually 2-lobed or bifid, imbricate, free or united with stamens into a tube. ANDROECIUM: stamens 5, free or united. Anthers 2-locular opening lengthwise. Connective often dorsally thickened. Nectary present. GYNOECIUM: superior to inferior, 2-3-locular. Style simple or 2-3-fid at apex. Ovules 2 in each loculus, pendulous.

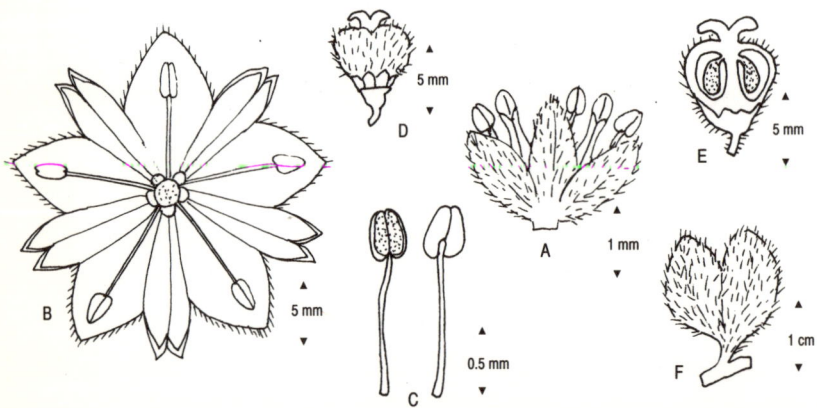

Key: view from side (A) above (B) of single flower, stamens (C), gynoecium in full (D), in longitudinal section (E), a fruit (F) of *Dichapetalum gelonoides*.

1. ***Dichapetalum gelonoides***, balu nakuta (S),
    (T I:254 & VI:47)), N, 5, shrub to small tree.
    Leaves: variable, usually oval to oblong, acute
      to tapering base, acuminate to acute apex,
      margin **few large teeth** at apex,
      sometimes hairy beneath.
    Trunk: young parts of branchlets
      pubescent.
    Flowers: pale green, small;
      I-small, axillary clusters.
    Fruits: grey-green, globose,
      compressed, pubescent.
    Site: rain forest understory; W.

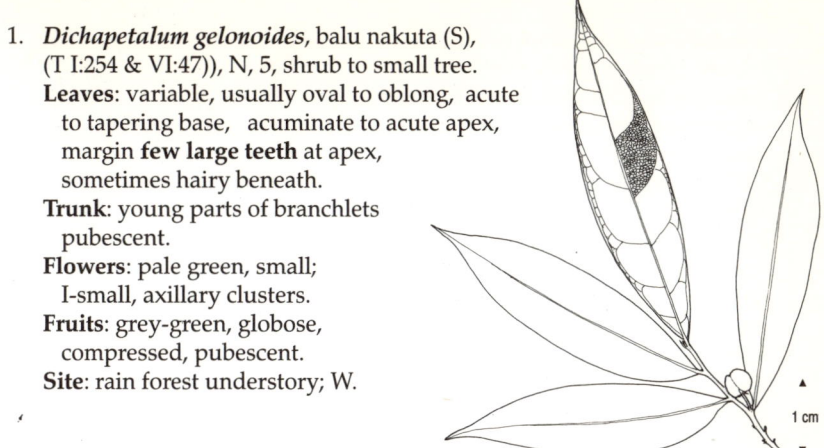

1 cm

40. DILLENIACEAE

FAMILY DESCRIPTION - **Habit**: trees, shrubs, rarely herbs or lianas. **Leaves**:
spiral, simple. **Stipules**: absent or when present wing-like and adnate to the
petiole, deciduous. **Flowers**: bisexual or unisexual, actinomorphic, solitary, in
racemes or cymes. **Fruits**: follicle or berry-like. Seeds mostly with a crested or
laciniate aril.

FLOWER PARTS - CALYX: usually 5, free, imbricate and persistent. COROLLA:
usually 5, free, imbricate, crumpled in bud. ANDROECIUM: stamens numerous,
free or variously united into bundles at base. Anthers opening lengthwise or by
apical pores. GYNOECIUM: superior. Carpels several, free, rarely one. Ovules one
or more per loculus. Styles free.

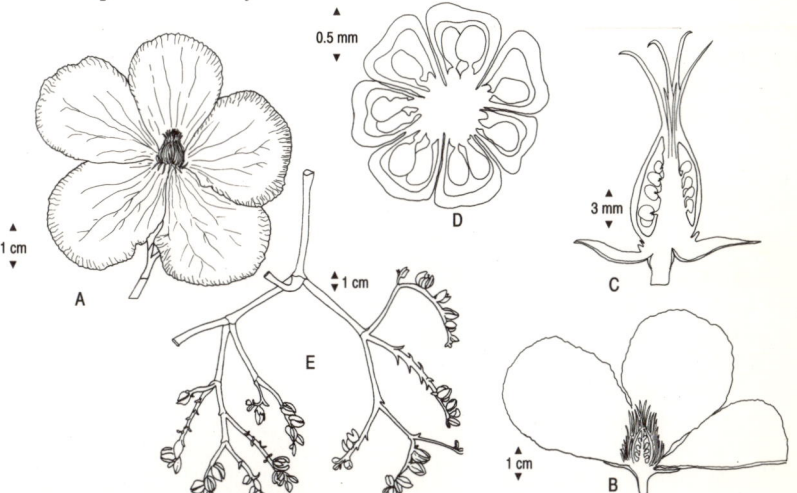

Key: full flower (A), half flower (B), longitudinal section (C) of gynoecium,
transverse section (D) of ovary of ***Dillenia suffruticosa***. Inflorescence of
Schumacheria castaneifolia (E).

153

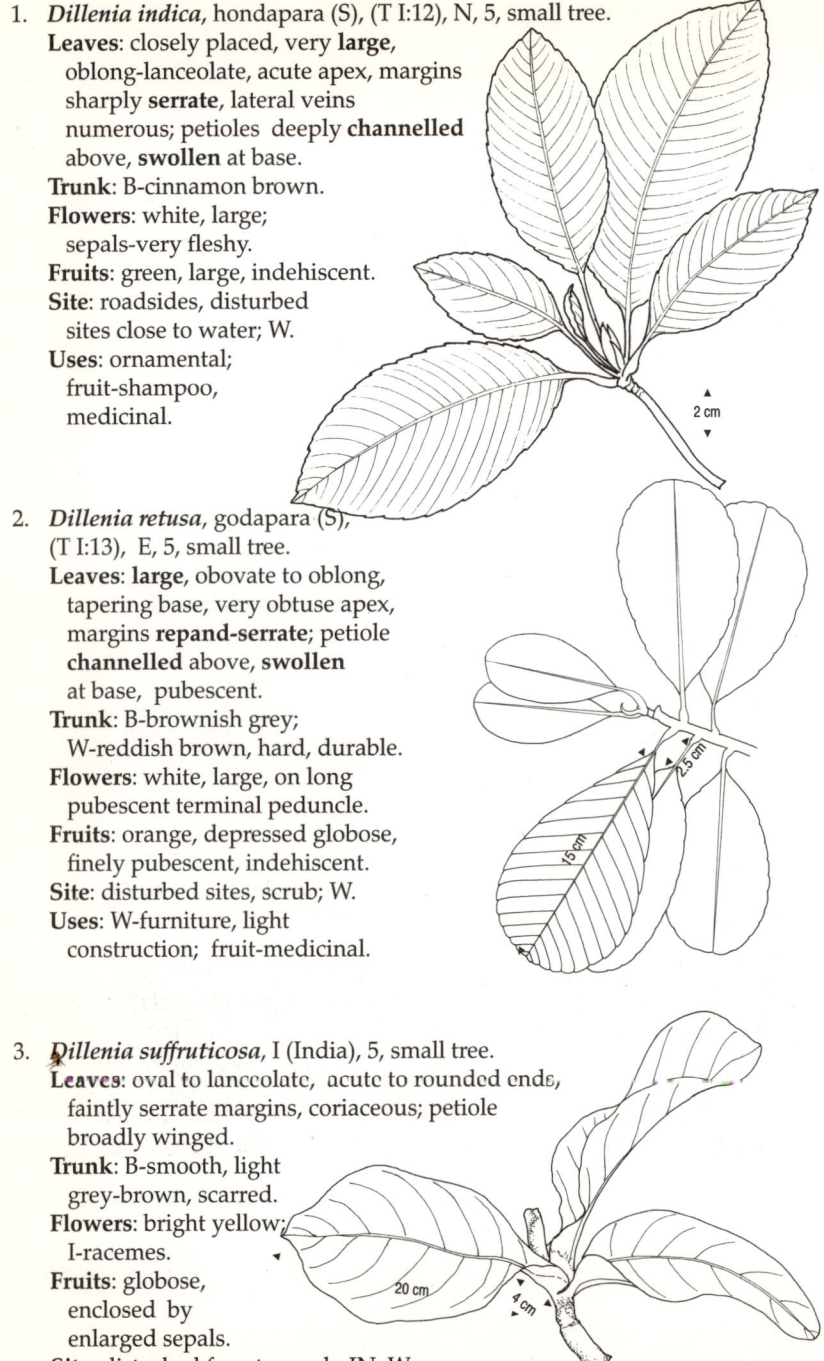

1. ***Dillenia indica***, hondapara (S), (T I:12), N, 5, small tree.
   Leaves: closely placed, very **large**, oblong-lanceolate, acute apex, margins sharply **serrate**, lateral veins numerous; petioles deeply **channelled** above, **swollen** at base.
   Trunk: B-cinnamon brown.
   Flowers: white, large; sepals-very fleshy.
   Fruits: green, large, indehiscent.
   Site: roadsides, disturbed sites close to water; W.
   Uses: ornamental; fruit-shampoo, medicinal.

   2 cm

2. ***Dillenia retusa***, godapara (S), (T I:13), E, 5, small tree.
   Leaves: **large**, obovate to oblong, tapering base, very obtuse apex, margins **repand-serrate**; petiole **channelled** above, **swollen** at base, pubescent.
   Trunk: B-brownish grey; W-reddish brown, hard, durable.
   Flowers: white, large, on long pubescent terminal peduncle.
   Fruits: orange, depressed globose, finely pubescent, indehiscent.
   Site: disturbed sites, scrub; W.
   Uses: W-furniture, light construction; fruit-medicinal.

   2.5 cm
   15 cm

3. ***Dillenia suffruticosa***, I (India), 5, small tree.
   Leaves: oval to lanceolate, acute to rounded ends, faintly serrate margins, coriaceous; petiole broadly winged.
   Trunk: B-smooth, light grey-brown, scarred.
   Flowers: bright yellow; I-racemes.
   Fruits: globose, enclosed by enlarged sepals.
   Site: disturbed forest, scrub; IN, W.

   20 cm
   4 cm

4. **Dillenia triquetra**, diyapara (S), (T I:11), E, 8, tree.
   Leaves: broadly oblong oval, obtuse to blunt
   pointed apex, margins **coarsely serrate**; petiole
   channelled above in mature leaves;
   **prominent fleshy horseshoe-
   shaped cushion** on upper
   side of petiole of young leaves.
   Trunk: twigs smooth, brown,
   marked with **large leaf scars**.
   Flowers: white, few, large; sepals
   fleshy and persistent; I-small,
   racemes opposite leaves.
   Fruits: small, globular, enclosed
   by enlarged sepals.
   Site: secondary forest, scrub,
   disturbed areas; W.

5. **Schumacheria alnifolia**, (T I:10), E, 3,
   shrub.
   Leaves: broadly oval, lanceolate,
   margins **serrate**.
   Trunk: branches **stout**,
   divaricate.
   Flowers: yellow, small;
   I-small, axillary panicles.
   Fruits: indehiscent; seed
   surrounded at base by
   membranous aril.
   Site: rain forest gaps
   and fringes; LM, UW.

6. **Schumacheria castaneifolia**, kekiri wara (S), (T I:10), E, 8, tree.
   Leaves: **large**, broadly oval to ovate, pointed apex, margins
   sinuate **dentate**.
   Trunk: B-smooth, brownish,
   conspicuously scarred;
   W-hard, light brown.
   Flowers: yellow,
   numerous, sessile;
   I-large terminal panicles.
   Fruits: indehiscent; seed
   with a membranous
   aril at base.
   Site: rain forest gaps
   and fringes; W.
   Uses: W-poles, light construction.

41. DIPTEROCARPACEAE

FAMILY DESCRIPTION - **Habit**: trees with resinous wood. **Leaves**: spiral, simple. **Stipules**: present, caducous. **Flowers**: bisexual, actinomorphic, axillary panicles, racemes or cymes. Scented. **Fruits**: dry, indehiscent, samara or nutlet.

FLOWER PARTS - CALYX: 5, free or basally united and adnate to ovary, imbricate or valvate. Sepals usually enlarged in fruit and remain wing-like. COROLLA: 5, free or slightly united, contorted. ANDROECIUM: stamens 5 to numerous. Filaments free or basally united. Anthers 2-locular, basifixed and opening lengthwise or by an apical pore. GYNOECIUM: superior, 3-locular, 2 ovules per loculus, axile placentation. Style entire or 3-lobed.

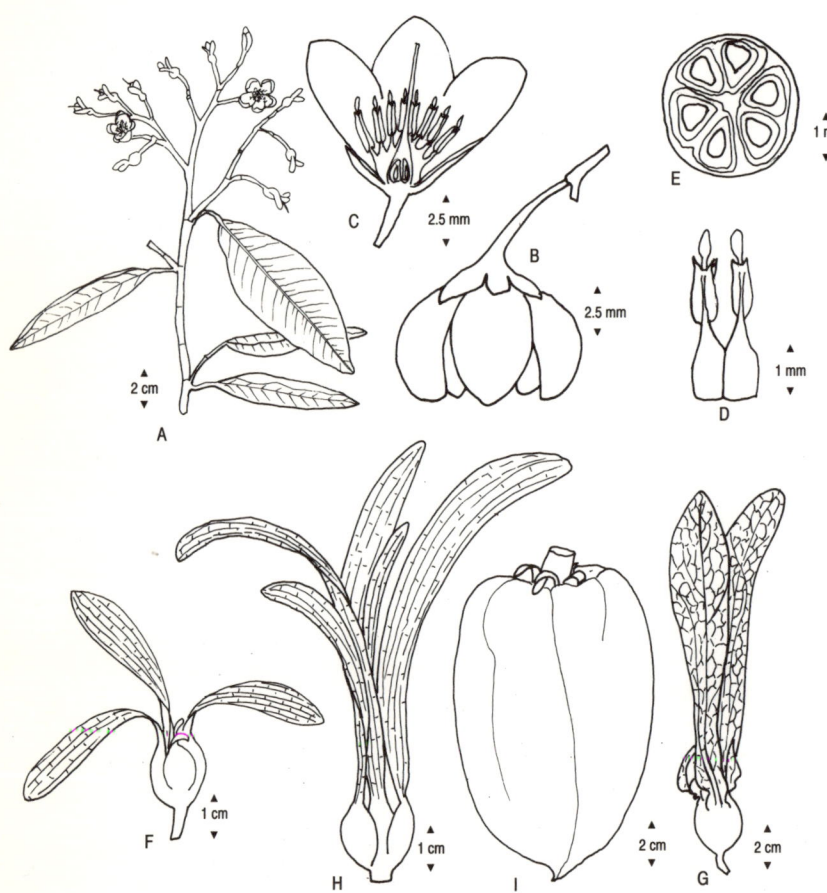

Key: inflorescence (A), full flower (B), half flower (C), stamens (D) and transverse section of ovary (E) of *Shorea cordifolia*. Fruits of *Shorea cordifolia* (F), *Dipterocarpus hispidus* (G), *Shorea stipularis* (H) and *Vateria copallifera* (I).

1. ***Dipterocarpus hispidus***, bu hora (S), (DF I:371), E, 45, tree.
   Leaves: **large**, oblong-ovate, cordate base, acuminate
   apex, lateral veins about 14-20 pairs, golden brown
   stiff hairs abundant beneath, sparse above; young
   leaves enclosed in conspicuous stipules bearing
   dense long golden brown hairs; large
   leaved spreading crown.
   Trunk: columnar; B-flaky,
   light orange to pale
   brown; twigs stout.
   Flowers: I-hardly
   branched raceme.
   Fruits: subglobose, wings
   2 long and 3 short.
   Site: rain forest canopy,
   along water courses; W.
   Uses: W-construction
   timber, plywood.

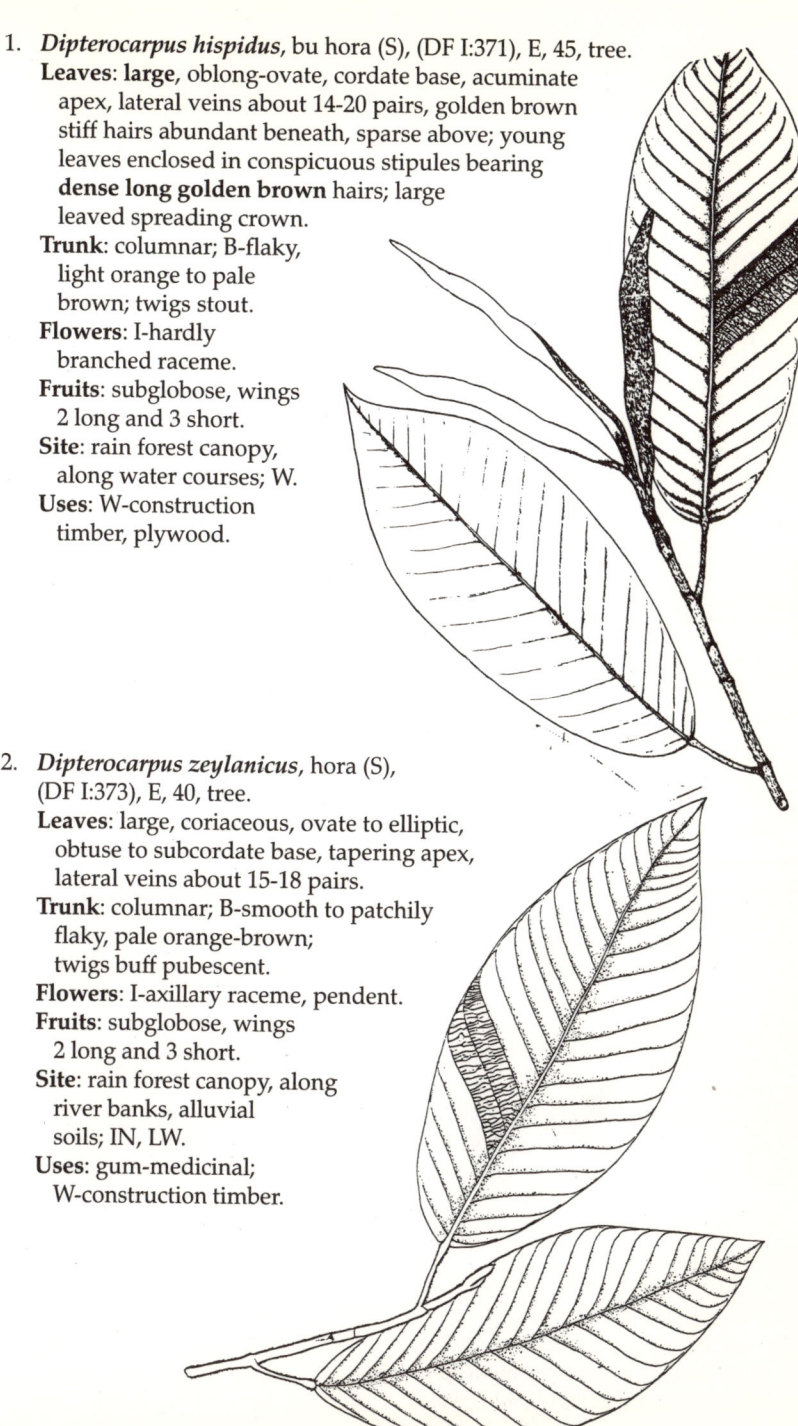

41

DIPTEROCARPACEAE

2. ***Dipterocarpus zeylanicus***, hora (S),
   (DF I:373), E, 40, tree.
   Leaves: large, coriaceous, ovate to elliptic,
   obtuse to subcordate base, tapering apex,
   lateral veins about 15-18 pairs.
   Trunk: columnar; B-smooth to patchily
   flaky, pale orange-brown;
   twigs buff pubescent.
   Flowers: I-axillary raceme, pendent.
   Fruits: subglobose, wings
   2 long and 3 short.
   Site: rain forest canopy, along
   river banks, alluvial
   soils; IN, LW.
   Uses: gum-medicinal;
   W-construction timber.

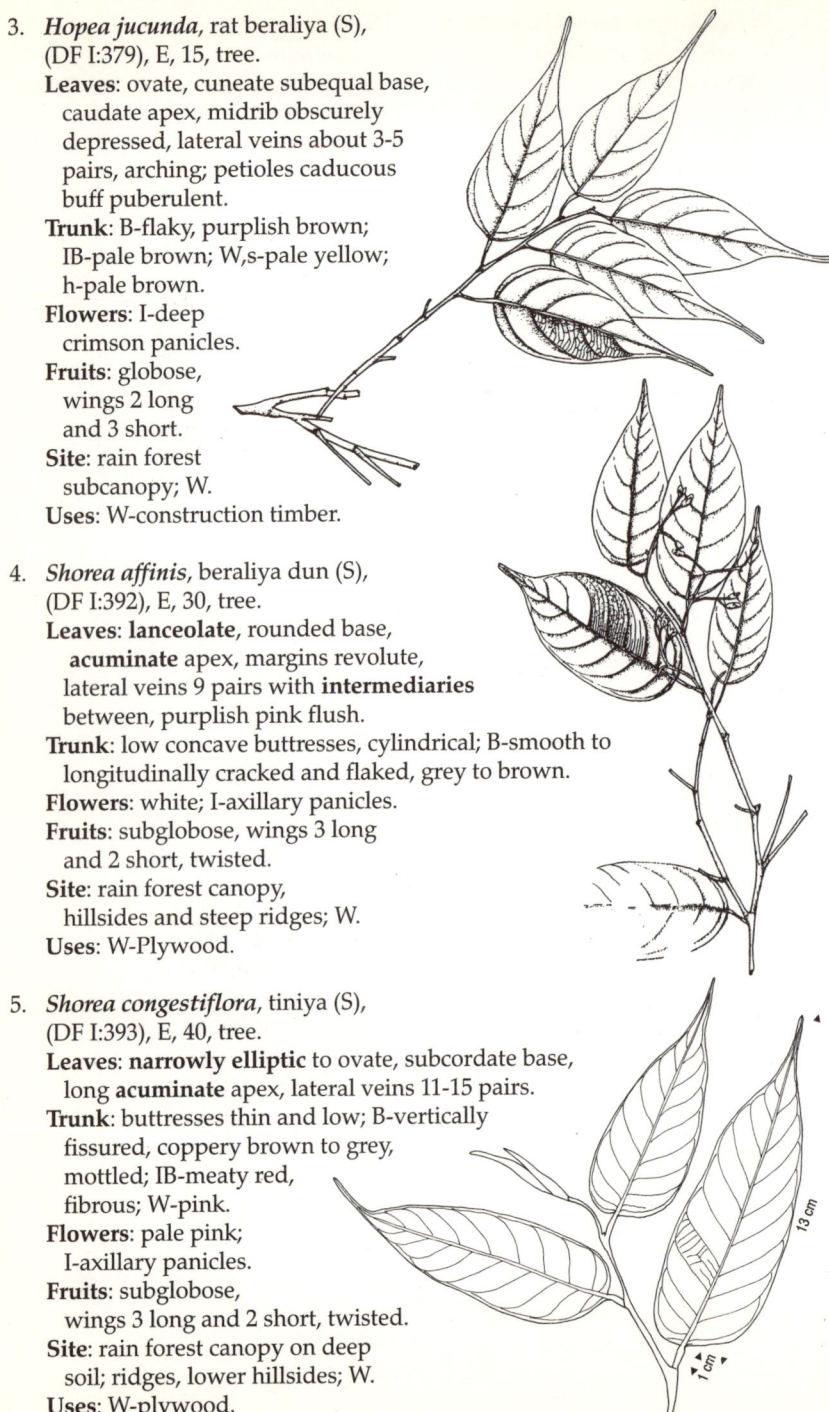

3. *Hopea jucunda*, rat beraliya (S),
   (DF I:379), E, 15, tree.
   Leaves: ovate, cuneate subequal base,
   caudate apex, midrib obscurely
   depressed, lateral veins about 3-5
   pairs, arching; petioles caducous
   buff puberulent.
   Trunk: B-flaky, purplish brown;
   IB-pale brown; W,s-pale yellow;
   h-pale brown.
   Flowers: I-deep
   crimson panicles.
   Fruits: globose,
   wings 2 long
   and 3 short.
   Site: rain forest
   subcanopy; W.
   Uses: W-construction timber.

4. *Shorea affinis*, beraliya dun (S),
   (DF I:392), E, 30, tree.
   Leaves: **lanceolate**, rounded base,
   acuminate apex, margins revolute,
   lateral veins 9 pairs with **intermediaries**
   between, purplish pink flush.
   Trunk: low concave buttresses, cylindrical; B-smooth to
   longitudinally cracked and flaked, grey to brown.
   Flowers: white; I-axillary panicles.
   Fruits: subglobose, wings 3 long
   and 2 short, twisted.
   Site: rain forest canopy,
   hillsides and steep ridges; W.
   Uses: W-Plywood.

5. *Shorea congestiflora*, tiniya (S),
   (DF I:393), E, 40, tree.
   Leaves: **narrowly elliptic** to ovate, subcordate base,
   long **acuminate** apex, lateral veins 11-15 pairs.
   Trunk: buttresses thin and low; B-vertically
   fissured, coppery brown to grey,
   mottled; IB-meaty red,
   fibrous; W-pink.
   Flowers: pale pink;
   I-axillary panicles.
   Fruits: subglobose,
   wings 3 long and 2 short, twisted.
   Site: rain forest canopy on deep
   soil; ridges, lower hillsides; W.
   Uses: W-plywood.

6. *Shorea cordifolia*, kotikan beraliya (S), (DF I:394), E, 20, tree.
   Leaves: **ovate** to lanceolate, rounded base, tapering apex, lateral veins 8-11 pairs, midrib **prominently elevated** on upper surface.
   Trunk: low concave butresses, low branching; B-thin flaky, chocolate brown; IB-pale brown; W-pale cream.
   Flowers: white; I-panicles.
   Fruits: ovoid, pointed, wings 3 long and 2 short, twisted.
   Site: rain forest subcanopy, hill slopes; W.
   Uses: W-plywood; fruit-vegetable.

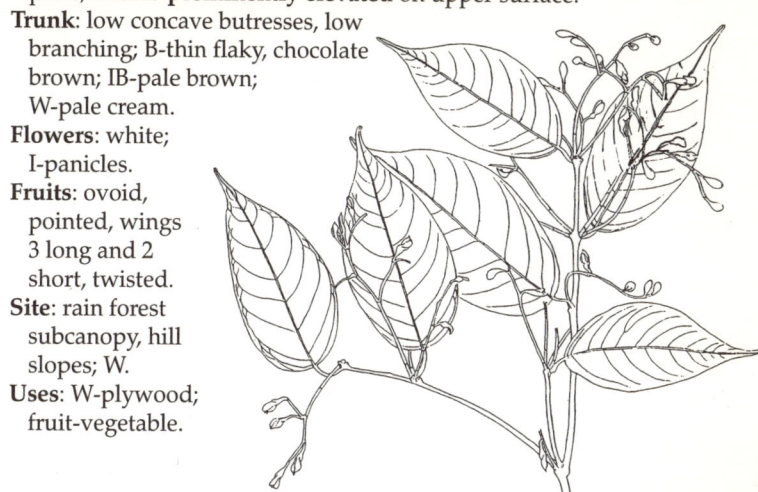

7. *Shorea disticha*, beraliya (S), (DF I:396), E, 40, tree.
   Leaves: **ovate-oblong** to elliptic, obtuse to subcordate base, tapering apex, lateral veins about 8-11 pairs, without intermediaries, intercostals densely scalariform.
   Trunk: B-yellow-brown, coming away in large patches; IB-pale brown; twigs **yellow-brown pubescent**; resinous.
   Flowers: white; I-axillary panicle.
   Fruits: ovoid, pointed, wings 3 long and 2 short, twisted.
   Site: rain forest canopy, deep soil, ridges, slopes; W.
   Uses: W-construction timber; resin-incense; fruits-vegetable.

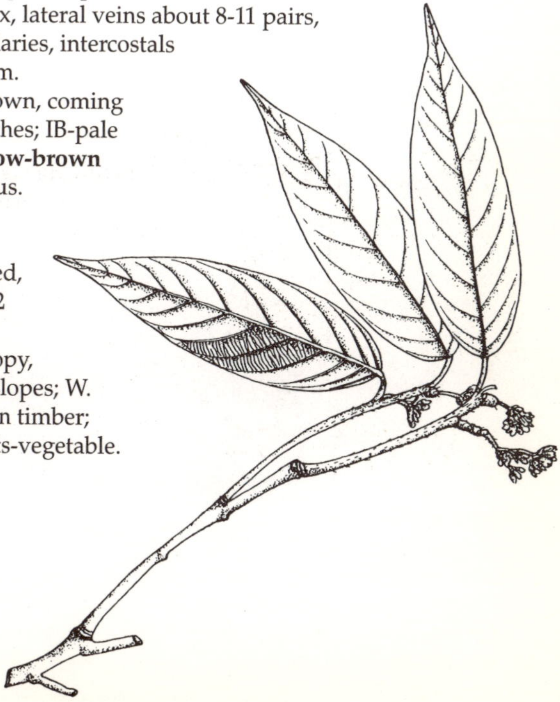

8. **Shorea dyeri**, yakahalu dun (S), (DF I:384), E, 40, tree.
   Leaves: lanceolate to sickle-shaped,
   cuneate to subequal base, tapering
   apex, lateral veins about 11-13 pairs,
   intercostals obscure,
   densely scalariform.
   Trunk: buttresses; B-thinly flaky, pale
   tawny; young parts buff pubescent.
   Flowers: I-terminal or
   axillary panicles.
   Fruits: small ovoid, apiculate, wings
   3 long narrowly spathulate,
   and 2 short, not twisted.
   Site: rain forest
   canopy,
   skeletal soils
   on ridges and
   slopes; LW.
   Uses: W-heavy
   construction.

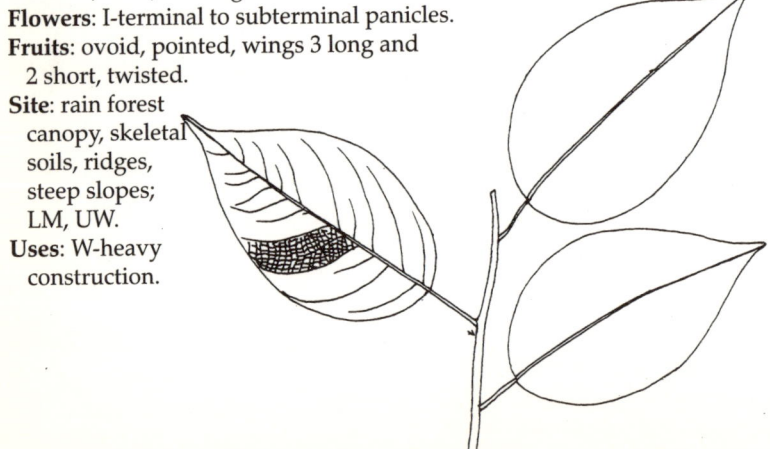

1 cm

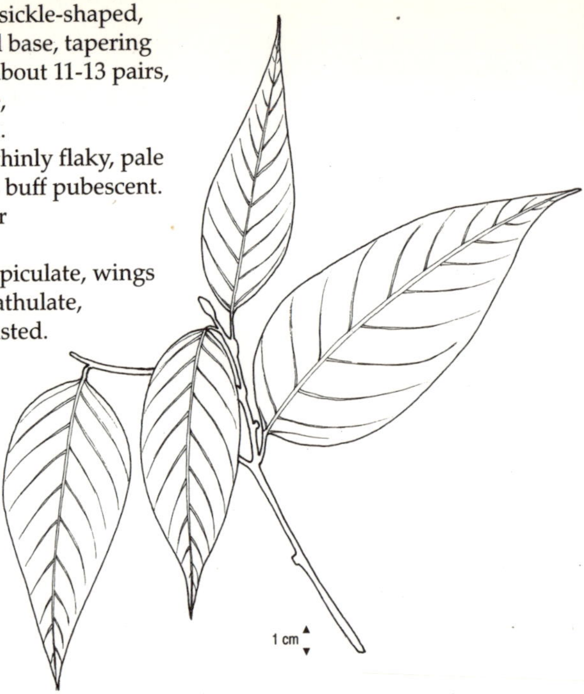

9. **Shorea gardneri**, rath dun (S)/koongili maram (T), (DF I:397), E, 35, tree.
   Leaves: ovate, **revolute**, base obtuse to cordate on unfurling, lateral veins
   about 12 pairs, with a few intermediaries.
   Trunk: buttresses, often wavy; B-cracked, flaky, purplish brown;
   IB-fibrous, thick, orange-brown; W,s-dark yellow
   brown, hard; h-orange-brown.
   Flowers: I-terminal to subterminal panicles.
   Fruits: ovoid, pointed, wings 3 long and
   2 short, twisted.
   Site: rain forest
   canopy, skeletal
   soils, ridges,
   steep slopes;
   LM, UW.
   Uses: W-heavy
   construction.

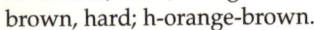

10. *Shorea megistophylla*, honda beraliya (S),
    (DF I:398), E, 40, tree.
    Leaves: **large, oblong-elliptic**, acuminate apex,
      lateral veins 13-18 pairs, intercostals
      scalariform; flush **red**.
    Trunk: tall, straight, cylindrical, often
      pronounced buttresses;
      B-coming away
      in large pieces, patchy,
      yellow-brown; IB-pale
      brown; W-pale yellow.
    Flowers: pale whitish
      pink; I-panicle.
    Fruits: subglobose,
      wings 3 long and
      2 short, twisted.
    Site: rain forest canopy, lower
      slopes and waterways; W.
    Uses: W-light construction; resin-incense;
      fruit-vegetable.

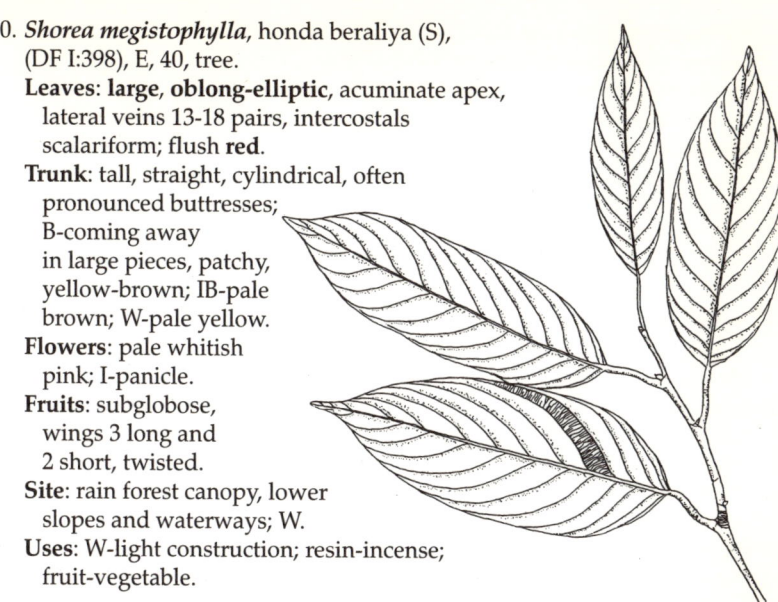

41

11. *Shorea oblongifolia*, beraliya (S),
    (DF I:386), E, 40, tree.
    Leaves: oblong, cordate base, tapering apex,
      subrevolute margins, lateral veins 13-16
      pairs, intercostals scalariform.
    Trunk: often crooked, small buttresses;
      B-irregularly thinly flaky, tawny
      brown.
    Flowers: cream to pinkish; I-axillary
      to ramiflorous panicles.
    Fruits: ovoid to ellipsoid,
      apiculate, wings 3 long and
      2 short, not twisted.
    Site: rain forest canopy,
      alluvial valleys,
      lower slopes; W.
    Uses: W-construc-
      tion timber.

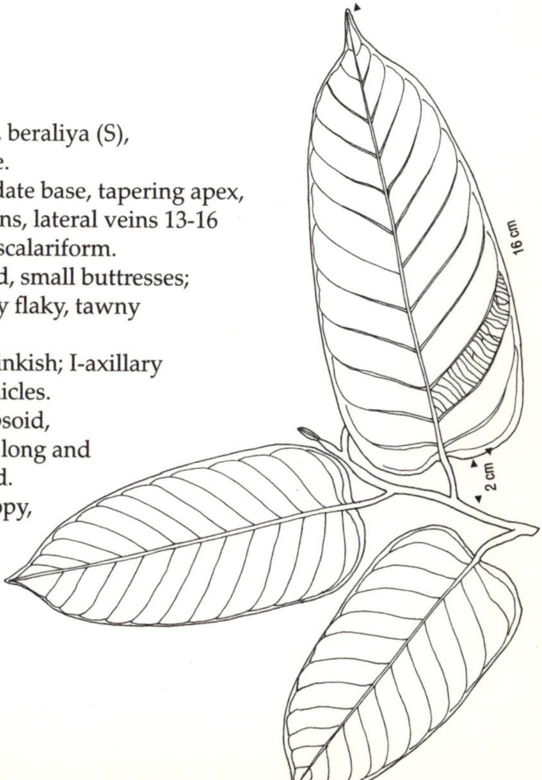

16 cm

2 cm

DIPTEROCARPACEAE

12. *Shorea stipularis*, nawada (S), (DF I:390), E, 45, tree.

Leaves: lanceolate to broadly oblong-ovate, short pointed ends, lateral veins 14-17 pairs; petioles geniculate;

Trunk: straight, cylindrical, low rounded buttresses; B-becoming regularly fissured in flakes, pale chocolate brown; IB-red-brown; W-pale yellow, **resinous**; young twigs with conspicuous **persistent large ovate** stipules.

Flowers: I-axillary panicles.

Fruits: stout, wings 2 short and 3 long, not twisted.

Site: rain forest canopy; W.

Uses: resin-incense; W-construction timber; B-arrests fermentation.

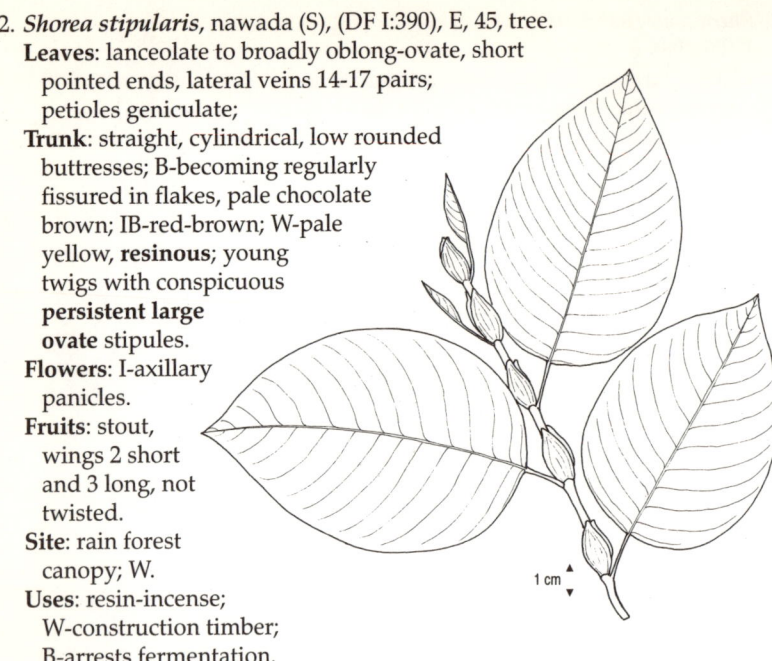

1 cm

13. *Shorea trapezifolia*, tiniya dun (S), (DF I:401), E, 40, tree.

Leaves: **lanceolate to elliptic**, acuminate apex, lateral veins about 11-14 pairs.

Trunk: prominent buttresses; B-becoming **regularly vertically** fissured, patchy light brown-grey; IB-pale brown; W-pale yellow; young twigs **pale yellow-brown pubescent**.

Flowers: white; I-panicles.

Fruits: subglobose, wings 3 long and 2 short, twisted.

Site: rain forest canopy; deep soil ridges, lower slopes; LM, UW.

Uses: W-light construction, plywood.

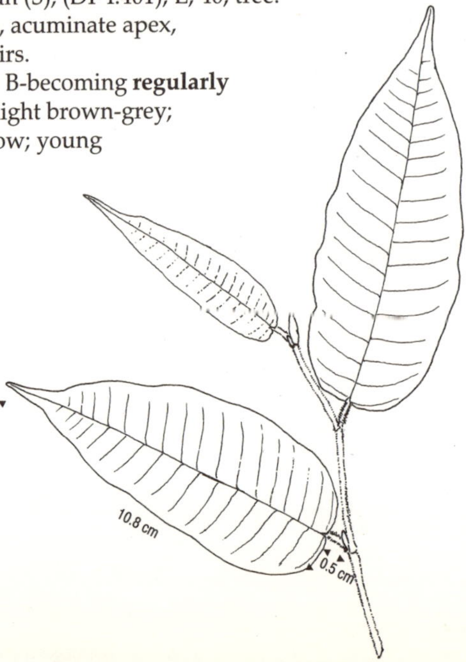

10.8 cm

0.5 cm

14. *Shorea worthingtonii*, beraliya (S), (DF I:402), E, 30, tree.
Leaves: **elliptic to ovate**, margins **subrevolute**, lateral veins 6-9 pairs; dark green foliage.
Trunk: rounded buttresses; B-becoming cracked, flaked, pale brown; IB-pale brown; W-yellow.
Flowers: white; I-axillary panicles.
Fruits: subglobose, wings 3 long and 2 short, twisted.
Site: rain forest canopy, rocky skeletal soils on ridges, hill tops; W.
Uses: W-timber construction.

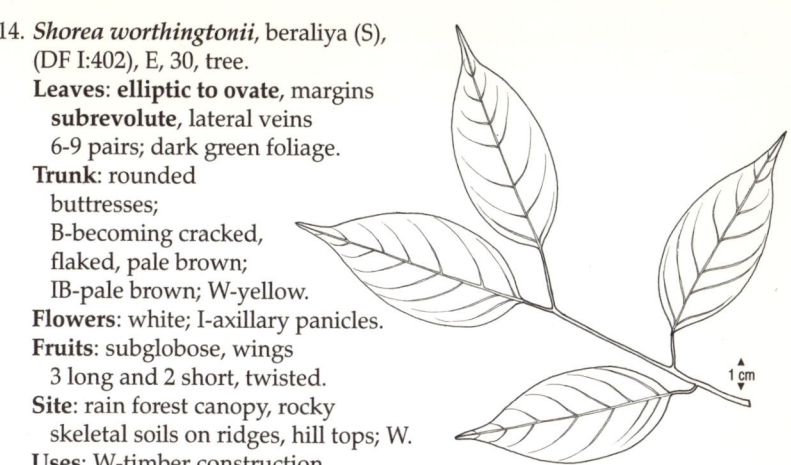

41

15. *Shorea zeylanica*, dun (S)/koongili (T), (DF I:403), E, 40, tree.
Leaves: elliptic to lanceolate, cuneate **prominently revolute** base, caudate apex.
Trunk: low concave butresses; B-yellow-brown, irregularly flaky; IB-pale brown; W-pale yellow.
Flowers: white; I-lax pendent panicles.
Fruits: ovoid, pointed, wings 3 long and 2 short, twisted.
Site: rain forest canopy, ridges and slopes; LM, UW.
Uses: W-construction timber.

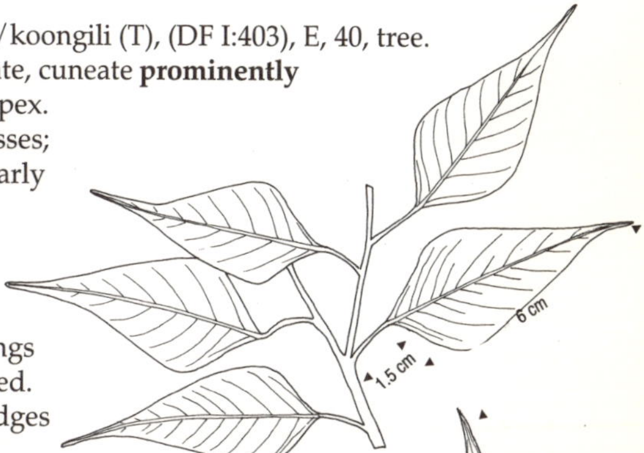

16. *Stemonoporus acuminatus*, (DF I:406), E, 25, tree.
Leaves: lanceolate, cuneate base, tapering apex, lateral veins about **10-12** pairs, intercostals densely **reticulate**; petiole geniculate.
Trunk: often distorted.
Flowers: I-axillary.
Fruits: globose, thick corky fibrous pericarp, short calyx lobes, no wings.
Site: montane and rain forest subcanopy; exposed hills, ridges; M, LM.

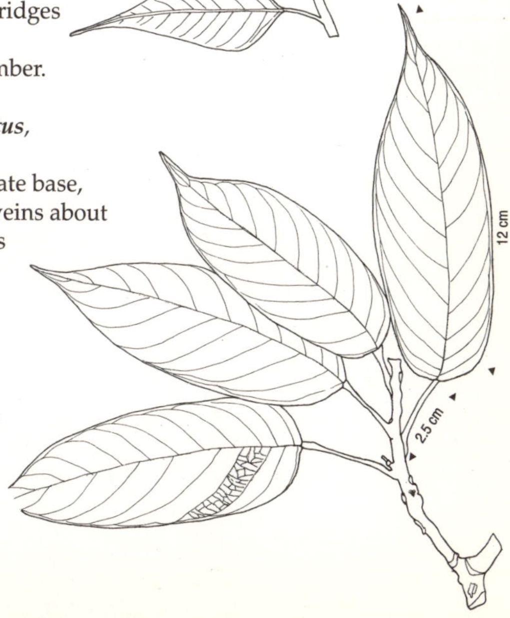

17. *Stemonoporus canaliculatus*, (DF I:408), E, 15, tree.
Leaves: elliptic-lanceolate, cuneate base, tapering apex,
    lateral veins **11-20**, arched, forming looped
    intramarginal vein, intercostals
    raised below, reticulate.
Trunk: stems often flop over,
    layering at base.
Flowers: solitary, axillary.
Fruits: globose, subacute,
    thick corky fibrous
    pericarp, short calyx
    lobes, no wings.
Site: rain forest
    understory, thin
    skeletal soils; W.

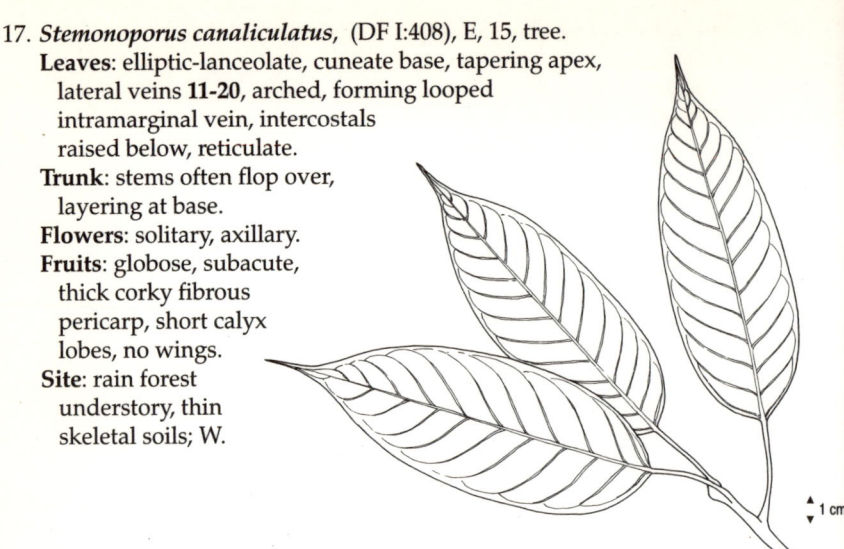

18. *Stemonoporus gardneri*, (DF I:412), E, 20, tree.
Leaves: ovate, obtuse to subcordate base, tapering apex, lateral veins **9-13**
    pairs, intercostals densely **reticulate**.
Trunk: many ascending twisted branches; twigs prominently **leaf-scarred**;
    young parts pale, sparsely pubescent.
Flowers: pale lemon-yellow; I-lax paniculate.
Fruits: globose, smooth, short calyx lobes, no wings.
Site: montane and rain forest subcanopy; M, LM.

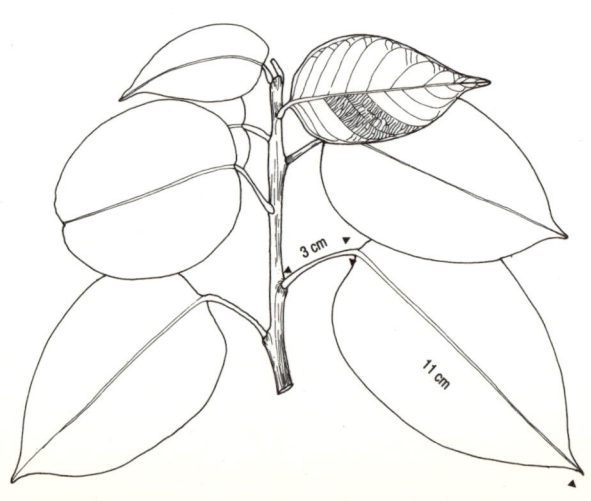

19. *Vateria copallifera*, hal (S), (DF I:419), E, 35, tree.
   Leaves: broadly to narrowly oblong, obtuse to cordate base, abruptly
   tapering apex, **coriaceous**; petiole **geniculate**.
   Trunk: twigs densely **tomentose**.
   Flowers: cream; I-stout branching panicles.
   Fruits: very large, no wings, fibrous pericarp.
   Site: rain forest canopy; alluvial floodplains; often in
   forest relics surrounded by cultivation; W.
   Uses: W-light construction; B-arresting fermentation;
   fruit-flour (helapa, pittu); resin-varnish.

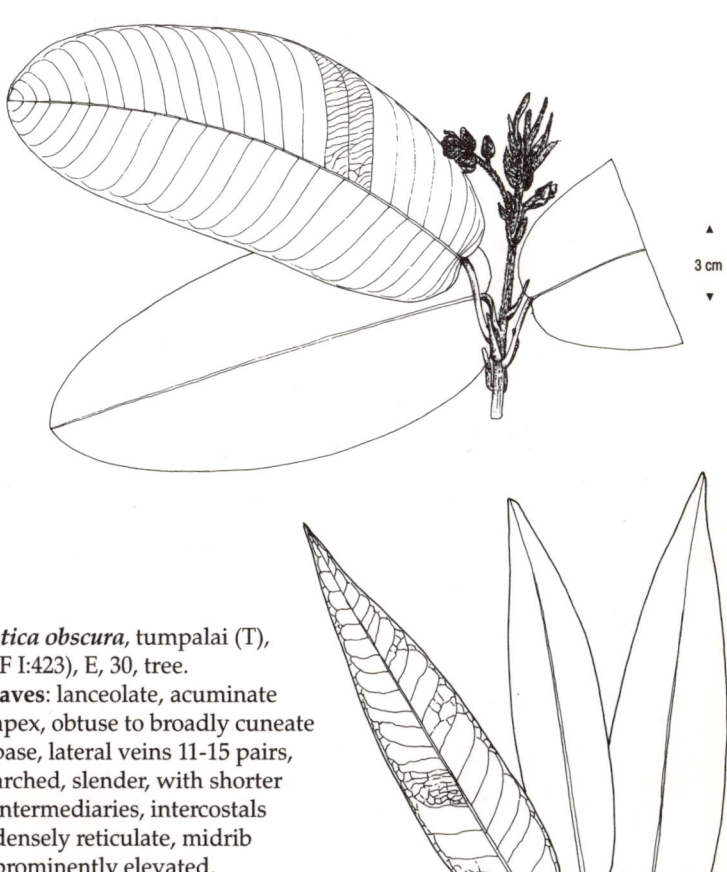

3 cm

41

DIPTEROCARPACEAE

20. *Vatica obscura*, tumpalai (T),
   (DF I:423), E, 30, tree.
   Leaves: lanceolate, acuminate
   apex, obtuse to broadly cuneate
   base, lateral veins 11-15 pairs,
   arched, slender, with shorter
   intermediaries, intercostals
   densely reticulate, midrib
   prominently elevated.
   Trunk: B-pale grey, smooth,
   hoop-marked; young parts
   densely pubescent.
   Flowers: I-many-flowered,
   axillary panicles.
   Fruits: globose, subequal short wings.
   Site: forest canopy; riverine forest; DL.

42. EBENACEAE

FAMILY DESCRIPTION - Habit: trees or shrubs. Sometimes with black heartwood. **Leaves**: spiral, rarely opposite, simple, entire. **Stipules**: absent. **Flowers**: mostly unisexual, (plants often dioecious), actinomorphic, solitary or in cymes. **Fruits**: berry, rarely dehiscent. Seeds large.

FLOWER PARTS - CALYX: 3-7, free, persistent, often accrescent in fruit, imbricate. COROLLA: 3-5, basally united, contorted. ANDROECIUM: stamens 2-numerous, epipetalous or borne on the receptacle. Filaments free or united in pairs. Anthers 2- locular, dehiscing lengthwise. GYNOECIUM: superior, 3-to many-locular, 1-2 pendulous ovules per loculus. Style often divided.

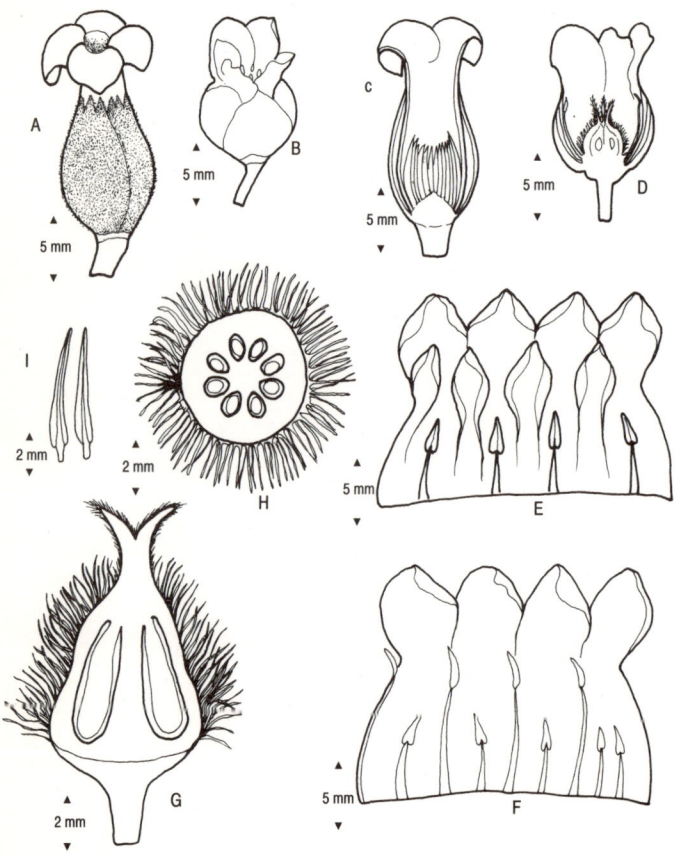

Key: full male (A) and bisexual (B) flowers, half of male (C) and bisexual (D) flowers, opened out corolla of female flower (E) and male flower (F), longitudinal section of gynoecium (G), transverse section of ovary (H) and stamens (I) of **Diospyros discolor**.

1. **_Diospyros acuminata_**, (DF III:4), E, 5, small tree.
   Leaves: lanceolate to ovate, rounded to acute
   base, acuminate apex, veins reticulate **above**
   and below, lateral veins about 7 pairs.
   Trunk: B-**smooth to finely fissured**;
   IB-pale reddish to light brown;
   W-white; young branchlets
   densely tomentose.
   Flowers: white with silky hairs,
   1-3 together, axillary.
   Fruits: globose, sparsely pilose;
   calyx not cup-shaped,
   3-lobed.
   Site: rain forest
   understory; W.

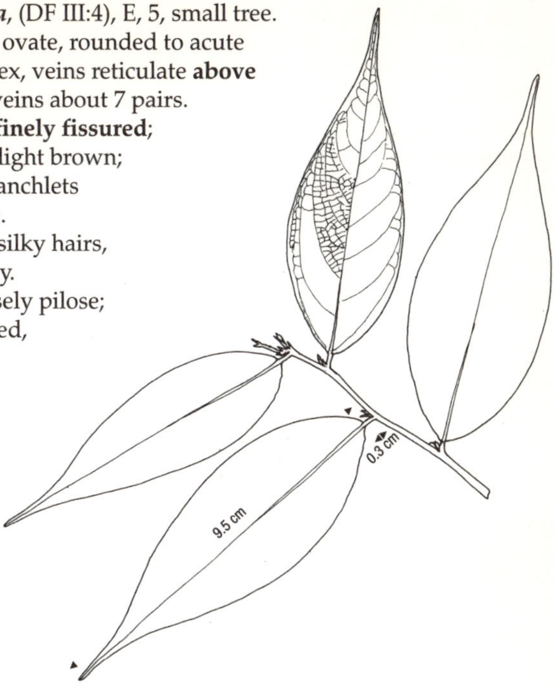

2. **_Diospyros ebenum_**, ebony (E)/ kaluwara (S)/karunkali (T),
   (DF III:16), N, 20, tree.
   Leaves: oblong-oval, **obscurely acuminate to rounded**
   apex, lateral veins conspicuous, about 5-8 pairs.
   Trunk: B-**black to grey-black**,
   peeling in **rectangular**
   pieces; IB-light
   brown-yellowish darkening
   on exposure; W,s-pale;
   h-black to brown; branchlets
   sometimes sparsely pilose.
   Flowers: yellow to whitish,
   sessile; I-male compressed
   cyme of 3-15
   flowers; I-female
   axillary, solitary.
   Fruits: globose; calyx
   woody green cup.
   Site: monsoon and
   intermediate forest
   canopy; DL, IN.
   Uses: W-luxury wood for furniture,
   carving, turnery; fruit-medicinal, fish poison.

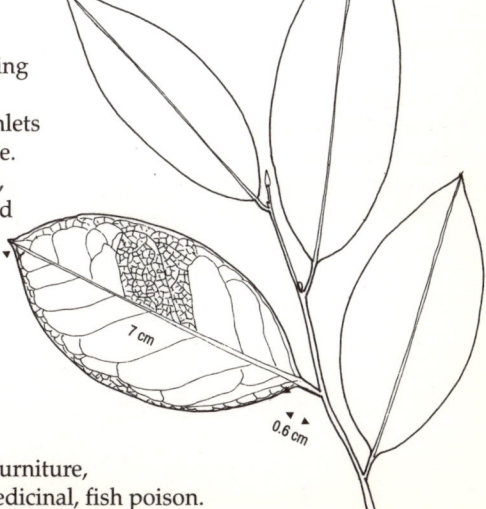

3. **_Diospyros ferrea_**, kalu habaraliya (S), (DF III:9), N, 7, tree.
   Leaves: obovate-oblong to spathulate, base tapered, lateral
   veins 5-6 pairs, tertiaries reticulate beneath.
   Trunk: B-**smooth to fissured, rough, blackish**
   to dark grey; IB-reddish brown.
   Flowers: white, 1-3 in leaf axils.
   Fruits: yellow to brown-red;
   smooth, glossy; calyx
   cup-shaped.
   Site: monsoon forest
   understory, scrub; DL.

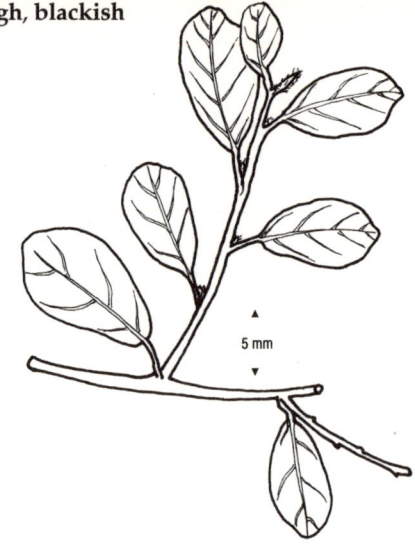

5 mm

4. **_Diospyros hirsuta_**, (DF III:39), E, 6, tree.
   Leaves: ovate-lanceolate to elliptic, base tapered, pointed apex, widely
   spaced, prominent, arched lateral veins with **intermediaries** between.
   Trunk: B-**finely fissured, black**; IB-reddish.
   Flowers: yellow, sessile to subsessile
   between 2 small bracteoles; I-male
   dense axillary clusters; female
   flowers axillary, solitary.
   Fruits: ellipsoid-ovoid, densely
   coppery hairy; calyx
   lobes broad with
   thick reflexed tips.
   Site: rain forest
   understory, along
   waterways; W.

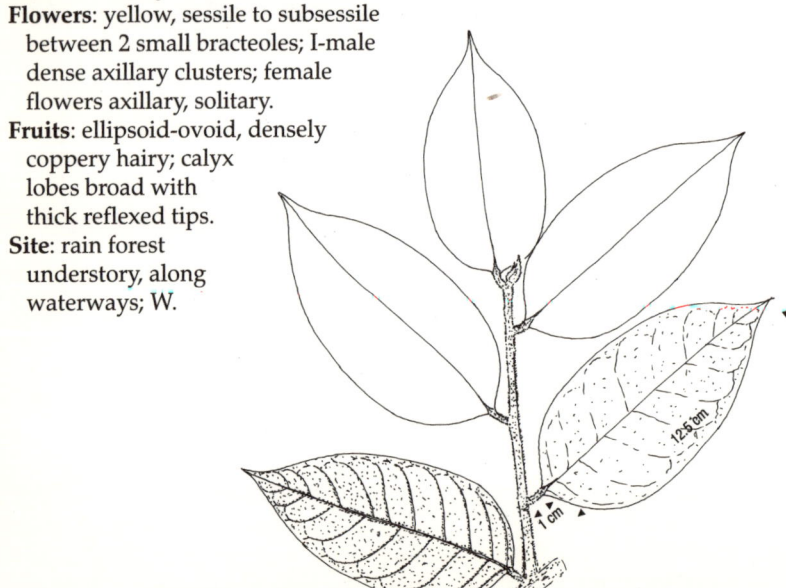

12.5 cm

1 cm

5. *Diospyros insignis*, porawa mara (S), (DF III:43), N, 6, tree.
   Leaves: ovate-elliptic, tapering base, shortly
   acuminate apex, lateral veins widely
   spaced; flush pinkish.
   Trunk: B-peeling in **large thin flakes**,
   black with **brown** beneath; IB-pale
   brown to pinkish; W -whitish.
   Flowers: white to yellowish;
   I-male, axillary, clustered,
   sessile, 3-20; I-female, sessile, 1-6.
   Fruits: green, depressed globose,
   glabrous, obscure rounded ribs;
   calyx cup thick, enlarged,
   recurved, woody.
   Site: rain forest understory; W.

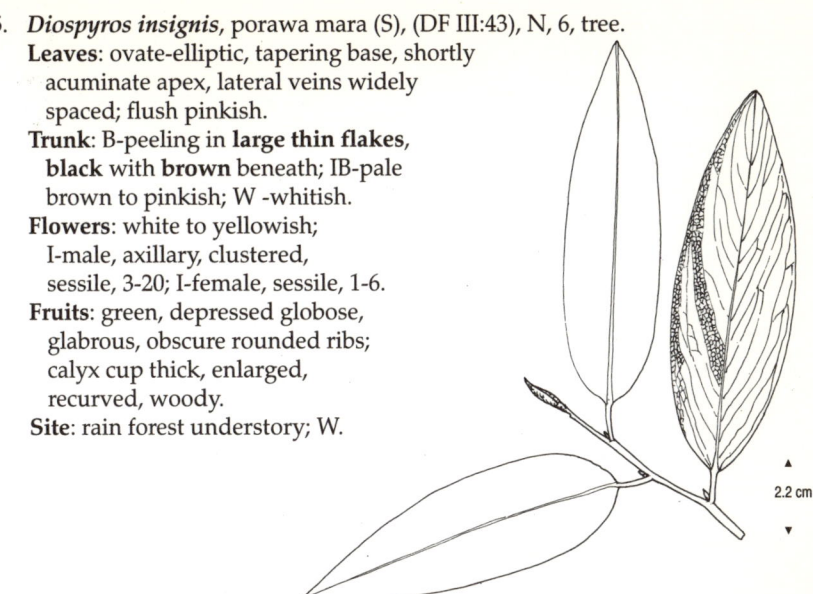

2.2 cm

42

6. *Diospyros malabarica*, timbiri (S)/panichchai (T), (DF III:27), N, 14, tree.
   Leaves: dark green, glossy, oblong to lanceolate-oblong, tapering base,
   lateral veins 8-10 pairs, densely **reticulate** intercostals.
   Trunk: B-**rough, peeling in scaly strips**, **black**; IB-beefy
   red; W-hard, heavy; branchlets yellowish,
   sparsely tomentose.
   Flowers: yellow, pilose.
   Fruits: globose, often
   depressed on top,
   mealy glands
   red-orange, lesions
   often exuding red
   gum; pulp sweet
   and transparent;
   calyx cup woody,
   reflexed.
   Site: monsoon and
   intermediate forest
   subcanopy, along
   waterways; DL, IN.
   Uses: W-boat building,
   construction; W, fruit,
   flowers-medicinal.
   Fruit-for treating fishing
   nets and lines to make
   them durable.

EBENACEAE

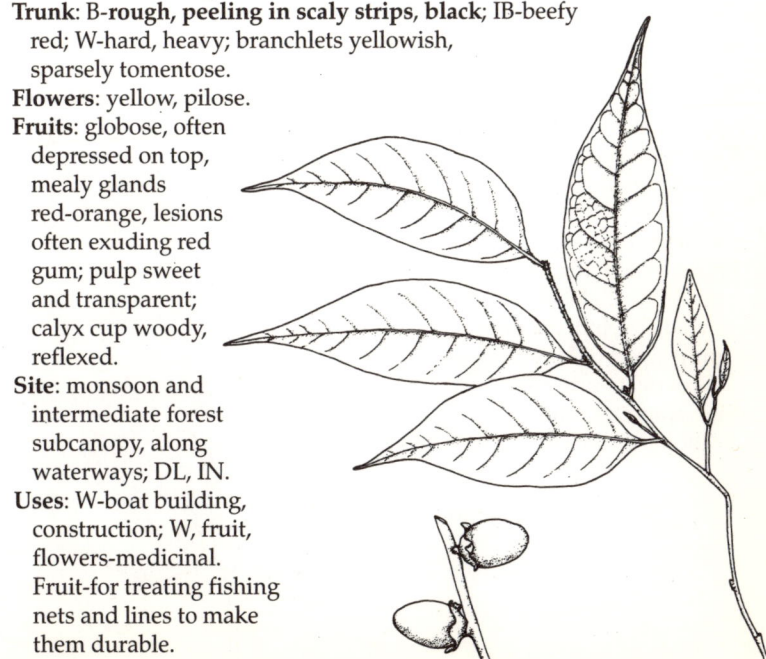

7. **Diospyros melanoxylon**, coromandel ebony (E)/ kadumberiya (S)/karum tumbi (T), (DF III:37), N, 20, tree.
   Leaves: elliptic to oblong, acute base, lateral veins about 8-9 pairs.
   Trunk: B-rough, deeply fissured, blackish;W,s-durable, h-variegated brown and black, durable; twigs densely rusty woolly.
   Flowers: yellowish white; I-male, 3-7, in dichasial cymes; female axillary, solitary.
   Fruits: yellowish, globose, pointed.
   Site: monsoon forest canopy, savannah; DL.
   Uses: leaves-wrappers for beedi; W-furniture.

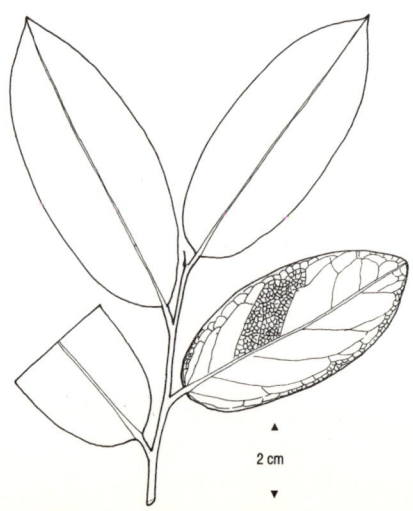

6.5 cm

8. **Diospyros montana**, vakkana (T), (DF III:20), N, 14, tree.
   Leaves: oblong to subovate, rounded to truncate base, obtusely acuminate apex, lateral veins 6-8 pairs, arched.
   Trunk: B-smooth to peeling, grey to yellowish; IB-yellow; exudate-watery; spines at base of branches stiff, divaricate.
   Flowers: white or yellow; I-male, axillary panicles or cymes; female axillary, solitary.
   Fruits: yellow turning red-brown, globose, ovoid, smooth, glabrous; pulp sweet or bitter; calyx lobes ovate-orbicular.
   Site: monsoon forest canopy, scrub; DL.

2 cm

9. ***Diospyros oocarpa***, kalu kadumberiya (S)/ vellai karunkkali (T), (DF III:12), N, 30, tree.

Leaves: ovate to oval, rounded base, acute apex, lateral veins approximately 10 pairs, obscure.

Trunk: straight, sometimes fluted; B-**smooth** to **rough**, **peeling** in large irregular pieces, **dark brown** to **grey**; IB-yellowish to pale brown; W,s-pale pinkish; h-brown with black striations.

Flowers: male yellow, female greenish-yellow, sometimes solitary; I-male and female sessile cymes.

Fruits: black, oblong-ovoid, top flat to rounded, glabrous; calyx slightly enlarged, disc-shaped.

Site: monsoon and intermediate forest canopy, scrub; DL, IN.

Uses: W-turnery, decorative work.

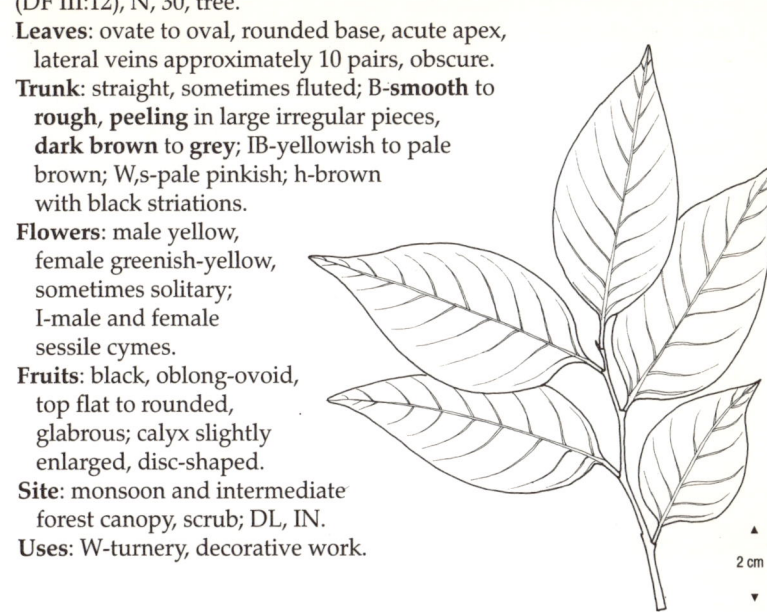

2 cm

42

10. ***Diospyros ovalifolia***, kunumella (S)/ vedukunari (T), (DF III:15), N, 15, tree.

Leaves: oblanceolate to oblong, base tapered, above glossy, lateral veins 6-7 pairs, reticulate intercostals.

Trunk: B-**superficial** longitudinal **black ridges and brown fissures**; IB-pale reddish brown; W-brown to dirty white.

Flowers: male yellow, 3-7 in sessile clusters on woody twigs; female solitary, or 3-9, axillary.

Fruits: subsessile, depressed globose; calyx lobes shallow.

Site: monsoon and intermediate forest subcanopy, scrub; DL, IN.

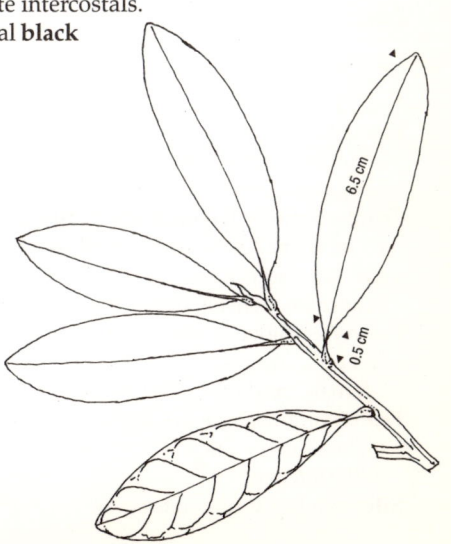

6.5 cm

0.5 cm

EBENACEAE

11. *Diospyros racemosa*, kaluwella (S)/vellai
    thoveri (T), (DF III:22), N, 20, tree.
    Leaves: elliptic to lanceolate, base acute to
        rounded, shortly acuminate apex, veins on
        both sides densely **reticulate**.
    Trunk: B-**black to dark brown**, rather
        smooth, **hardly** peeling; IB-reddish;
        branchlets glabrous.
    Flowers: yellow; I-male,
        3-12 in simple to complex
        cymes; I-female,
        solitary or in cymes.
    Fruits: yellowish, smooth, ovoid
        to ellipsoid with flattened apex.
    Site: rain forest subcanopy, understory; W.

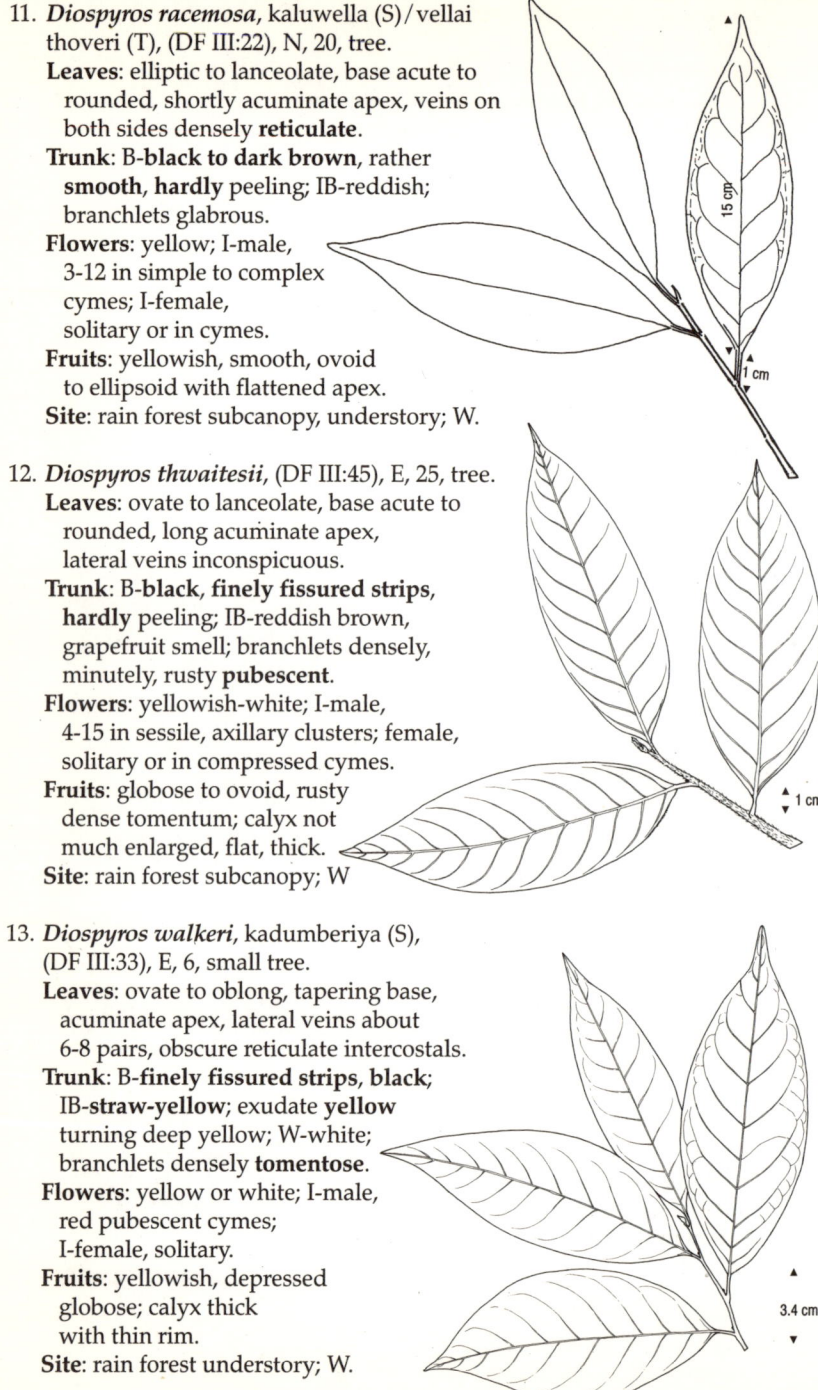

12. *Diospyros thwaitesii*, (DF III:45), E, 25, tree.
    Leaves: ovate to lanceolate, base acute to
        rounded, long acuminate apex,
        lateral veins inconspicuous.
    Trunk: B-**black**, **finely fissured strips**,
        hardly peeling; IB-reddish brown,
        grapefruit smell; branchlets densely,
        minutely, rusty **pubescent**.
    Flowers: yellowish-white; I-male,
        4-15 in sessile, axillary clusters; female,
        solitary or in compressed cymes.
    Fruits: globose to ovoid, rusty
        dense tomentum; calyx not
        much enlarged, flat, thick.
    Site: rain forest subcanopy; W

13. *Diospyros walkeri*, kadumberiya (S),
    (DF III:33), E, 6, small tree.
    Leaves: ovate to oblong, tapering base,
        acuminate apex, lateral veins about
        6-8 pairs, obscure reticulate intercostals.
    Trunk: B-**finely fissured strips**, **black**;
        IB-**straw-yellow**; exudate **yellow**
        turning deep yellow; W-white;
        branchlets densely **tomentose**.
    Flowers: yellow or white; I-male,
        red pubescent cymes;
        I-female, solitary.
    Fruits: yellowish, depressed
        globose; calyx thick
        with thin rim.
    Site: rain forest understory; W.

43. ELAEOCARPACEAE

FAMILY DESCRIPTION - **Habit**: trees and shrubs. **Leaves**: spiral, simple. **Stipules**: present. **Flowers**: usually bisexual, actinomorphic, racemes or cymes. **Fruits**: capsule or drupe.

FLOWER PARTS - CALYX: 4-5, free or united, valvate. COROLLA: 4-5 or often none, free, rarely basally united, often apically fringed, usually valvate. ANDROECIUM: numerous, free, arising from a disk or enlarged receptacle forming an androgynophore. Connective often conspicuously prolonged. GYNOECIUM: superior, 2-many, rarely 1-locular, with 2-many ovules per loculus. Placentation axile.

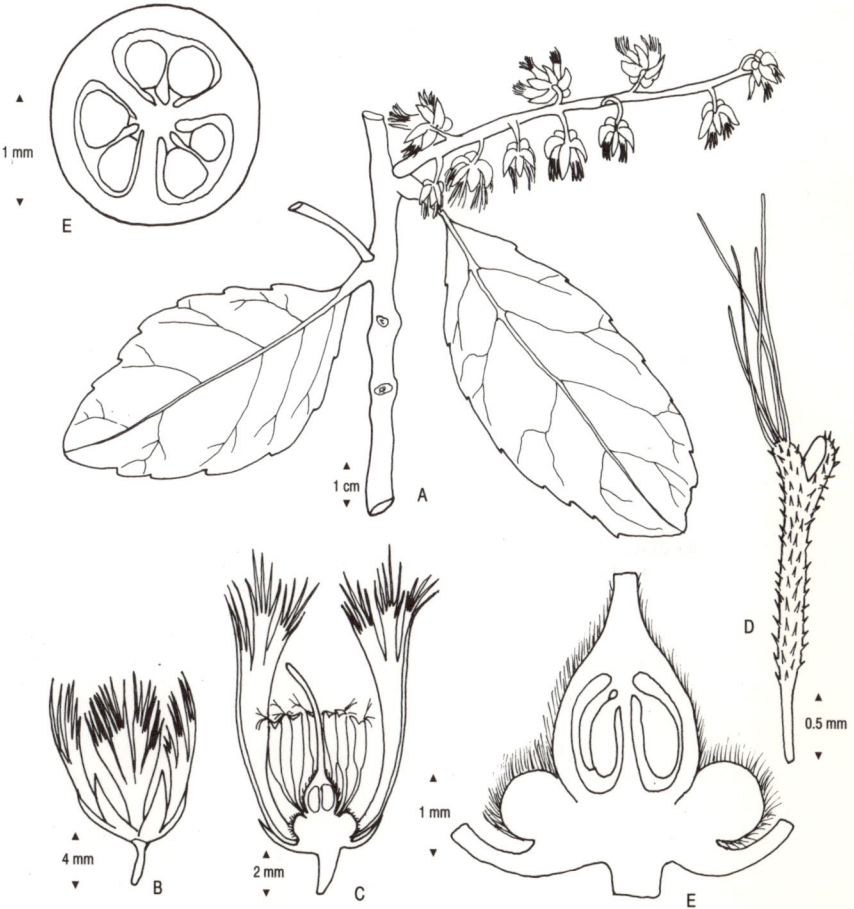

Key: inflorescence (A), full flower (B), half flower (C), stamen (D), longitudinal (E) and transverse (F) sections of ovary of *Elaeocarpus serratus*.

1. *Elaeocarpus amoenus*, titta weralu (S), (T I:185), N, 15, tree.
    Leaves: variable, oblong-oval, tapering ends, margins
       scalloped to serrate, **glandular pits** in axils of lateral veins beneath.
    Trunk: much branched; young parts finely adpressed puberulous.
    Flowers: white, petals deeply cut into linear segments;
       I-drooping, short spreading racemes.
    Fruits: smooth, nearly globular drupe.
    Site: rain forest subcanopy; LM, IN, UW.
    Uses: seeds-necklaces.

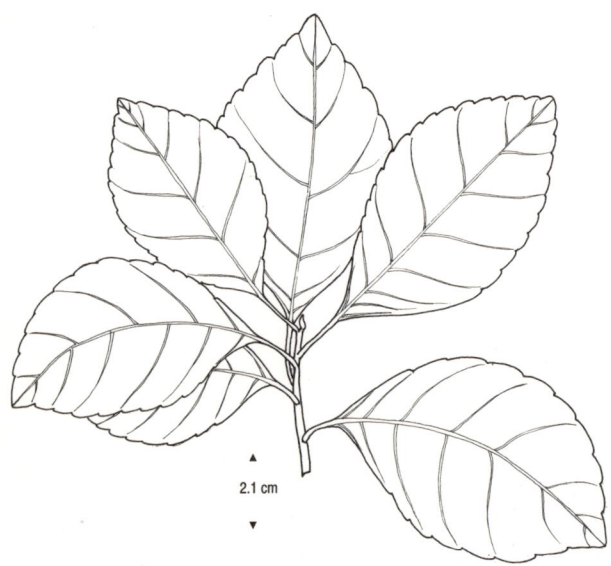

2.1 cm

2. *Elaeocarpus glandulifer*, titta weralu (S), (T I:187), E, 20, tree.
    Leaves: lanceolate, acute base, tapering apex, margins
       coarsely **serrate**, lateral veins beneath
       prominently oblique with **large**
       glandular pits in axils;
       petiole **very long**.
    Trunk: young parts glabrous.
    Flowers: white;
       I-drooping, terminal
       or axillary racemes.
    Fruits: ovoid, smooth,
       glaucous drupe.
    Site: montane and rain
       forest canopy;
       M, LM, UW.

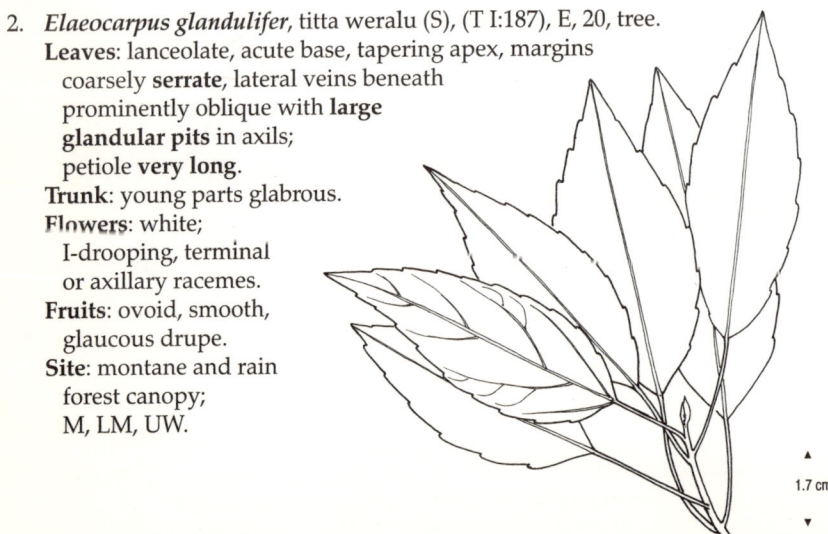

1.7 cm

3. *Elaeocarpus serratus*, weralu (S), (T I:184), N, 15, small tree.
   Leaves: oval-obovate, acute base, rounded to obtuse apex, margins
   shallowly **serrate**, beneath glandular thickenings on vein axils.
   Trunk: W-light yellowish white;
   young parts pubescent.
   Flowers: white; I-numerous,
   axillary racemes.
   Fruits: dull yellow-green,
   ovoid, smooth, bluntly
   pointed drupe.
   Site: home
   gardens; IN, W.
   Uses: fruit-edible.

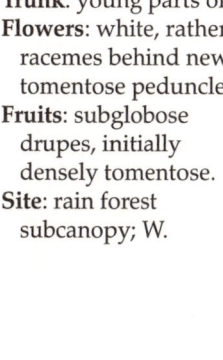

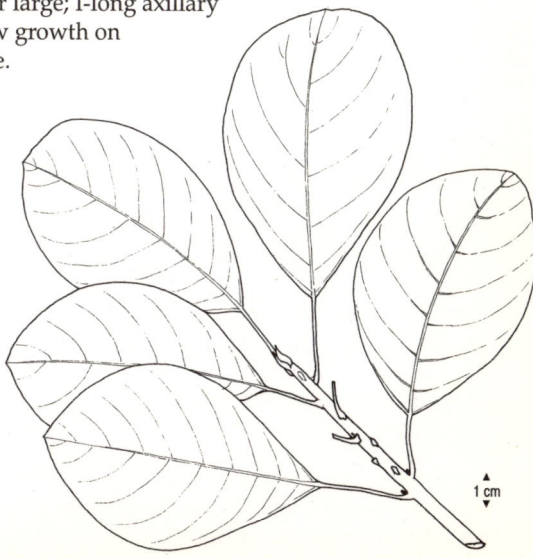

43

ELAEOCARPACEAE

4. *Elaeocarpus subvillosus*, gal weralu (S), (T I:186), N, 15, small tree.
   Leaves: oval, narrow to rounded base, shortly acuminate, margins
   shallowly **spiny-serrate**, pubescent beneath.
   Trunk: young parts of twigs finely tomentose.
   Flowers: white, rather large; I-long axillary
   racemes behind new growth on
   tomentose peduncle.
   Fruits: subglobose
   drupes, initially
   densely tomentose.
   Site: rain forest
   subcanopy; W.

1 cm

44. ERICACEAE

FAMILY DESCRIPTION - Habit: small trees or shrubs, rarely lianas. **Leaves**: spiral, opposite or whorled, simple. **Stipules**: absent. **Flowers**: usually bisexual, actinomorphic or slightly zygomorphic, often in racemes. **Fruits**: loculicidal or septicidal capsule, berry or drupe. Seeds numerous, sometimes winged.

FLOWER PARTS- CALYX: 4-7, more or less united, valvate or imbricate. COROLLA: 4-7, united into a campanulate or urceolate tube, or sometimes free, lobes convolute or imbricate. ANDROECIUM: stamens double number of corolla lobes, rarely the same number, free. Filaments sometimes flattened or dilated and basally united forming a tube, straight or s-curved. Anthers 2-celled, frequently appendaged, opening by a terminal pore or lengthwise. Nectary disc intrastaminal, surrounding and often attached to the ovary. GYNOECIUM: superior or inferior, carpels 4-10, typically 5. Placentation axile or basally axile and apically parietal, with numerous ovules in each loculus.

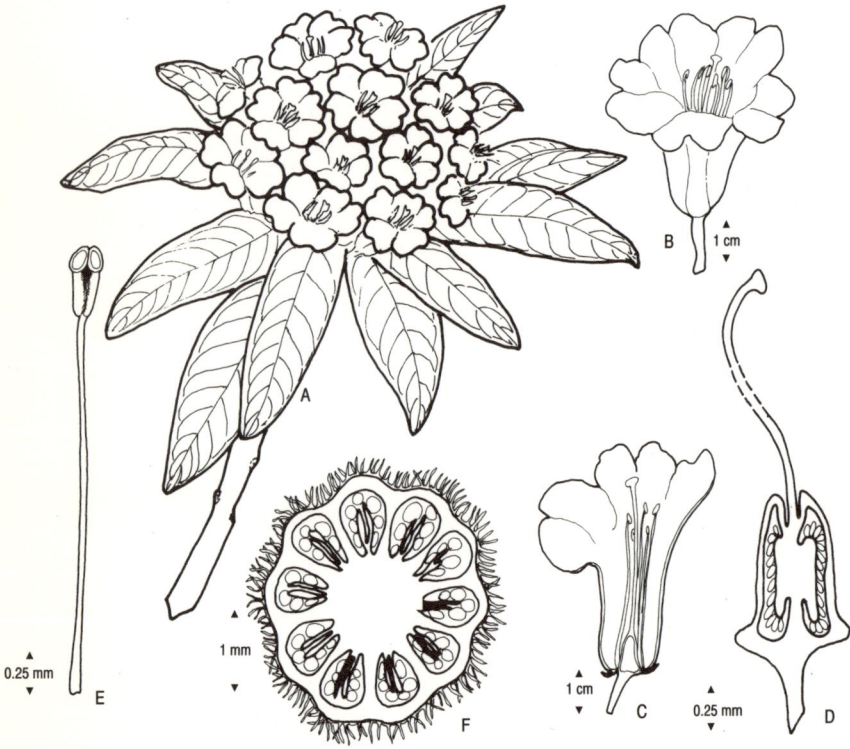

Key: inflorescence (A), full flower (B), half flower (C), longitudinal section of gynoecium (D), single stamen (E) and transverse section of ovary (F) of *Rhododendron arboreum* var. *zeylanicum*.

1. *Gaultheria rudis*, wal kapuru (S), (T III:62), N, 2, shrub.
   Leaves: numerous, oblong to lanceolate, rounded base, bluntly pointed
   apex, margins **serrate**, stiff, whitish,
   veins reticulate, scattered glands on
   lower surface; wintergreen' smell.
   Trunk: B-orange-brown; twigs
   pink; much branched.
   Flowers: numerous,
   small, globose;
   I-pubescent,
   axillary cymes.
   Fruits: purplish-blue, small
   pubescent capsule;
   enclosed by fleshy,
   ovoid, shiny,
   smooth calyx.
   Site: montane forest understory,
   wet patana grassland; M.
   Uses: leaves-medicinal.

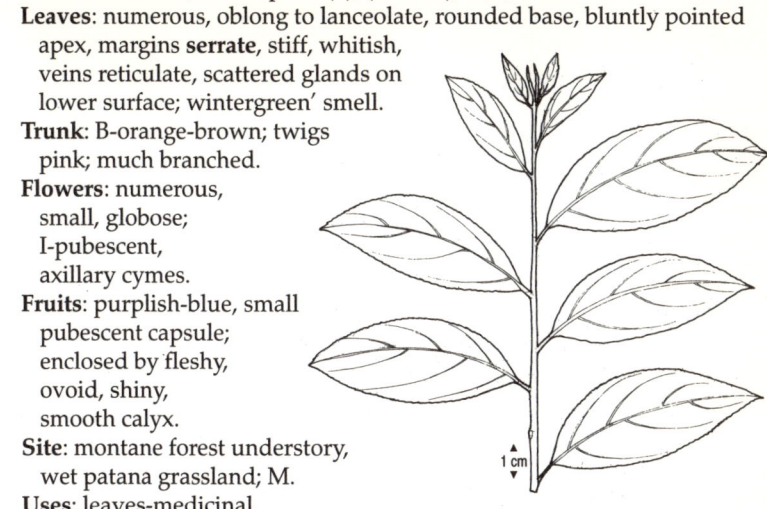

1 cm

44

2. *Rhododendron arboreum* var. *zeylanicum*, ma ratmal (S),
   (T III:63), E, 3, small tree or shrub.
   Leaves: terminally crowded, oblong-oval to lanceolate, tapering base,
   acute apex, margins **revolute**, densely **pubescent** beneath.
   Trunk: B-deeply furrowed, dark grey; branches stout, twisted;
   twigs conspicuously leaf-scarred.
   Flowers: **dark crimson**, **large**, conspicuous; I-closely
   placed short terminal head-like raceme.
   Fruits: capsule, very hard, woody.
   Site: montane forest gaps
   and fringes, wet patana
   grassland; M.
   Uses: leaves-medicinal.

ERICACEAE

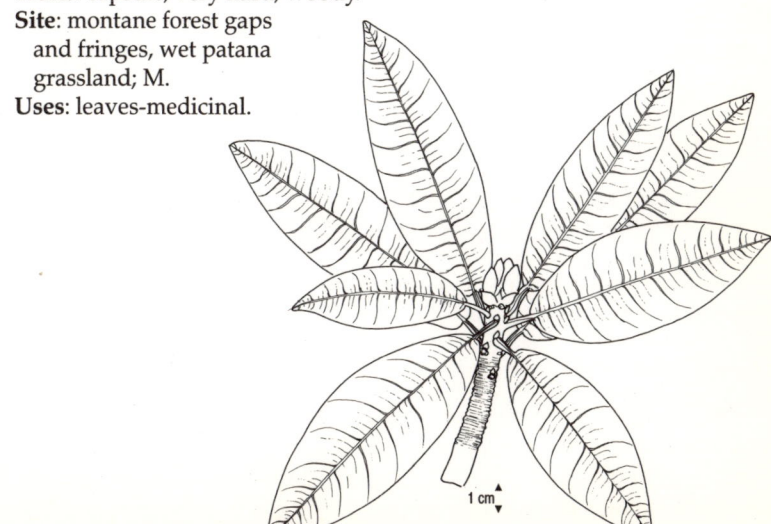

1 cm

45. ERYTHROXYLACEAE

FAMILY DESCRIPTION - **Habit**: trees or shrubs. **Leaves**: spiral, rarely opposite, simple, entire. **Stipules**: intrapetiolar. **Flowers**: usually unisexual, actinomorphic, solitary or axillary clusters. **Fruits**: 1-seeded drupe.

FLOWER PARTS - CALYX: 5, tubular, lobes imbricate or valvate. COROLLA: 5, free, imbricate or convolute, with bifid, ligulate appendages or callosities on inner surface. ANDROECIUM: usually 10, basally united into a tube. Anthers 2- lobed, dehiscing longitudinally. GYNOECIUM: superior, 3-locular with 2 of them usually sterile, 1-2 pendulous ovules per loculus. Styles 3, free or united. Stigma club-shaped, capitate or the stigmatic surface obliquely depressed.

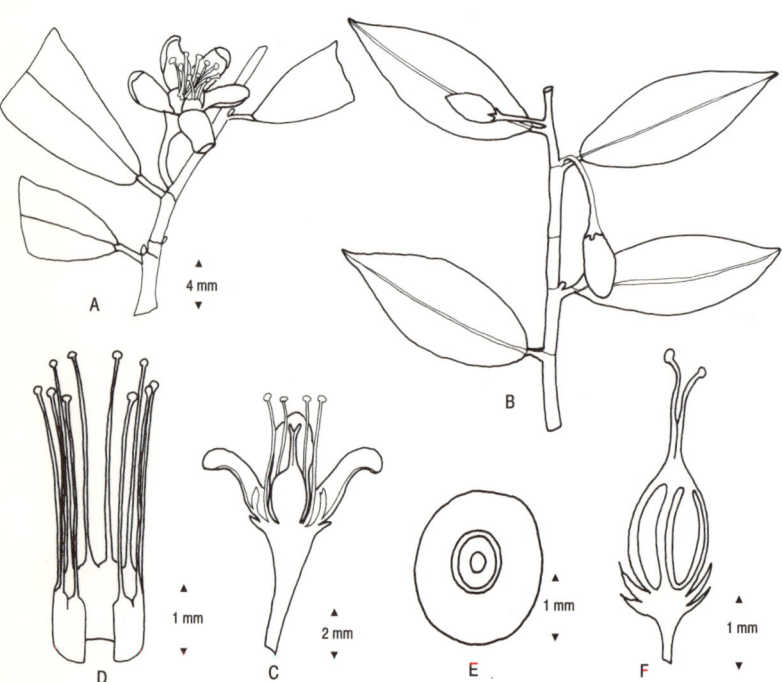

Key: twigs bearing flower (A) and mature fruits (B); half flower (C), androecium showing basally fused stamens (D), longitudinal section of gynoecium (E), transverse section of ovary (F) of *Erythroxylum zeylanicum*.

1. ***Erythroxylum monogynum***, deva daram (T), (T I:190), N, 5, small tree.
   Leaves: oval to obovate, base tapering, very obtuse apex, veins reticulate.
   Trunk: B-very rough, dark brown;
   W,h-hard, heavy, dark brown,
   durable; much branched;
   resinous.
   Flowers: greenish white,
   1-4 together, axillary.
   Fruits: bright scarlet,
   smooth, pointed, oblong
   drupe; surrounded by
   persistent calyx at base.
   Site: monsoon and
   secondary forest,
   scrub; DL.
   Uses: resin-boat wood
   preservative;
   leaves, stem-medicinal.

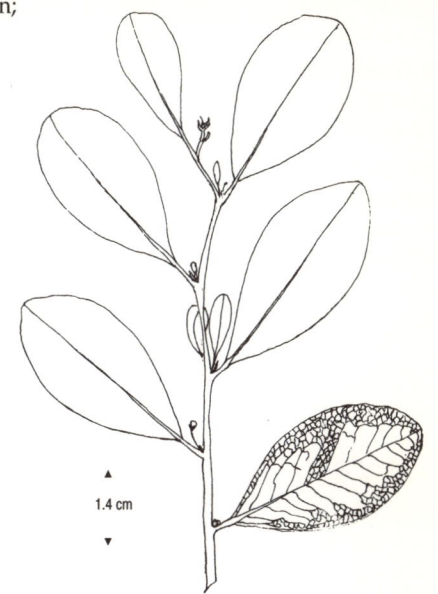

1.4 cm

45

ERYTHROXYLACEAE

2. ***Erythroxylum zeylanicum***, bata kirilla (S)/chiruchemannatti (T),
   (T I:191 & VI:34), E, 3, shrub.
   Leaves: linear-lanceolate to oval-lanceolate,
   tapering base, acuminate to obtuse apex.
   Trunk: much branched, twiggy; B-pale.
   Flowers: solitary, axillary.
   Fruits: red, grooved, pointed, drupe.
   Site: monsoon and intermediate
   forest understory; mostly
   by streams. DL, W.
   Uses: leaves-medicinal.

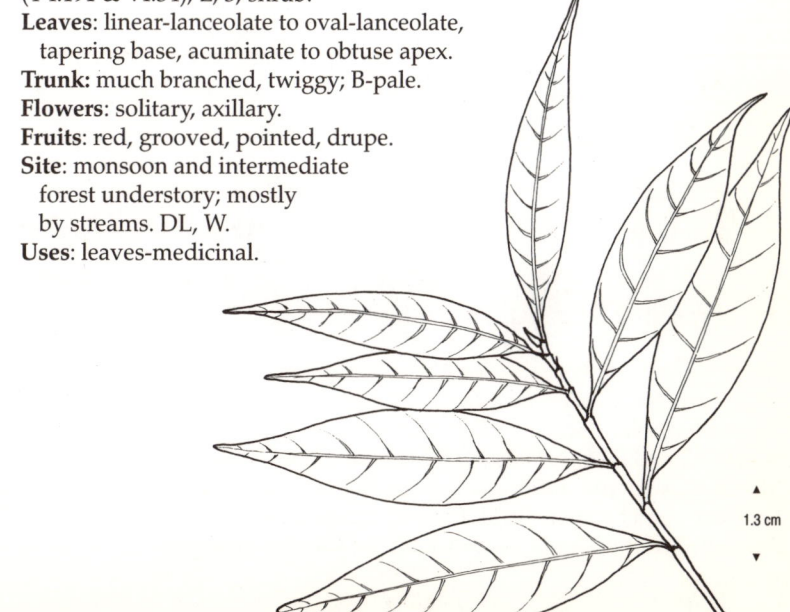

1.3 cm

46. EUPHORBIACEAE

FAMILY DESCRIPTION - Habit: trees, shrubs, lianas or herbs. Some with milky latex. **Leaves**: spiral, sometimes opposite or whorled, simple or compound. **Stipules**: large, represented by small glands, or absent. **Flowers**: unisexual, (plants monoecious or rarely dioecious), usually actinomorphic, cymose inflorescences. **Fruits**: capsule or drupe. Seeds often with conspicuous caruncles.

FLOWER PARTS - CALYX: reduced or absent, 5 when present, valvate or imbricate. COROLLA: reduced or absent, 5 when present, valvate or imbricate. ANDROECIUM: stamens 5-numerous. Filaments basally united. Anthers with longitudinal slits or apical pores. Nectary disc present. Rudimentary ovary present in male flowers. GYNOECIUM: superior, usually 3-locular, with free or basally united styles. 1-2 pendulous ovules per loculus.

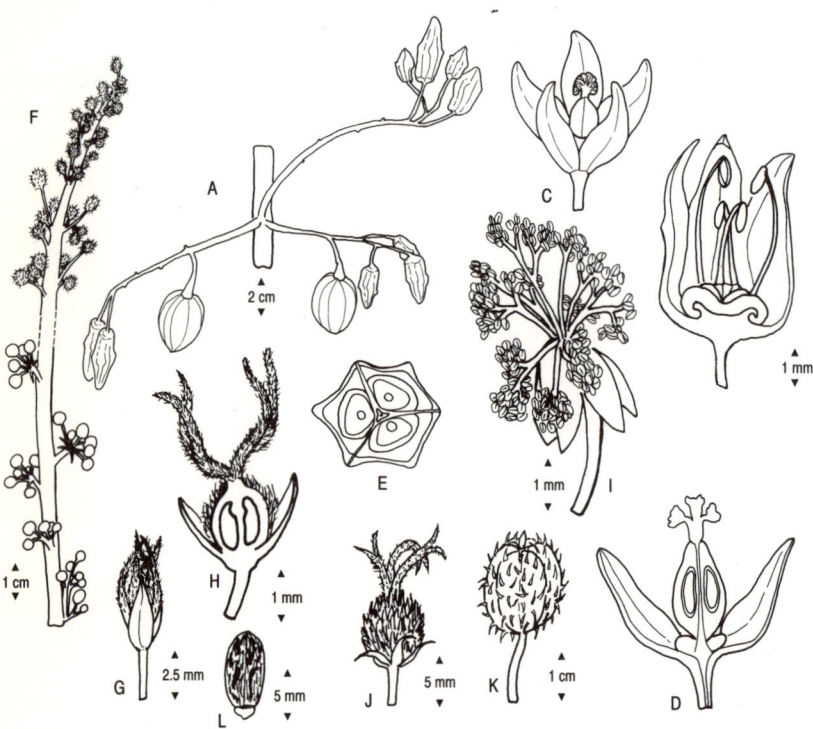

Key: inflorescence bearing male and female (A) flowers, half a male (B), full female (C) and half a female (D) flowers, and transverse section of ovary (E) of **Manihot esculenta.** Inflorescence bearing male and female flowers (F), full female (G) and half female (H) flowers, a male flower (I), a young fruit (J), fully developed fruit (K) and mature seed (L) of *Ricinus communis.*

1. *Acalypha hispida*, red hot cat tail (E), I (India), 3, shrub.
   Leaves: green, oblong, margins
   weakly serrate.
   Trunk: many stems.
   Flowers: I-female dark
   red, velvety long 'tails'.
   Fruits: very small
   capsules.
   Site: hedges, gardens,
   parks; widespread.
   Uses: ornamental.

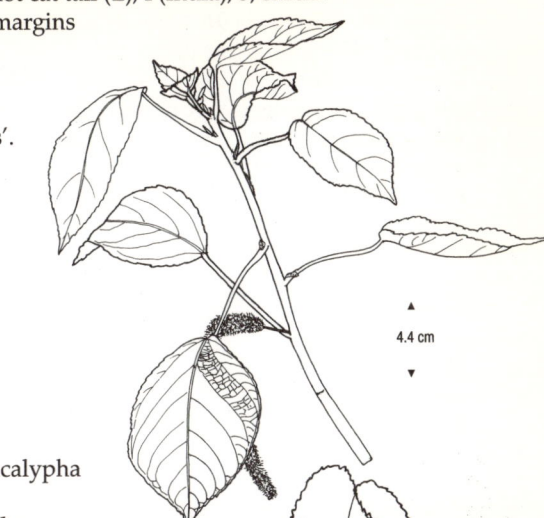

4.4 cm

2. *Acalypha wilkesiana*, acalypha
   (E), I (Fiji), 3, shrub.
   Leaves: **variegated red-bronze-**
   green, **large**, margins
   scalloped.
   Trunk: many stems.
   Flowers: inconspicuous;
   I-female brownish,
   drooping 'tails'.
   Fruits: very small capsules.
   Site: hedges, gardens,
   parks; widespread.
   Uses: ornamental.

2.7 cm

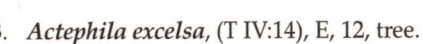

3. *Actephila excelsa*, (T IV:14), E, 12, tree.
   Leaves: very **variable**,
   oblong-linear to broadly
   oval, tapering base.
   Trunk: B-lenticellate, dark
   brown, **leaf-scarred**.
   Flowers: small,
   axillary clusters.
   Fruits: dark green,
   3-lobed capsule on
   long stalk; persistent
   calyx at base.
   Site: montane and rain
   forest understory; M, LM, W.

1 cm

4. *Agrostistachys coriacea*, beru (S), (T IV:56 & VI:265), E, 4, shrub.
    Leaves: **crowded terminally**, obovate to oblong,
    tapering base, shortly acuminate to
    obtuse apex, **stiff**.
    Trunk: branches stout, leaf-scarred;
    young parts **resinous**.
    Flowers: female solitary;
    I-male 1-4 clusters
    in lax racemes;
    Fruits: small,
    3-lobed capsule.
    Site: rain forest
    understory;
    M, LM, UW.
    Uses: leaves-roof
    thatching.

5. *Agrostistachys hookeri*, maha beru (S),
    (T IV:55), E, 5, small tree.
    Leaves: rather **crowded**, very **large**, **linear-oblanceolate**,
    tapering base, shortly acuminate apex, margins
    shallowly serrate toward apex.
    Trunk: very slightly branched; stems
    thick, smooth, leaf-scarred.
    Flowers: I-lax, elongated racemes.
    Fruits: 3-lobed capsule.
    Site: rain forest understory,
    rocky ridgetops, disturbed
    sites, waterways; W.

6. *Aleurites moluccana*, tel kekuna (S),
    (T VI:263), I (Moluccas), 20, tree.
    Leaves: **terminally**
    clustered, ovate to
    lobed; **glands** on petiole.
    Trunk: B-pale greyish;
    young parts densely hairy.
    Flowers: white; I-terminal
    panicles.
    Fruits: fleshy, glabrous.
    Site: cultivated, now
    naturalized; W.
    Uses: W-plywood; seeds-oil.

7. *Antidesma alexiteria*, hin embilla (S),
   (T IV:44), N, 5, small tree.
   Leaves: **numerous, small**, oval, acute
      base, acuminate apex, shiny.
   Trunk: much branched;
      twigs verticillate.
   Flowers: green; I-spikes,
      solitary or few.
   Fruits: Red, small, ovoid,
      lop-sided, capsules.
   Site: monsoon,
      intermediate, and rain
      forest understory;
      DL, IN, W.

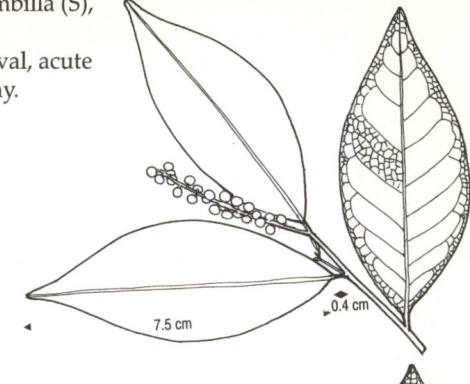

7.5 cm   0.4 cm

46

EUPHORBIACEAE

8. *Antidesma bunius*, karawala kebella (S),
   (T IV:43), N, 10, tree.
   Leaves: rather large, lanceolate to obovate,
      tapering base, slightly acuminate apex;
      needle-like stipules, hairy, deciduous.
   Trunk: B-greyish brown; young
      parts rusty pubescent.
   Flowers: reddish, numerous, rather
      lax; I-terminal spikes.
   Fruits: red-black, globose-ovoid,
      smooth.
   Site: rain forest understory; W.
   Uses: fruit-edible.

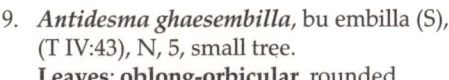

2.7 cm

9. *Antidesma ghaesembilla*, bu embilla (S),
   (T IV:43), N, 5, small tree.
   Leaves: **oblong-orbicular**, rounded
      ends, often notched apex,
      densely **pubescent** beneath.
   Trunk: B-grey; young parts
      rusty **pubescent**.
   Flowers: sessile; I-reddish,
      spikes in terminal panicles.
   Fruits: reddish purple,
      nearly globose, smooth.
   Site: rain forest understory; W.
   Uses: leaves, fruit-edible.

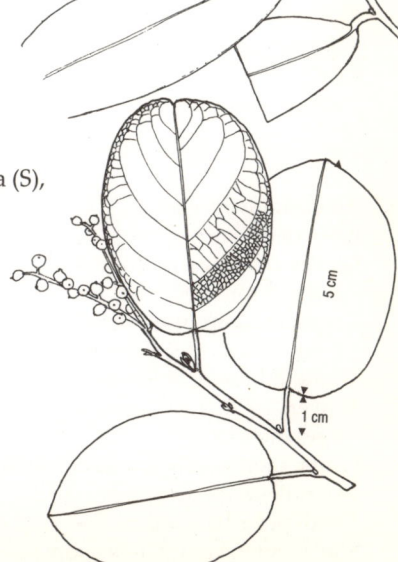

5 cm   1 cm

10. *Antidesma pyrifolium*, (T IV:45), E, 10, tree.
    Leaves: lanceolate to oblong, acute base,
      shortly acuminate to caudate apex.
    Trunk: young parts **pubescent**.
    Flowers: pale pinkish yellow;
      I-lax, axillary, often
      basally branched.
    Fruits: nearly globose,
      purple pulp.
    Site: montane and rain
      forest understory;
      M, LM, W.

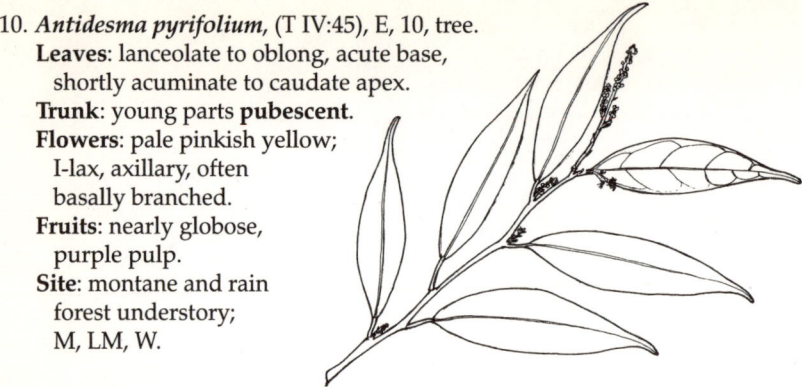

11. *Aporusa cardiosperma*, kampotta, kambokka (S), (T IV:39), E, 15, tree.
    Leaves: numerous, **crowded**, **large**, broadly oval-rotund, cordate to
      rounded base, rounded apex, thick, **leathery**, **shiny**, dark green.
    Trunk: W-hard, durable; branchlets
      thick, leaf-scarred.
    Flowers: in 1-3's; I-numerous,
      closely crowded spikes;
      trees dioecious.
    Fruits: brownish orange,
      large, broadly ovoid,
      slightly tapering to
      blunt-pointed, smooth;
      pericarp thick, spongy,
      tardily dehiscent
      into 3 valves.
    Site: rain forest understory,
      degraded forest; W.
    Uses: ornamental; W-heavy
      construction; fruit-edible.

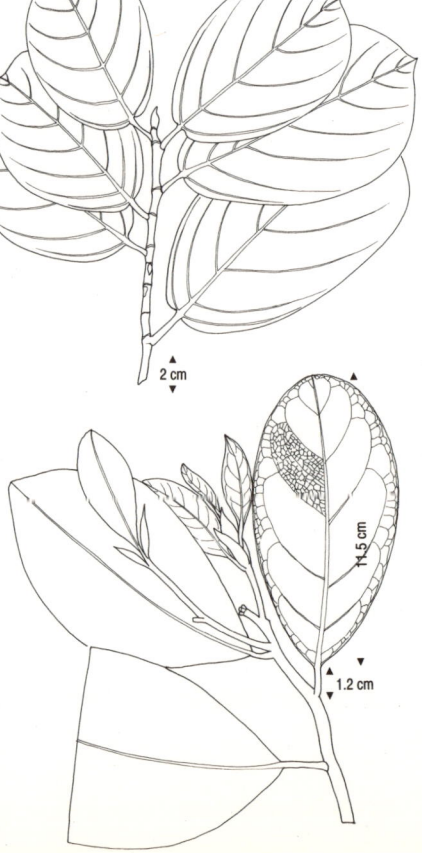

12. *Aporusa fusiformis*, (T IV:41),
    E, 10, tree.
    Leaves: broadly oval to rotund,
      acute to rounded base, rounded
      apex, margins **often revolute**,
      thick, coriaceous; flush purple.
    Trunk: B-brown; twigs stout.
    Flowers: yellowish; I-spikes;
      trees dioecious.
    Fruits: short nearly sessile clusters,
      spindle-shaped, capped
      with persistent stigma.
    Site: montane forest subcanopy; M.

13. ***Aporusa lanceolata***, hin kebella (S), (T IV:40), E, 15, tree.
   Leaves: oblong-lanceolate, acute base,
      caudate to acuminate apex.
   Trunk: branches slender.
   Flowers: sessile, very small, yellow;
      I-spikes; trees dioecious.
   Fruits: small, nearly sessile,
      axillary clusters, globose,
      two-valved, yellow pulp.
   Site: rain forest
      understory; W.
   Uses: fruit-edible.

14. ***Aporusa lindleyana***, kebella (S)/vittil (T),
   (T IV:40 & VI:262), N, 10, tree.
   Leaves: ovate-oval, rounded
      base, slightly acuminate apex,
      margins **undulate**, shiny.
   Trunk: B-smooth, brown.
   Flowers: sessile; I-spikes;
      trees dioecious.
   Fruits: nearly globose,
      pointed style,
      pericarp thin,
      indehiscent;
      yellow pulp.
   Site: rain forest
      understory; W.
   Uses: leaves, fruit-edible.

15. ***Blachia umbellata***, kosatta (S), (T IV:53), N, 3,
   shrub to small tree.
   Leaves: numerous, oval, acute base, slightly
      acuminate apex, shiny dark green.
   Trunk: B-smooth, finely lenticellate;
      branches slender.
   Flowers: green; I-on
      long terminal
      peduncles; plants
      monoecious.
   Fruits: glabrous, lobed,
      very bluntly keeled.
   Site: rain forest
      understory; MC.

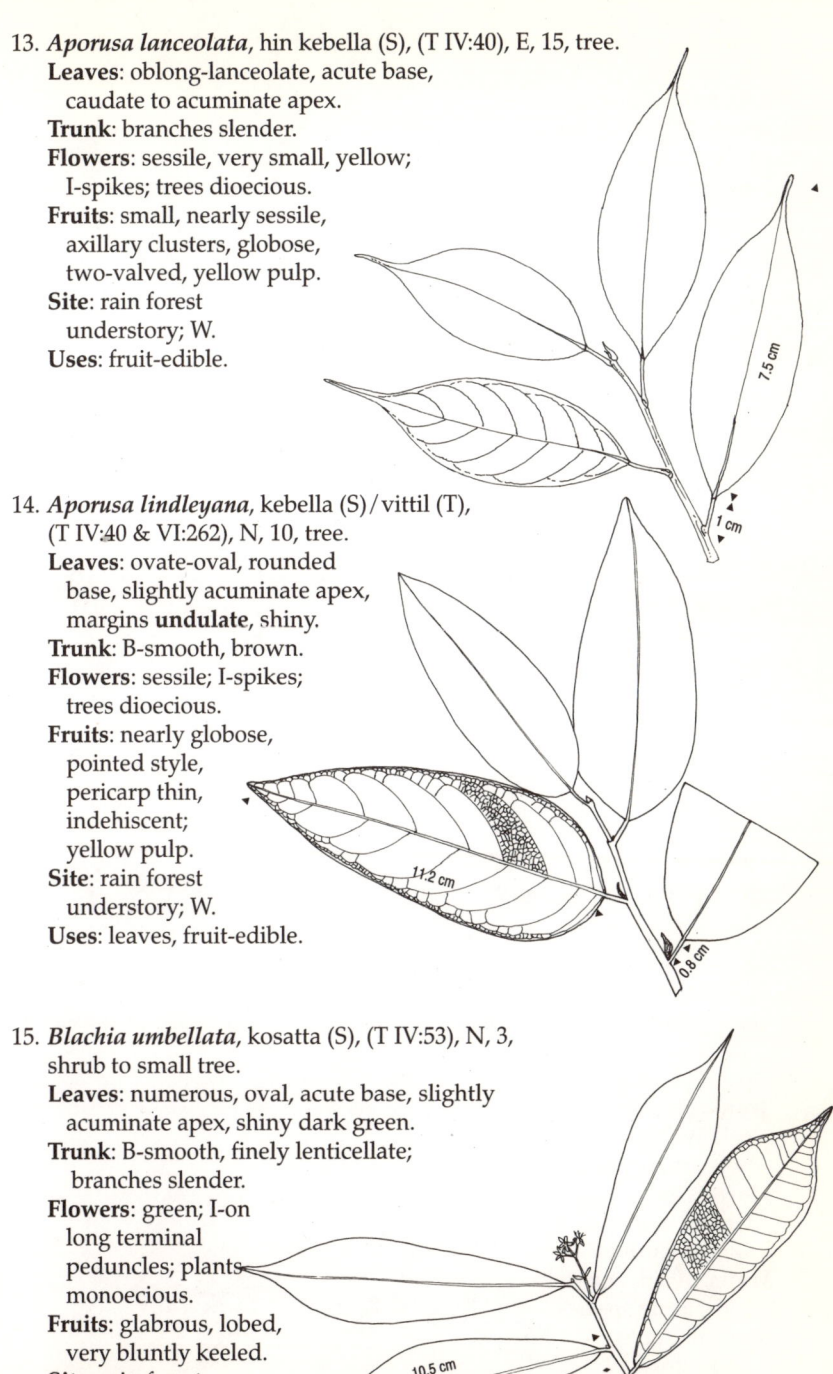

46

EUPHORBIACEAE

16. *Breynia retusa*, wal murunga (S), (T IV:33 & VI:261), N, 2, shrub.

Leaves: **2-ranked**; numerous, rotund-oval; **aromatic**.

Trunk: B-smooth, grey; twigs slender, **angular**.

Flowers: with young leaves, few, solitary, on underside of branches; male lemon-yellow; female pale green.

Fruits: depressed globose, faintly 3-lobed berry, pericarp fleshy orange-red; flattened calyx at base.

Site: monsoon, intermediate, rain forest understory; DL, IN, W.

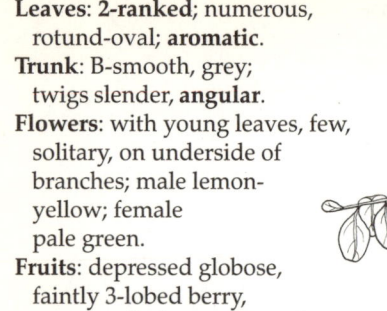

1 cm

17. *Breynia vitis-idaea*, gas kayila (S)/ manipulnati (T), (T IV:34), N, 3, shrub.

Leaves: **2-ranked**; numerous, oval, acute ends.

Trunk: B-yellow-grey; branches long, **horizontal**; twigs **angular**.

Flowers: yellow; male very small in clusters; female solitary; plants monoecious.

Fruits: dull red berry, small, globose, smooth; small calyx at base.

Site: rain forest understory; W.

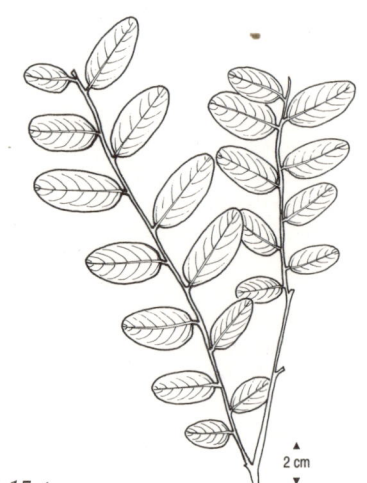

2 cm

18. *Bridelia moonii*, pat kela (S), (T IV:11), E, 15, tree.

Leaves: numerous, oblong-oval, rounded base, obtuse to acuminate apex, lateral veins **11-15** pairs, parallel, whitish beneath.

Trunk: B-smooth, pale yellowish grey; young twigs **rusty pubescent**.

Flowers: green tinged red; I-small, axillary, spicate clusters.

Fruits: ovoid, tapering ends, acute apex.

Site: rain forest understory; W.

Uses: W-termite resistant, building construction.

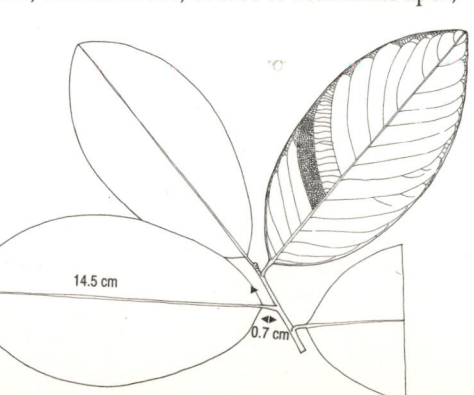

14.5 cm

0.7 cm

19. *Bridelia retusa*, kata kela (S)/mul-venkai (T), (T IV:10), N, 10, tree.
   Leaves: numerous, oblong-oval, rounded base, rounded apex, lateral veins **16-20** pairs, parallel, numerous intermediaries, whitish beneath.
   Trunk: same as *B. moonii*.
   Flowers: same as *B. moonii*.
   Fruits: purplish black, globose pulpy drupe; persistent calyx.
   Site: monsoon and intermediate forest subcanopy; DL, IN, W.
   Uses: W-house construction; roots, stem, bark-medicinal.

20. *Chaetocarpus castanocarpus*, gal hedawaka (S)/sadavaku (T), (T IV:74), N, 20, tree.
   Leaves: oval-ovate, tapering base, acuminate to acute apex.
   Trunk: W-hard, splitting; twigs slender.
   Flowers: globose clusters on short peduncles.
   Fruits: bright red, oblong-globose capsule, densely covered with needle-like prickles, woody valves.
   Site: rain forest subcanopy and understory; W.
   Uses: W-fuelwood.

21. *Chaetocarpus coriaceus*, pol hedawaka (S), (T IV:75), E, 20, tree.
   Leaves: broadly oval to suborbicular, suddenly tapering, shortly acuminate to subacute apex.
   Trunk: W-hard, splitting.
   Flowers: pinkish, axillary clusters.
   Fruits: ovoid-globose, bluntly 3-angled, glabrous capsule covered by close pointed warts.
   Site: rain forest subcanopy; W.
   Uses: W-fuelwood.

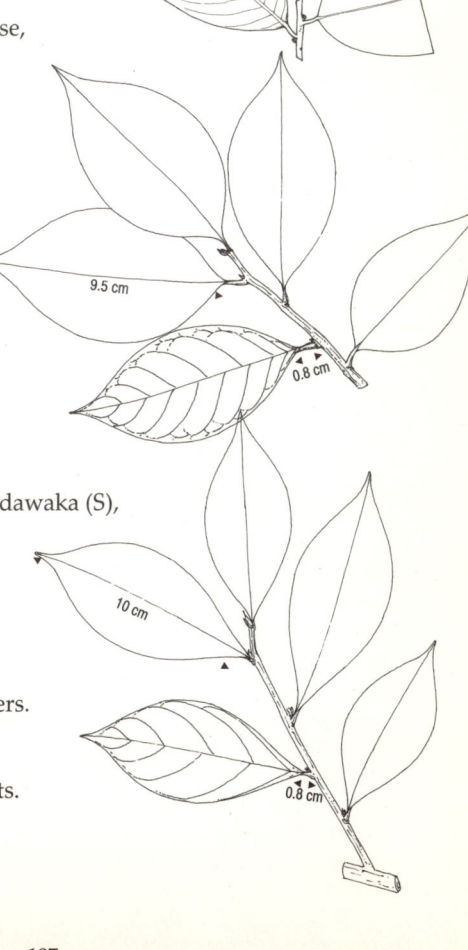

46

EUPHORBIACEAE

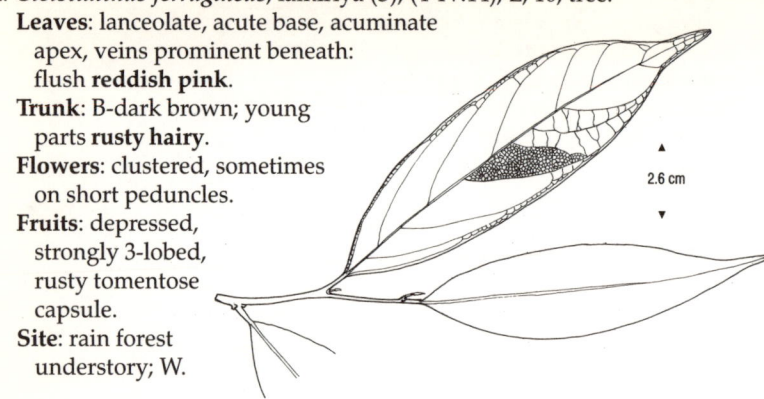

22. *Cleistanthus ferrugineus*, lamiriya (S), (T IV:14), E, 10, tree.
Leaves: lanceolate, acute base, acuminate
apex, veins prominent beneath:
flush **reddish pink**.
Trunk: B-dark brown; young
parts **rusty hairy**.
Flowers: clustered, sometimes
on short peduncles.
Fruits: depressed,
strongly 3-lobed,
rusty tomentose
capsule.
Site: rain forest
understory; W.

2.6 cm

23. *Codiaeum variegatum*, croton (E,S), I (Pacific Is.), 3, shrub.
Leaves: **terminally whorled**,
variegated **red-pink-yellow-
orange-bronze** with green, very
variable, narrow to broad,
margins wavy or twisted.
Trunk: many-stemmed.
Flowers: inconspicuous; I-spikes.
Fruits: green, small capsules.
Site: gardens, roadsides;
widespread.
Uses: ornamental; hedges.

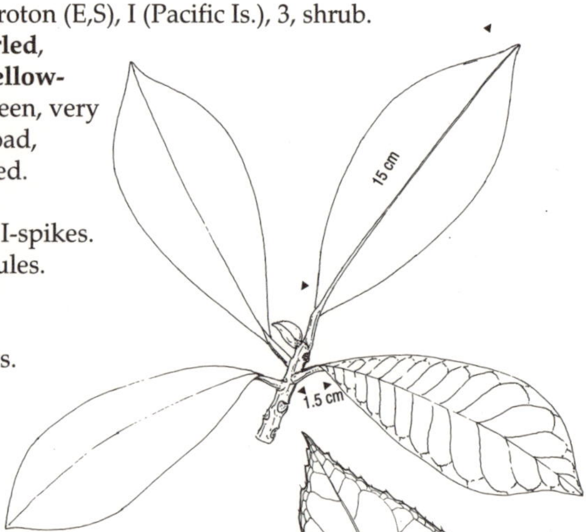

15 cm

1.5 cm

24. *Croton aromaticus*, wal
keppetiya (S),
(T IV:47), N, 2, shrub.
Leaves: ovate, subcordate base, margins irregularly
scalloped, old leaves red; **aromatic**; petiole
apex has pair of small **circular glands**.
Trunk: B-smooth, pale grey; young
parts with **stellate** hairs.
Flowers: greenish white, numerous,
in stellate-hairy short panicles.
Fruits: nearly globose, barely 3-lobed
capsule, rough, scanty hair.
Site: secondary forest, scrub;
DL, IN, W.
Uses: B-secretion (lac) from scale
insect collected for lacquer
(keppetiya resin).

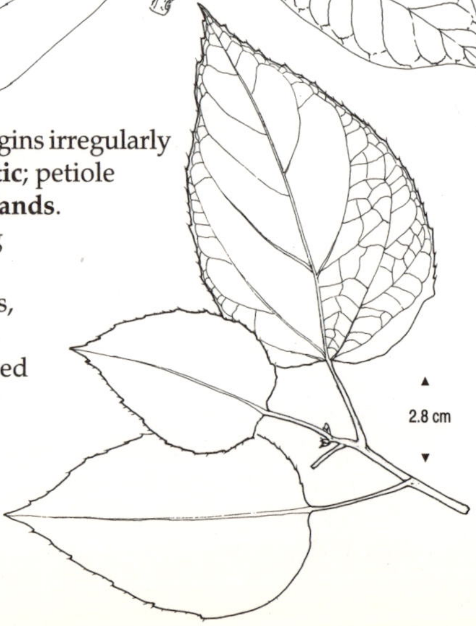

2.8 cm

25. *Croton officinalis*, (T IV:49), N, 3, shrub to small tree.
Leaves: oblong-lanceolate, rounded base, margins **shallowly serrate**, base faintly 3-veined, scattered hairs beneath; petiole apex has pair of **small glands**.
Trunk: B-whitish; scattered stellate hairs on young parts.
Flowers: white; I-racemes short, very lax.
Fruits: 3-lobed, stellate-hairy capsule, crowned with long persistent styles.
Site: monsoon forest, scrub; DL, IN.

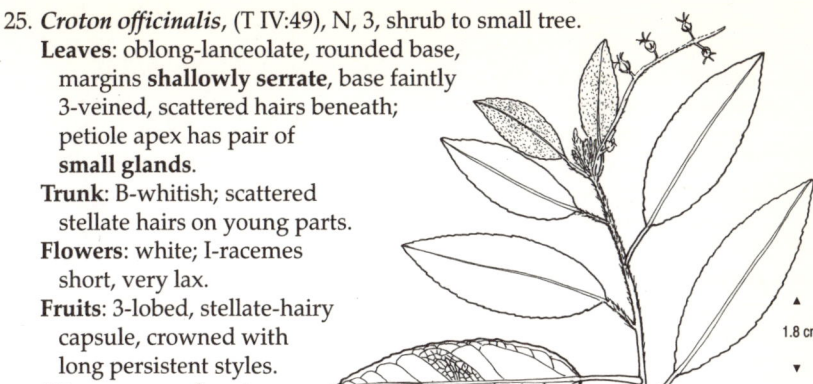

1.8 cm

26. *Dimorphocalyx glabellus*, weliwenna (S)/ tentukki (T), (T IV:54), E, 5, small tree.
Leaves: lanceolate-oblong, tapering base, acuminate apex, dull dark green, margins entire to slightly dentate.
Trunk: B-cinnamon-grey; twigs slender.
Flowers: white.
Fruits: pubescent capsule surrounded by persistent calyx.
Site: monsoon forest understory, scrub; DL.
Uses: whole plant- medicinal.

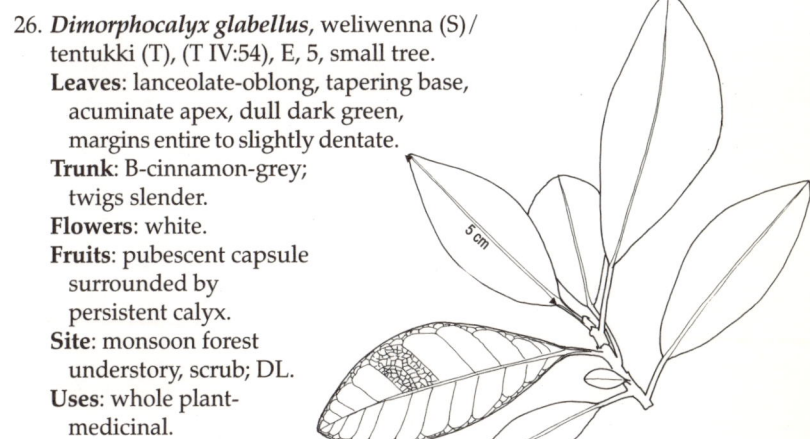

5 cm

27. *Drypetes sepiaria*, wira (S)/virai (T), (T IV:36), N, 5, small tree.
Leaves: broadly oval-oblong, base cordate to rounded, rounded to obtuse-retuse apex.
Trunk: often **gnarled, twisted or fluted**, rigid, much-branched, puberulous.
Flowers: I-in bracteolate, axillary clusters or short racemes.
Fruits: crimson, small, fusiform drupe.
Site: monsoon forest, scrub; DL.
Uses: fruit-edible; W-fuelwood.

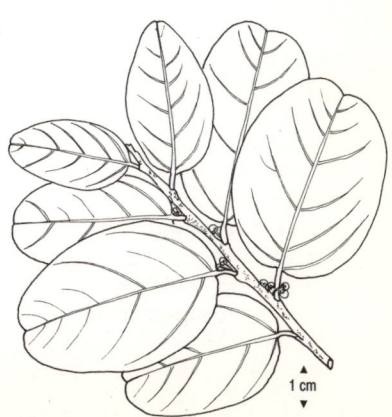

1 cm

46

EUPHORBIACEAE

28. *Euphorbia antiquorum*, daluk (S)/ chatura kalli (T), (T IV:4), N, 3, small tree.
   Leaves: very small, fleshy, deciduous, sessile, on summit of each elevation; stipular **spines** persistent, divaricate.
   Trunk: stout, cylindrical; B-thick, very rough, corrugated, brown; young branches whorled, green, fleshy, jointed, 3-winged, coarsely scalloped; latex **milky**.
   Flowers: greenish yellow; I-heads in small stalked cymes, in 3's.
   Fruits: 3-lobed capsule.
   Site: rocky and stony places; DL.
   Uses: whole plant-medicinal.

12 cm

29. *Euphorbia pulcherrima*, poinsettia (E), (T IV:1), I (Mex.), 2, shrub.
   Leaves: terminally crowded; elliptic to lanceolate.
   Trunk: twigs green, cylindrical, jointed; latex **white**.
   Flowers: small, inconspicuous; bracts **scarlet, pink or white**, large, **conspicuous**, resemble leaves; I-shaggy heads.
   Fruits: terminal clusters of capsules.
   Site: home gardens, parks; IN, W.
   Uses: ornamental; hedges.

3.5 cm

9.3 cm

30. *Euphorbia tirucalli*, nawahandi (S)/ kalli (T), (T IV:5), I (Trop. Africa), 2, shrub.
   Leaves: many, small, oval to lanceolate.
   Trunk: branches thornless, cylindrical; latex **white**, copious.
   Flowers: inconspicuous.
   Fruits: capsule.
   Site: gardens, scrub; DL, IN, W.
   Uses: hedges; ornamental; roots, stem, sap-medicinal.

2 cm

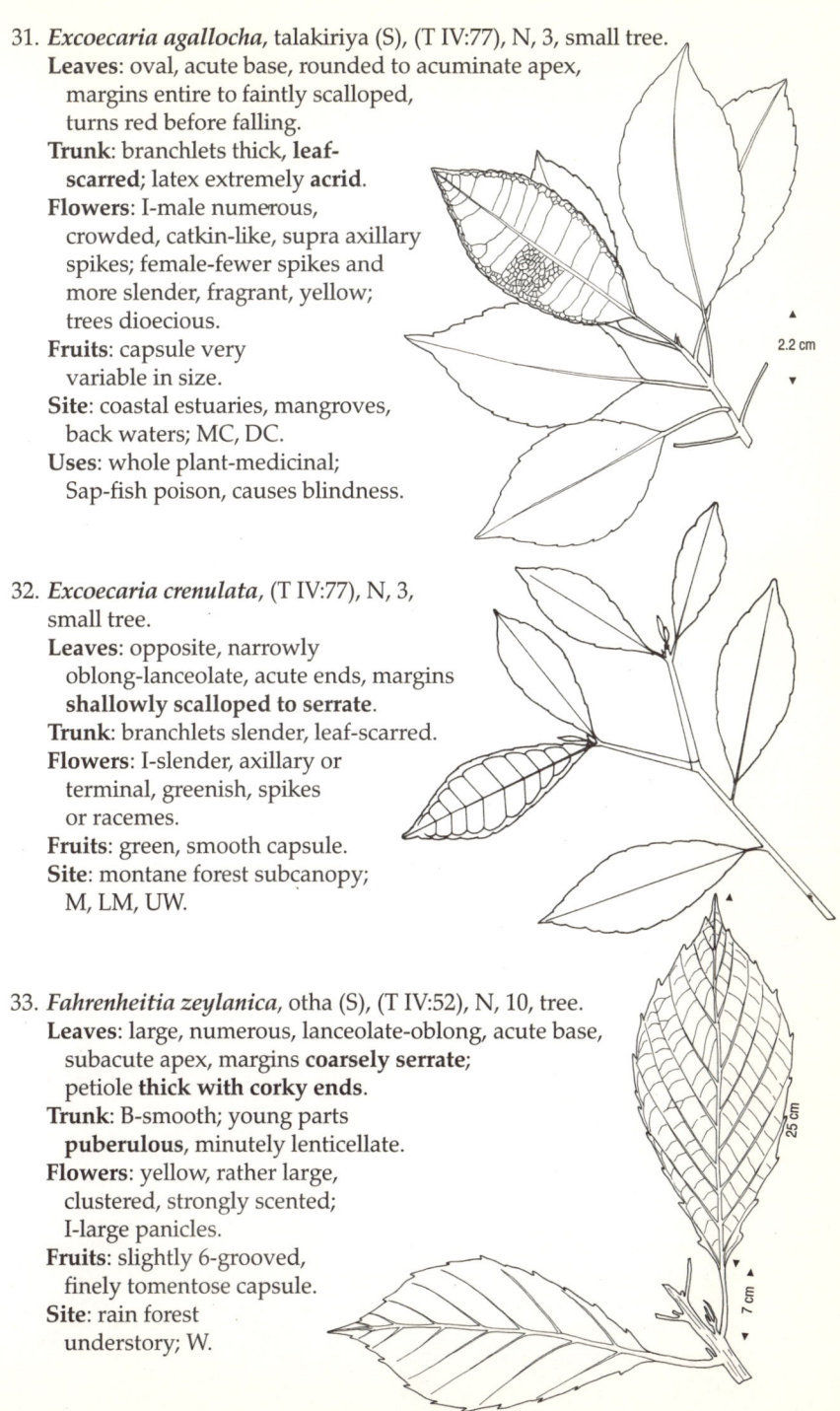

31. *Excoecaria agallocha*, talakiriya (S), (T IV:77), N, 3, small tree.
    Leaves: oval, acute base, rounded to acuminate apex,
        margins entire to faintly scalloped,
        turns red before falling.
    Trunk: branchlets thick, **leaf-
        scarred**; latex extremely **acrid**.
    Flowers: I-male numerous,
        crowded, catkin-like, supra axillary
        spikes; female-fewer spikes and
        more slender, fragrant, yellow;
        trees dioecious.
    Fruits: capsule very
        variable in size.
    Site: coastal estuaries, mangroves,
        back waters; MC, DC.
    Uses: whole plant-medicinal;
        Sap-fish poison, causes blindness.

2.2 cm

32. *Excoecaria crenulata*, (T IV:77), N, 3,
    small tree.
    Leaves: opposite, narrowly
        oblong-lanceolate, acute ends, margins
        shallowly scalloped to serrate.
    Trunk: branchlets slender, leaf-scarred.
    Flowers: I-slender, axillary or
        terminal, greenish, spikes
        or racemes.
    Fruits: green, smooth capsule.
    Site: montane forest subcanopy;
        M, LM, UW.

33. *Fahrenheitia zeylanica*, otha (S), (T IV:52), N, 10, tree.
    Leaves: large, numerous, lanceolate-oblong, acute base,
        subacute apex, margins **coarsely serrate**;
        petiole **thick with corky ends**.
    Trunk: B-smooth; young parts
        puberulous, minutely lenticellate.
    Flowers: yellow, rather large,
        clustered, strongly scented;
        I-large panicles.
    Fruits: slightly 6-grooved,
        finely tomentose capsule.
    Site: rain forest
        understory; W.

25 cm

7 cm

46

EUPHORBIACEAE

34. *Flueggea leucopyrus*, hin katu pila (S)/ mudpulanti (T),
　　(T IV:33), N, 2, shrub.
　　Leaves: **small**, obovate to **rotund**.
　　Trunk: B-white to grey; branches **straggly**; twigs leafy,
　　　horizontally divaricate, ending in spines.
　　Flowers: I-green, pedunculate, axillary clusters.
　　Fruits: white, globose, smooth,
　　　3-celled berry.
　　Site: scrub, monsoon
　　　forest; DL.
　　Uses: leaves-
　　　medicinal.

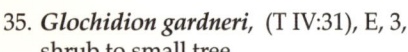

1 cm

35. *Glochidion gardneri*, (T IV:31), E, 3,
　　shrub to small tree.
　　Leaves: oval, tapering,
　　　sometimes unequal
　　　base, shortly acuminate
　　　to sickle-shaped apex.
　　Trunk: branchlets
　　　slender, angular.
　　Flowers: Yellow,
　　　small clusters.
　　Fruits: orange-red, much
　　　depressed, smooth,
　　　6-lobed capsule,
　　　crowned with slender style.
　　Site: rain forest understory; LM, W.

7.5 cm

0.4 cm

36. *Glochidion moonii*, bu hunukirilla (S),
　　(T IV:32), E, 3, shrub to small tree.
　　Leaves: lanceolate-oval, acute ends,
　　　acuminate, veins conspicuously
　　　reticulate, **hairy**.
　　Trunk: branchlets more
　　　or less **tomentose**.
　　Flowers: pale yellow, numerous;
　　　male on long hairy peduncles;
　　　female sessile; solitary
　　　or in axillary fascicles.
　　Fruits: pubescent, strongly
　　　3-lobed capsule, topped
　　　by persistent style.
　　Site: rain forest understory;
　　　LM, W.

2 cm

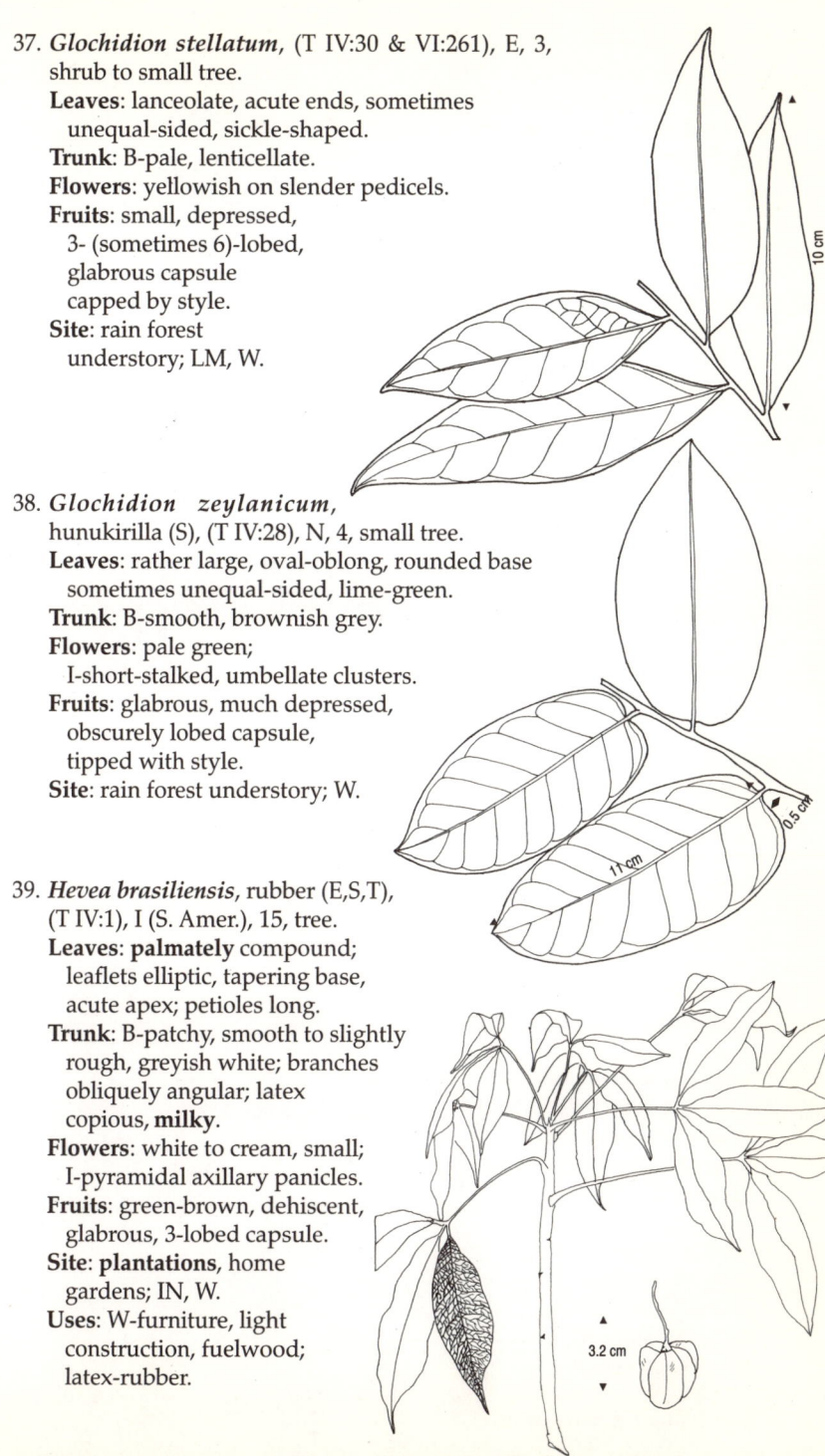

37. *Glochidion stellatum*, (T IV:30 & VI:261), E, 3,
   shrub to small tree.
   Leaves: lanceolate, acute ends, sometimes
      unequal-sided, sickle-shaped.
   Trunk: B-pale, lenticellate.
   Flowers: yellowish on slender pedicels.
   Fruits: small, depressed,
      3- (sometimes 6)-lobed,
      glabrous capsule
      capped by style.
   Site: rain forest
      understory; LM, W.

38. *Glochidion zeylanicum*,
   hunukirilla (S), (T IV:28), N, 4, small tree.
   Leaves: rather large, oval-oblong, rounded base
      sometimes unequal-sided, lime-green.
   Trunk: B-smooth, brownish grey.
   Flowers: pale green;
      I-short-stalked, umbellate clusters.
   Fruits: glabrous, much depressed,
      obscurely lobed capsule,
      tipped with style.
   Site: rain forest understory; W.

39. *Hevea brasiliensis*, rubber (E,S,T),
   (T IV:1), I (S. Amer.), 15, tree.
   Leaves: **palmately** compound;
      leaflets elliptic, tapering base,
      acute apex; petioles long.
   Trunk: B-patchy, smooth to slightly
      rough, greyish white; branches
      obliquely angular; latex
      copious, **milky**.
   Flowers: white to cream, small;
      I-pyramidal axillary panicles.
   Fruits: green-brown, dehiscent,
      glabrous, 3-lobed capsule.
   Site: **plantations**, home
      gardens; IN, W.
   Uses: W-furniture, light
      construction, fuelwood;
      latex-rubber.

10 cm

46

EUPHORBIACEAE

0.5 cm

11 cm

3.2 cm

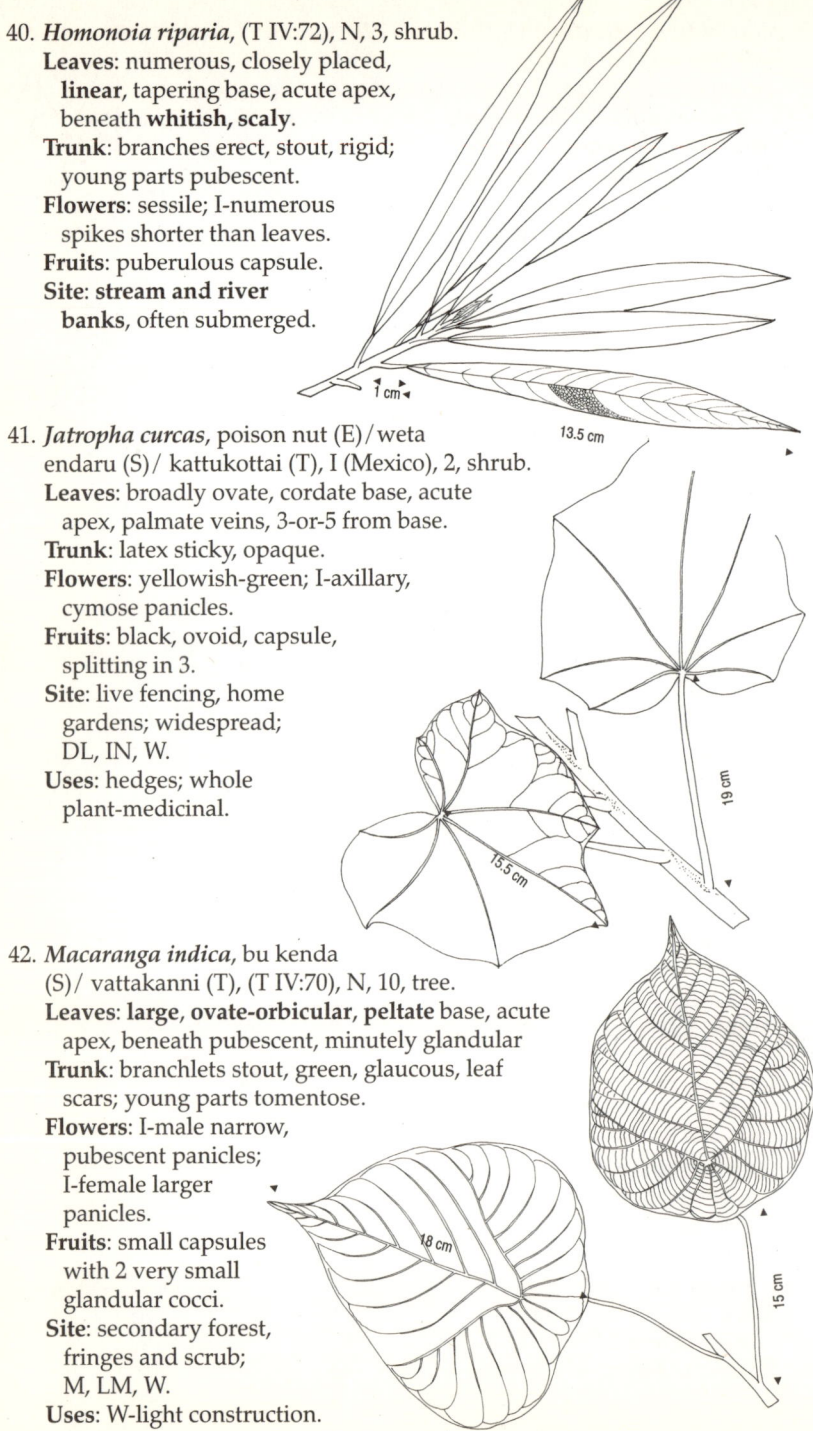

40. *Homonoia riparia*, (T IV:72), N, 3, shrub.
Leaves: numerous, closely placed,
linear, tapering base, acute apex,
beneath **whitish, scaly**.
Trunk: branches erect, stout, rigid;
young parts pubescent.
Flowers: sessile; I-numerous
spikes shorter than leaves.
Fruits: puberulous capsule.
Site: **stream and river
banks**, often submerged.

1 cm

13.5 cm

41. *Jatropha curcas*, poison nut (E)/weta
endaru (S)/ kattukottai (T), I (Mexico), 2, shrub.
Leaves: broadly ovate, cordate base, acute
apex, palmate veins, 3-or-5 from base.
Trunk: latex sticky, opaque.
Flowers: yellowish-green; I-axillary,
cymose panicles.
Fruits: black, ovoid, capsule,
splitting in 3.
Site: live fencing, home
gardens; widespread;
DL, IN, W.
Uses: hedges; whole
plant-medicinal.

19 cm

15.5 cm

42. *Macaranga indica*, bu kenda
(S)/ vattakanni (T), (T IV:70), N, 10, tree.
Leaves: **large**, **ovate-orbicular**, **peltate** base, acute
apex, beneath pubescent, minutely glandular
Trunk: branchlets stout, green, glaucous, leaf
scars; young parts tomentose.
Flowers: I-male narrow,
pubescent panicles;
I-female larger
panicles.
Fruits: small capsules
with 2 very small
glandular cocci.
Site: secondary forest,
fringes and scrub;
M, LM, W.
Uses: W-light construction.

18 cm

15 cm

43. *Macaranga peltata*, kenda (S)/ vattakanni (T), (T IV:70), N, 10, tree.
   Leaves: very **large**, **broadly oval**, **peltate** base, acuminate, acute apex,
      veins prominent beneath; petiole **very long**.
   Trunk: B-smooth, light grey; branchlets stout,
      waxy shine, green, conspicuously
      leaf-scarred.
   Flowers: greenish; I-numerous
      panicles.
   Fruits: small, globose, glabrous,
      glandular-warted,
      sticky capsules.
   Site: rain forest gaps
      and fringes, secondary
      forest; W.
   Uses: leaves-wrapper for
      steaming local sweet
      (helapa); juice-medicinal;
      W-light construction.

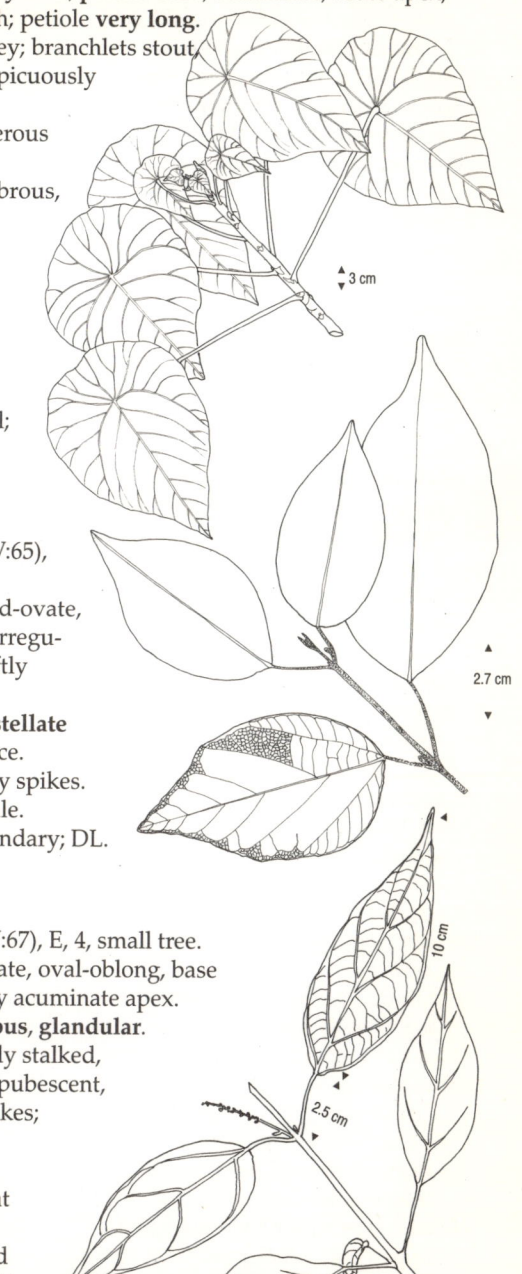

44. *Mallotus eriocarpus*, (T IV:65),
      E, 2, shrub.
   Leaves: variable, rhomboid-ovate,
      rounded base, margins irregu-
      larly **faintly dentate**, softly
      pubescent.
   Trunk: branches slender, **stellate**
      yellow-brown pubescence.
   Flowers: I-reddish, axillary spikes.
   Fruits: thickly hairy capsule.
   Site: monsoon forest, secondary; DL.

45. *Mallotus fuscescens*, (T IV:67), E, 4, small tree.
   Leaves: opposite or alternate, oval-oblong, base
      acute to rounded, shortly acuminate apex.
   Trunk: young parts **glabrous**, **glandular**.
   Flowers: pale green, shortly stalked,
      crowded; I-male rufous-pubescent,
      axillary and terminal spikes;
      I-female lax racemes on
      long peduncles.
   Fruits: glabrous, somewhat
      glandular capsules.
   Site: rain forest fringes and
      gaps, secondary forest; W.

46

EUPHORBIACEAE

46. *Mallotus philippensis*, hamparilla (S)/kapila (T), (T IV:68), N, 10, tree.
    Leaves: variable, ovate to linear-lanceolate, round to acute base,
      pale beneath, base strongly **3**-veined, dotted with **minute**
      crimson glands; **small glands** also
      at the apex of pubescent petiole.
    Trunk: B-smooth; young parts
      yellow-brown pubescent.
    Flowers: sessile; I-male several erect
      terminal spikes; I-female
      stalked, short racemes.
    Fruits: capsule with rounded
      lobes, smooth, covered by
      minute, red scales.
    Site: monsoon, intermediate
      and rain forest; DL, IN, W.
    Uses: whole plant-medicinal; fruit-dye.

47. *Mallotus rhamnifolius*, marai tium (T),
    (T IV:66), N, 4, shrub to small tree.
    Leaves: opposite or alternate, oblong-oval,
      rounded base, acute apex, prominently
      3-veined beneath with **scattered**
      yellow scales.
    Trunk: B-smooth, yellowish;
      young parts **slightly yellow-**
      brown-pubescent.
    Flowers: male whitish, I-male
      spikes; female solitary.
    Fruits: finely stellate-
      pubescent, globose.
    Site: rain forest fringes
      and gaps, scrub; W.

48. *Mallotus tetracoccus*, bu kenda (S)/
    mullu polavu (T), (T IV:65 & VI:267), N, 12, tree.
    Leaves: numerous, variable, oblong-oval to rhomboidal, narrowed to
      rounded peltate base, sometimes toothed, dark
      green above, **silvery green** beneath;
      pair of **large glands** near leaf base.
    Trunk: B-smooth, yellowish grey;
      branchlets stout; young parts rough,
      brownish stellate hairs.
    Flowers: brownish; I-terminal,
      stellate-pubescent, panicles.
    Fruits: capsule with woolly spines,
      capped with black pubescent styles.
    Site: rain forest gaps and fringes,
      secondary forest, scrub; W.

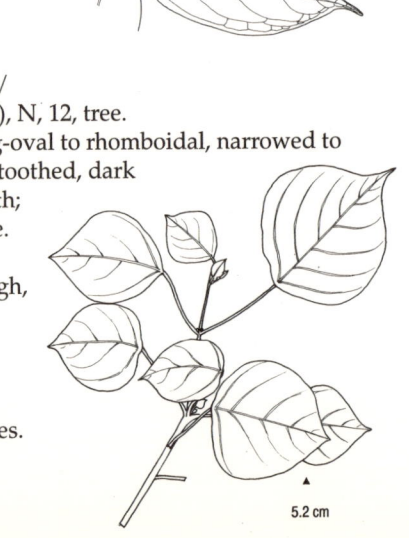

49. *Mallotus walkerae*, (T IV:66), E, 4, small tree.
   Leaves: opposite or spiral, variable, rhomboid-oblong,
      obtuse to acute base, margins **entire
      to serrate**, **minute glands** beneath.
   Trunk: young parts **glabrous,
      glandular, sticky**.
   Flowers: greenish white; female-
      solitary; I-male short spikes.
   Fruits: 3-lobed capsule with
      soft deciduous spines.
   Site: forest gaps and fringes,
      secondary forest, scrub; DL, IN, W.

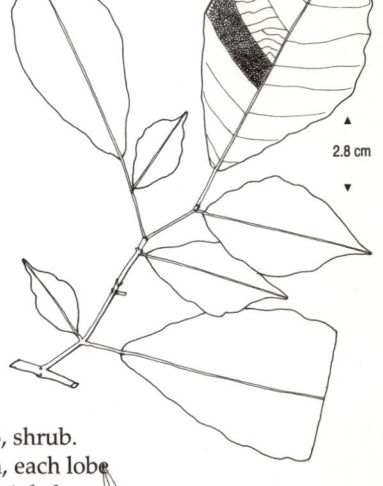

2.8 cm

50. *Manihot esculenta*, manioc (E)/
   manyokka (S), (T IV:1), I (S. Amer.), 3, shrub.
   Leaves: **palmately-lobed**, dark green, each lobe
      oblanceolate with an acute apex; petiole long.
   Trunk: stems slender, glabrous, erect, **leaf
      scars prominent**; B-silvery grey to dark
      brown streaked purple.
   Flowers: very small; I-terminal
      panicles.
   Fruits: capsule with 6
      longitudinal angled
      wings.
   Site: home gardens,
      chenas; widespread.
   Uses: roots, young
      leaves-edible.

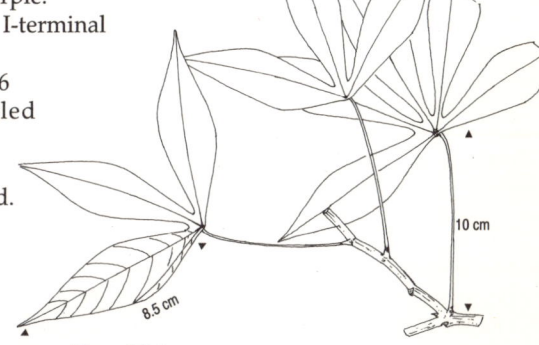

10 cm

8.5 cm

51. *Manihot glaziovii*, ceara rubber (E)/
   gas manyokka (S), I (Africa), 6, tree.
   Leaves: palmately lobed, 3-5,
      peltate base, whitish beneath.
   Trunk: large; B-purple-brown,
      papery, peeling; latex milky.
   Flowers: very small;
      I-terminal panicles.
   Fruits: globose capsule,
      without wings or ridges.
   Site: secondary forest;
      IN, W.
   Uses: shade and live fencing.

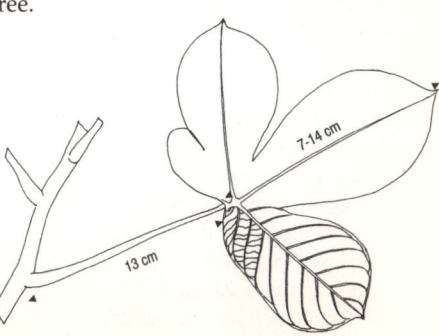

7-14 cm

13 cm

46

EUPHORBIACEAE

52. *Mischodon zeylanicus*, tammenna (S)/ tampanai (T), (T IV:38), N, 15, tree.
Leaves: numerous, **whorled in 4's**, often drooping, linear-oblong-lanceolate, narrow rounded base, **notched** apex, flush **reddish**.
Trunk: B-rough, brown; young parts finely pubescent, bluntly quadrangular.
Flowers: pale yellowish green or pinkish.
Fruits: dark green, glabrous capsule, back of 3 lobes bluntly keeled.
Site: monsoon forest subcanopy, scrub; DL.
Uses: W-construction.

5.6 cm

53. *Phyllanthus acidus*, star gooseberry (E)/ nelli (S), I (India), 3, shrub to small tree.
Leaves: **2-ranked**, ovate, acute ends.
Trunk: branches stout, leaf-scarred, branchlets resemble pinnate leaves.
Flowers: I-usually below leaves, drooping clusters.
Fruits: roundish, slightly ribbed fleshy.
Site: home gardens; DL, IN, W.
Uses: fruit-edible.

24-29 cm

2.5-3.5 cm

2 cm

54. *Phyllanthus emblica*, nelli (S)/toppinelli (T), (T IV:19), N, 10, tree.
Leaves: **2-ranked**, numerous, closely placed, overlapping, **linear to strap-shaped**, rounded base; deciduous.
Trunk: B-thin, grey; numerous bosses of leaf-bearing branchlets that resemble pinnate leaves. crown **feathery-like**.
Flowers: greenish yellow; male small, numerous; female few, nearly sessile; I-male axillary clusters.
Fruits: pale green-yellow, globose, fleshy berry.
Site: savannah and dry patana; exposed places; W.
Uses: whole plant-medicinal; W-construction; fruit-edible.

15 cm

55. *Phyllanthus indicus*, karaw (S), (T IV:27 & VI:259), N, 12, tree.
   Leaves: lanceolate-oval, acute base, subacute
   apex, whitish beneath; deciduous.
   Trunk: B-whitish; W-white, tough.
   Flowers: small; I-axillary umbellate
   clusters at base of growth.
   Fruits: green, globose, faintly
   3-lobed, smooth capsule.
   Site: rain forest understory; W.
   Uses: W-construction.

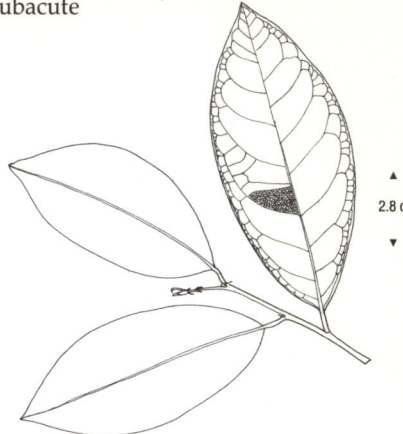

2.8 cm

56. *Phyllanthus myrtifolius*, ganga werella (S), (T IV:22), E, 2, shrub.
   Leaves: numerous **in 2-3's on suppressed**
   branchlets, lanceolate-linear.
   Trunk: stems numerous, irregular; branchlets
   resemble pinnate leaves; B-vertically fissured;
   young parts **finely pubescent**.
   Flowers: I-several together in lax clusters.
   Fruits: purplish red to greenish, small,
   depressed, slightly 3-lobed, capsule.
   Site: along water courses
   in forest; gardens; IN, W.
   Uses: hedges; ornamental.

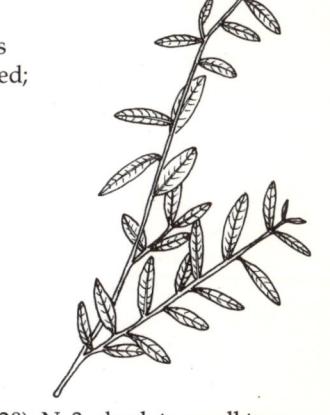

57. *Phyllanthus polyphyllus*, kuratiya (S), (T IV:20), N, 3, shrub to small tree.
   Leaves: numerous, closely spaced, often overlapping,
   linear-oblong, rounded base, apex
   with a short spiny tip; **sessile**.
   Trunk: young parts **striate;** branchlets
   resemble pinnate leaves.
   Flowers: male numerous, 2-3
   together; female larger, solitary.
   Fruits: dry, depressed, 3-lobed
   capsule, capped by
   persistent style.
   Site: scrub; DL.

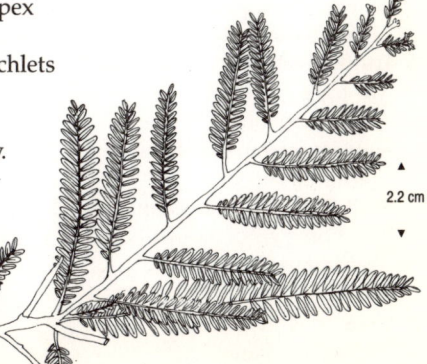

2.2 cm

58. *Phyllanthus reticulatus*, wel kayila (S)/pula (T), (T IV:19 &
     VI:259), N, 3, shrub.
   Leaves: lanceolate-oblong, obtuse or acute.
   Trunk: branches lenticellate; young
      parts **finely pubescent;** branchlets
      resemble pinnate leaves.
   Flowers: male in clusters of
      2-6; female solitary.
   Fruits: purplish black, fleshy,
      depressed globose,
      smooth, shiny, berry.
   Site: scrub; DL.
   Uses: fruit-edible; whole
      plant-medicinal;
      twigs-brushing teeth.

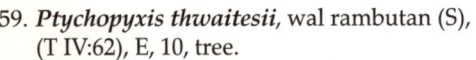

59. *Ptychopyxis thwaitesii*, wal rambutan (S),
     (T IV:62), E, 10, tree.
   Leaves: oval, acute base, shortly acuminate
      apex, **hairy on veins beneath**.
   Trunk: branchlets stout,
      yellow-brown tomentose.
   Flowers: crimson-red; I-large,
      glandular, pubescent, terminal
      panicles; plants dioecious.
   Fruits: red, globose, beaked,
      thickly set with stalked
      spherical glands.
   Site: rain forest understory; W.

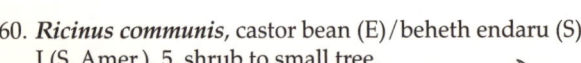

60. *Ricinus communis*, castor bean (E)/beheth endaru (S),
     I (S. Amer.), 5, shrub to small tree.
   Leaves: **large, palmately lobed, star-shaped**,
      margins **serrate**.
   Trunk: B-smooth, light brown;
      IB-light grey; twigs
      ringed at nodes.
   Flowers: green, small; I-erect
      terminal clusters.
   Fruits: 3-lobed, spiny capsule.
   Site: home gardens, scrub and
      secondary forest;
      widespread.
   Uses: ornamental; whole
      plant-medicinal;
      seeds-castor oil.

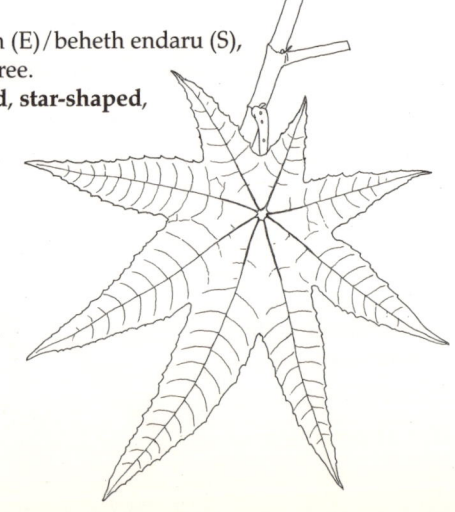

61. **Sauropus androgynus**, mella dum kola (S), (T IV:16), N, 3, small shrub.
   Leaves: ovate-lanceolate, obtuse base, acute,
     whitish beneath; **nearly sessile**.
   Trunk: branches smooth,
     green, elongated.
   Flowers: yellowish green;
     I-small drooping clusters.
   Fruits: pale white to greenish
     yellow, large, pendulous,
     globose, fleshy.
   Site: rain forest
     understory;
     secondary
     forest; LM, UW.
   Uses: leaves,
     fruit-vegetable.

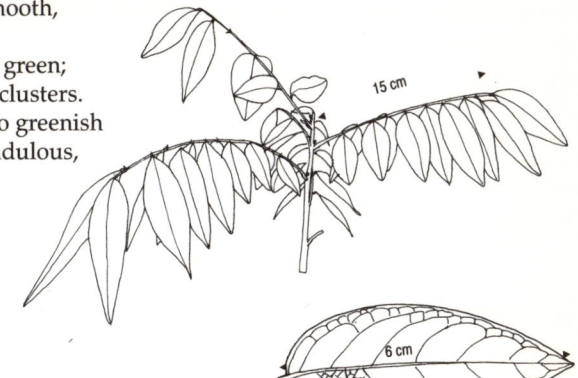

62. **Suregada lanceolata**, potpattai (T), (T IV:73), N, 5, small tree.
   Leaves: very variable, obovate to lanceolate, acute base, acuminate to
     obtuse apex, margins **sometimes spinous-serrrate**.
   Trunk: B-smooth, grey, annular scars; young parts glabrous.
   Flowers: pale yellow; I-clusters, sometimes in short racemes.
   Fruits: glabrous, rough capsule.
   Site: rain forest understory; W.

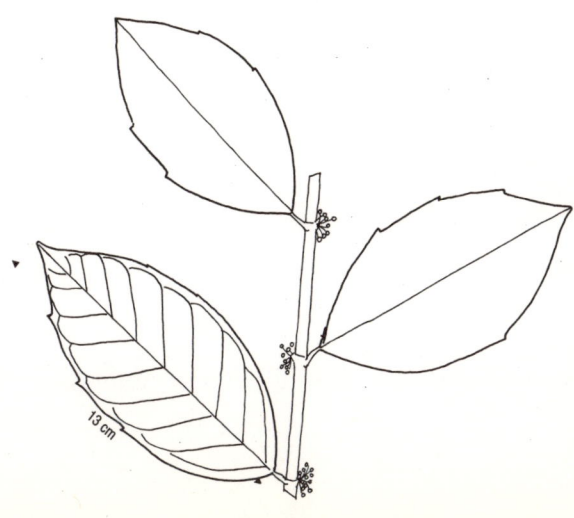

47. FLACOURTIACEAE

FAMILY DESCRIPTION - Habit: trees or shrubs. **Leaves**: spiral, rarely opposite or whorled. Simple, sometimes gland-dotted. **Stipules**: present. **Flowers**: bisexual, sometimes unisexual, (plants monoecious or dioecious), actinomorphic, cymes. **Fruits**: usually a berry, less often a loculicidal capsule or drupe. Seeds often arillate.

FLOWER PARTS - CALYX: 2-15, free, imbricate or basally united. COROLLA: 2-15 or sometimes absent, free, imbricate, not distinct from calyx. ANDROECIUM: stamens numerous, rarely few, free or in bundles. Anthers 2-locular, opening lengthwise. Staminodes sometimes present. Nectary disc present, sometimes prolonged connective. GYNOECIUM: superior, carpels 2-10, united, with 1 loculus. Placentation parietal. Placenta intruded into loculus sometimes. Styles and stigmas as many as placentas.

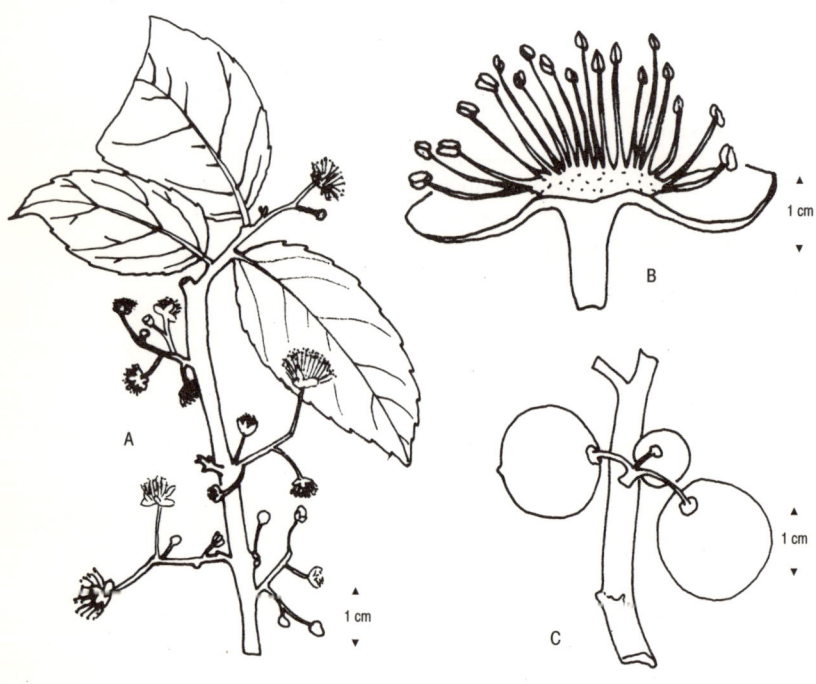

Key: inflorescence (A), half male flower (B), developing fruits (C) of *Flacourtia indica*.

1. *Flacourtia indica*, Uguressa (S)/Katakali, Mulanninchi (T),
   (T I:73 & VI:15), N, 4, shrub to small tree.
   Leaves: elliptic to obovate, obtuse apex,
   margins **crenate**; tapering base.
   Trunk: much-branched; trunk and main
   branches with **rigid branched
   spines;** younger branches
   with axillary simple spines.
   Flowers: greenish, very small,
   I-axillary racemes,
   shorter than leaves.
   Fruits: purplish, globose,
   smooth berry.
   Site: monsoon and rain
   forest understory, scrub;
   cultivated in homegardens; DL, W.
   Uses: ornamental; whole plant medicinal;
   fruit-edible.

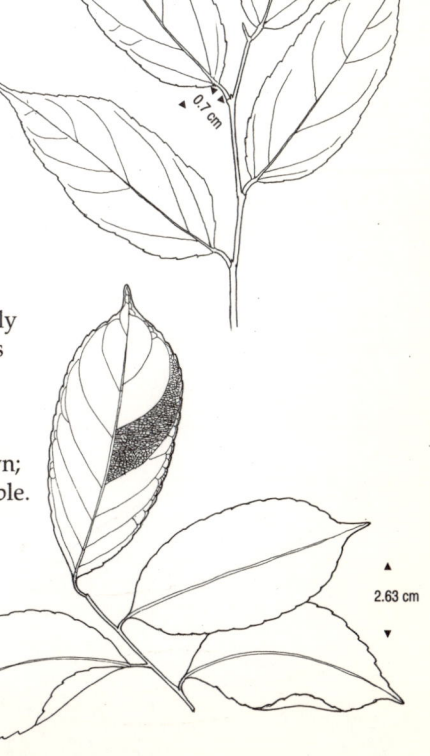

1 cm

2. *Flacourtia inermis*, lovi (S), (T I:73),
   I (Malesia), 5, small tree.
   Leaves: margins **scalloped to serrate**.
   Trunk: branches **spiny**.
   Flowers: I-axillary clusters.
   Fruits: red, pulpy, acid berry.
   Site: home gardens; widespread.
   Uses: ornamental; fruit-edible.

9.5 cm

0.7 cm

3. *Homalium zeylanicum*, liyan (S),
   (T II:239), N, 30, tree.
   Leaves: oval, narrowed base, shortly
   acuminate to acute apex, margins
   scalloped to serrrate, **purplish
   red**; reddish flush.
   Trunk: straight; B-rough, breaking
   irregularly, white; W,s-pale brown;
   h-dark brown, hard, heavy, durable.
   Flowers: greenish white, very
   numerous; I-dense corymbs
   in pendulous panicles.
   Fruits: capsule.
   Site: rain forest canopy; W.
   Uses: W-light to medium
   construction.

2.63 cm

47

FLACOURTIACEAE

4. **Hydnocarpus octandra**, wal divul (S), (T I:76), E, 20, tree.
   Leaves: oval, acute unequal base, shortly caudate, obtuse apex, glandular.
   Trunk: B-brown; young shoots **pubescent**.
   Flowers: greenish white; I-axillary clusters.
   Fruits: large, globose; pericarp thick, hard, tomentose berry.
   Site: rain forest subcanopy; W.

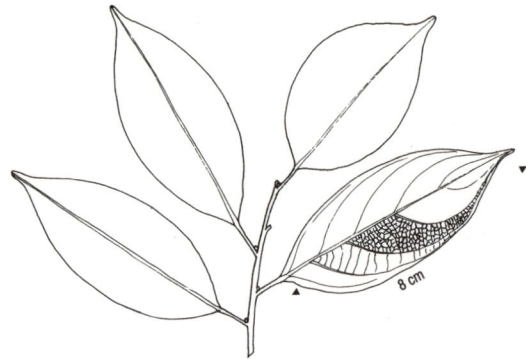

5. **Hydnocarpus venenata**, makulu (S)/ makul (T), (T I:75), E, 20, tree.
   Leaves: lanceolate, tapered base, subacute apex, margins **serrate**,
     pubescent on veins beneath.
   Trunk: B-smooth, whitish; W-hard, yellow; young shoots **pubescent**.
   Flowers: male flowers brownish white, solitary or in 2's, axillary, densely
     pubescent; I-female small, crowded, short, axillary panicles.
   Fruits: globose, tomentose berry with a blunt beak.
   Site: riverine forest canopy; along water courses in rain forest; DL, IN, W.
   Uses: fruits-fish poison; seeds-medicinal.

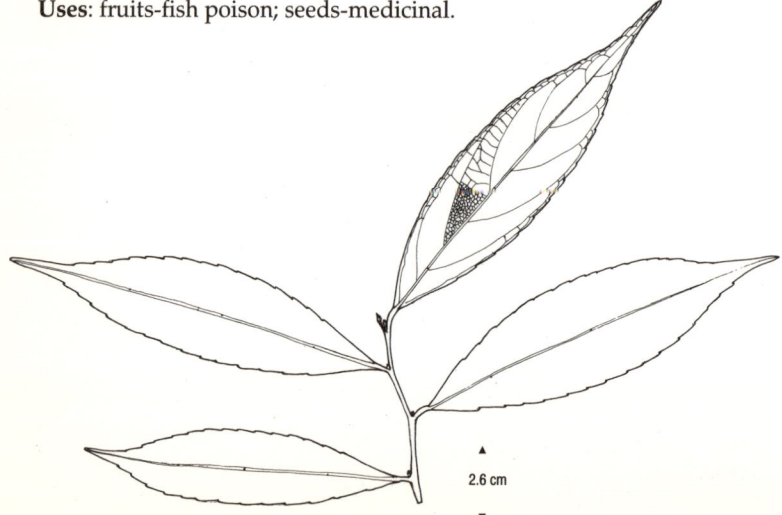

6. *Scolopia acuminata*, katu kenda (S), (T I:70), N, 10, tree.
    Leaves: oval, tapering base, very acuminate to
       obtuse apex, margins **shallowly dentate**.
    Trunk: B-smooth, rufous-grey; branches
       straight, **simple spines** ·
       when young.
    Flowers: pinkish white,
       numerous, orange
       papillae; I-panicles
       shorter than leaves.
    Fruits: green, globose
       or ovoid, fleshy berry.
    Site: monsoon, intermediate
       and rain forest understory;
       DL, IN, W.

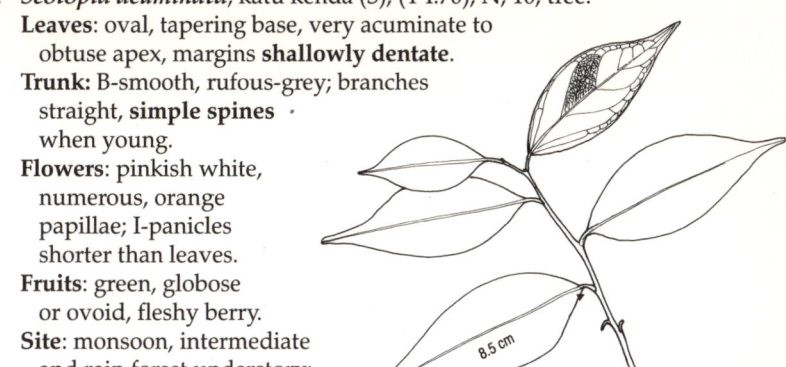

7. *Scolopia crassipes*, katu kenda (S), (T I:71), E, 10, tree.
    Leaves: lanceolate to oval, tapering base, obtuse apex,
       margins entire or slightly scalloped.
    Trunk: branchlets **spiny**.
    Flowers: white, bracts
       conspicuous; I-crowded,
       dense, axillary racemes.
    Fruits: green, fleshy,
       subglobose, pointed
       berry.
    Site: montane forest
       canopy; LM, M.

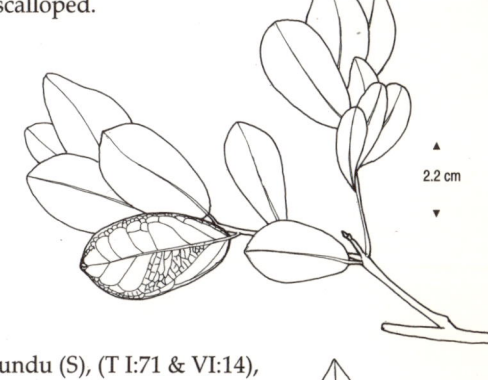

8. *Scolopia schreberi*, katu kurundu (S), (T I:71 & VI:14),
    E, 10, tree.
    Leaves: very variable, oblong-ovate,
       cordate to rounded base, acute to
       obtuse apex, margins **faintly
       scalloped**.
    Trunk: B-grey; Branches slender,
       compound axillary spines.
    Flowers: white; I-simple
       racemes shorter
       than leaves.
    Fruits: bright red,
       fleshy, ovoid berry.
    Site: monsoon,
       intermediate, and
       rain forest understory;
       DL, IN, W.

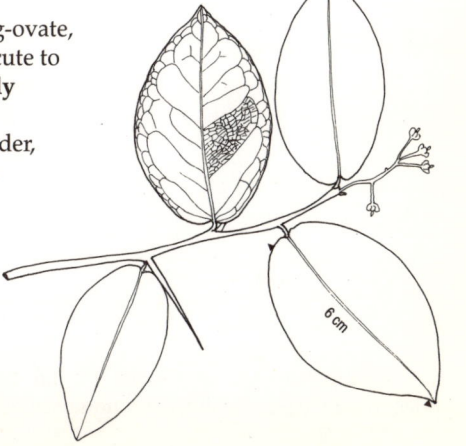

48. GOODENIACEAE

FAMILY DESCRIPTION - **Habit**: shrubs or herbs, rarely trees. Some poisonous. **Leaves**: spiral, rarely opposite or whorled, simple. **Stipules**: absent. **Flowers**: bisexual, actinomorphic or zygomorphic, solitary or in heads, racemes or cymes. **Fruits**: usually a capsule, sometimes a drupe or nut. Seeds usually flat, sometimes winged.

FLOWER PARTS - CALYX: 5, basally united and adnate to ovary, rarely free. COROLLA: 5, united,1-2-lipped, lobes valvate. ANDROECIUM: 5, epipetalous or free. Filaments united basally, surrounding style. Anthers opening lengthwise, connivent or united, forming a cylinder. GYNOECIUM: inferior, 1-4-locular with 1 to numerous ovules per loculus. Placentation axile. Style simple or 2-3-fid. Stigma indusiate.

Key: Inflorescence (A), full flower (B), half flower (C), longitudinal section of gynoecium (D) and transverse section of ovary (E) of *Scaevola sericea*.

1. *Scaevola sericea*, takkada (S), (T III:54 & VI:174), N, 3, shrub.
    Leaves: Numerous, large, obovate, tapering base, rounded apex,
       margins **slightly dentate; sessile.**
    Trunk: branches very stout, shiny, pale green.
    Flowers: white to green-yellow, large;
       I-short, flat, dichotomous, axillary cymes.
    Fruits: white, oblong to globose,
       somewhat lobed drupe, succulent
       with a persistent calyx.
    Site: coastal secondary forest,
       scrub; MC.

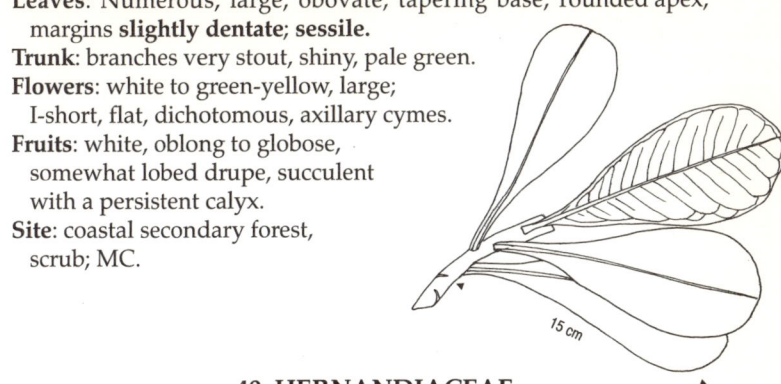

15 cm

49. HERNANDIACEAE

FAMILY DESCRIPTION - Habit: trees, shrubs or lianas. **Leaves**: spiral, simple
or palmately compound. **Stipules**: absent. **Flowers**: bisexual or unisexual, (plants
monoecious or rarely dioecious), actinomorphic, cymes. **Fruits**: usually an
achene, incompletely enveloped by accrescent involucre derived from 2-3 united
bracteoles or sometimes a 2-4- winged samara.

FLOWER PARTS - PERIANTH: sometimes calyx and corolla are
indistinguishable. 2-10 subequal segments in one or two whorls. ANDROECIUM:
stamens 3-5, often 4. Anthers 2-locular, dehiscing longitudinally. Staminodes
often present. GYNOECIUM: inferior with one carpel bearing one pendulous ovule.
Style elongate.

49

HERNANDIACEAE

A

B

1 mm

C   1 mm

D   1 mm

1 cm

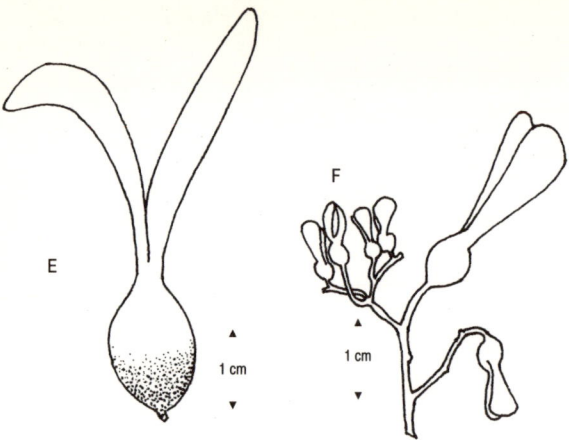

Key: Inflorescence (A), full bisexual flower (B), full male flower (C), ovary in longitudinal section (D), cluster of fruits (E) single fruit (F) of *Gyrocarpus americanus*.

1. **Gyrocarpus americanus**, hima (S)/tanakku (T), (DF VI:118), N, 12, tree.
   Leaves: spiral, terminally crowded, very broadly ovoid-rhomboid, acuminate-acute apex, **3** main veins, dense soft **pubescent** below.
   Trunk: B-smooth, **shiny, greenish white**; W-light, soft, greyish white; branches stout, leaf-scarred; young parts puberulous.
   Flowers: cream-coloured, small; male numerous; female few; I-globular clusters in much branched cymes.
   Fruits: ovoid, furrowed apex, wrinkled below, linear-spathulate wings.
   Site: scrub, rocky outcrops; DL, IN.
   Uses: W-boat building.

50. ICACINACEAE.

FAMILY DESCRIPTION - Habit: trees, shrubs or lianas. **Leaves**: spiral, simple, entire or toothed. **Stipules**: absent. **Flowers**: bisexual, rarely unisexual, (plants dioecious), actinomorphic, panicles. **Fruits**: often 1-seeded drupe, rarely a samara.

FLOWER PARTS - CALYX: 4-5, tubular, lobes imbricate or rarely valvate. COROLLA: 4-5, free or united, valvate, rarely absent. ANDROECIUM: stamens same number as petals. Filaments free or epipetalous, often hairy below the anthers. Anthers 2-locular, sometimes deeply 4-lobed, dehiscing longitudinally. GYNOECIUM: superior, 1-locular, rarely 3-5-locular. Ovule solitary, pendulous. Style usually short.

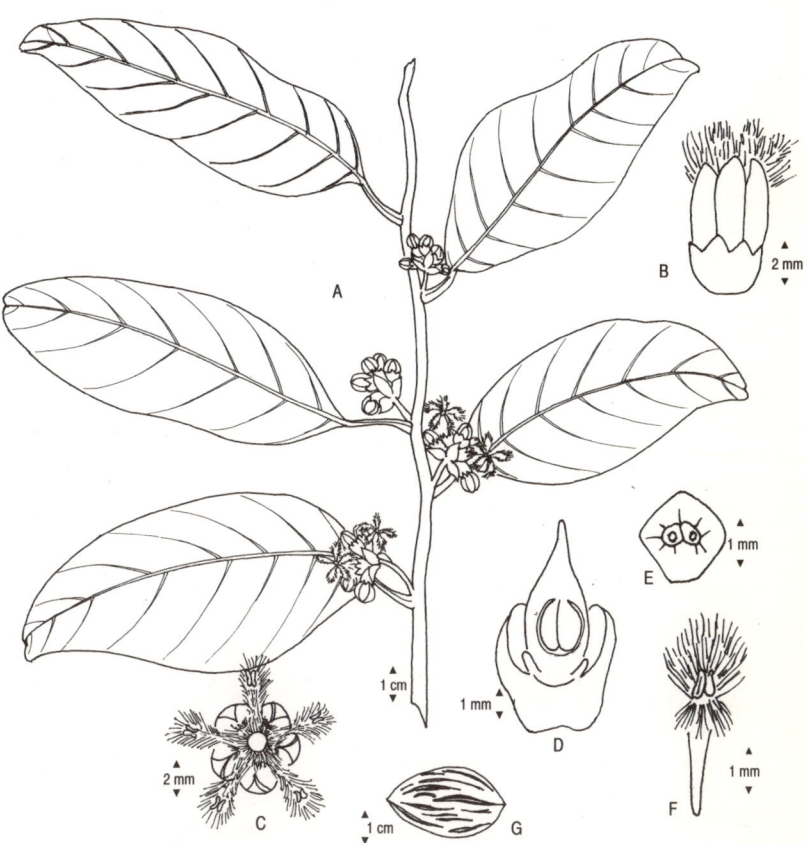

Key: Inflorescence-bearing twig (A), full flower in side view (B) and from above (C), ovary in longitudinal (D) transverse (E) sections, a single stamen (F) and seed (G) of *Stemonurus apicalis*.

1. *Apodytes dimidiata*, (T I:262), E, 10, tree.
   Leaves: variable, broadly oval-lanceolate to lanceolate, tapering base, rounded to acuminate apex, margins **entire, revolute**.
   Trunk: B-smooth, grey; young parts **puberulent**.
   Flowers: white; I-small, lax, terminal panicles.
   Fruits: black, lop-sided, compressed, shiny drupe with persistent style.
   Site: montane forest subcanopy; M.

2. *Gomphandra coriacea*, (T I:261 & VI:48), E, 10, tree.
   Leaves: variable, broadly oval to narrowly oblong-lanceolate, coriaceous; long petioles.
   Trunk: no description.
   Flowers: I-larger than *S. tetrandrus*, panicles sessile, extra-axillary.
   Fruits: drupe.
   Site: montane forest understory; M.

3. *Gomphandra tetrandra*, (T I:261 & VI:48), N, 7, shrub to small tree.
   Leaves: variable, broadly oval to narrowly oblong-lanceolate, tapering base, **acuminate to caudate** to obtuse apex.
   Trunk: young parts **puberulous**.
   Flowers: pale green, very small; I-usually in 3's, axillary, paniculate clusters.
   Fruits: white, oblong, blunt ended drupe.
   Site: rain forest understory; W.

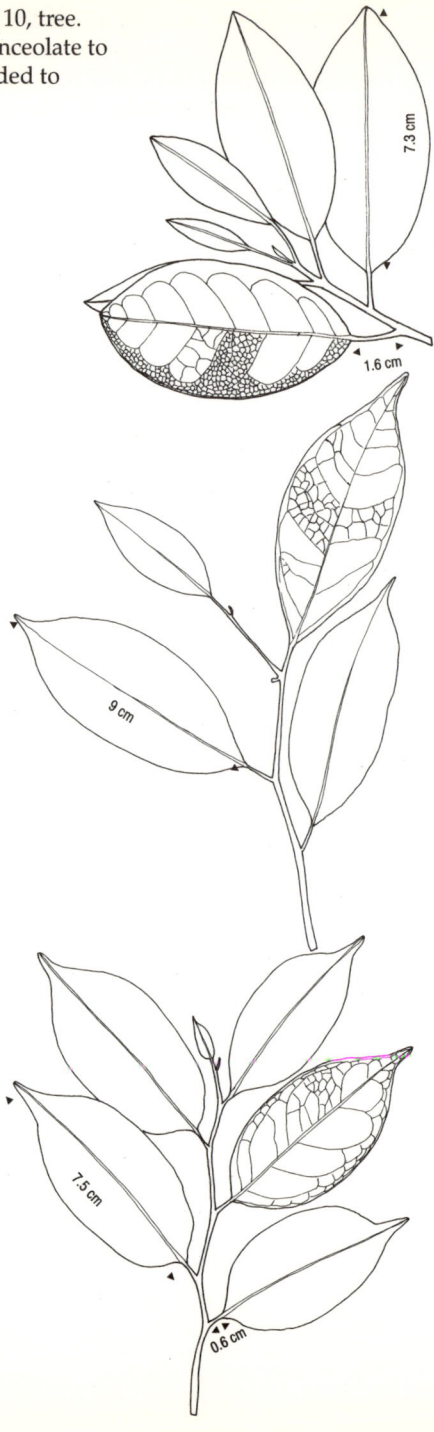

7.3 cm

1.6 cm

9 cm

7.5 cm

0.6 cm

4. **Stemonurus apicalis**, uru honda (S), (T I:260), N, 20, tree.
   Leaves: lanceolate-oblong, acute base,
       shortly acuminate apex, margins
       entire; flush **coppery red**.
   Trunk: straight; B-smooth; lower
       branches **drooping**.
   Flowers: purplish green;
       I-sessile, heads sur-
       rounded by 4 bracts.
   Fruits: lower half purplish
       green, upper half white,
       ovoid-oblong, smooth drupe
       with persistent calyx.
   Site: rain forest subcanopy; W.
   Uses: W-construction timber.

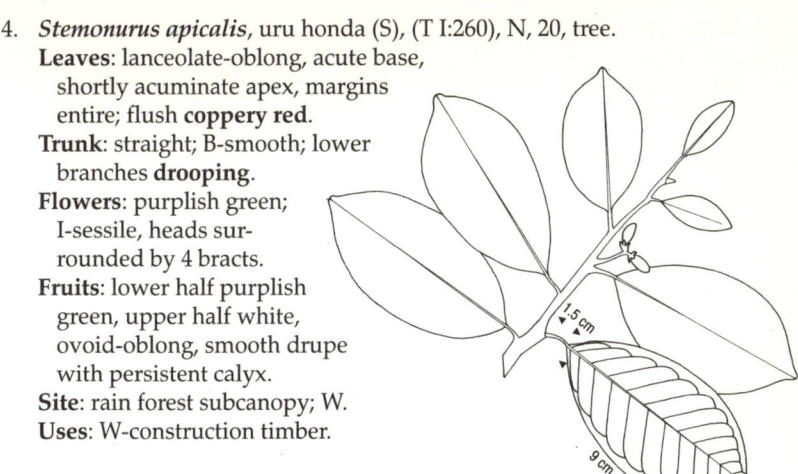

51. LAURACEAE

FAMILY DESCRIPTION - Habit: trees, shrubs or semi-parasites. **Leaves**:
spiral, rarely whorled or opposite, usually entire. **Stipules**: absent. **Flowers**: small,
greenish or yellowish, bisexual or unisexual (plants monoecious or dioecious),
actinomorphic, cymes or racemes. **Fruits**: 1-seeded berry or drupe, often
enclosed in persistent cup.

FLOWER PARTS - PERIANTH: calyx and corolla often similar. Usually 6, basally
united. The tube usually persists as a cupule at the base of fruit. ANDROECIUM:
typically in 4 whorls, each with 3 stamens. 4th whorl reduced to staminodes.
Filaments sometimes glandular at base. Anthers open by valves from base
upwards. GYNOECIUM: superior, rarely inferior. 1-carpellary and 1-locular. Ovule
solitary and pendulous.

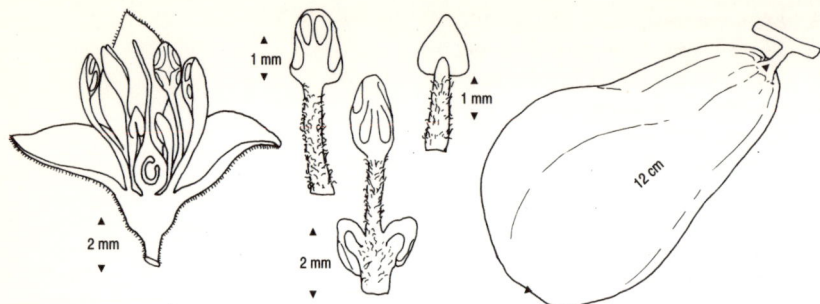

Key: twig bearing inflorescence (A), full flower (B), half flower (C), different types of stamens (D, E, F) and fruit (G) of *Persea americana*.

1. *Actinodaphne albifrons*, (T III:448), E, 12, tree.
    Leaves: **opposite or in 3**'s, large, elliptic, acute base,
        acuminate apex, **3** veins at base, 4 pairs of
        arched laterals, beneath **tomentose white**;
        flush covered with long white hairs.
    Trunk: B-thick, smooth, grey to
        light brown; branchlets
        minutely tomentose.
    Flowers: I-yellowish brown,
        axillary clusters, tomentose.
    Fruits: ovoid drupe with
        shallowly cup-shaped
        persistent perianth.
    Site: montane and rain forest
        subcanopy; M, LM, UW.

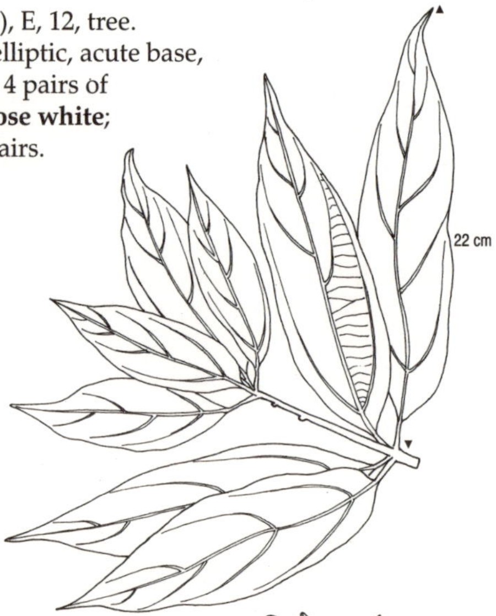

2. *Actinodaphne molochina*, (T III:445), E, 9, tree.
    Leaves: **4-6 in a whorl**, variable, usually small,
        broadly obovate-oval, acute ends, beneath **rufous
        tomentose**, stiff, lateral veins arched.
    Trunk: B-whitish; branchlets stout;
        young parts **rufous tomentose**.
    Flowers: brownish yellow; I-small,
        axillary clusters or umbels.
    Fruits: globose, pointed drupe;
        persistent perianth.
    Site: montane forest understory; M.

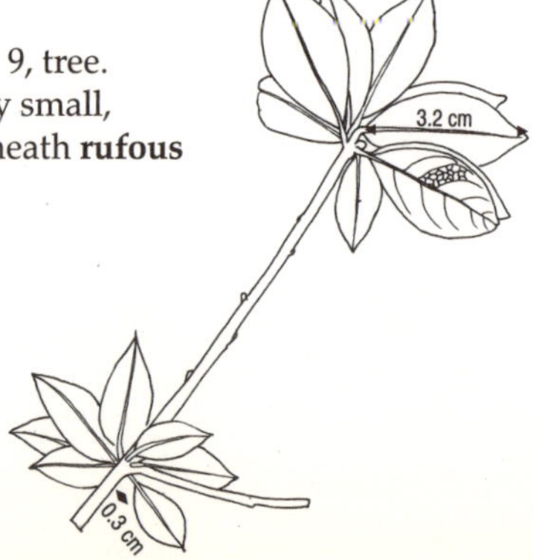

3. *Actinodaphne speciosa*, elephant ears (E)/
   pol katu (S), (T III:448), E, 12, tree.
   Leaves: opposite or in 3's, large
       oval-rotund, acute to obtuse apex,
       somewhat **3**-veined at base with
       arched laterals, beneath red-brown
       woolly; young leaves velvety.
   Trunk: slightly branched; B-thick,
       smooth, grey; W-yellowish, heavy,
       smooth; branchlets densely tomentose.
   Flowers: yellowish brown;
       I-axillary clusters, tomentose.
   Fruits: ovoid drupe with a shallowly
       cup-shaped, persistent perianth.
   Site: montane forest subcanopy; M.

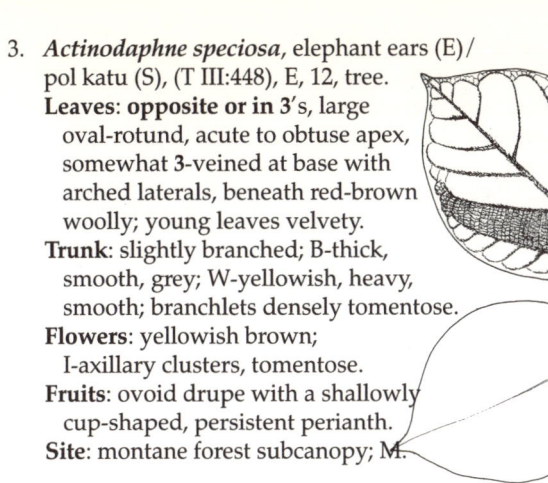

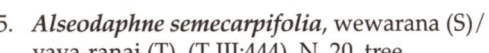

4. *Actinodaphne stenophylla*, (T III:446), E, 10, tree.
   Leaves: whorled in 4-7's, linear-oblong, acute base,
       obtuse apex, beneath **greyish white**.
   Trunk: B-yellowish grey; young
       parts **pubescent**.
   Flowers: pale yellow, bracts silky;
       I-crowded, small axillary clusters.
   Fruits: purple, globose drupe; shallowly
       cup-shaped persistent perianth.
   Site: dry patana grassland,
       intermediate and montane
       forest subcanopy and
       understory; IN, LM, W.

5. *Alseodaphne semecarpifolia*, wewarana (S)/
   yava-ranai (T), (T III:444), N, 20, tree.
   Leaves: whorled, often crowded, oblong-obovate, tapering
       base, acute or rounded apex, whitish beneath.
   Trunk: much-branched; B-thick, vertically furrowed, yellowish brown;
       W,h-pale greyish orange, heavy, durable; branchlets **whorled**.
   Flowers: yellowish green;
       I-lax axillary panicles
       longer than leaves.
   Fruits: oblong-ovoid,
       blunt, smooth drupe.
   Site: monsoon forest
       canopy, scrub; DL, IN.
   Uses: W-heavy
       construction,
       resistant to termites.

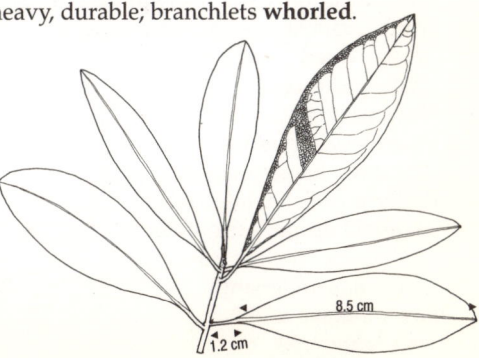

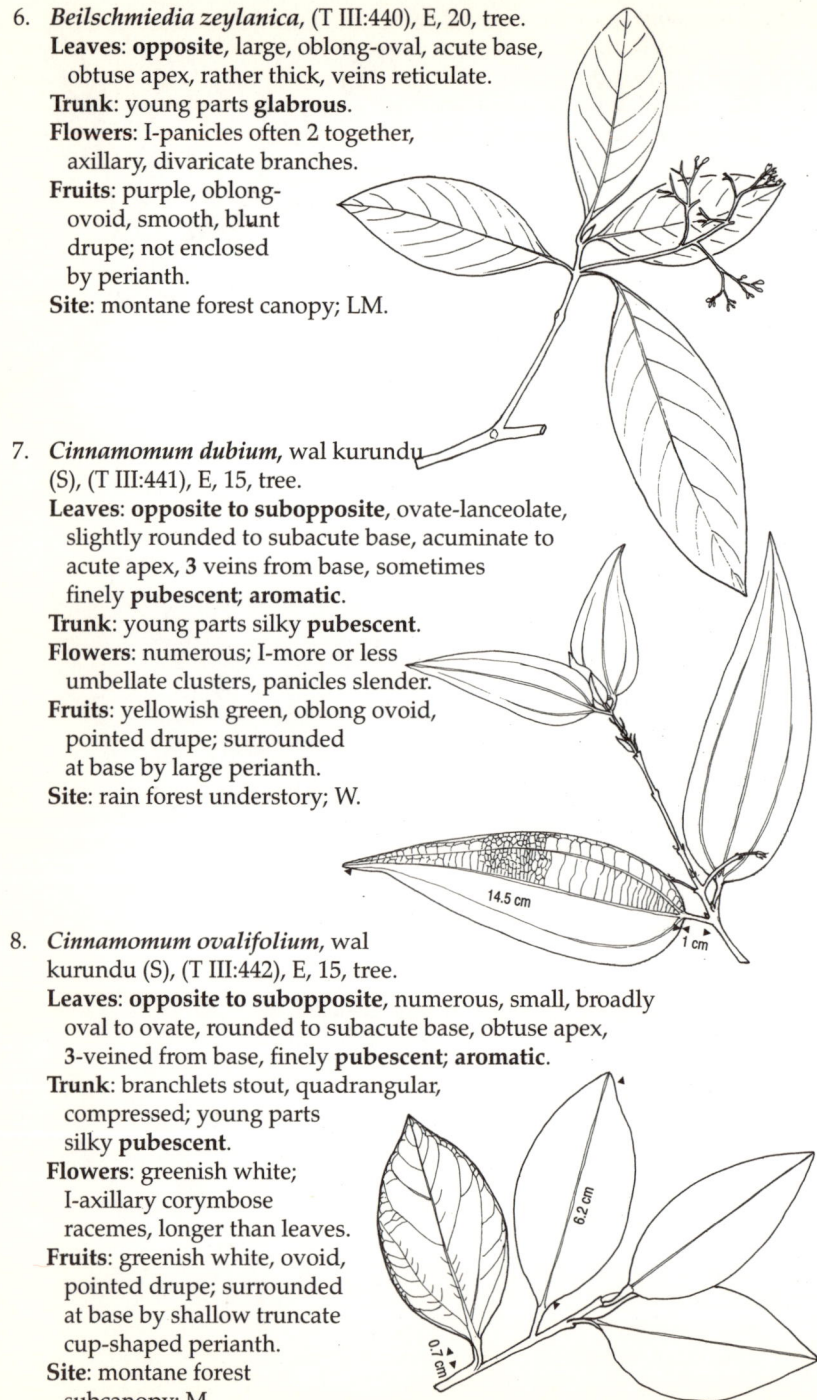

6. *Beilschmiedia zeylanica*, (T III:440), E, 20, tree.
    Leaves: **opposite**, large, oblong-oval, acute base,
      obtuse apex, rather thick, veins reticulate.
    Trunk: young parts **glabrous**.
    Flowers: I-panicles often 2 together,
      axillary, divaricate branches.
    Fruits: purple, oblong-
      ovoid, smooth, blunt
      drupe; not enclosed
      by perianth.
    Site: montane forest canopy; LM.

7. *Cinnamomum dubium*, wal kurundu
    (S), (T III:441), E, 15, tree.
    Leaves: **opposite to subopposite**, ovate-lanceolate,
      slightly rounded to subacute base, acuminate to
      acute apex, **3 veins from base**, sometimes
      finely **pubescent; aromatic**.
    Trunk: young parts silky **pubescent**.
    Flowers: numerous; I-more or less
      umbellate clusters, panicles slender.
    Fruits: yellowish green, oblong ovoid,
      pointed drupe; surrounded
      at base by large perianth.
    Site: rain forest understory; W.

14.5 cm

1 cm

8. *Cinnamomum ovalifolium*, wal
    kurundu (S), (T III:442), E, 15, tree.
    Leaves: **opposite to subopposite**, numerous, small, broadly
      oval to ovate, rounded to subacute base, obtuse apex,
      3-veined from base, finely **pubescent; aromatic**.
    Trunk: branchlets stout, quadrangular,
      compressed; young parts
      silky **pubescent**.
    Flowers: greenish white;
      I-axillary corymbose
      racemes, longer than leaves.
    Fruits: greenish white, ovoid,
      pointed drupe; surrounded
      at base by shallow truncate
      cup-shaped perianth.
    Site: montane forest
      subcanopy; M.

6.2 cm

0.7 cm

9. **Cinnamomum verum**, kurundu (S)/ karuva (T), (T III:440), N, 20, tree.
   Leaves: **opposite to subopposite**, variable, oval to lanceolate, subacute
   base, shortly acuminate to obtuse apex, **3-5** veins from base with fine
   reticulate tertiary veins; **aromatic**; flush **bright pink**.
   Trunk: B-rather thick, reddish; twigs
   often compressed.
   Flowers: pale yellow; I-numerous
   lax panicles, longer than leaf.
   Fruits: dark purple, oblong-ovoid,
   minutely pointed, dry to
   slightly fleshy drupe.
   Site: rain forest subcanopy,
   home gardens, plantations; W.
   Uses: B-cinnamon; Leaves-cinnamon
   oil; whole plant-medicinal.

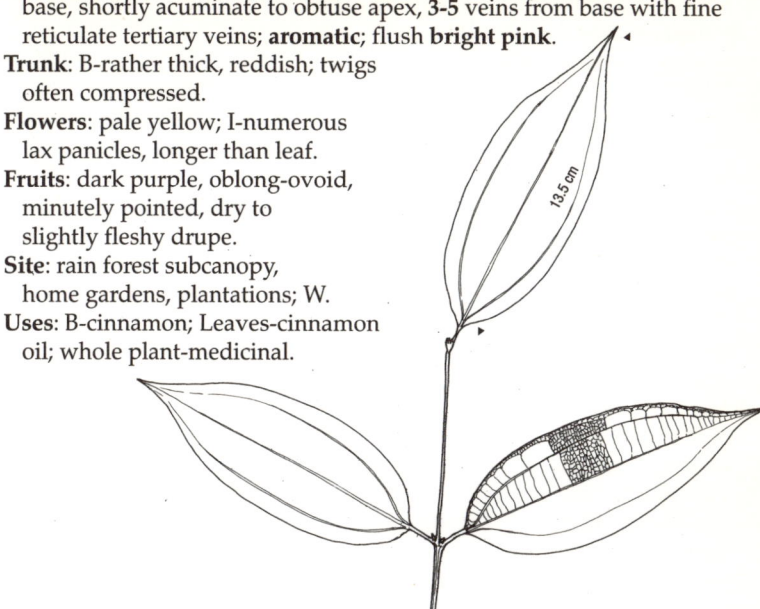

51

10. **Cryptocarya wightiana**, gulu mora (S), (T III:438), N, 30, tree.
    Leaves: alternate, oblong-oval to lanceolate, rounded
    to subacute base, acute apex, finely pubescent,
    beneath whitish, coriaceous, stiff.
    Trunk: W-pale brown, yellow, hard, heavy;
    young parts **rusty-pubescent**.
    Flowers: numerous; I-spreading, axillary,
    pubescent panicles, shorter than leaves.
    Fruits: purplish black, globose,
    smooth, shining drupe.
    Site: rain forest subcanopy; LM, W.
    Uses: W-rafters, construction.

LAURACEAE

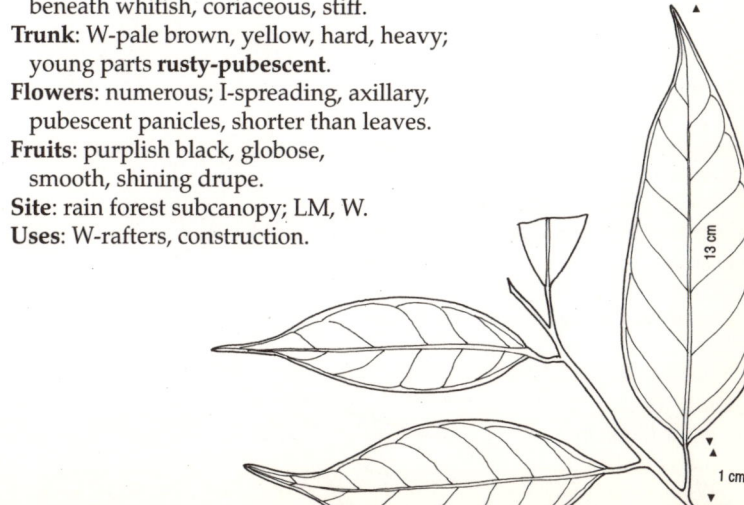

11. *Litsea gardneri*, talan (S), (T III:453), E, 30, tree.
    Leaves: alternate, large, oval-oblong, acute base,
        obtuse apex, beneath minutely **white
        tomentose**, many arched lateral
        veins; flush pale violet.
    Trunk: B-leaf-scarred, grey;
        young parts **pubescent**.
    Flowers: pale yellow, few, sessile;
        I-umbels, in axillary racemes
        shorter than leaves.
    Fruits: ovoid drupe; shallowly
        cup-shaped perianth.
    Site: montane
        and rain forest
        canopy;
        LM, UW.
    Uses: construction.

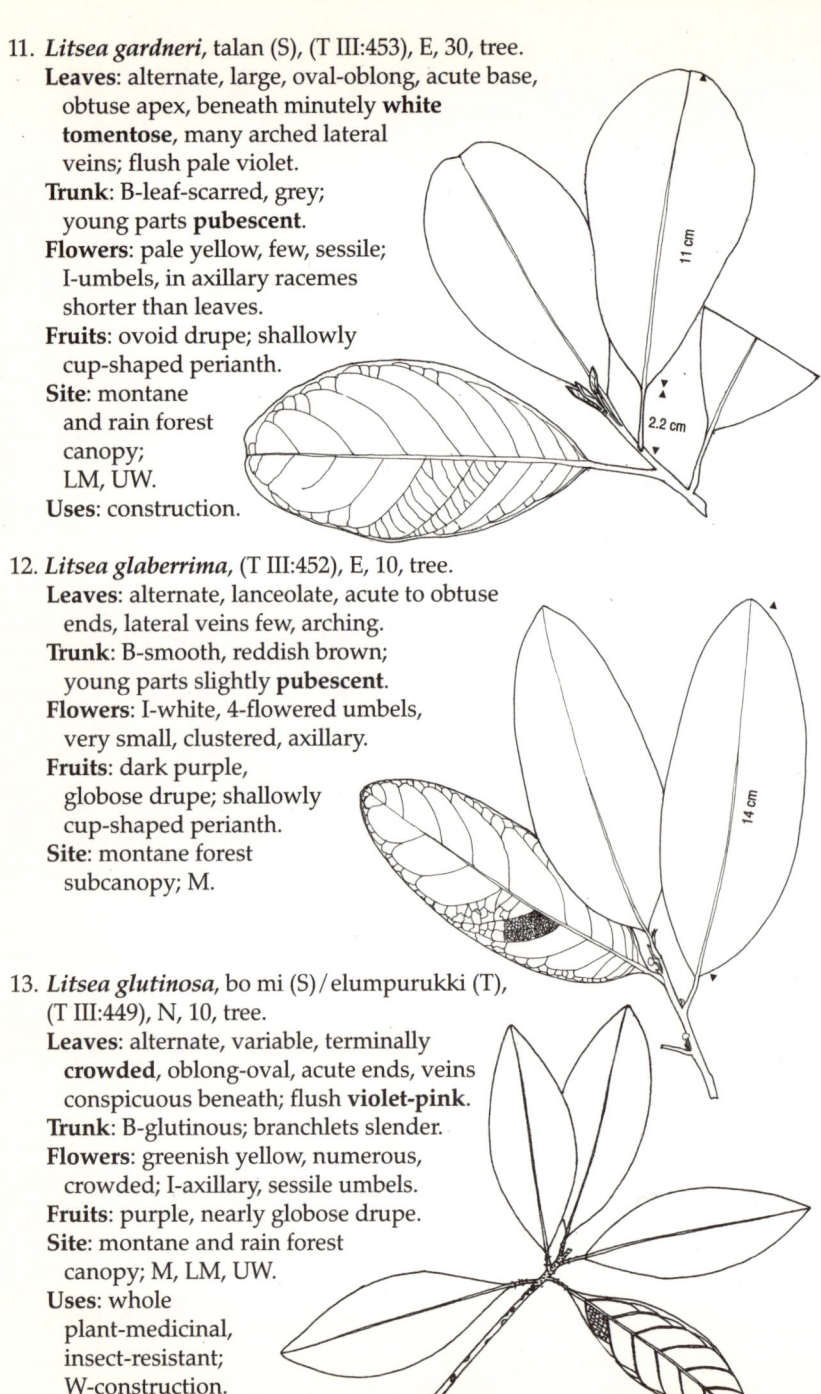

12. *Litsea glaberrima*, (T III:452), E, 10, tree.
    Leaves: alternate, lanceolate, acute to obtuse
        ends, lateral veins few, arching.
    Trunk: B-smooth, reddish brown;
        young parts slightly **pubescent**.
    Flowers: I-white, 4-flowered umbels,
        very small, clustered, axillary.
    Fruits: dark purple,
        globose drupe; shallowly
        cup-shaped perianth.
    Site: montane forest
        subcanopy; M.

13. *Litsea glutinosa*, bo mi (S)/elumpurukki (T),
    (T III:449), N, 10, tree.
    Leaves: alternate, variable, terminally
        crowded, oblong-oval, acute ends, veins
        conspicuous beneath; flush **violet-pink**.
    Trunk: B-glutinous; branchlets slender.
    Flowers: greenish yellow, numerous,
        crowded; I-axillary, sessile umbels.
    Fruits: purple, nearly globose drupe.
    Site: montane and rain forest
        canopy; M, LM, UW.
    Uses: whole
        plant-medicinal,
        insect-resistant;
        W-construction.

14. *Litsea iteodaphne*, (T III:452), E, 10, shrub to tree.
    Leaves: alternate, variable, narrowly
        oval-lanceolate, rounded to acute
        base, subacute apex, veins
        inconspicuous, beneath paler.
    Trunk: B-rather rough; young
        parts **finely silky hairy**.
    Flowers: greenish white,
        in 4's, nearly sessile; I-small
        umbels, solitary, or
        clustered, axillary.
    Fruits: purplish red,
        ovoid drupe; cup-
        shaped perianth.
    Site: montane and rain
        forest subcanopy and
        understory; M, W.

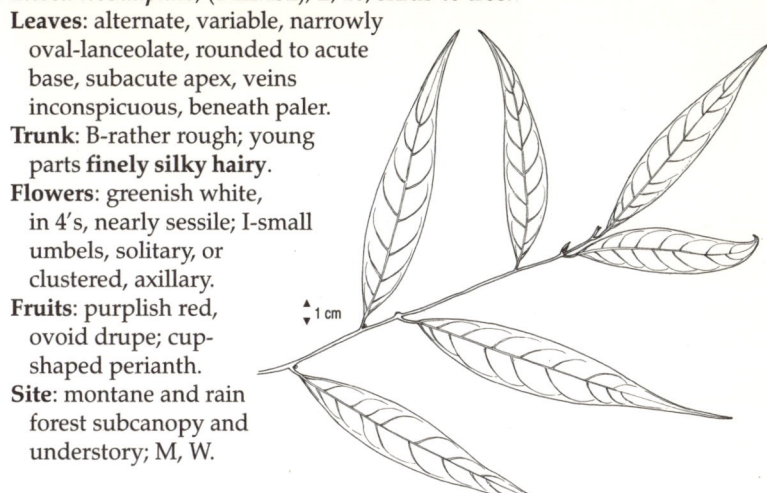

51

15. *Litsea longifolia*, rat keliya (S),
    (T III:450 & VI:248), E, 3, small tree.
    Leaves: alternate, oval, acute to rounded
        base, beneath **rufous pubescent**.
    Trunk: B-rough, grey; young
        parts **brown tomentose**.
    Flowers: I-umbels small, clustered
        in axils of fallen leaves,
        short peduncles.
    Fruits: globose drupe; on
        shallow cup-shaped
        perianth.
    Site: montane and rain
        forest understory
        and fringes; LM, W.
    Uses: B-medicinal.

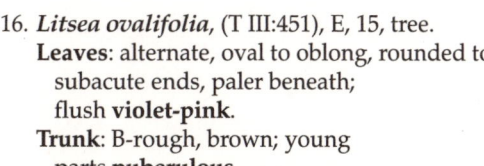

LAURACEAE

16. *Litsea ovalifolia*, (T III:451), E, 15, tree.
    Leaves: alternate, oval to oblong, rounded to
        subacute ends, paler beneath;
        flush **violet-pink**.
    Trunk: B-rough, brown; young
        parts **puberulous**.
    Flowers: pale yellow; I-small, 4-flowered
        umbels, axillary, clustered.
    Fruits: purple, ovoid, or globose drupe.
    Site: montane forest canopy; M.

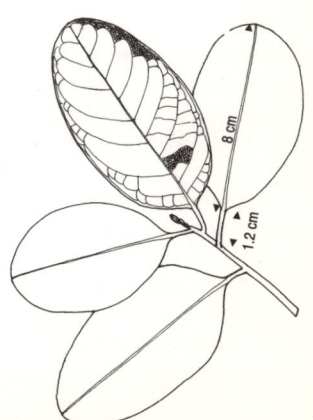

17. *Litsea quinqueflora*, lena idda (S), (T III:449), N, 10, tree.
Leaves: alternate, numerous, terminally **crowded**, large, oval, acute ends, beneath finely **yellowish tomentose**.
Trunk: thin, grey; W-hard, heavy, close-grained; branchlets stout, **yellow tomentose**.
Flowers: pale yellow, numerous; I-umbels, rather large, on drooping peduncles.
Fruits: depressed-globose drupe; on shallow cup-shaped perianth.
Site: montane and rain forest canopy and subcanopy; LM, W.
Uses: W-construction.

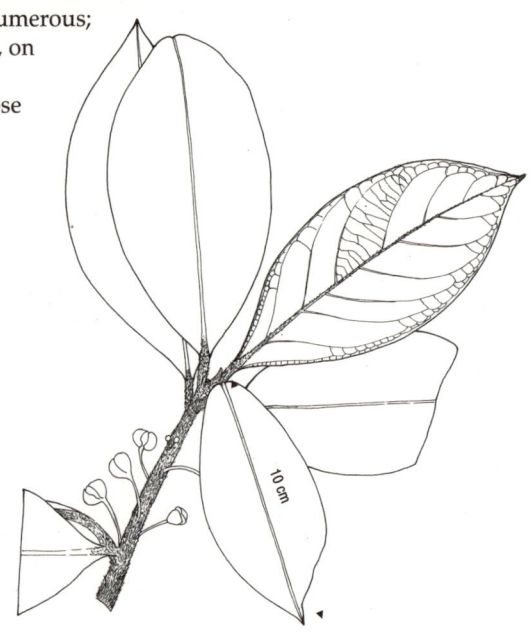

18. *Neolitsea cassia*, dawul kurundu (S), (T III:454), E, 15, small tree.
Leaves: alternate, numerous, **terminally crowded**, lanceolate, tapering ends, margins **undulate**, 3 veins at base.
Trunk: B-thick, smooth, grey; W-light, hard, pale orange.
Flowers: I-4 to 5-flowered umbels, small, clustered.
Fruits: dark purple, ovoid, pointed or globose drupe; shallow cup-shaped perianth.
Site: montane and rain forest understory; M, LM, UW.
Uses: W-panelling; leaves-mucilaginous extract used in preparation of local sweet (asmi); B, leaves-medicinal.

19. *Neolitsea fuscata*, kudu dawula (S), (T III:453), E, 15, tree.
   Leaves: alternate, lanceolate-oval, acute base,
      subacute apex, **3** veins at base, beneath
      finely **white tomentose**.
   Trunk: B-smooth; young parts
      finely **rufous tomentose**.
   Flowers: yellowish; I-small umbels,
      nearly sessile, axillary clusters.
   Fruits: globose drupe, on large,
      thickened, flat perianth.
   Site: montane forest
      canopy; M.

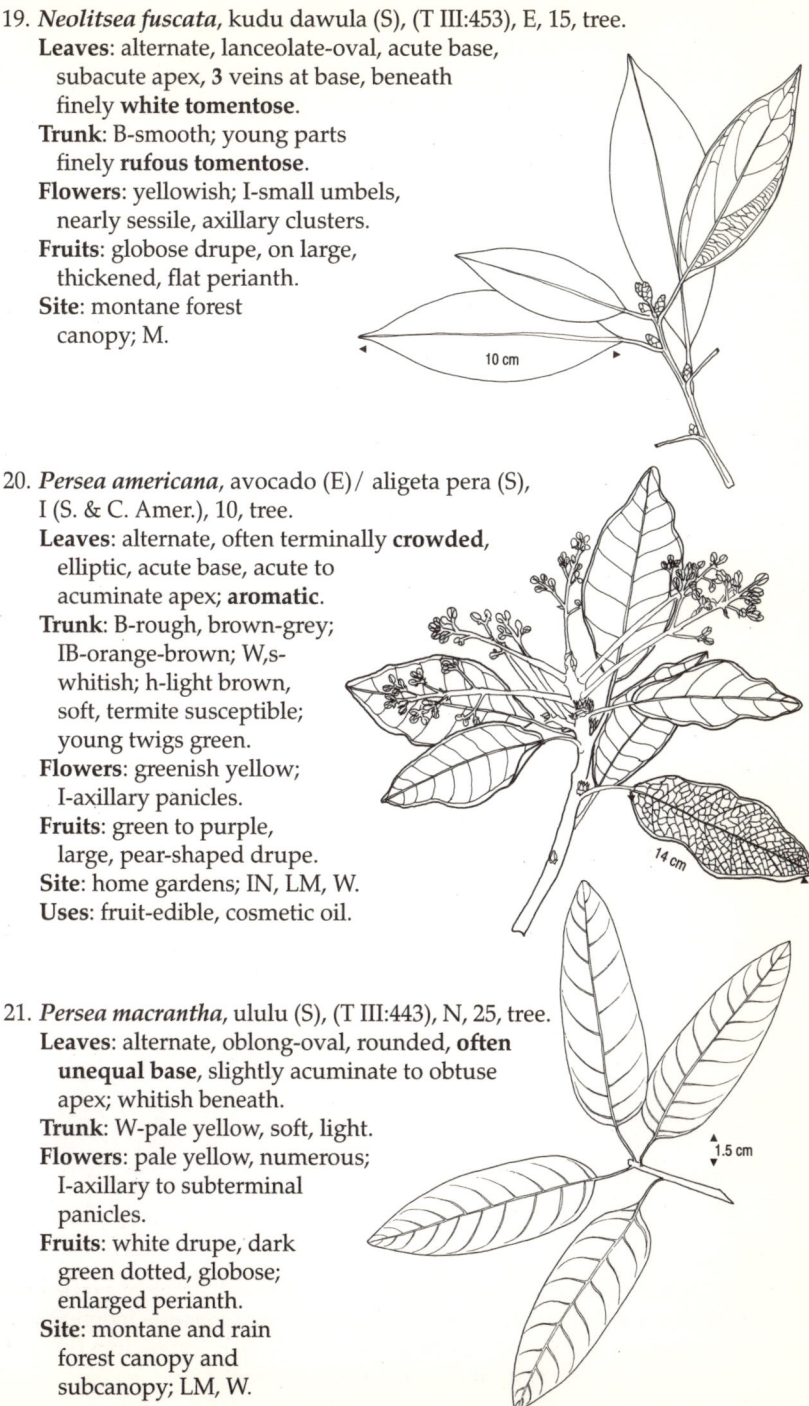

10 cm

20. *Persea americana*, avocado (E)/ aligeta pera (S),
   I (S. & C. Amer.), 10, tree.
   Leaves: alternate, often terminally **crowded**,
      elliptic, acute base, acute to
      acuminate apex; **aromatic**.
   Trunk: B-rough, brown-grey;
      IB-orange-brown; W,s-
      whitish; h-light brown,
      soft, termite susceptible;
      young twigs green.
   Flowers: greenish yellow;
      I-axillary panicles.
   Fruits: green to purple,
      large, pear-shaped drupe.
   Site: home gardens; IN, LM, W.
   Uses: fruit-edible, cosmetic oil.

14 cm

51

LAURACEAE

21. *Persea macrantha*, ululu (S), (T III:443), N, 25, tree.
   Leaves: alternate, oblong-oval, rounded, **often
      unequal base**, slightly acuminate to obtuse
      apex; whitish beneath.
   Trunk: W-pale yellow, soft, light.
   Flowers: pale yellow, numerous;
      I-axillary to subterminal
      panicles.
   Fruits: white drupe, dark
      green dotted, globose;
      enlarged perianth.
   Site: montane and rain
      forest canopy and
      subcanopy; LM, W.

1.5 cm

52. LECYTHIDACEAE

FAMILY DESCRIPTION - Habit: trees or shrubs. **Leaves**: spiral, simple, entire or toothed. **Stipules**: small and caducous, or absent. **Flowers**: bisexual, actinomorphic or zygomorphic, solitary or in panicles, fascicles or racemes sometimes arising from old wood. **Fruits**: capsule with distal operculum, often very large, or a drupe or berry. Seeds often nut-like, winged or often with funicular aril.

FLOWER PARTS - CALYX: 4-6, united, lobes valvate or slightly imbricate. COROLLA: 4-6, free or united into a campanulate tube. ANDROECIUM: numerous stamens in whorls, basally united on a ring. This ring may be symmetrical or asymmetrical and extended on one side into a flat ligule, sometimes curved over the ovary as a hood. Sometimes outer stamens modified into staminodes and united forming a corona. Anthers basifixed or rarely adnate, opening by a longitudinal slit. GYNOECIUM: inferior, 2-6-locular. Ovules 1 to many per loculus. Placentation axile, style simple.

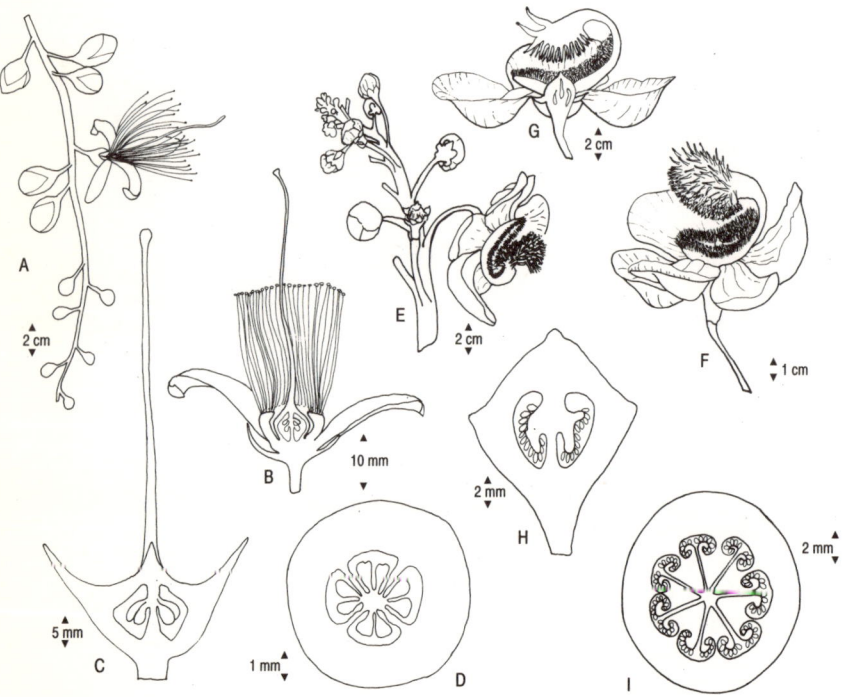

Key: part of inflorescence (A), half flower (B), longitudinal section of gynoecium (C) and transverse section of ovary (D) of *Barringtonia racemosa*. Inflorescence (E), full flower (F), half flower (G), longitudinal section of gynoecium (H) and transverse section of ovary (I) of *Couroupita guianensis*.

1. ***Barringtonia acutangula***, ela midella(S) / adampu(T),
   (DF III:197), N, 10, tree.
   Leaves: elliptic, tapering base, acuminate apex,
     margin **serrate to scalloped**, slightly
     hairy; **nearly sessile**.
   Trunk: B-rough, brownish grey,
     W-white, heavy, even-grained.
   Flowers: cream with bright
     crimson stamens; I-pendulous
     terminal racemes.
   Fruits: bluntly quadrangular,
     ovoid drupe.
   Site: around **tanks, water ways,
     flood plains**; DL.
   Uses: whole plant-medicinal.

2. ***Barringtonia asiatica***, diya mudilla(S), (DF III:196), N, 15, tree.
   Leaves: large, **crowded** terminally, oblong-ovate,
     tapering base, marginal vein; **sessile**.
   Trunk: twigs with large leaf scars.
   Flowers: creamy white;
     I-terminal raceme.
   Fruits: large, quadran-
     gular, ovoid drupe;
     pericarp thick,
     spongy, fibrous.
   Site: home gardens,
     roadsides,
     seashores; MC.
   Uses: ornamental.

3. ***Barringtonia racemosa***, diya mudilla(S),
   (DF III:197), N, 15, tree.
   Leaves: terminally **clustered**, oval to
     obovate, tapering base; nearly **sessile**.
   Trunk: B-grey, promi nently leaf-scarred.
   Flowers: pink to red; I-long
     terminal racemes.
   Fruits: ovoid drupe.
   Site: **estuaries, rivers,
     lakes**; MC, W.
   Uses: root, fruit,
     seed-medicinal.

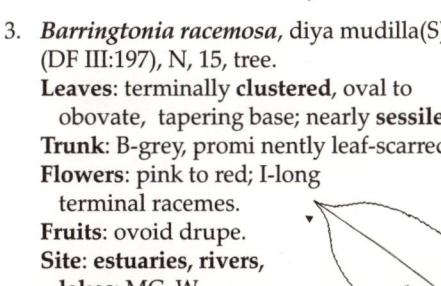
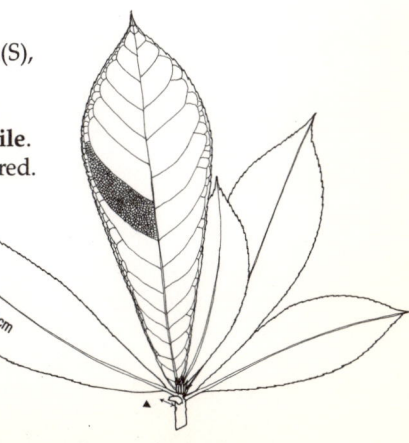

4. *Careya arborea*, kahata(S)/kachaddai(T), (DF III:200), N, 15, tree.
   Leaves: **large**, broadly ovate, tapering base, rounded
     apex, margin **scalloped**, thick, shiny; **sessile**.
   Trunk: B-rough, dark grey; W,h-dark reddish brown,
     heavy, even-grained, hard, durable;
     twigs prominently leaf-scarred.
   Flowers: pale green, large, sessile;
     I-crowded on terminal spikes.
   Fruits: green, glabrous drupe
     with persistent calyx.
   Site: savannah, dry
     patana grassland; IN, W.
   Uses: fire-resistant; B-tannin;
     IB-fibre; Stem, flower,
     fruit-medicinal;
     fruit-vegetable.

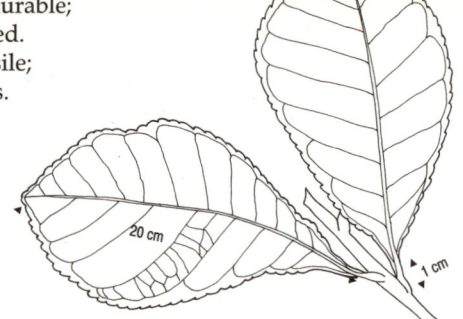

5. *Couroupita guianensis*, cannon ball tree(E)/ sal(S),
   (DF III:201), I(S. Amer.), 30, tree.
   Leaves: terminally **crowded**, oblong-ovate, petiole short.
   Trunk: B-smooth.
   Flowers: whitish pink; I-conspicuous racemes on trunk.
   Fruits: brown, **conspicuous**, large, globose berry.
   Site: **gardens**, roadsides, temples; IN, W.
   Uses: ornamental; flowers-temple offering.

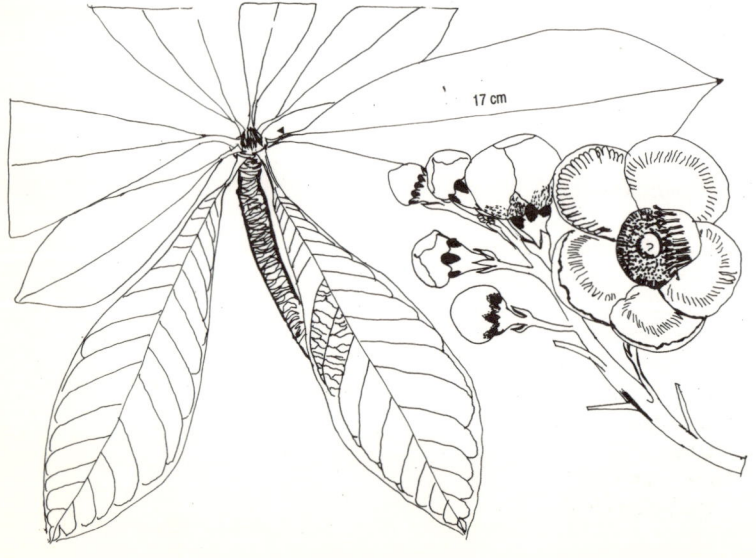

53. LEGUMINOSAE

FAMILY DESCRIPTION - **Habit**: trees, shrubs, herbs or lianas. **Leaves**: usually spiral, compound or simple by suppression of leaflets. Petiole and leaflets with basal pulvini often controlling orientation and sleep movements. **Stipules**: present. **Fruits**: legumes, sometimes follicular, or indehiscent and samaroid, or a drupe.

FLOWER PARTS - CALYX: typically 5, when tubular with valvate or rarely imbricate lobes, or often bilabiate. COROLLA: typically 5, free or the two anterior ones basally united. A hypanthium sometimes present. ANDROECIUM: stamens usually 10 or numerous, free, monadelphous or diadelphous. Anthers 2-lobed dehiscing by longitudinal slits, infrequently by pores. GYNOECIUM: superior, 1-carpel, 1 loculus. Ovules 1 to many, marginal placentation.

MIMOSOIDEAE, (Mimosa subfamily)

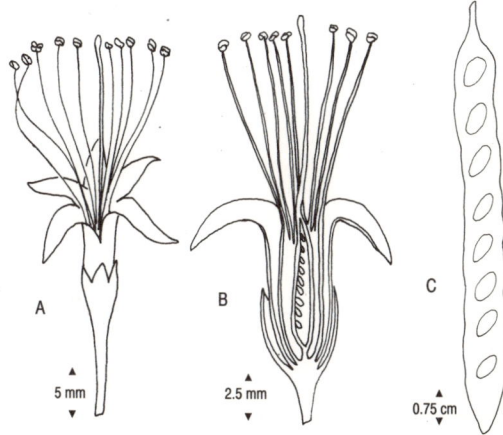

A   5 mm   B   2.5 mm   C   0.75 cm

Key: full bisexual flower (A), half flower (B) and legume (C) of Albizia saman.

1. ***Acacia decurrens***, feathery wattle(E)/ yakada sabbukku(S); (W:203), I(Australia), 15, tree.
   Leaves: bipinnate; leaflets very narrow, paired with no terminal, acute ends.
   Trunk: B-dark grey; W,s-light brown; h-reddish brown, resinous.
   Flowers: yellow.
   Fruits: legumes, constricted between seeds.
   Site: tea plantations; M, LM.
   Uses: shade tree, windbreaks in tea plantations; W-fuelwood.

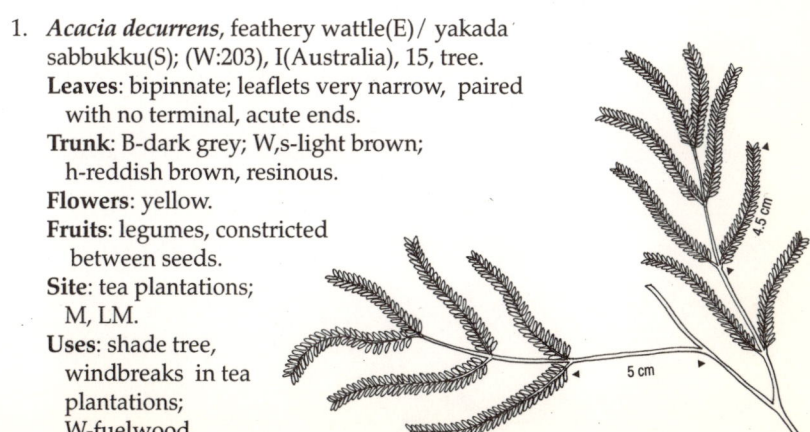

2. *Acacia leucophloea,* katu andara(S)/ velvel(T), (DF I:486), I(India), 10, tree.
   Leaves: bipinnate; leaflets paired with no terminal, oblong-linear,
   rounded ends; sessile; glands between the bases
   of lower leaflets; paired **straight** stipular
   thorns, pilose, at base of rachis.
   Trunk: thorns glossy, black; B-smooth,
   irregularly scaly, pale yellowish white;
   IB-dirty pale yellow; W, h-hard, tough.
   Flowers: white; I-globose
   heads in terminal panicles.
   Fruits: legumes strap-shaped,
   slightly fleshy, straight,
   flattened; seeds oval, flat.
   Site: forest plantations; DC, DL.
   Uses: B-flavouring and clarifier
   for arrack, dyeing sails,
   mats, ropes; fruits-edible,
   fodder; W-heavy construction.

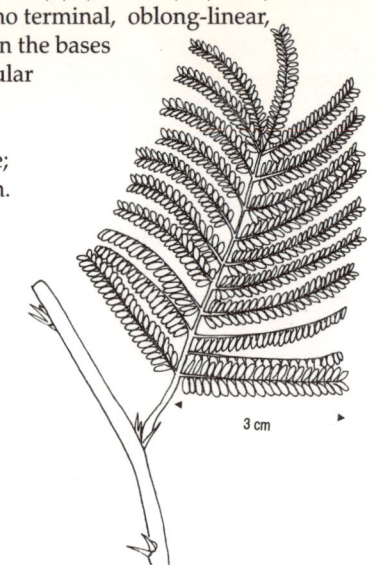

3 cm

3. *Acacia mangium,* I(Australia), 20, tree.
   Leaves: dimorphic; bipinnate in seedlings; on mature branches petiole
   and rachis form **parallel-veined flat winged** phyllodes.
   Trunk: straight; B-rough, logitudinally furrowed,
   brown; W-hard, strong, durable.
   Flowers: white or cream; I-loose spikes.
   Fruits: blackish brown legumes,
   intertwined into irregular,
   spiral clusters.
   Site: forest plantations; LM, W.
   Uses: W-furniture, fuelwood,
   pulp, construction.

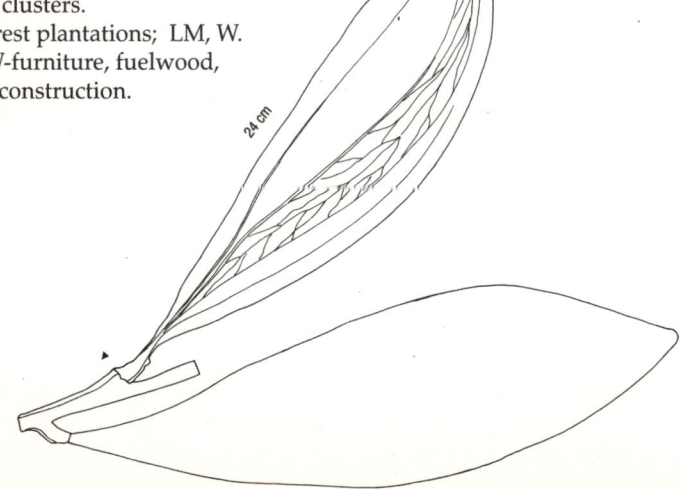

24 cm

4. *Acacia melanoxylon*, Australian blackwood(E),
   (DF I:493), I(Australia), 20, tree
   Leaves: **dimorphic**; bipinnate in seedlings; on
      maturing, rachis and petioles form flat
      parallel-veined phyllodes.
   Trunk: B-dark brown.
   Flowers: inconspicuous.
   Fruits: curled legumes.
   Site: forest and tea
      plantations; M, LM.
   Uses: shade tree; W-construction,
      fuelwood.

8 cm

5. *Acacia planifrons*, umbrella thorn(E)/ odai(T), (DF I:491), N, 6, tree.
   Leaves: bipinnate; leaflets **strap-shaped**, rounded to truncate base,
      obtuse apex, glandless; petioles of leaflets short;
      paired stipular **spines dimorphic**, one
      hooked and conical the other
      straight and slender.
   Trunk: B-deeply fissured, grey;
      IB-white, outer green; W-onion
      smell when cut, pale yellow.
   Flowers: creamish white;
      I-pedunculate head.
   Fruits: fleshy, flattened
      coiled legumes.
   Site: thorn scrub; DL.
   Uses: W-fuelwood;
      fruits-fodder.

2 cm

6. *Adenanthera pavonina*, jumble bead(E)/
   madatiya(S), (DF I:470), I(India), 20, tree.
   Leaves: bipinnate; leaflets alternate.
   Trunk: B-smooth to fissured, grey;
      IB-light brown; W,s-light
      brown; h-red, hard, durable.
   Flowers: pale yellow, small,
      numerous; I-erect racemes.
   Fruits: dark brown, coiled legumes;
      seeds shiny, **scarlet**, lens-shaped.
   Site: gardens, roadsides; IN, W.
   Uses: W-cabinet work, fuelwood;
      seeds-ornamental, medicinal,
      used for weighing gold.

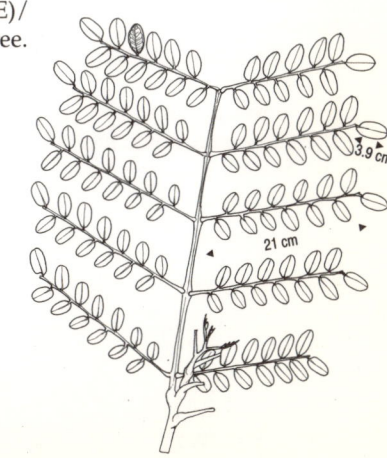

3.9 cm

21 cm

7. *Albizia falcataria*, rata mara(S), (DF I:503), I(Malesia), 20, tree.
   Leaves: bipinnate; leaflets paired with no terminal,
   obliquely oblong, **large disc-shaped
   gland** at rachis base.
   Trunk: B-smooth, silvery grey;
   branchlets minutely
   pubescent, tiny lenticels.
   Flowers: greenish yellow to
   cream; I-panicles spike-like.
   Fruits: legumes straight,
   strap-shaped, flat, narrow
   wing on ventral margin.
   Site: forest plantations; LM, W.
   Uses: coffee and tea shade;
   reforestation, soil
   stabilization and
   improvement; W-light
   construction, tea boxes.

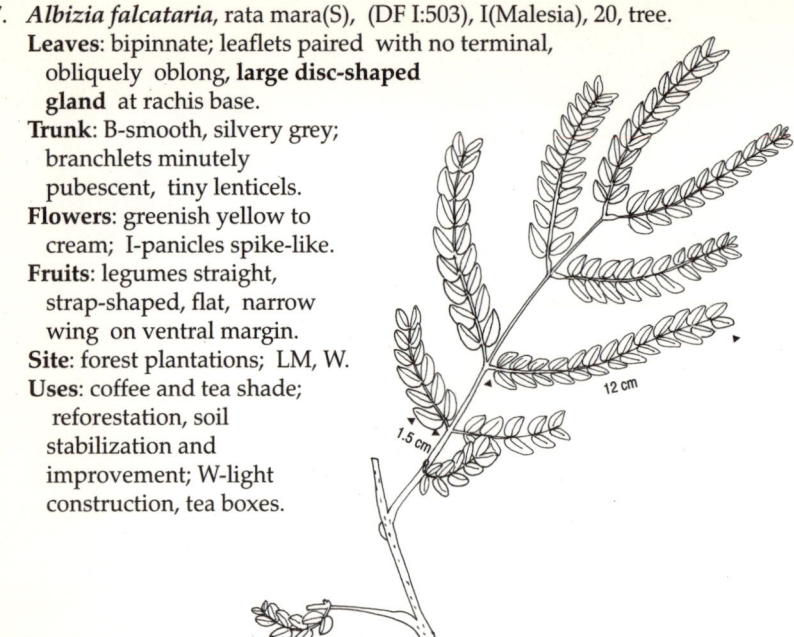

8. *Albizia lebbeck*, lebbek(E)/mara(S)/ vakai(T), (DF I:502),
   I(India), 15, small tree.
   Leaves: bipinnate; leaflets paired with no terminal;
   disc-shaped gland at rachis base.
   Trunk: B-smooth to fissured, grey;
   IB-pink; W,s-whitish; h-light yellow,
   coarse-grained.
   Flowers: cream, fragrant, clustered;
   I-axillary heads, several together.
   Fruits: **straw-coloured**, flat legumes.
   Site: forest plantations; DL, IN, W.
   Uses: B-tannin, medicine;
   W-fuelwood, posts, light
   construction.

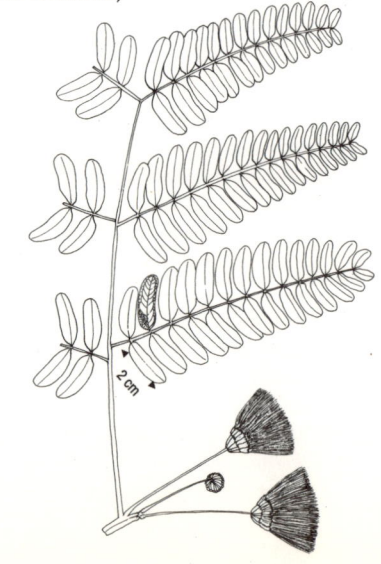

9. ***Albizia odoratissima***, suriya mara(S)/ karu vakai(T),
(DF I:499), N, 30, tree.
   Leaves: bipinnate; leaflets paired with no
   terminal, oval **gland** near rachis base.
   Trunk: B-smooth, yellowish grey,
   soft; IB-light red with white
   marbling; young parts
   tomentose.
   Flowers: greenish white,
   sessile; I-globose heads,
   3-6 clustered at nodes.
   Fruits: red-brown,
   strap-shaped legumes;
   flat, thin, oblong.
   Site: widespread.
   Uses: reforestation-
   quick growing,
   soil improvement;
   W-construction;
   B, leaves-medicinal.

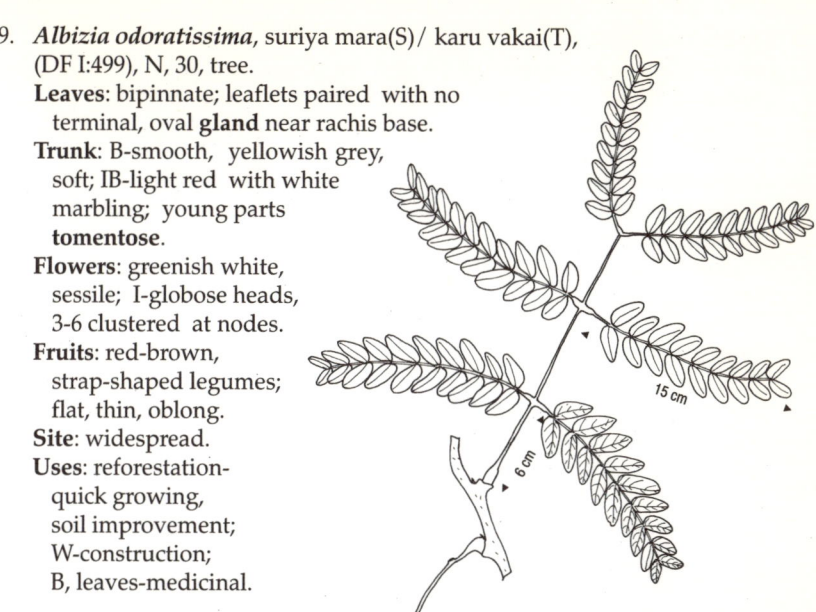

53

10. ***Albizia saman***, rain tree(E)/ pare mara(S)/ enak vakai(T),
(DF I:477), I(C. Amer.), 25, tree.
   Leaves: bipinnate; leaflets paired with no terminal,
   diamond-shaped; glands at base
   of leaflets; crown spreading.
   Trunk: B-rough, grey; IB-pink; W,s-yellowish;
   h-dark chocolate-streaked
   soft, light, durable.
   Flowers: pinkish;
   I-umbels on
   hairy stalks.
   Fruits: brown to
   black, flattened
   legumes; reddish
   brown seeds.
   Site: avenues,
   gardens;
   widespread.
   Uses: shade tree;
   fruit-sweet;
   W-furniture,
   interior
   construction.

LEGUMINOSAE

11. *Archidendron bigemina*, kalatiya(S), (DF I:507), E, 10, tree.
    Leaves: bipinnate; leaflets paired with no
    terminal, lanceolate to nearly oblanceolate,
    gland at middle of rachis.
    Trunk: B-smooth, dark brown;
    IB-reddish; W-white, soft,
    slight garlic smell.
    Flowers: cream-coloured,
    sessile; I-terminal,
    pseudo-umbels on
    panicle.
    Fruits: dirty brown,
    inside bright brown
    flattened legumes;
    seeds glossy, dark blue.
    Site: rain forest fringes; W.

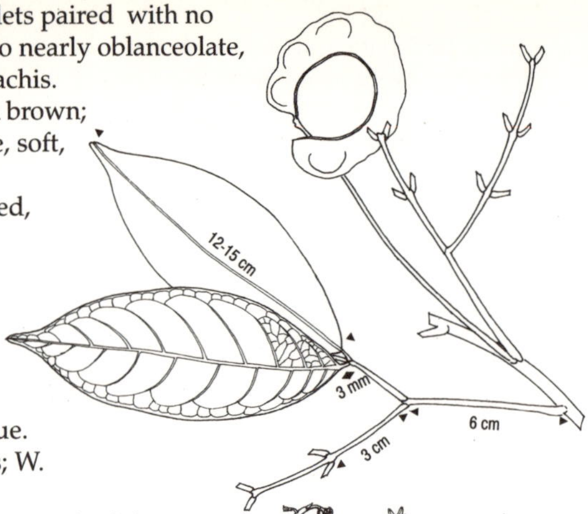

12. *Archidendron subcoriacea*, mi mini
    mara(S), (DF I:505), N, 4, small tree.
    Leaves: bipinnate; leaflets paired with
    no terminal, acute to obtuse apex,
    lateral veins prominent,
    laxly pubescent below.
    Trunk: twisted and gnarled;
    B-dark brown to grey;
    roughish, peeling,
    quadrangular;
    branchlets angular,
    densely rusty pubescent.
    Flowers: white.
    Fruits: legumes flattened,
    curled up.
    Site: montane forest
    subcanopy; M, LM.

13. *Calliandra guildingii*, I(Caribbean), 4, shrub to small tree.
    Leaves: bipinnate; leaflets paired
    with no terminal, **velvety
    pubescent**.
    Trunk: much-branched.
    Flowers: stamens tufted, basally
    white, terminaly pink;
    I-axillary heads, fluffy.
    Fruits: legumes.
    Site: gardens, parks;
    LM, W.
    Uses: ornamental.

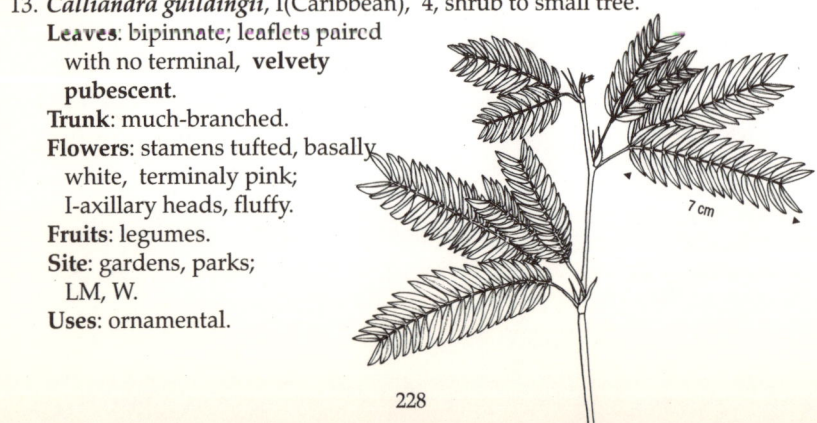

14. ***Dichrostachys cinerea***, andara(S)/vindatallai(T),
    (DF I:466), N, 4, shrub to small tree.
    Leaves: bipinnate; leaflets paired with no terminal,
        linear-lanceolate, base truncate, subacute apex.
    Trunk: many, spreading branches, spiny at
        ends; B-yellowish grey, fissured.
    Flowers: yellow, pink to violet-pink,
        turning white; I-spikes,
        2-3 or solitary.
    Fruits: dark brown legumes;
        linear, glabrous.
    Site: thorn scrub; DL.
    Uses: leaves-fodder; B-fibre;
        stem, root-medicinal.

15. ***Leucaena leucocephala***, ipil ipil(E,S), tagavai(T),
    (DF I:472), I(C. Amer.), 5, small tree.
    Leaves: bipinnate; leaflets paired with no
        terminal, narrowly oblong-lanceolate.
    Trunk: B-smooth, lenticellular, grey;
        IB-light green; W,s-light yellow;
        h-dark brown, heavy, hard.
    Flowers: whitish, clustered;
        I-numerous, globose heads.
    Fruits: legumes clustered, narrowed
        ends; seeds shiny brown.
    Site: roadsides, plantations; DL, IN, W.
    Uses: W-fuelwood; leaves-fodder.

16. ***Pithcellobium dulce***, Madras thorn(E)/
    andara(S)/ karka puli(T), (DF I:495), I(C. Amer.), 10, tree.
    Leaves: bipinnate; leaflets paired with no terminal;
        paired **spines** (stipules) at leaf
        base; flush pink-brown.
    Trunk: B-smooth to furrowed, light grey,
        prominently scarred; IB-light brown;
        W,s-yellow; h-reddish brown, soft,
        heavy, durable; gum red-brown.
    Flowers: creamy white, clustered;
        I-small, globose heads,
        axillary or terminal.
    Fruits: pink to brown, coiled legumes;
        pink to whitish pulp; shiny, black seeds.
    Site: roadsides, home gardens; DL, IN, W.
    Uses: shade tree, hedges; pulp-edible, lemonade;
        B-tannin, yellow dye; gum-mucilage; fruit-fodder.

FABOIDEAE (Bean Subfamily)

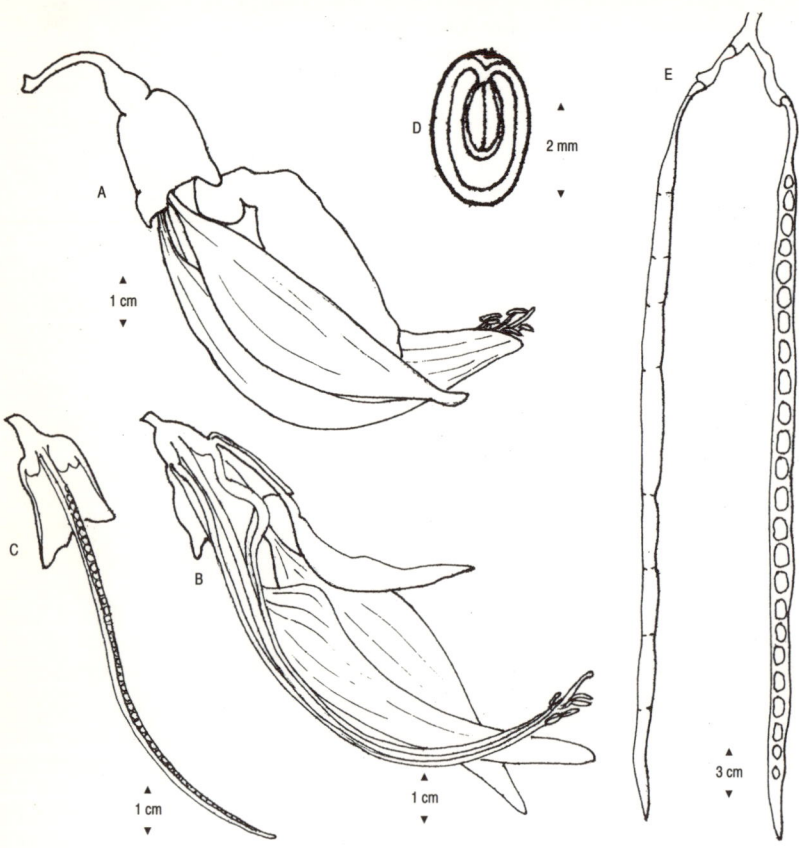

Key: full flower (A), half flower (B), longitudinal section of gynoecium (C),
transverse section of ovary (D), and legume (E) of *Sesbania grandiflora*.

17. **Butea monosperma**, flame of the forest(E)/ gas kela(S)/parasu(T), (DF VII:266)), N, 15, tree.

Leaves: trifoliolate; terminal largest, rhomboid-orbicular, unequal, veins reticulate, beneath **finely tomentose**; rachis pubescent.

Trunk: crooked, irregular; B-thick, rough, exfoliating, ash-coloured; W-resinous; young shoots **densely pubescent**; sap **ruby-red**.

Flowers: **orange-scarlet**, large, drooping, 2-3 together, silvery silky hairs; I-cauliflorous racemes.

Fruits: pendulous legumes, leathery, finely pubescent; seeds reddish brown, flat, oval, broad, smooth.

Site: monsoon and intermediate forest; DL.

Uses: ornamental; whole plant-medicinal; resin-incense.

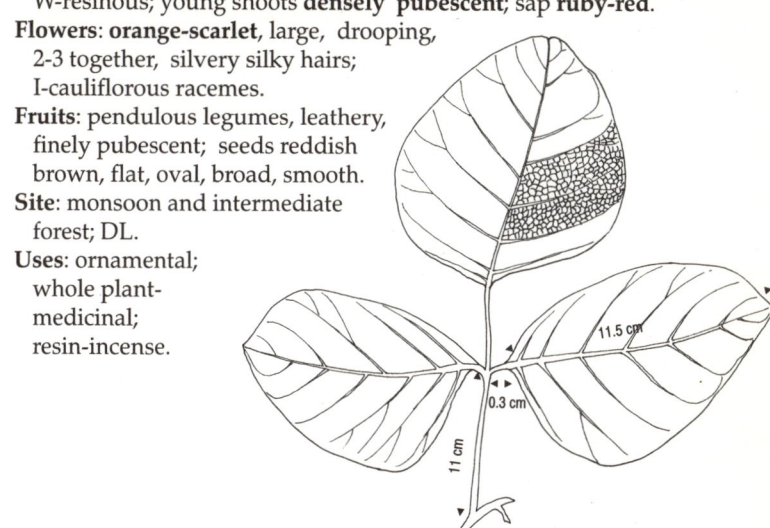

53

18. **Desmodium umbellatum**, (T II:47), N, 2, shrub.

Leaves: trifoliolate; terminal large, oval, obtuse ends, lateral veins prominent, pubescent, whitish beneath.

Trunk: young branches **densely pubescent**.

Flowers: I-densely silky, axillary umbels.

Fruits: legumes slightly curved, 3-6 broadly oblong joints.

Site: monsoon, intermediate and rain forest gaps and fringes; DL, IN, W.

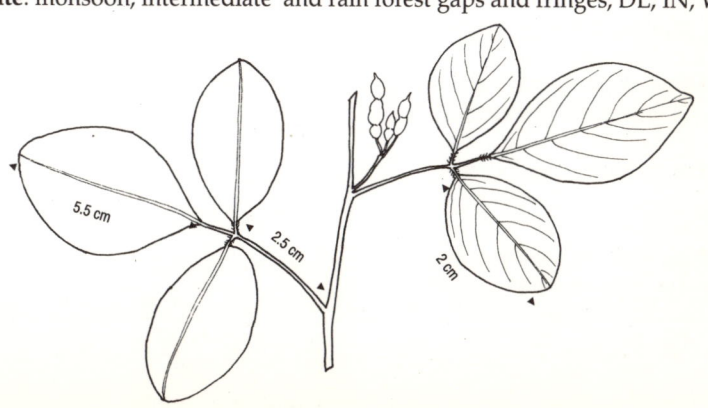

19. *Erythrina fusca*, yak erabadu(S), (DF VII:245), N, 10, tree.
    Leaves: trifoliolate; broadly oval, silvery white
        beneath; stipules small, roundish.
    Trunk: large pyramidal **corky spines**
        B-brown.
    Flowers: crimson, wings and keel
        purple, rather large; I-closely
        crowded racemes.
    Fruits: curved, densely downy,
        legume with a sharp beak.
    Site: along water courses; DL, IN, W.

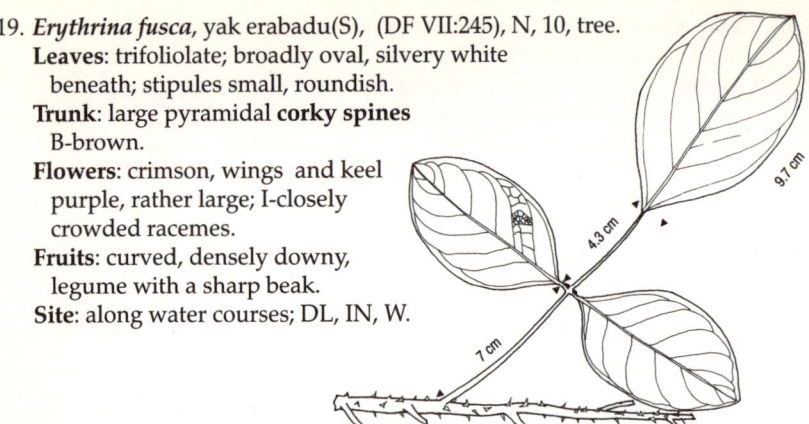

20. *Erythrina variegata*, dadap(E)/erabadu(S)/
    mullu murukku(T), (DF VII:251), N, 10, tree.
    Leaves: trifoliolate; terminal largest, roundish ovate; large, deciduous.
    Trunk: **black prickles**; B-smooth grey.
    Flowers: scarlet, wings and keel
        crimson, numerous, in 2-3's,
        closely crowded.
    Fruits: legumes cylindrical,
        sharp curved beak.
    Site: roadsides, gardens,
        forests; DL, IN, W.
    Uses: shade tree, live
        fencing; leaves,
        B-medicinal.

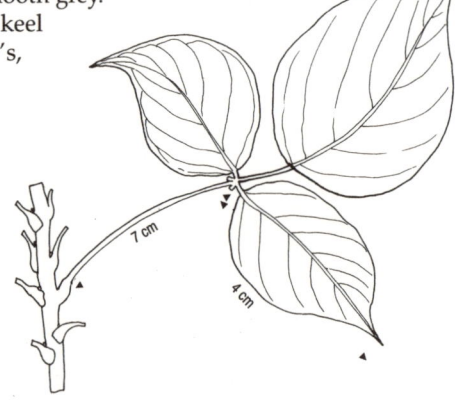

21. *Flemingia macrophylla*, (DF VII:362), N, 1-2, shrub.
    Leaves: trifoliolate; narrowly rhomboid-lanceolate, base rounded,
        tapering acute apex; rachis **hairy, sometimes winged**.
    Trunk: young branches angular, **adpressed** hairy.
    Flowers: pale pink, crowded; I-dense
        oblong sessile racemes, 2-3
        in axil, bracts small, silky.
    Fruits: blunt, inflated, finely
        pubescent, oblong-ovoid
        legume.
    Site: monsoon, intermediate
        and rain forest understory
        and scrub near water;
        DL, IN, W.

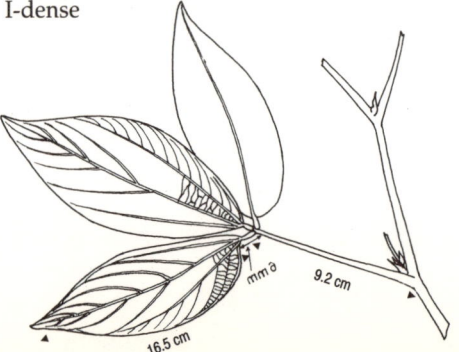

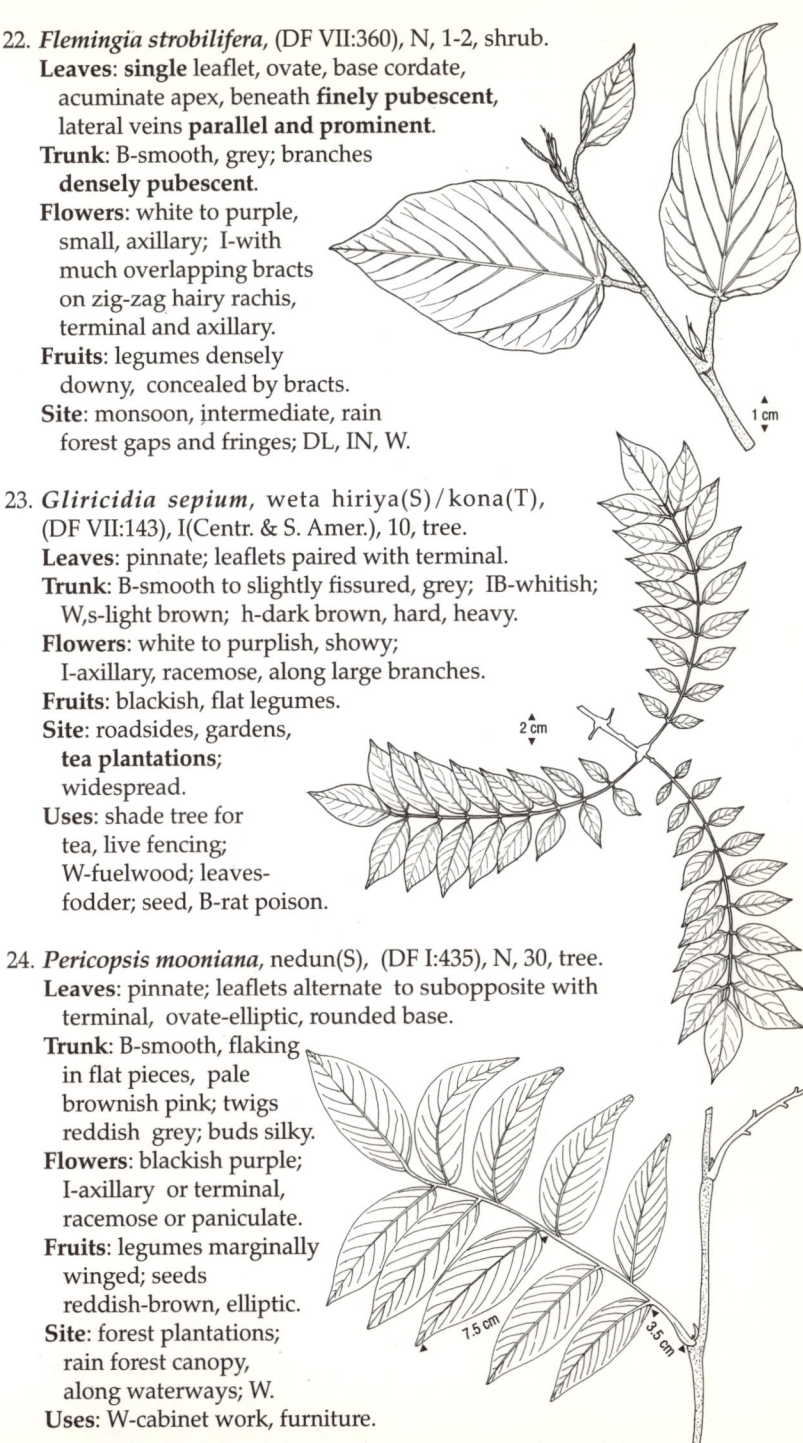

22. *Flemingia strobilifera*, (DF VII:360), N, 1-2, shrub.
Leaves: **single** leaflet, ovate, base cordate,
acuminate apex, beneath **finely pubescent**,
lateral veins **parallel and prominent**.
Trunk: B-smooth, grey; branches
densely pubescent.
Flowers: white to purple,
small, axillary; I-with
much overlapping bracts
on zig-zag hairy rachis,
terminal and axillary.
Fruits: legumes densely
downy, concealed by bracts.
Site: monsoon, intermediate, rain
forest gaps and fringes; DL, IN, W.

1 cm

23. *Gliricidia sepium*, weta hiriya(S)/kona(T),
(DF VII:143), I(Centr. & S. Amer.), 10, tree.
Leaves: pinnate; leaflets paired with terminal.
Trunk: B-smooth to slightly fissured, grey; IB-whitish;
W,s-light brown; h-dark brown, hard, heavy.
Flowers: white to purplish, showy;
I-axillary, racemose, along large branches.
Fruits: blackish, flat legumes.
Site: roadsides, gardens,
tea plantations;
widespread.
Uses: shade tree for
tea, live fencing;
W-fuelwood; leaves-
fodder; seed, B-rat poison.

2 cm

24. *Pericopsis mooniana*, nedun(S), (DF I:435), N, 30, tree.
Leaves: pinnate; leaflets alternate to subopposite with
terminal, ovate-elliptic, rounded base.
Trunk: B-smooth, flaking
in flat pieces, pale
brownish pink; twigs
reddish grey; buds silky.
Flowers: blackish purple;
I-axillary or terminal,
racemose or paniculate.
Fruits: legumes marginally
winged; seeds
reddish-brown, elliptic.
Site: forest plantations;
rain forest canopy,
along waterways; W.
Uses: W-cabinet work, furniture.

7.5 cm

3.5 cm

53

LEGUMINOSAE

25. *Pongamia pinnata*, magul karanda(S)/ punku(T), (DF VII:225), N, 15, tree.

Leaves: pinnate; leaflets paired with terminal, oval to lanceolate, base rounded, acuminate apex, shiny.

Trunk: B-smooth, soft, grey; W-yellowish white, hard, not durable.

Flowers: greenish pink-white, calyx purplish brown; I-racemes 2 together.

Fruits: woody oval-oblong, glabrous legume, pointed, flat and indehiscent.

Site: monsoon, intermediate, rain forest, waterways; DL, IN, W.

Uses: shade tree, hedges, ornamental; young twigs-as tooth brushes; whole plant-medicinal.

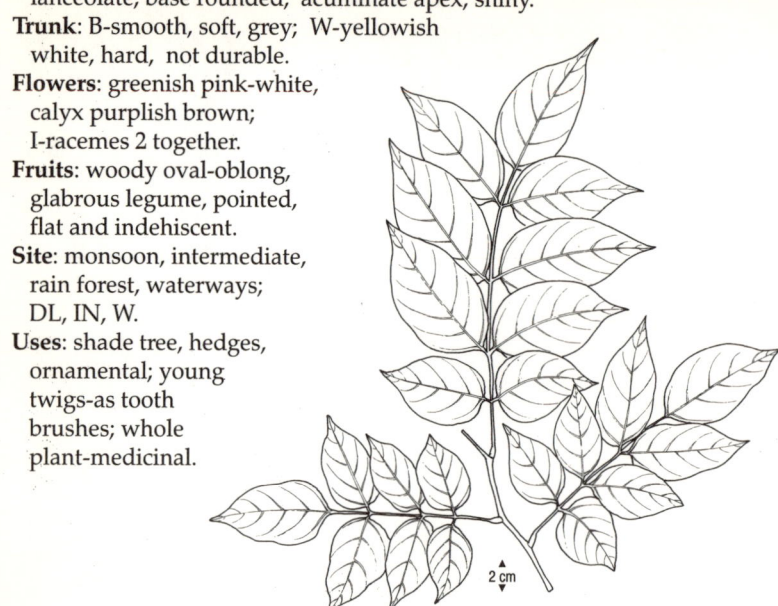

2 cm

26. *Pterocarpus indicus*, Andaman red wood(E)/ wal ehela(S)/kengei(T), (DF VII:234), I(India), 25, tree.

Leaves: pinnate; leaflets **alternate** with terminal.

Trunk: B-finely fissured, light brown; IB-light brown to reddish streaked; W-reddish, hard; **red gummy** exudate.

Flowers: yellow, small, numerous, fragrant.

Fruits: brown, circular, flattened legumes, with winged margin.

Site: roadsides, gardens; widespread; IN, W.

Uses: ornamental, shade, live fencing; W-furniture, construction, flooring.

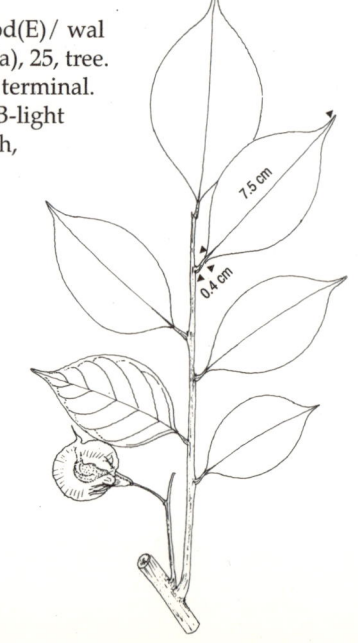

7.5 cm

0.4 cm

27. ***Pterocarpus marsupium***, gammalu(S) /
    venkai(T), (DF VII:233), N, 20, tree.
    Leaves: pinnate; leaflets **alternate**,
    acute base, **bilobed** apex,
    numerous parallel lateral veins.
    Trunk: crooked; B-thick, yellowish
    grey, W-reddish brown, dark,
    durable, heavy, hard; resin
    dark red; spreading crown.
    Flowers: bright yellow,
    lax; I-panicles.
    Fruits: nearly circular
    legume with beaked
    apex, winged margins.
    Site: savannah, intermediate
    forest canopy; DL, IN.
    Uses: W-furniture, turnery, cabinet work;
    resin-incense, exudate medicinal.

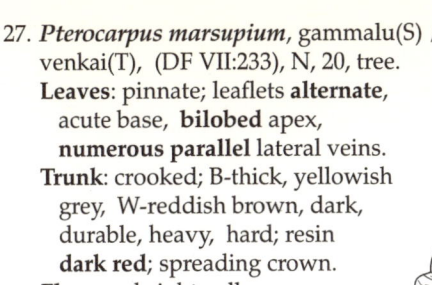

7 cm

53

LEGUMINOSAE

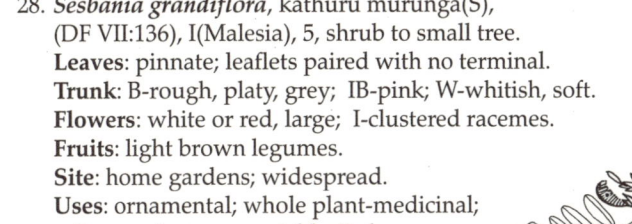

10-18 cm

28. ***Sesbania grandiflora***, kathuru murunga(S),
    (DF VII:136), I(Malesia), 5, shrub to small tree.
    Leaves: pinnate; leaflets paired with no terminal.
    Trunk: B-rough, platy, grey; IB-pink; W-whitish, soft.
    Flowers: white or red, large; I-clustered racemes.
    Fruits: light brown legumes.
    Site: home gardens; widespread.
    Uses: ornamental; whole plant-medicinal;
    flowers, leaves-vegetable; B-dye.

29. ***Sophora tomentosa***, mudu murunga(S),
    (DF I:439), N, 3, shrub.
    Leaves: pinnate; leaflets **alternate to
    opposite**, oblong to orbicular,
    subcordate to cuneate base, obtuse
    apex, **densely pubescent** beneath.
    Trunk: branchlets grey, **velvety pubescent**.
    Flowers: yellow to white; I-erect racemes.
    Fruits: coriaceous, velvety, downy
    legume, constricted between
    seeds; seeds brown, ellipsoid.
    Site: coastal scrub, calcareous
    sand; DC, MC.
    Uses: ornamental.

14 cm

12 cm

2 cm

30. *Tadehagi triquetrum*, baloliya(S), (T II:49), N, 3, shrub.
   Leaves: variable size, linear-lanceolate, subcordate base,
      tapering acute apex; stipules lanceolate, acute,
      brown; petiole **broadly winged**.
   Trunk: branches few, erect, 3-angled,
      hairy on angles only.
   Flowers: bright violet, often white,
      small; I-lax, slender, erect, spike-like
      terminal and axillary racemes;
      bracts with few bristles.
   Fruits: linear-oblong, scarcely
      indented, erect legume,
      hairy, as broad as
      long, 4-7 joints.
   Site: rain forest fringes,
      secondary forest; IN, W.

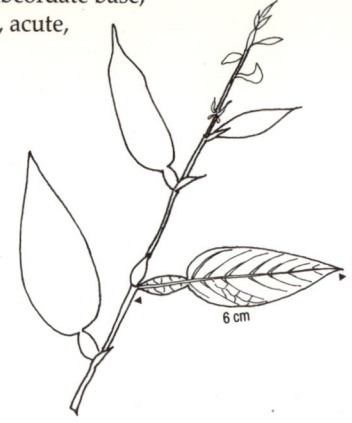

CAESALPINIOIDEAE (Cassia subfamily)

Key: inflorescence (A), half flower (B), longitudinal section of gynoecium (C) of
Cassia fistula.

236

31. ***Bauhinia racemosa***, maila(S)/atti(T), (DF VII:39), N, 4, small tree.
    Leaves: **small**, **truncate** base, **2 oval lobes**, 9 main veins
        from base, whitish beneath.
    Trunk: B-rough, furrowed, blackish;
        W-heavy, hard, pale brown to grey.
    Flowers: yellowish white;
        I-erect lax terminal racemes.
    Fruits: pendulous, sickle-shaped,
        glabrous legume, tapering
        base, blunt apex; 12-20 seeded.
    Site: monsoon forest, scrub; DL.
    Uses: B-rope; W-poles; leaves,
        stem, flowers-medicinal.

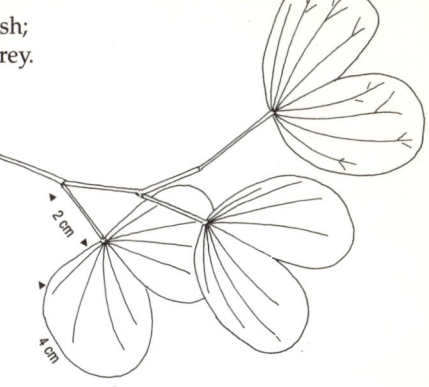

32. ***Bauhinia tomentosa***, yellow
    bauhinia(E)/ kaha petan(S)/tiruvati(T), (DF VII:38), N, 4, small tree.
    Leaves: **2 oval lobes**, **truncate** base, 7 main veins
        from base, **pubescent and whitish** beneath.
    Trunk: B-longitudinally furrowed, yellowish
        grey; W,h-hard, dark red, heavy, durable.
    Flowers: sulphur-yellow with a
        dark purple blotch at base,
        conspicuous, bell-shaped;
        I-small, terminal racemes.
    Fruits: finely pubescent,
        pointed legume,
        narrowed base;
        8-12 seeded.
    Site: monsoon forest,
        scrub, home gardens; DL.
    Uses: ornamental; whole
        plant-medicinal.

33. ***Bauhinia variegata***, kobo leela(S),
    (DF VII:43), I(Eastern Asia), 4, small tree.
    Leaves: **large**, **2 lobes**, **truncated** at
        base, **12-13** main veins from base.
    Trunk: B-smooth, grey brown.
    Flowers: white to mauve,
        conspicuous orchid-like.
    Fruits: pendulous, flattened
        legumes; seeds many.
    Site: roadsides, gardens; DL, W.
    Uses: ornamental; root, flower,
        B-medicinal.

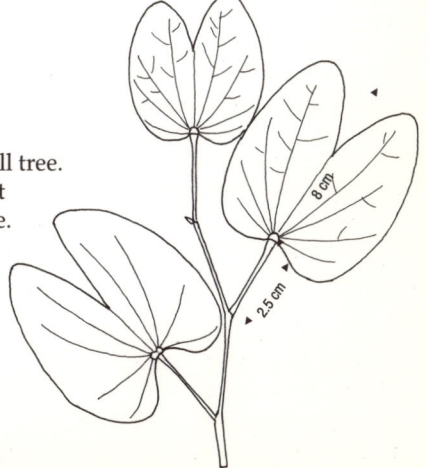

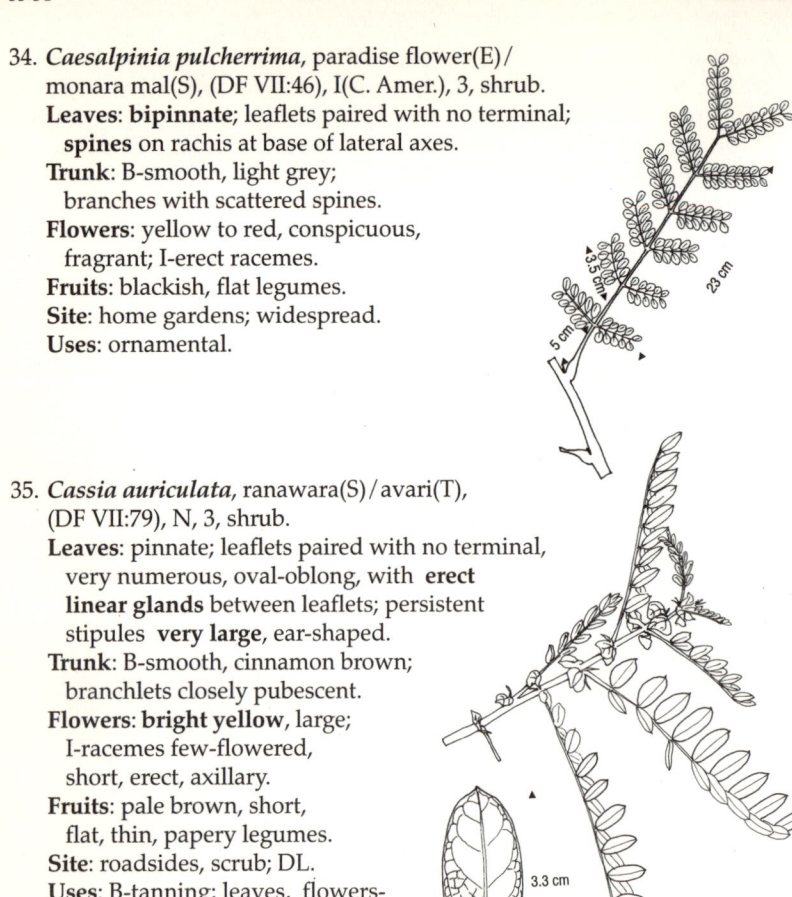

34. *Caesalpinia pulcherrima*, paradise flower(E)/
monara mal(S), (DF VII:46), I(C. Amer.), 3, shrub.
Leaves: **bipinnate**; leaflets paired with no terminal;
spines on rachis at base of lateral axes.
Trunk: B-smooth, light grey;
branches with scattered spines.
Flowers: yellow to red, conspicuous,
fragrant; I-erect racemes.
Fruits: blackish, flat legumes.
Site: home gardens; widespread.
Uses: ornamental.

35. *Cassia auriculata*, ranawara(S)/avari(T),
(DF VII:79), N, 3, shrub.
Leaves: pinnate; leaflets paired with no terminal,
very numerous, oval-oblong, with **erect
linear glands** between leaflets; persistent
stipules **very large**, ear-shaped.
Trunk: B-smooth, cinnamon brown;
branchlets closely pubescent.
Flowers: **bright yellow**, large;
I-racemes few-flowered,
short, erect, axillary.
Fruits: pale brown, short,
flat, thin, papery legumes.
Site: roadsides, scrub; DL.
Uses: B-tanning; leaves, flowers-
herbal tea; roots, stem-medicinal.

36. *Cassia fistula*, Indian laburnum(E)/ ehela(S)/
tiru kontai(T), (DF VII:62), N, 15, tree.
Leaves: pinnate; leaflets paired with no terminal.
Trunk: B-smooth to scaly, grey-brown;
W-reddish, hard, heavy.
Flowers: **golden-yellow**, conspicuous;
I-long-stalked, drooping,
terminal clusters.
Fruits: black, long, cylindrical
legumes; pulp dark brown.
Site: roadsides, home gardens,
monsoon forest subcanopy; DL.
Uses: ornamental; whole
plant-medicinal; B-tanning;
W-cabinet work, tools, posts.

37. *Cassia javanica*, pink cassia(E), (DF VII:67),
   I(Tropical Asia), 15, tree.
   Leaves: pinnate; leaflets paired with no terminal.
   Trunk: B-smooth grey; IB-light brown; W,s-whitish, soft.
   Flowers: **pink**, conspicuous, fragrant; I-panicles.
   Fruits: dark brown, long,
      slender, cylindrical legumes;
      seeds shiny and flattened,
      brown, 20-100 horizontally
      placed; malodorous.
   Site: roadsides, gardens; IN, W.
   Uses: ornamental; shade.

38. *Cassia roxburghii*, red cassia(E)/
   ratu wa(S)/ vakai(T), (DF VII:65), N, 15, tree.
   Leaves: pinnate; leaflets paired with no terminal,
      oblong obtuse, **unequal** base, **notched** apex,
      pubescent beneath; umbrella-shaped crown.
   Trunk: branches drooping, dark brown,
      young parts **densely pubescent;** B-cracked;
      W,h-very hard, reddish brown;
   Flowers: **pink or salmon**;
      I-supra axillary racemes.
   Fruits: black, cylindrical,
      straight, woody legumes.
   Site: monsoon forest subcanopy,
      secondary forest; DL.
   Uses: ornamental.

39. *Cassia siamea*, Siam cassia(E)/wa(S)/ vakai(T),
   (DF VII:71), N, 20, tree.
   Leaves: pinnate; leaflets paired with no terminal.
   Trunk: B-smooth grey; IB-light brown;
      W,s-light brown; h-dark brown,
      hard, termite-susceptible.
   Flowers: **bright yellow**, conspicuous;
      I-erect terminal corymbose racemes.
   Fruits: dark brown, flat, long,
      narrow, pubescent
      legumes; 8-15-seeded.
   Site: monsoon forest canopy,
      close to water; roadsides,
      gardens; DL, IN.
   Uses: ornamental, shade; B-tanning;
      W-furniture, posts, fuelwood.

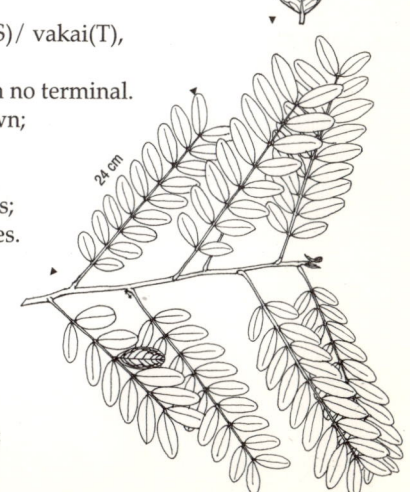

40. *Cassia spectabilis*, kaha kona(S)/murjal kona(T),
    (DF VII:69), I(S. America), 5, small tree.
    Leaves: pinnate; leaflets paired with no terminal,
      ovate to lanceolate; umbrella-shaped crown.
    Trunk: B-vertical lines of lenticels;
      W-heavy, tough.
    Flowers: **yellow**, fragrant;
      I-terminal racemes.
    Fruits: black, long
      legumes with many
      closely packed seeds.
    Site: roadsides, home
      gardens; widespread.
    Uses: ornamental; W-tool handles.

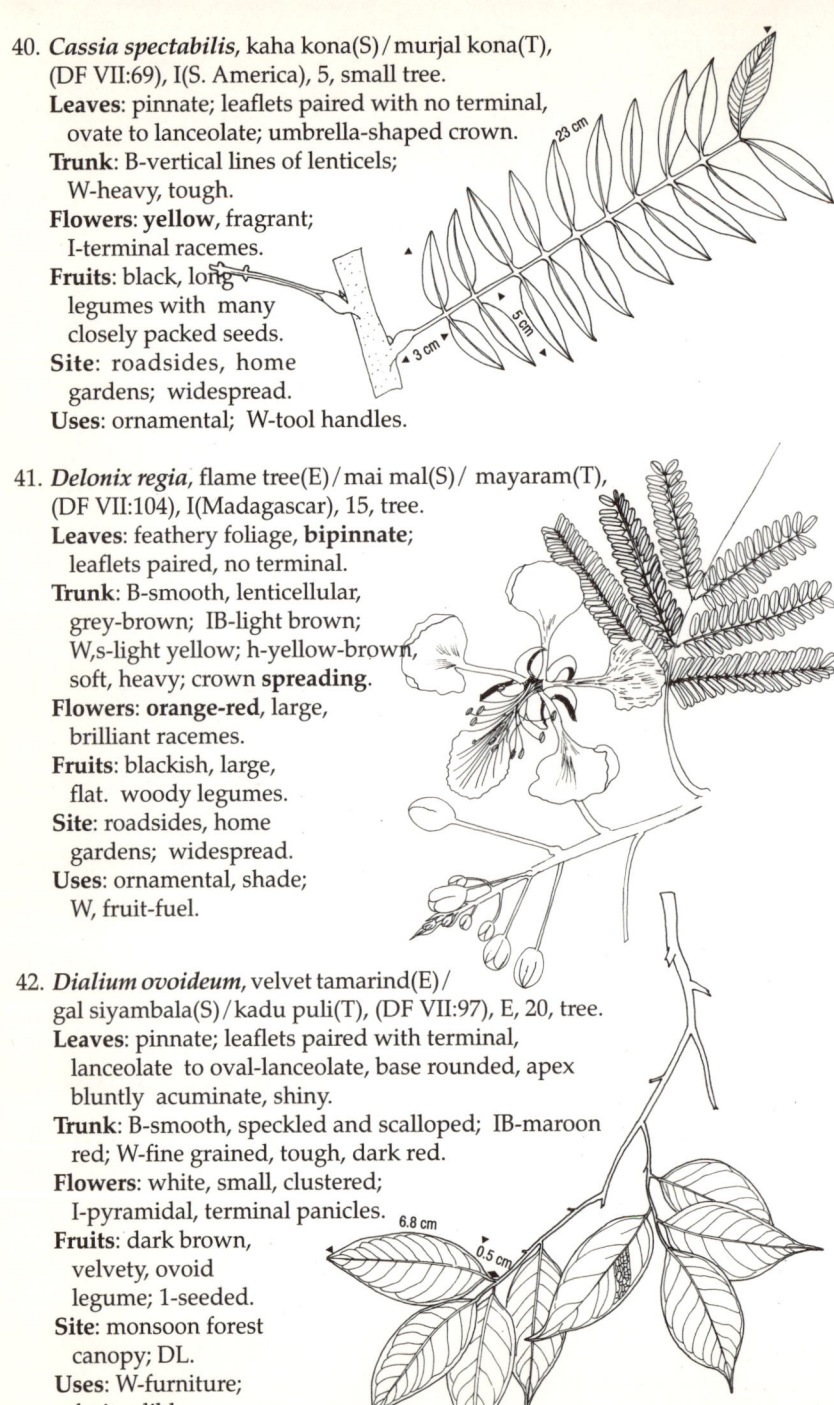

41. *Delonix regia*, flame tree(E)/mai mal(S)/ mayaram(T),
    (DF VII:104), I(Madagascar), 15, tree.
    Leaves: feathery foliage, **bipinnate**/
      leaflets paired, no terminal.
    Trunk: B-smooth, lenticellular,
      grey-brown; IB-light brown;
      W,s-light yellow; h-yellow-brown,
      soft, heavy; crown **spreading**.
    Flowers: **orange-red**, large,
      brilliant racemes.
    Fruits: blackish, large,
      flat. woody legumes.
    Site: roadsides, home
      gardens; widespread.
    Uses: ornamental, shade;
      W, fruit-fuel.

42. *Dialium ovoideum*, velvet tamarind(E)/
    gal siyambala(S)/kadu puli(T), (DF VII:97), E, 20, tree.
    Leaves: pinnate; leaflets paired with terminal,
      lanceolate to oval-lanceolate, base rounded, apex
      bluntly acuminate, shiny.
    Trunk: B-smooth, speckled and scalloped; IB-maroon
      red; W-fine grained, tough, dark red.
    Flowers: white, small, clustered;
      I-pyramidal, terminal panicles.
    Fruits: dark brown,
      velvety, ovoid
      legume; 1-seeded.
    Site: monsoon forest
      canopy; DL.
    Uses: W-furniture;
      fruit-edible.

43. *Humboldtia laurifolia*, gal karanda(S), (DF VII:94),
    N, 5, shrub to small tree.
    Leaves: pinnate with **large stipules at base**; leaflets
    paired, no terminal, flush **white**, **drooping**.
    Trunk: branches **spreading**; **drooping**, **hollow**
    internodes inhabited by ants.
    Flowers: white, honey-scented;
    I-axillary racemes, bracts at
    base green, rounded,
    at apex pink.
    Fruits: glabrous, veiny, much
    compressed, pointed legume.
    Site: rain forest understory; W.

44. *Parkinsonia aculeata*, Jerusalem
    thorn(E), (DF VII:44), I(C. Amer.), 7, small tree.
    Leaves: **bipinnate**, primary raches very
    short, secondary rachis flattened;
    leaflets **small**, **deciduous**.
    Trunk: B-smooth, yellow-green-blue;
    IB-green; W,s-yellow; h-reddish.
    Flowers: yellow; I-loose lateral racemes.
    Fruits: brown legume,
    constricted between seeds.
    Site: naturalized in thorn scrub;
    roadsides, home gardens; DC.
    Uses: ornamental; W-fuelwood.

45. *Peltophorum pterocarpum*, yellow flame tree(E), (DF VII:57), N, 15, tree.
    Leaves: feathery foliage, **bipinnate**; leaflets paired, no terminal.
    Trunk: B-smooth to furrowed, lenticellular, light grey to brown;
    IB-light brown; W,s-whitish; twigs densely rusty hairy.
    Flowers: rusty yellow, showy; I-paniculate.
    Fruits: reddish to brown legume;
    winged, flat, broad; 1-4 seeded.
    Site: roadsides, home
    gardens; widespread.
    Uses: ornamental, shade;
    W-fuelwood; B-medicinal.

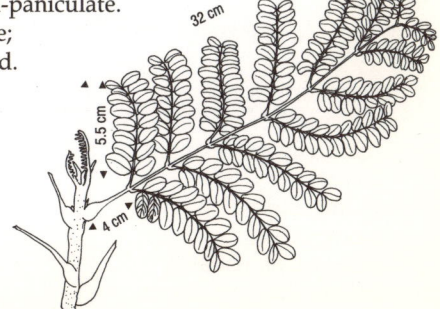

46. *Saraca asoca*, asoka(E,S,T), (DF VII:99), N, 4, small tree.

Leaves: large, spreading, pinnate; leaflets paired, no terminal, linear, tapering base, acute apex, flush **drooping**.

Trunk: B-reddish grey, cracked; W-pale red, soft; branches spreading.

Flowers: pale orange-scarlet, numerous; I-large, dense axillary and terminal corymbose panicles, sweet-scented.

Fruits: legumes with large, tapering ends, compressed, glabrous.

Site: monsoon and rain forest understory near waterways, gardens; DL.

Uses: ornamental; leaves, B-medicinal.

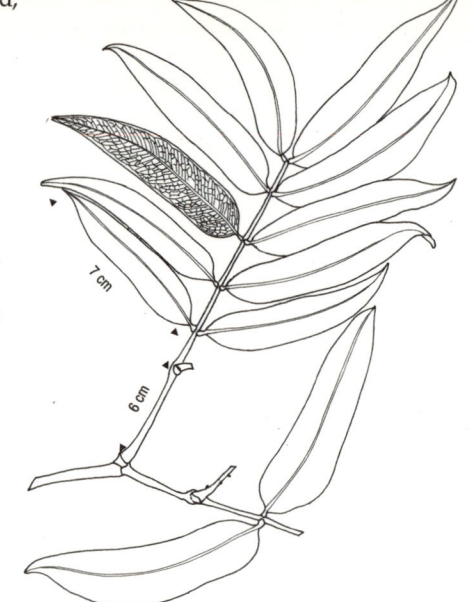

7 cm

6 cm

47. *Tamarindus indica*, tamarind(E)/siyambala(S)/ puli(T), (DF VII:96), I(Trop. Africa), 25, tree.

Leaves: feathery, pinnate; leaflets paired, no terminal.

Trunk: B-rough, fissured, grey to brown; IB-brownish; W,s-light yellow; h-dark purplish.

Flowers: pale yellow, tinged with red, showy; I-lateral and terminal clusters.

Fruits: grey-brown, thick, rough legumes; pulp dark brown.

Site: roadsides, home gardens; DL, IN, W.

Uses: ornamental; whole plant-medicinal; W-furniture, fuelwood, charcoal; pulp-edible, flavouring.

1 cm

54. LOGANIACEAE

FAMILY DESCRIPTION - **Habit**: trees, shrubs, herbs or lianas. **Leaves**: opposite, rarely spiral, simple, entire or lobed. **Stipules**: present or absent. **Flowers**: bisexual, actinomorphic, solitary or terminal cymes. Bracts small, or rarely large and petaloid. **Fruits**: septicidal capsule, rarely a berry or drupe. Seeds sometimes winged.

FLOWER PARTS - CALYX: 4-5, free or united, valvate or imbricate. COROLLA: 4-5, united, variously shaped, lobes valvate, imbricate or contorted in bud. ANDROECIUM: stamens 4-5, epipetalous. Anthers 2-locular, basifixed or versatile, opening lengthwise. GYNOECIUM: superior or sometimes half- inferior, 2-3-locular. Style apically 2-lobed, seldom branched. Ovules 1 to many per loculus with axile placentation.

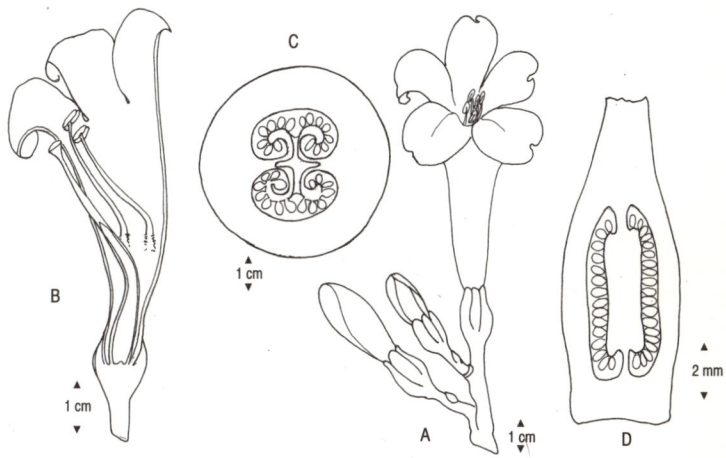

Key: part of an inflorescence (A), half flower (B), transverse (C) and longitudinal (D) sections of the gynoecium of *Fagraea obovata.*

1. *Fagraea ceilanica*, etamburu (S), (T III:170), N, 4, shrub or small tree.
    Leaves: **large**, oval-obovate, tapering base, **rounded** apex, margins entire, wavy, dark green and shiny above; **nearly sessile**, stipules absent.
    Trunk: B-rough, warted, vertically cracked, greyish brown.
    Flowers: cream-white to yellowish, sweet scented, large; I-cymes.
    Fruits: broadly ovoid, bluntly pointed, smooth, shining, fleshy berry; seeds numerous.
    Site: montane and rain forest understory, sometimes epiphytic; LM, W.
    Uses: ornamental, medicinal.

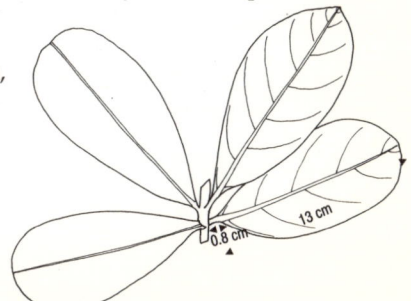

2. ***Strychnos nux-vomica***, strychnine tree (E)/ goda kaduru (S)/
kanchurai (T), (T III:175), N, 20, tree.

Leaves: broadly oval, **unequal-sided**,
acute to rounded base, acuminate
apex, **5-veined**, laterals often
faint; stipules absent.

Trunk: twigs yellowish
grey, numerous, **opposite.**

Flowers: greenish white;
numerous, pubescent;
I-terminal paniculate cymes.

Fruits: orange-red, slightly rough,
shiny berry; seeds button-shaped
covered with pinkish-grey pulp.

Site: seeondary forest, scrub; DL.

Uses: whole plant-medicinal;
W-furniture, termite resistant;
seeds-poisonous, source of strychnine.

55. LYTHRACEAE

FAMILY DESCRIPTION - Habit: trees, shrubs or herbs. **Leaves**: opposite, rarely whorled or spiral. Simple. **Stipules**: vestigial or absent. **Flowers**: bisexual, actinomorphic, rarely zygomorphic, with conspicuous hypanthium which is sometimes spurred or subtended by epicalyx. **Fruits**: usually a capsule, dehiscing variously. Seeds sometimes winged.

FLOWER PARTS - CALYX: 4, 6 or 8, marginal on hypanthium, valvate. COROLLA: 4,6 or 8 or absent, free and arising from rim or within upper end of hypanthium. Petals crumpled in bud. ANDROECIUM: stamens twice as many as calyx or corolla, in 2 whorls inserted in hypanthium, or numerous and centrifugal. Anthers with longitudinal slits. Nectary disc present. GYNOECIUM: superior, carpels 2-6, locules 2-6, rarely 1. Two to many ovules per loculus. Placentation axile.

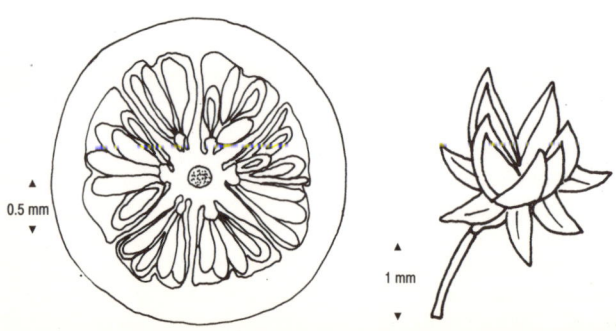

Key: part of inflorescence (A), longitudinal section of gynoecium (B), transverse section of ovary (C) and mature dehisced fruit (D) of *Lagerstroemia speciosa*.

1. *Lagerstroemia speciosa*, queen of flowers (E)/murutha (S)/
    pu muruthu (T), (T II:228 & VI:129), N, 10, tree.
    Leaves: **2-ranked**, large, oval to lanceolate,
        rounded base, acute apex, margins
        entire to scalloped, lateral veins
        prominent; deciduous.
    Trunk: B-pale, smooth to flaking; W-brownish
        red, hard, light; branches spreading.
    Flowers: **light mauve to bright
        rose-pink**; I-**conspicuous**, large panicle.
    Fruits: broadly ovoid, hard,
        woody, pointed, splitting
        into 6 valves, persistent
        cup-like calyx;
        seeds 6, winged.
    Site: forest waterways,
        roadsides, parks, home
        gardens; widespread.
    Uses: ornamental;
        whole plant-medicinal;
        W-furniture, construction.

56. MAGNOLIACEAE

FAMILY DESCRIPTION - Habit: trees or shrubs. **Leaves**: spiral, simple, entire. **Stipules:** large, enclosing terminal bud, deciduous. **Flowers**: bisexual, actinomorphic, usually solitary, terminal or axillary. **Fruits**: follicle, samara or berry.

FLOWER PARTS - PERIANTH: sepals and petals not always differentiated. Sepals often 3. Petals 6 to many, cyclically or spirally arranged. ANDROECIUM: stamens numerous, often strap-shaped, free, spirally arranged on the basal portion of floral axis. Anthers 2-celled, dehiscing longitudinally. GYNOECIUM: superior, sessile or borne on a gynophore. Carpels 2 to many, free, spirally arranged on elongated axis. Locules one per carpel. Placentation marginal with 1 to many ovules.

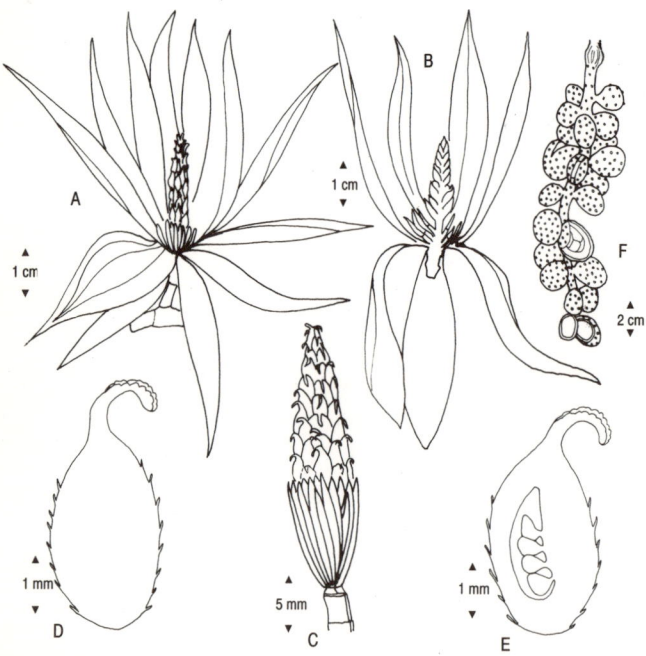

Key: full flower (A), half flower (B), androecium with spirally arranged stamens at the base and gynoecium with spirally arranged carpels at the upper end of the axis (C), a single carpel (D), a carpel in longitudinal section (E) and mature fruit showing an aggregate of follicles (F) of *Michelia champaca*.

1. ***Michelia champaca***, champak tree (E)/gini sapu (S)/
   chambuga (T), (DF VI:155), I (India), 30, tree.
   Leaves: elliptic or ovate, **glossy**, light green,
   large terminal leaf buds; stipules present.
   Trunk: B-smooth, grey brown; W,h-yellow-brown
   to dark brown, but purple when freshly cut;
   twigs leaf-scarred.
   Flowers: yellow,
   fragrant, large
   pointed buds,
   waxy petals.
   Fruits: pale
   brown, aggregate
   of warty follicles;
   seeds coral-pink,
   angular.
   Site: home
   gardens,
   roadsides,
   forest plantations; IN, W.
   Uses: ornamental; whole
   plant-medicinal; W-furniture,
   construction; flowers-temple offering.

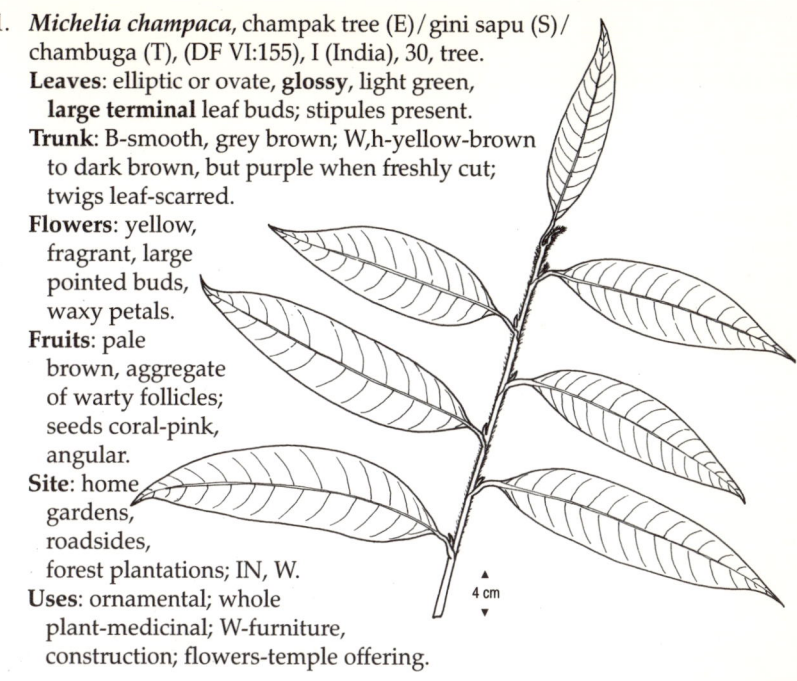

4 cm

56

MAGNOLIACEAE

2. ***Michelia nilagirica***, wal sapu (S), (DF VI:153), N, 15, tree.
   Leaves: lanceolate to oval-lanceolate, tapering ends, above bright green,
   beneath **fine white-pubescent**; stipules **densely silky**, deciduous.
   Trunk: B-smooth, thick, grey; W-pale brown,
   durable, strong; twigs ring-scarred;
   young parts silky.
   Flowers: sulphur-yellow,
   solitary, pale.
   Fruits: pale
   yellow,
   tinged
   purple,
   aggregate
   of follicles;
   seeds bright
   scarlet.
   Site: montane
   forest canopy; M.
   Uses: W-furniture,
   heavy construction.

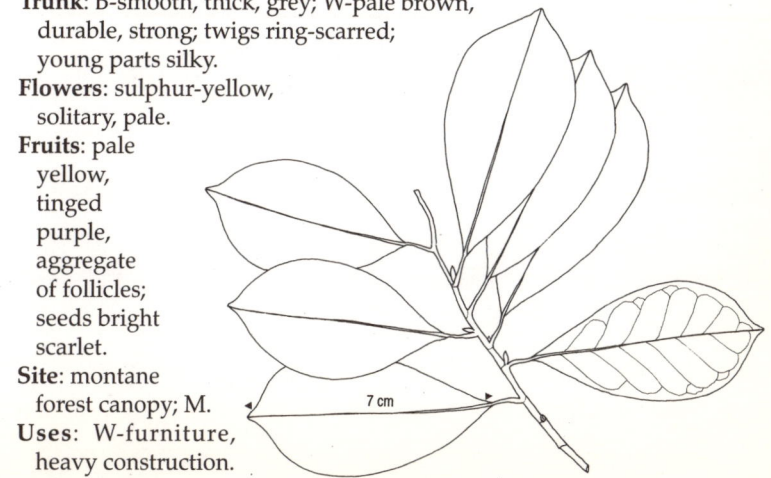

7 cm

57. MALVACEAE

FAMILY DESCRIPTION - Habit: trees, shrubs or herbs. **Leaves**: spiral, simple, entire or variously lobed. **Stipules**: usually present. **Flowers**: bisexual, often with epicalyx, actinomorphic, solitary or in cymes. **Fruits**: loculicidal capsule, schizocarp, rarely berry or samara.

FLOWER PARTS - Calyx: 5, sometimes basally united, valvate. COROLLA: 5, free, often basally adnate to androecium, convolute or rarely imbricate. ANDROECIUM: stamens numerous, united into a tube. Anthers 1-locular, opening lengthwise. GYNOECIUM: superior, 2 to many carpels, often 5. Styles as many as carpels and basally united. Loculi as many as carpels. Placentation axile, 1 to many ovules per loculus.

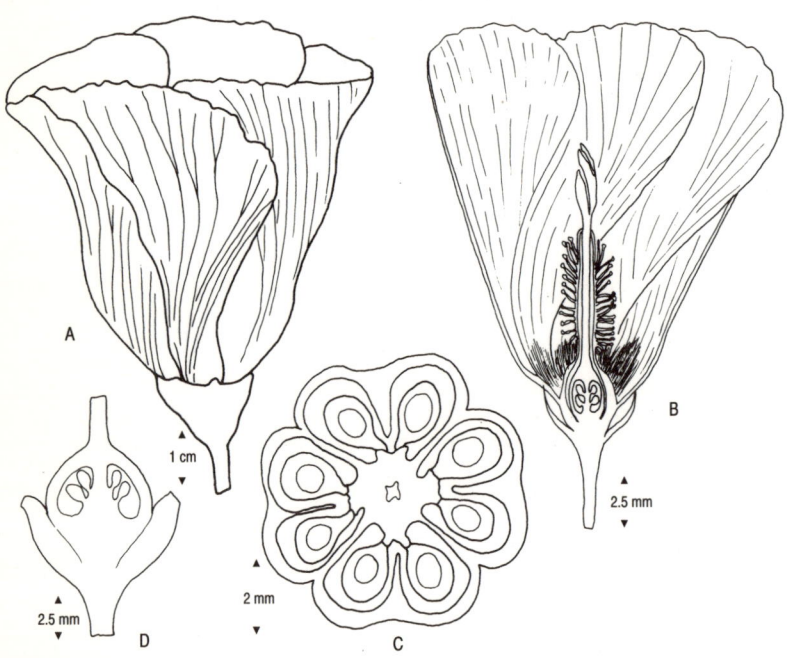

Key: full flower (A), half flower (B), transverse section (C) and longitudinal section (D) of ovary of *Thespesia populnea*.

1. **Hibiscus eriocarpus**, (T I:152 & VI:29),
   N, 3, shrub to small tree.
   Leaves: large, roundish, cordate base,
   palmately 3-lobed, lobes acute,
   margins **coarsely dentate**, hairy.
   Trunk: much branched;
   B-smooth; young
   parts **tomentose**.
   Flowers: pale pink to white
   with deep purple centre,
   large, hairy, conspicuous,
   axillary and solitary.
   Fruits: globose, depressed
   or pointed, densely
   hairy capsule.
   Site: monsoon forest
   understory; DL.
   Uses: ornamental.

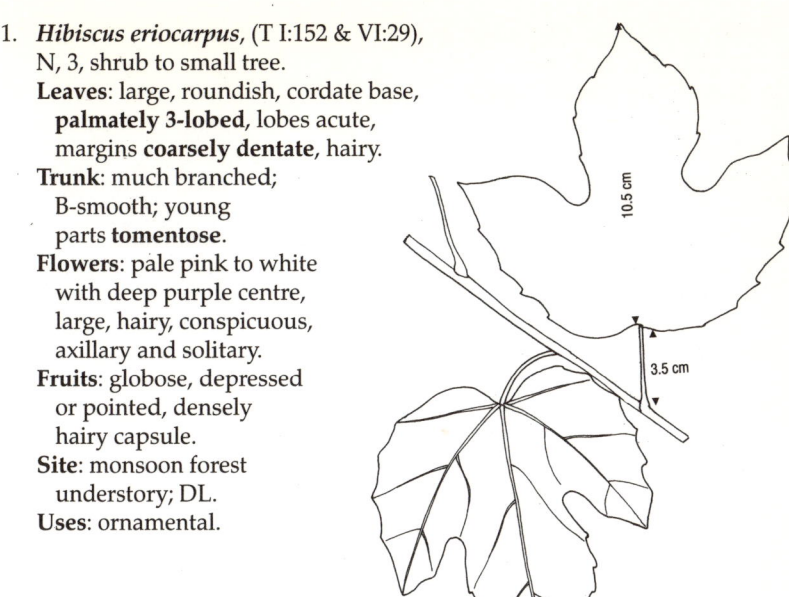

57

2. **Hibiscus rosa-sinensis**, shoe flower (E)/ wada mal (S)/
   nir paratthi (T), (T VI:30), I (Malesia), 3, shrub.
   Leaves: ovate, dark green, acuminate apex,
   margins **coarsely serrate**.
   Trunk: B-smooth,
   grey; IB-green.
   Flowers: red to white,
   large, solitary,
   conspicuous,
   funnel-shaped.
   Fruits: brown,
   small capsule.
   Site: home gardens;
   widespread.
   Uses: ornamental;
   leaves, root,
   flower
   medicinal.

MALVACEAE

3. *Hibiscus tiliaceus*, belli patta (S),
   (T I:157), N, 4, small tree.
   Leaves: **heart-shaped**, **9 or 11**
      main veins from base, beneath
      star-shaped hairs; petiole long.
   Trunk: B-smooth, grey,
      fibrous; young
      shoots **pubescent**.
   Flowers: yellow with
      crimson centre, large,
      funnel-shaped.
   Fruits: grey-green,
      elliptic, hairy capsule.
   Site: swampy thickets,
      coastal scrub; MC, DL.
   Uses: whole plant-medicinal;
      B-peeled for rope, fish-nets, mats.

4. *Thespesia populnea*, suriya (S)/pu varathu (T),
   (T I:158), I (Malesia), 10, tree.
   Leaves: **heart-shaped**, shiny green, **palmately** 7-veined;
      petioles long. IB-fibrous, yellowish; W,s-light brown;
      h-chocolate brown; young twigs with peltate scales.
   Flowers: pale yellow with purple centres, single, large, bell-shaped.
   Fruits: dark grey, rounded, flattened, dry capsule; seeds woolly.
   Site: coastal thickets; MC, DC.
   Uses: ornamental; live fencing; green manure;
      whole plant-medicinal; W-furniture, boats.

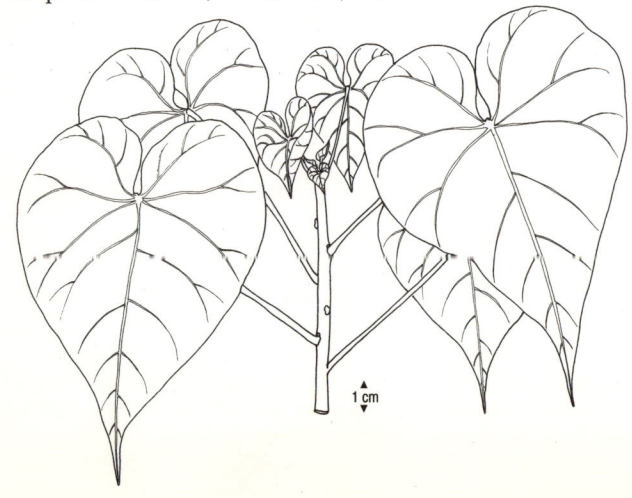

58. MELASTOMATACEAE

FAMILY DESCRIPTION - Habit: trees, shrubs, herbs, or lianas. Often with 4-angled stems. **Leaves**: opposite, rarely whorled, simple, often with 3-9 prominent nearly parallel veins. **Stipules**: absent. **Flowers**: bisexual, actinomorphic except the androecium, cymes. **Fruits**: loculicidal capsule or berry.

FLOWER PARTS - CALYX: 3-10, or a rim on hypanthium, sometimes united into a hood. COROLLA: 3-10, free, convolute in bud. ANDROECIUM: number same as petals or double, often dimorphic. Filaments free, often geniculate and twisted to one side of the flower. Anthers 2-locular, basifixed and dehiscing by a single terminal pore, rarely by 2 slits. GYNOECIUM: mostly inferior, 2- to many-locular, rarely 1-locular, Style simple. Ovules numerous per loculus, with axile, rarely basal or parietal placentation.

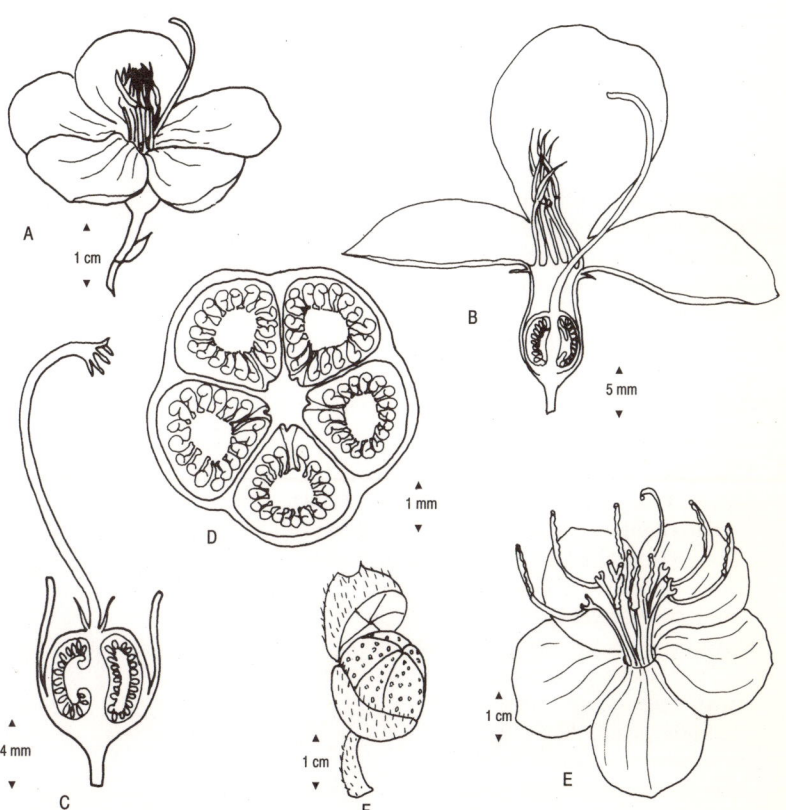

Key: full flower (A), half flower (B), longitudinal section of gynoecium (C) and transverse section of ovary (D) of *Osbeckia octandra*. Full flower (E) and mature fruit (F) of *Melastoma malabathricum*.

1. **Axinandra zeylanica**, pol hunna (S), (DF VI:239), E, 10, tree.
   Leaves: large, lanceolate-oblong, **subcordate** base, acuminate to acute apex, veins prominent beneath with **intramarginal pair**.
   Trunk: B-smooth, pale brown; branches numerous, straight, **drooping** almost to base; twigs with **4 stipular wings** at nodes.
   Flowers: white, very small, disagreeable odour; I-lax, slender, stalked, axillary and terminal racemes.
   Fruits: blackish brown, ovoid, woody, dehiscent capsule; seeds winged.
   Site: rain forest subcanopy; UW.

2. **Lijndenia capitellata**, pini baru (S), (DF VI:203), E, 3, shrub.
   Leaves: ovate, rounded base, long-pointed to obtuse apex, beneath pale, **3-veined**; flush purplish.
   Trunk: B-smooth, pale brown; branchlets slender, **cylindrical**.
   Flowers: white, very small; I-axillary heads.
   Fruits: blue-black, globose berry.
   Site: rain forest understory; W.

3. **Melastoma malabathricum**, ma bowitiya (S), (DF VI:159), N, 3, shrub.
   Leaves: lanceolate, acute ends, margins **faintly ciliate-denticulate**, **5-veined** with outermost close to margin, beneath **finely softly** hairy.
   Trunk: B-fibrous, yellowish; branchlets cylindrical, **red bristly** hairs.
   Flowers: violet-mauve, few, conspicuous, silvery scales; I-terminal paniculate cymes.
   Fruits: black, soft, coriaceous, broadly ovoid, dehiscing irregularly in a circle exposing 5 large segments of purplish black or white, pulpy placenta covered with minute seeds.
   Site: rain forest gaps and fringes, secondary forest and scrub; W.
   Uses: fruit-edible.

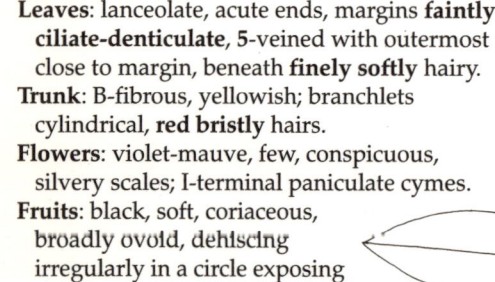

4. ***Memecylon angustifolium***, blue mist (E)/kora kaha (S),
   (DF VI:217), N, 2, shrub.
   Leaves: **linear to linear-**
      lanceolate, tapering base,
      caudate-acuminate to rounded
      apex, stiff, shiny above,
      veins **inconspicuous**.
   Trunk: branches slender,
      erect; twigs **quadrangular**.
   Flowers: bright purple-blue,
      I-umbellate or paniculate.
   Fruits: blackish purple, small berry.
   Site: monsoon and rain forest
      understory; along waterways; DL, W.
   Uses: ornamental.

5. ***Memecylon capitellatum***,
   weli kaha (S)/katti kaya (T),
   (DF VI:213), E, 3, shrub.
   Leaves: **oblong to rhomboid-oval**,
      base acute, obtuse apex, bright
      light green, shiny, lateral veins
      inconspicuous, **intramarginal**
      pair faint, petiole short.
   Trunk: twigs brown, **cylindrical**.
   Flowers: violet-blue, large, sessile,
      few together; I-small heads.
   Fruits: black-purple, globose
      berry, crowned
      with large calyx.
   Site: monsoon and rain
      forest understory; DL, IN, W.
   Uses: leaves, stem-medicinal;
      W-tool handles.

6. ***Memecylon parvifolium***, (DF VI:232), E, 10, tree.
   Leaves: broadly obovate, base tapering, **obtuse**
      to rounded apex, veins **invisible**; dark
      green, shiny above; young flush reddish.
   Trunk: B-grey, finely cracked; W-yellow,
      hard, heavy; twigs **quadrangular**.
   Flowers: white to pinkish red,
      few, small, nearly sessile.
   Fruits: black berry, tipped with calyx.
   Site: montane forest subcanopy; M.

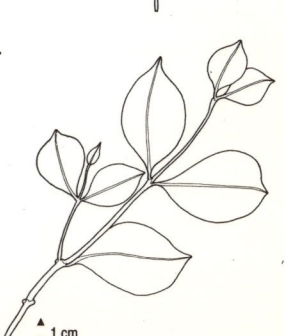

7. *Memecylon rostratum,* kuretiya (S),
   (DF VI:224), E, 4, small tree.
   Leaves: **small**, lanceolate to oval,
   base tapering, caudate-acuminate
   to obtuse apex, **faint** veins.
   Trunk: W-yellow, smooth, hard;
   branchlets numerous; twigs
   **compressed or nearly
   quadrangular**.
   Flowers: pale blue, very
   small; I-stalked
   umbels in leaf axils
   or leafless nodes.
   Fruits: small, globose berry.
   Site: rain forest understory; W.

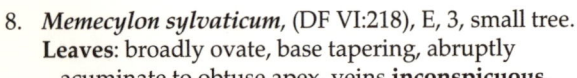

8. *Memecylon sylvaticum,* (DF VI:218), E, 3, small tree.
   Leaves: broadly ovate, base tapering, abruptly
   acuminate to obtuse apex, veins **inconspicuous**.
   Trunk: twigs **cylindrical**.
   Flowers: deep violet petals,
   white to pink calyx, large,
   nearly sessile; I-cymes small,
   crowded at leafless nodes.
   Fruits: ovoid, dark purple
   berry, crowned with calyx.
   Site: monsoon and rain
   forest understory; DL,W.

9. *Memecylon umbellatum,* kora kaha (S)/
   kaya (T), (DF VI:210), N, 4, shrub or small tree.
   Leaves: oval to oblong, acute base, slightly pointed
   to obtuse apex, lateral veins hardly visible,
   intramarginal pair, faintly **3** veined at base.
   Trunk: B-smooth, whitish;
   twigs **cylindrical**.
   Flowers: brilliant blue,
   small; I-umbellate
   cymes at leafless nodes.
   Fruits: berry, inconspicuous
   calyx at apex.
   Site: monsoon and rain forest
   understory; DL, IN, W.
   Uses: leaves-yellow dye;
   fruit-edible; ornamental;
   roots-medicinal; W-tool
   handles.

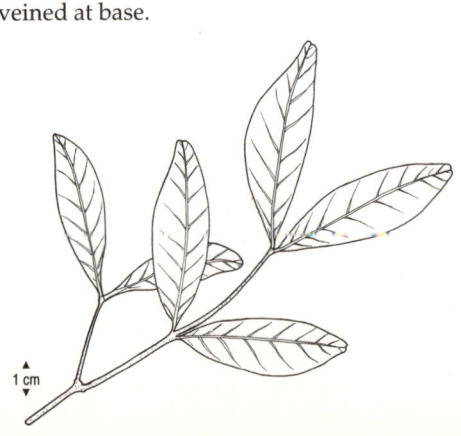

10. *Memecylon varians*, (DF VI:222), E, 4, small tree.
  Leaves: ovate-lanceolate, tapering base,
    caudate-acuminate to acute
    apex, thick, veins
    hardly visible.
  Trunk: twigs **quadrangular**.
  Flowers: pale blue-white,
    nearly sessile; I-small
    axillary heads.
  Fruits: berry capped
    with small calyx rim.
  Site: montane and rain
    forest subcanopy;
    M, LM, UW.

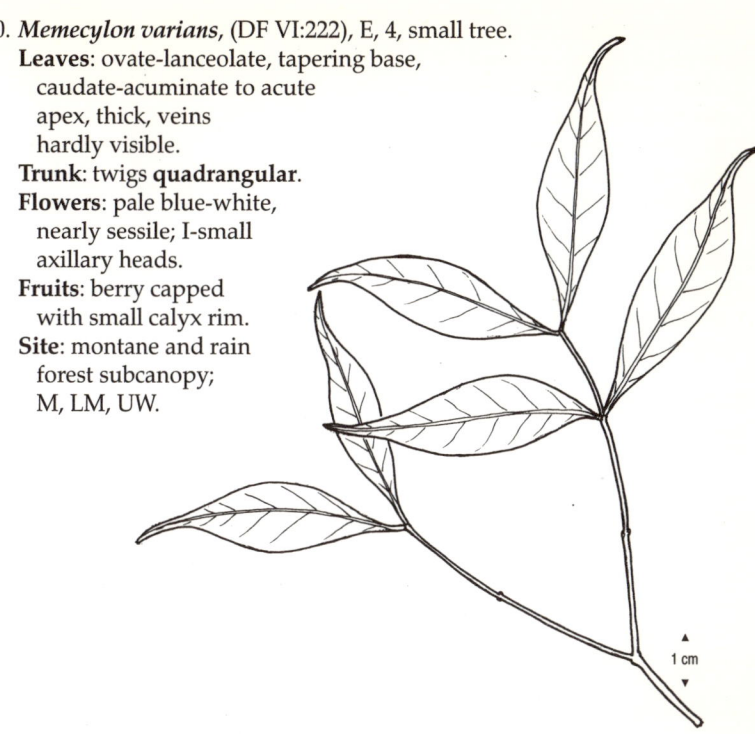

1 cm

11. *Osbeckia aspera*, bowitiya (S), (DF VI:166), N, 3, shrub.
  Leaves: variable, oblong-lanceolate, rounded base,
    acute slightly twisted apex,
    margins **finely dentate to
    scalloped**, 5 main veins
    from base, lateral veins
    faint; hairy on both surfaces.
  Trunk: twigs **quadrangular**.
  Flowers: mauve to brilliant
    crimson, often large,
    showy; hypanthium
    covered by bristly
    hairs; I-short cymes.
  Fruits: ovoid, truncate,
    rough, prickly capsule.
  Site: forest understory,
    secondary scrub;
    widespread.

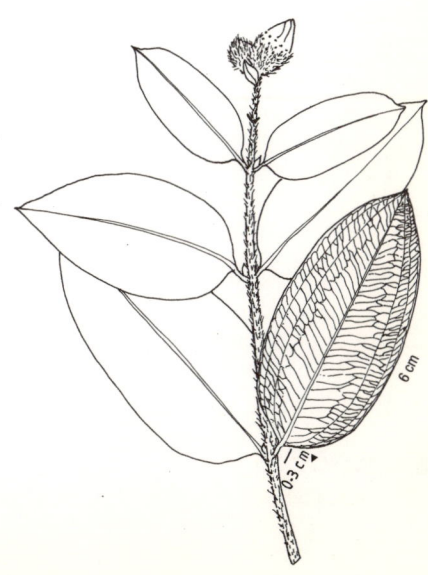

6 cm

0.3 cm

12. *Osbeckia lanata*, (DF VI:174), E, 3, shrub.
  Leaves: numerous, **crowded**, **small**,
    rotund to oval, rounded base,
    emarginate apex, margins **revolute**,
    3 main veins from base that are
    depressed on upper surface,
    beneath with **long brownish** hairs.
  Trunk: much-branched crown,
    flat-topped; twigs with
    copious **brownish wool**.
  Flowers: rich mauve, many,
    bracts leaf-like, reddish brown
    hairy, solitary or 3's, sessile.
  Fruits: capsule.
  Site: montane forest understory; M.

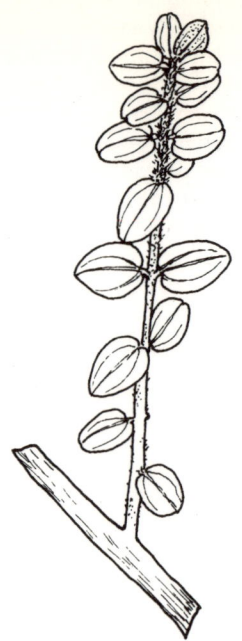

13. *Osbeckia octandra*, hin bowitiya (S),
  (DF VI:163), N, 2, shrub.
  Leaves: narrowly lanceolate to oblong,
    tapering base, margins **ciliate**, **faintly
    serrate**, 3 main veins from base,
    beneath scattered long hairs;
    petiole flat, hairy.
  Trunk: B-flaking in fibrous pieces,
    pale reddish brown; branches erect,
    twigs prominent; internodes
    quadrangular, very **bristly**.
  Flowers: pale purple, numerous,
    short, hairy pedicels; I-terminal
    cymes, corymbose or paniculate.
  Fruits: capsule.
  Site: montane and rain
    forest gaps and fringes,
    secondary forest, scrub
    and grasslands; M, W.
  Uses: leaves,
    roots-medicinal;
    ornamental.

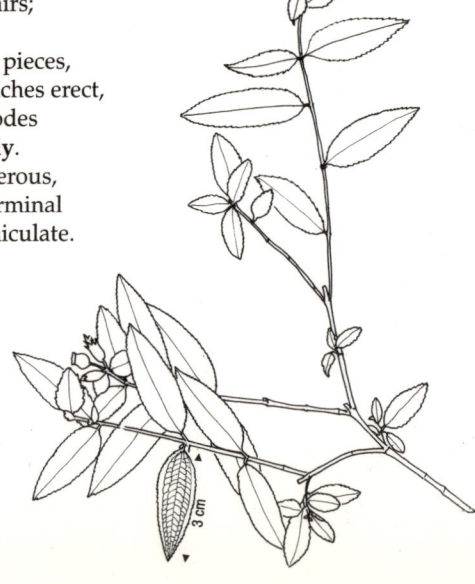

59. MELIACEAE

FAMILY DESCRIPTION - Habit: trees or shrubs, sometimes with hard, sweet-scented wood. **Leaves**: spiral, rarely decussate, pinnate to bipinnate, unifoliolate or simple. Leaflets entire, sometimes spiny. **Stipules**: absent. **Flowers**: bisexual or unisexual, (plants dioecious, polygamous or monoecious), actinomorphic, in spikes to dense panicles. **Fruits**: capsule, berry or drupe. Seeds winged and attached to a woody central axis, with corky outer layers, or with fleshy sarcotesta or aril, or a combination of both, or none of these.

FLOWER PARTS - CALYX: usually 4 or 5, sometimes reduced, usually on top of a tube, imbricate. COROLLA: 3-14 in 1 or 2 whorls, sometimes basally united, imbricate, convolute or valvate. ANDROECIUM: stamens 8-10, rarely numerous, mostly with united filaments and often anthers sessile in the tube. Anthers 2-locular, opening lengthwise. Nectary disc present around ovary. GYNOECIUM: superior, usually 2-6-locular. Ovules 1 to many, per loculus. Style 1 or none, stigma disciform or capitate.

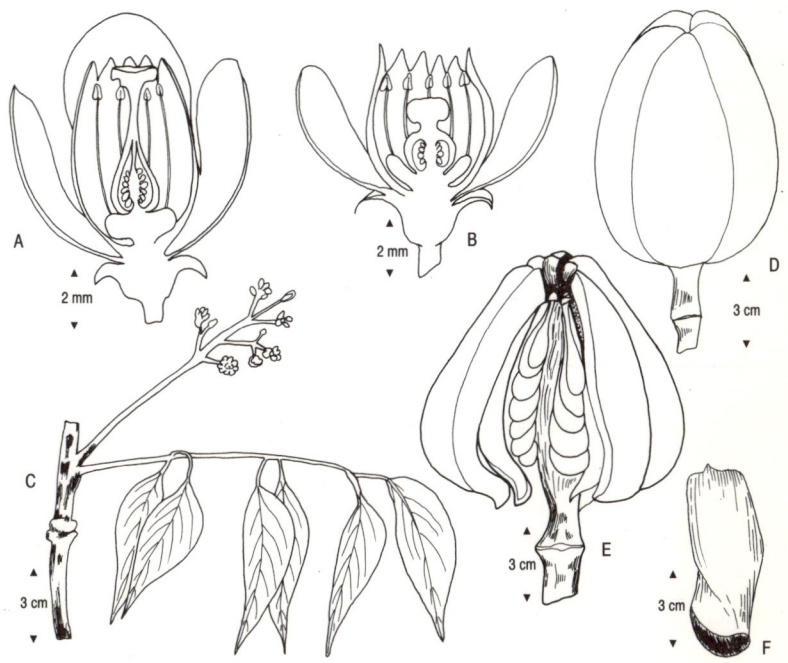

Key: male (A) and female (B) half flowers, inflorescence (C), mature undehisced (D) and dehisced fruits (E), and winged seed (F) of *Swietenia macrophylla*.

1. ***Aglaia apiocarpa***, (T I:245), N, 5, small tree.
   Leaves: pinnate, variable in size; leaflets paired
      with terminal, oval-lanceolate, acute
      to tapering base, pointed apex,
      coppery scaly beneath.
   Trunk: young parts with
      minute dark brown scales.
   Flowers: yellow; I-lax,
      brown-scaly, panicles.
   Fruits: orange, pear-shaped
      to globose, smooth berry.
   Site: montane forest understory; LM.

2. ***Aglaia elaeagnoidea***, puwangu (S)/
   kannakompu (T), (T I:246), N, 10, tree.
   Leaves: pinnate; leaflets paired with terminal,
      oval, acute or tapering base, obtuse
      apex, margins somewhat **undulate**.
   Trunk: B-thin, smooth, dull orange
      grey; W-hard, brown, heavy; young
      parts **minutely scaly**.
   Flowers: yellow, male small
      and numerous, female few;
      I-spreading panicles.
   Fruits: pale orange,
      ovoid to pear-shaped,
      minutely rusty hairy
      berry; seeds 1 or 2.
   Site: monsoon forest
      subcanopy; DL.
   Uses: whole plant-medicinal.

3. ***Aphanamixis polystachya***,
   hingul (S), (T I:249), N, 20, tree.
   Leaves: pinnate; leaflets **unequal**
      at base, pointed to obtuse
      apex; flush **purple**.
   Trunk: B-smooth, thin;
      W-hard, dark red; branches
      drooping and straight;
      young parts **finely silky**.
   Flowers: sessile, I-male numerous,
      small, axillary panicles; female
      few, large, long drooping spikes.
   Fruits: red, globose, fleshy, smooth
      capsule; arils yellow, fleshy.
   Site: montane and rain forest
      canopy and subcanopy; LM, UW.
   Uses: root, stem, B-medicinal.

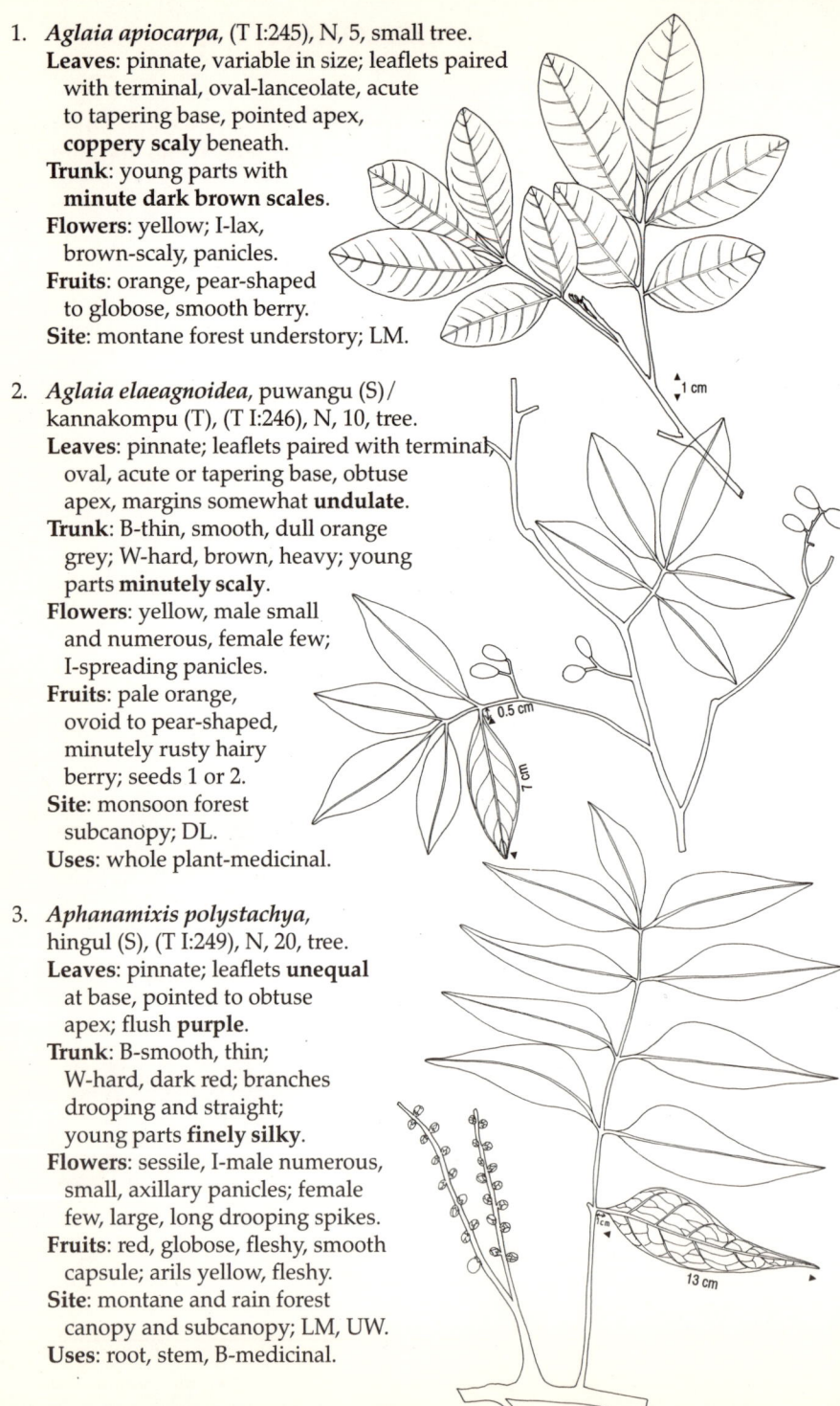

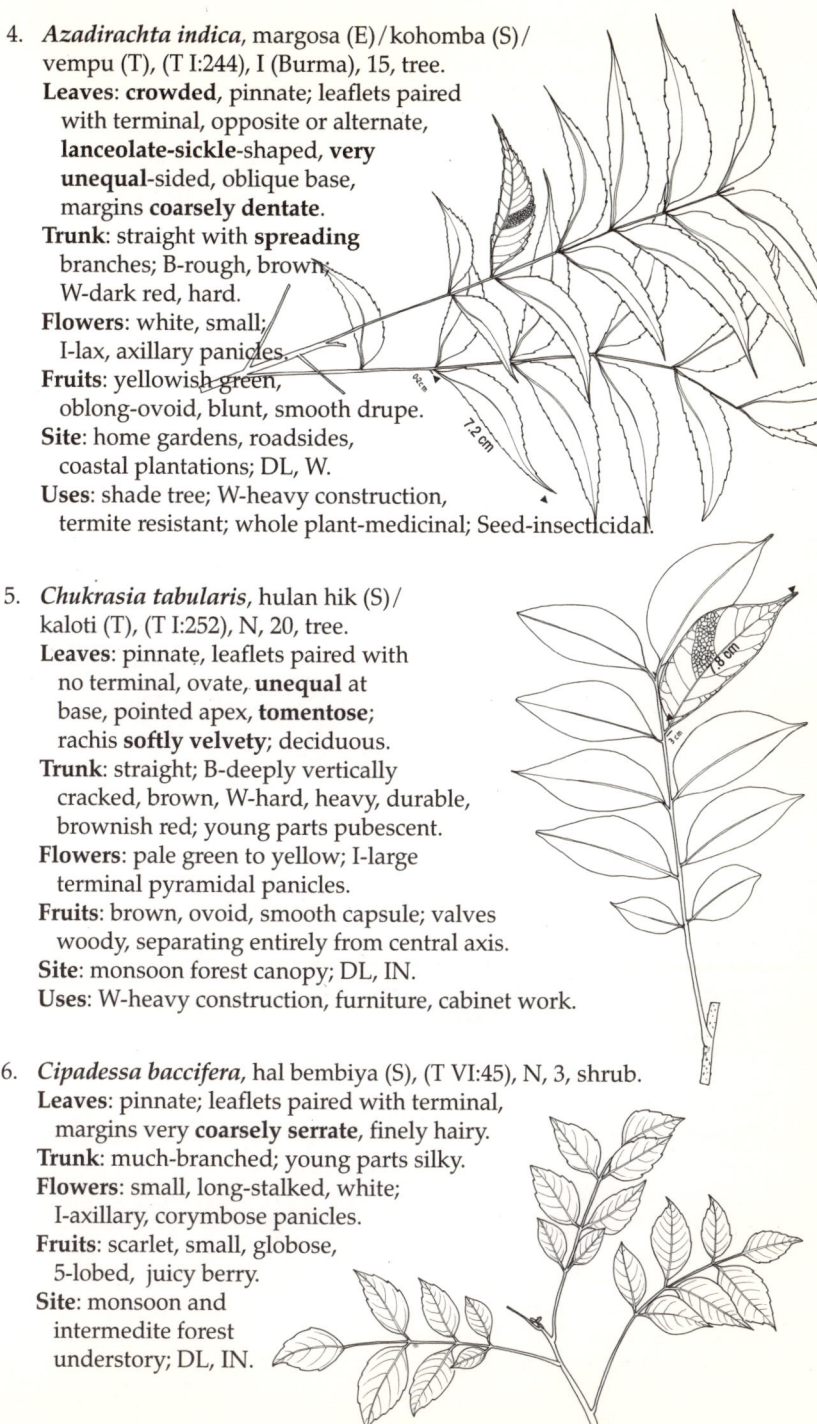

4. ***Azadirachta indica***, margosa (E)/kohomba (S)/
vempu (T), (T I:244), I (Burma), 15, tree.
   Leaves: **crowded**, pinnate; leaflets paired
   with terminal, opposite or alternate,
   lanceolate-sickle-shaped, **very
   unequal**-sided, oblique base,
   margins **coarsely dentate**.
   Trunk: straight with **spreading**
   branches; B-rough, brown,
   W-dark red, hard.
   Flowers: white, small;
   I-lax, axillary panicles.
   Fruits: yellowish green,
   oblong-ovoid, blunt, smooth drupe.
   Site: home gardens, roadsides,
   coastal plantations; DL, W.
   Uses: shade tree; W-heavy construction,
   termite resistant; whole plant-medicinal; Seed-insecticidal.

59

5. ***Chukrasia tabularis***, hulan hik (S)/
kaloti (T), (T I:252), N, 20, tree.
   Leaves: pinnate, leaflets paired with
   no terminal, ovate, **unequal** at
   base, pointed apex, **tomentose**;
   rachis **softly velvety**; deciduous.
   Trunk: straight; B-deeply vertically
   cracked, brown, W-hard, heavy, durable,
   brownish red; young parts pubescent.
   Flowers: pale green to yellow; I-large
   terminal pyramidal panicles.
   Fruits: brown, ovoid, smooth capsule; valves
   woody, separating entirely from central axis.
   Site: monsoon forest canopy; DL, IN.
   Uses: W-heavy construction, furniture, cabinet work.

MELIACEAE

6. ***Cipadessa baccifera***, hal bembiya (S), (T VI:45), N, 3, shrub.
   Leaves: pinnate; leaflets paired with terminal,
   margins very **coarsely serrate**, finely hairy.
   Trunk: much-branched; young parts silky.
   Flowers: small, long-stalked, white;
   I-axillary, corymbose panicles.
   Fruits: scarlet, small, globose,
   5-lobed, juicy berry.
   Site: monsoon and
   intermedite forest
   understory; DL, IN.

1 cm

7. *Dysoxylum ficiforme*, (T I:247), N, 30, tree.
   Leaves: pinnate; leaflets alternate, oval, acute,
      often **unequal**-sided base, **finely puberulous**.
   Trunk: straight; W-hard, heavy,
      reddish; young parts puberulous.
   Flowers: green, few; I-axillary panicles.
   Fruits: orange, depressed globose, deeply
      4 or 5 grooved, smooth capsule.
   Site: intermediate and rain
      forest canopy; IN, W.
   Uses: W-heavy construction.

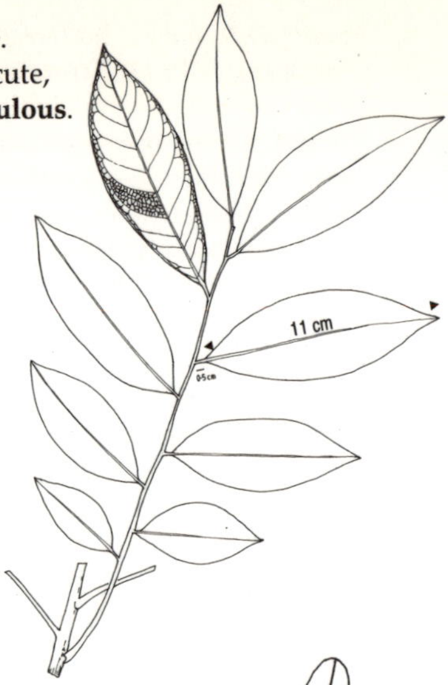

8. *Dysoxylum championii*, gona pana (S),
   (T I:248), E, 25, tree.
   Leaves: pinnate; leaflets obovate-
      oblong, tapering base, apex obtuse.
   Trunk: B-rough, grey; W-very hard,
      heavy, reddish.
   Flowers: yellow; I-axillary racemose panicles.
   Fruits: globose topear-shaped capsule,
      pointed, smooth; seeds
      half-covered by
      reddish aril.
   Site: montane and rain
      forest canopy;
      UM, LM, W.

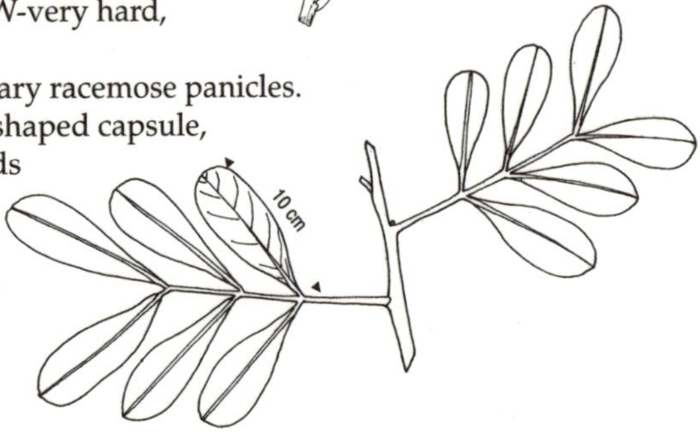

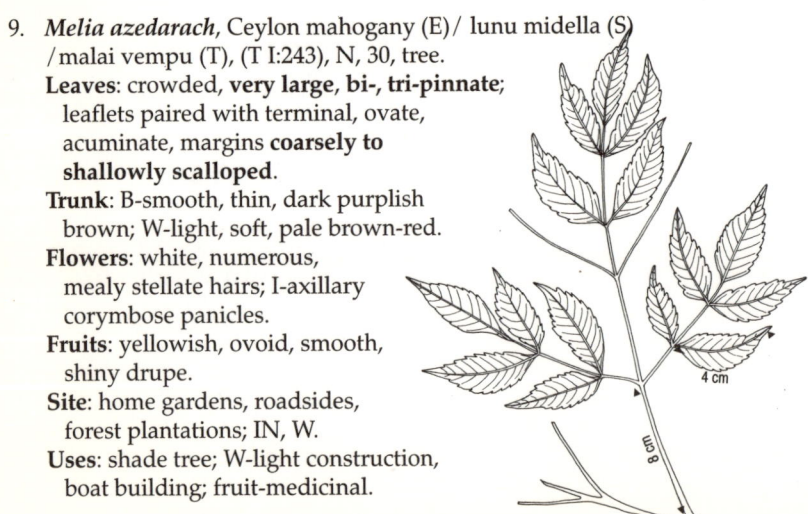

9. *Melia azedarach*, Ceylon mahogany (E)/ lunu midella (S)
   /malai vempu (T), (T I:243), N, 30, tree.
   Leaves: crowded, **very large**, **bi-, tri-pinnate**;
      leaflets paired with terminal, ovate,
      acuminate, margins **coarsely to
      shallowly scalloped**.
   Trunk: B-smooth, thin, dark purplish
      brown; W-light, soft, pale brown-red.
   Flowers: white, numerous,
      mealy stellate hairs; I-axillary
      corymbose panicles.
   Fruits: yellowish, ovoid, smooth,
      shiny drupe.
   Site: home gardens, roadsides,
      forest plantations; IN, W.
   Uses: shade tree; W-light construction,
      boat building; fruit-medicinal.

10. *Swietenia macrophylla*, Honduran (large-leaved) mahogany (E)/
    mahogani (S), I (Trop. Amer.), 20, tree.
    Leaves: pinnate; leaflets paired with no
      terminal, **unequal-sided** base.
    Trunk: buttressed; B-rough, fissured,
      light brown; IB-dark reddish;
      W,s-yellowish; h-pinkish to
      reddish, brown on exposure.
    Flowers: greenish yellow;
      I-short-stalked panicles.
    Fruit: erect, egg-shaped capsule,
      splitting upward; seeds winged.
    Site: intermediate and wet zone in forest
      plantations, home gardens; IN, W.
    Uses: W-construction, furniture.

11. *Swietenia mahagoni*, Cuban (small-leaved) mahogany (E)/
    mahogani (S), I (Trop. America), 20, tree.
    Leaves: pinnate; leaflets paired with no
      terminal, **unequal**-sided base.
    Trunk: B-smooth to scaly, dark
      reddish brown; IB-pink; W,s-whitish;
      h-turns dark red on exposure.
    Flowers: greenish
      yellow; I-lateral panicles.
    Fruits: dark brown, egg-shaped
      capsule; seeds winged.
    Site: intermediate zone
      in forest plantations; IN.
    Uses: W-all-purpose construction, furniture.

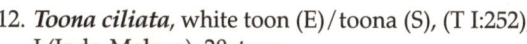

1 cm

1 cm

12. *Toona ciliata*, white toon (E)/toona (S), (T I:252),
    I (Indo-Malaya), 20, tree.
    Leaves: pinnate; leaflets paired with no
      terminal; flush **crimson;** deciduous.
    Trunk: B-deeply cracked, red-brown,
      hard; W-fragrant, reddish, durable.
    Flowers: white, small;
      I-axillary panicle.
    Fruits: woody capsule.
    Site: tea estates; M, LM.
    Uses: shade tree; W-construction,
      furniture, termite-resistant;
      ornamental; capsules-ornamental
      when dry.

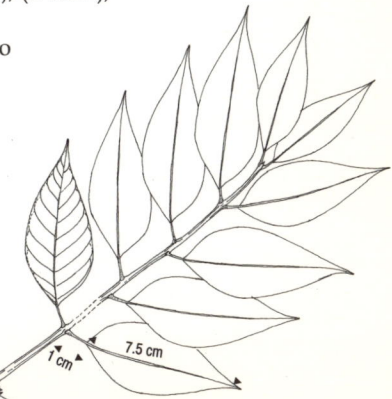

1 cm

7.5 cm

59

MELIACEAE

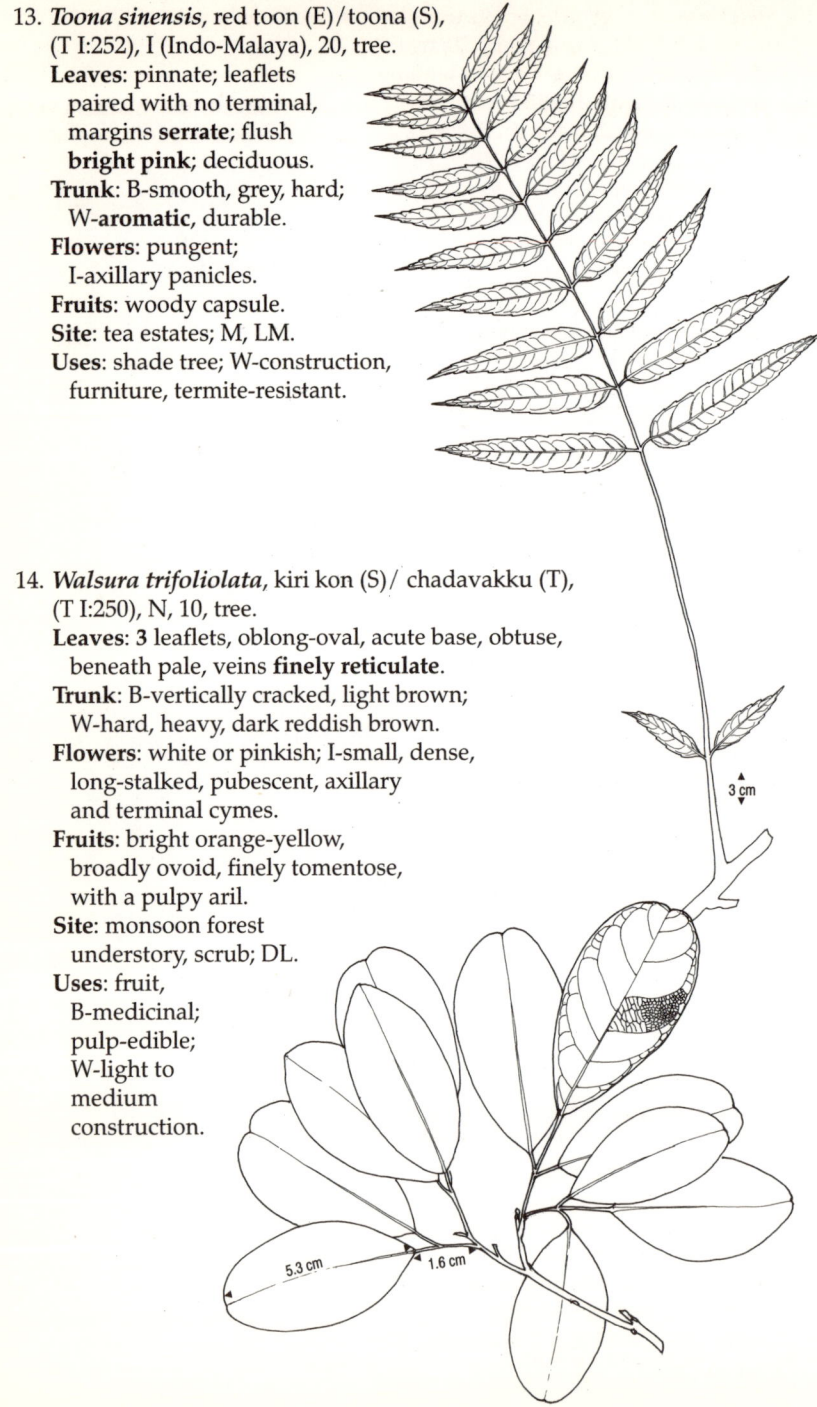

13. ***Toona sinensis***, red toon (E)/toona (S),
(T I:252), I (Indo-Malaya), 20, tree.
Leaves: pinnate; leaflets
paired with no terminal,
margins **serrate**; flush
bright pink; deciduous.
Trunk: B-smooth, grey, hard;
W-**aromatic**, durable.
Flowers: pungent;
I-axillary panicles.
Fruits: woody capsule.
Site: tea estates; M, LM.
Uses: shade tree; W-construction,
furniture, termite-resistant.

14. ***Walsura trifoliolata***, kiri kon (S)/ chadavakku (T),
(T I:250), N, 10, tree.
Leaves: 3 leaflets, oblong-oval, acute base, obtuse,
beneath pale, veins **finely reticulate**.
Trunk: B-vertically cracked, light brown;
W-hard, heavy, dark reddish brown.
Flowers: white or pinkish; I-small, dense,
long-stalked, pubescent, axillary
and terminal cymes.
Fruits: bright orange-yellow,
broadly ovoid, finely tomentose,
with a pulpy aril.
Site: monsoon forest
understory, scrub; DL.
Uses: fruit,
B-medicinal;
pulp-edible;
W-light to
medium
construction.

3 cm

5.3 cm

1.6 cm

60. MONIMIACEAE

FAMILY DESCRIPTION - Habit: trees, shrubs or lianas. **Leaves**: usually opposite, simple, often dotted with glands. **Stipules**: absent. **Flowers**: bisexual or unisexual, actinomorphic, solitary or in axillary cymes, usually with a hypanthium. **Fruits**: head of drupes or nuts, often enclosed by hypanthium.

FLOWER PARTS - Calyx: 4 in two whorls, fleshy. Corolla: 7-20 or more, or perianth not differentiated into calyx and corolla, sometimes reduced or absent. Androecium: numerous stamens, in 1 or 2 whorls. Filaments short, often flattened with or without nectaries at their base. Anthers 2-locular, opening lengthwise, or by valves from the base upwards. Gynoecium: superior, carpels free, several or rarely solitary. Each carpel unilocular with 1 ovule. Style short or elongated, stigma terminal, carpels sometimes sunk in receptacle. Placentation basal, erect or pendulous.

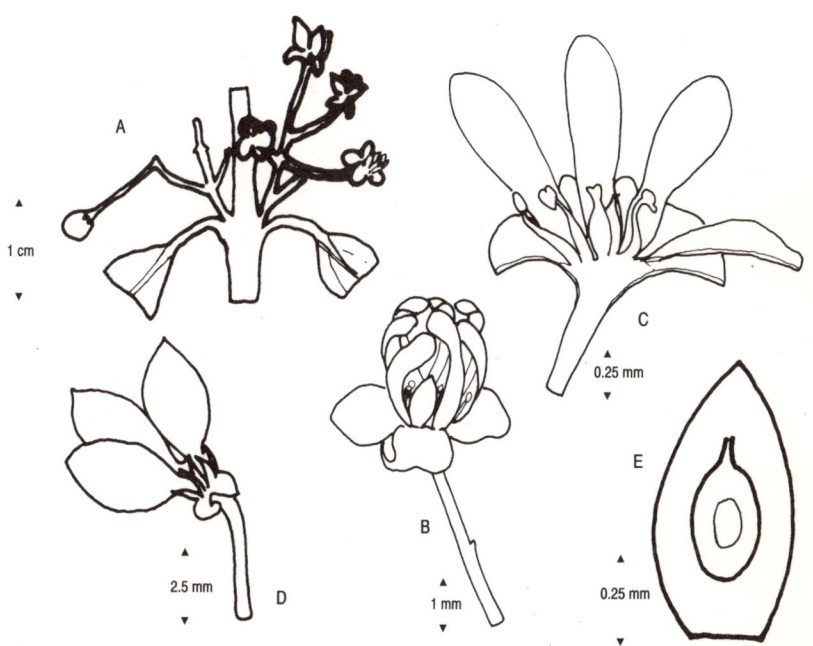

Key: inflorescence (A), full flower (B), half flower (C), aggregate of young drupes (D), and a longitudinal section of a drupe (E) of *Hortonia floribunda*.

1. *Hortonia angustifolia*, (T III:437), E, 5, small tree.
   Leaves: linear-lanceolate, pointed ends, faintly 3 veined at base.
   Trunk: branches slender.
   Flowers: pale yellow, slender, drooping; I-racemes with few flowers.
   Fruits: dark crimson, ovoid, compressed, obliquely pointed, pulpy.
   Site: rain forest near waterways; W.

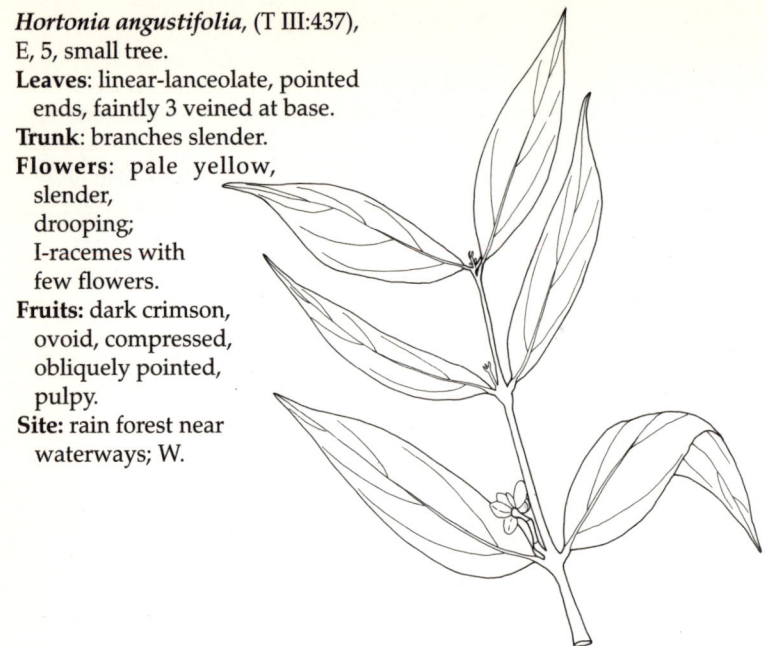

2. *Hortonia floribunda*, wawiya (S), (T III:437), E, 5, small tree.
   Leaves: lanceolate to oval, acute base.
   Trunk: branches long, glabrous.
   Flowers: yellow; I-nearly sessile axillary racemes and panicles.
   Fruits: dark crimson, ovoid, much compressed, obliquely pointed, very pulpy.
   Site: montane and rain forest understory and fringes; M, UW.

61. MORACEAE

FAMILY DESCRIPTION - Habit: trees, shrubs, herbs, climbers or stranglers with milky latex. **Leaves**: spiral, rarely opposite, simple. **Stipules**: present. **Flowers**: bisexual or unisexual, (plants monoecious or dioecious), actinomorphic, small, in axillary heads, disks or hollow receptacles. **Fruits**: drupes, often aggregated (*Morus*), or united together with perianths and axes (*Artocarpus*), or achenes within a fleshy receptacle (*Ficus*).

FLOWER PARTS - Calyx: usually 4, free or basally united, sometimes reduced or absent. Corolla: absent. Androecium: stamens same number as calyx. Filaments free. Anthers 2-locular, opening lengthwise. Ovary rudimentary or absent in male flowers. Gynoecium: superior or inferior, with 2 carpels, often one rudimentary. Each carpel unilocular with a solitary pendulous or rarely basal and erect ovule. Styles mostly 2, filiform.

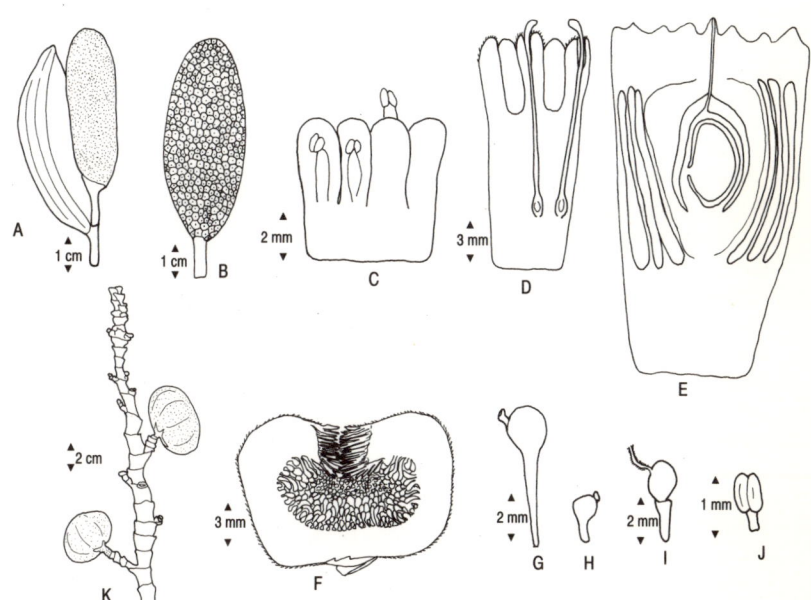

Key: male inflorescence (A), female inflorescence (B), a portion of the male inflorescence in section showing the stamens (C), a portion of the female inflorescence showing gynoecia (D), and a portion of the compound fruit showing fruitlet (E) of *Artocarpus heterophyllus*. Single inflorescence in section (F), gall flowers (G, H), female flower (I), male flower (J), and twig bearing inflorescences (K) of *Ficus hispida*.

1. *Artocarpus altilis*, bread fruit (E)/rata del,
   del (S)/ era pla (T), (DF III:219),
   I (New Guinea), 20, tree.
   Leaves: **shiny dark green**, very
     large, **lobed**, pointed hairy bud.
   Trunk: B-smooth with warty lenticels,
     brown; W, s-light yellow; h-golden, soft,
     light, termite-susceptible; latex milky.
   Flowers: numerous, minute;
     I- male thick, cylindrical cluster;
     female globose to oblong.
   Fruits: green, large with
     starchy white pulp.
   Site: home gardens; W.
   Uses: fruit-edible; shade tree;
     W-light construction.

2. *Artocarpus heterophyllus*, jak (E)/kos (S)/
   pla (T), (DF III:217), I(India), 25, tree.
   Leaves: **dark green**, elliptic to obovate,
     base cuneate, lateral veins **6-10 pairs**,
     with parallel intercostals; sapling
     leaves **1-3** pairs of lobes.
   Trunk: B-reddish brown;
     W-yellow; latex **milky**.
   Flowers: I-cauliflorous, cylindrical.
   Fruits: green, fleshy, large,
     composed of multiple segments.
   Site: home gardens; widespread.
   Uses: fruit-edible; leaves-fodder;
     W-construction, furniture,
     yellow dye; whole plant-medicinal.

3. *Artocarpus nobilis*, wal del (S)/arsini
   pla (T), (DF III:219), E, 25, tree.
   Leaves: **dark green**, stiff, obovate,
     shortly pointed apex, base cuneate,
     margins **wavy**, lateral veins **10-13 pairs**.
   Trunk: B-scaly, scarred,
     grey-orange; latex **milky**.
   Flowers: I-axillary, solitary or
     paired, cylindrical.
   Fruits: cylindrical, composed
     of multiple segments,
     persistent bracts.
   Site: intermediate and rain forest
     canopy and subcanopy,
     home gardens; IN, W.
   Uses: seeds-edible; W-construction.

4. ***Castilla elastica***, Panama rubber (E),
   (DF III:291), I (Central Amer.), 20, tree.
   Leaves: oblong to obovate, **cordate**
   to wedge-shaped base somewhat
   unequal, pointed apex; petiole
   short, stout, **golden** hairs.
   Trunk: B-light grey, lenticellate; twigs
   with adpressed golden hairs; latex milky.
   Flowers: I-male in groups of 4; female smaller.
   Fruits: orange-red, pulpy; seeds straw-coloured.
   Site: roadsides.
   Uses: latex-rubber; ornamental; shade tree.

5. ***Ficus benghalensis***, banyan (E)/nuga (S)/al (T),
   (DF III:251), I (India), 20, tree.
   Leaves: **spirally arranged**, ovate, base **cordate**,
   lateral veins **5-6** pairs, with **2-4** pairs at base,
   10 intercostals with regular reticulate veins,
   beneath **puberulous**.
   Trunk: rooting from
   spreading branches,
   large; latex **milky**.
   Flowers: I-globose,
   sessile, axillary, paired.
   Fruits: figs ripening orange
   to red, depressed globose.
   Site: roadsides; widespread.
   Uses: ornamental; sap, B-medicinal; aerial
   roots-brushing teeth; shade tree, sacred tree.

6. ***Ficus benjamina***, walu nuga (S), (DF III:256),
   I (India), 7, tree.
   Leaves: nearly 2-ranked, elliptic
   to lanceolate, base rounded,
   lateral veins **6-11** pairs, with
   secondary veins prominent.
   Trunk: B-grey, smooth;
   branches horizontal,
   interlocking; latex **milky**.
   Flowers: I-globose, sessile,
   axillary and paired.
   Fruit: figs ripening yellow
   to orange-red, globose.
   Site: gardens; W.
   Uses: ornamental; shade tree.

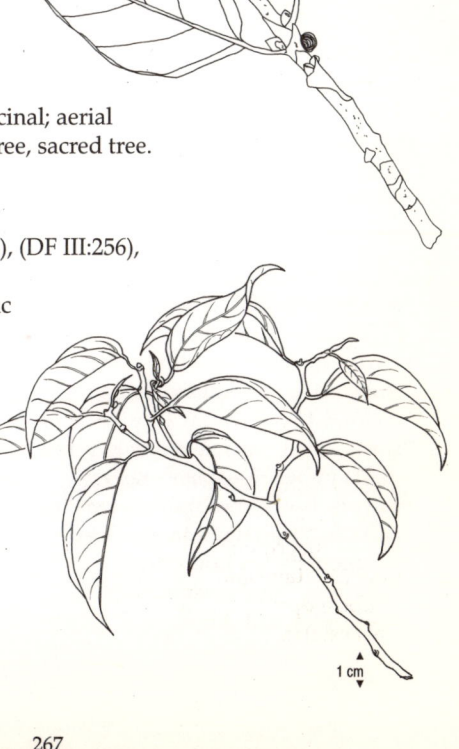

28 cm

1 cm

61

MORACEAE

7. *Ficus elastica*, India rubber tree (E),
   (DF III:261), I (India), 15, tree.
   Leaves: **spirally arranged**, elliptic to
     subovate, base wedge-shaped,
     lateral veins **15-20** pairs, with
     1-2 pairs at base, secondary veins
     well developed; stipules **red**.
   Trunk: copious **aerial roots**; B-warty
     with lenticels; latex **milky**.
   Flowers: I-obconical,
     axillary, paired.
   Fruits: figs ripening yellow.
   Site: home gardens,
     roadsides; widespread.
   Uses: ornamental, shade tree.

4 cm

8. *Ficus exasperata*, bu deliya (S),
   (DF III:274), N, 15, tree.
   Leaves: laxly spiral, opposite to
     nearly 2-ranked, varying elliptic,
     base wedge-shaped, margins **denticulate**
     to wavy, stiffly coriaceous, **coarsely** rough
     with white hairs, lateral veins **3-5** pairs,
     with **1-2** pairs at base, 2 basal glands.
   Trunk: latex **milky**.
   Flowers: I- sessile, subglobose, harshly
     scabrous, mostly solitary, axillary.
   Fruits: figs ripening yellow.
   Site: secondary forest canopy
     and subcanopy; DL, IN, W.
   Uses: leaves-sand paper.

10 cm

1 cm

9. *Ficus fergusonii*, kos gona (S)/al (T),
   (DF III:253), E, 15, tree.
   Leaves: spiral, ovate-elliptic, base
     subcordate to rounded, lateral
     veins **8-12** pairs with **2-3** pairs at base.
   Trunk: aerial roots near trunk; young
     parts **puberulous**; latex **milky.**
   Flowers: I-sessile subglobose, axillary, paired.
   Fruits: figs ripening orange to red.
   Site: montane and rain forest
     subcanopy and understory;
     tea estates; LM, M, W.

1 cm

1-10

10. *Ficus hispida*, kota dimbula (S), (DF III:277),
N, 5, small tree or shrub.
Leaves: **opposite**, oblong-elliptic,
acuminate apex, base **cordate** to
wedge-shaped, margins sometimes
scalloped, **4-9** pairs of curved
ascending lateral veins with
1-3 pairs at base, **rough**, pale
brown hispid; gland on petiole.
Trunk: laxly branched; B-smooth,
grey; twigs hollow; latex **milky**.
Flowers: I-globose, in clusters on leafless twigs,
sometimes close to the base of the trunk.
Fruits: figs ripening pale yellow,
depressed globose.
Site: secondary forest openings,
riverbanks; DL, IN.
Uses: whole plant-medicinal.

21 cm

2.8 cm

61

11. *Ficus microcarpa*, panu nuga (S), (DF III:258), N, 15, tree.
Leaves: spiral, 2-ranked, elliptic to obovate, lateral
veins **5-9** pairs at an acute angle with midrib,
1-2 pairs at base.
Trunk: developing **pillar roots** from
branches, **aerial roots** copious,
slender; B-grey; latex **milky**.
Flowers: I-subglobose, sessile,
axillary, paired.
Fruits: figs ripening pink to black.
Site: estuaries, river banks, rocky
outcrops, home gardens; widespread.
Uses: ornamental, shade tree.

1.4 cm

6.6 cm

MORACEAE

12. *Ficus nervosa*, kala maduwa (S),
(DF III:264), N, 25, tree.
Leaves: **spiral**, elliptic, base cuneate, lateral
veins **7-11** pairs with **1** pair at base; flush
pinkish brown; dense crown.
Trunk: buttressed; latex **scantily
white**, yellowish on exposure.
Flowers: sessile; I-pear-shaped,
axillary, paired or solitary.
Fruits: figs ripening red.
Site: montane and rain
forest canopy; LM, M, W.

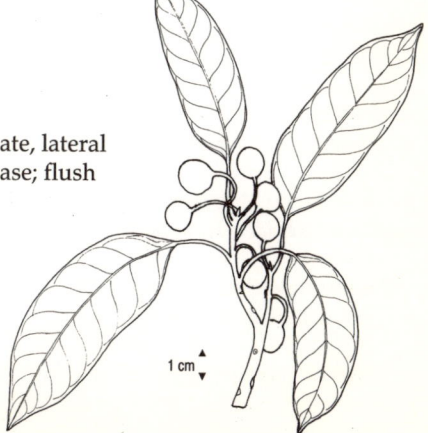

1 cm

13. *Ficus racemosa*, cluster fig (E) / attikka (S) / atti (T), (DF III:266), N, 20, tree.
   Leaves: elliptic, base wedge-shaped, lateral veins **4-8**
      pairs with **1** pair at base; stipules persistent on young
      twigs; crown often irregular and shabby.
   Trunk: buttressed;
      B-smooth to flaky, pinkish
      brown; latex copious, **milky**;
      young parts finely hairy.
   Flowers: I-subglobose to
      pear-shaped, large
      clusters on stems.
   Fruits: figs ripening red.
   Site: riverbanks; DL, IN.
   Uses: fruits-edible; whole
      plant-medicinal.

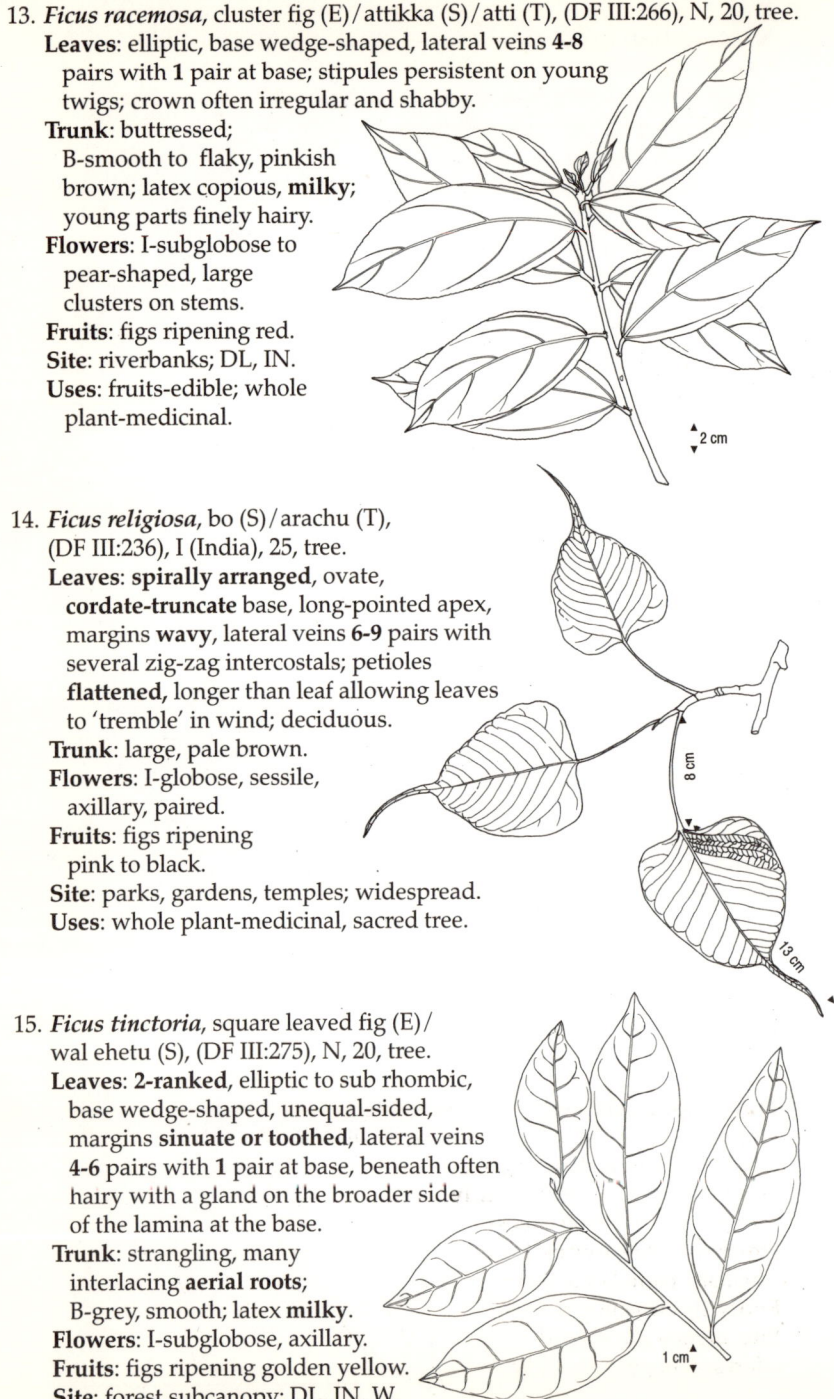

2 cm

14. *Ficus religiosa*, bo (S) / arachu (T),
   (DF III:236), I (India), 25, tree.
   Leaves: **spirally arranged**, ovate,
      cordate-truncate base, long-pointed apex,
      margins **wavy**, lateral veins **6-9** pairs with
      several zig-zag intercostals; petioles
      flattened, longer than leaf allowing leaves
      to 'tremble' in wind; deciduous.
   Trunk: large, pale brown.
   Flowers: I-globose, sessile,
      axillary, paired.
   Fruits: figs ripening
      pink to black.
   Site: parks, gardens, temples; widespread.
   Uses: whole plant-medicinal, sacred tree.

8 cm

13 cm

15. *Ficus tinctoria*, square leaved fig (E) /
   wal ehetu (S), (DF III:275), N, 20, tree.
   Leaves: **2-ranked**, elliptic to sub rhombic,
      base wedge-shaped, unequal-sided,
      margins **sinuate or toothed**, lateral veins
      4-6 pairs with **1** pair at base, beneath often
      hairy with a gland on the broader side
      of the lamina at the base.
   Trunk: strangling, many
      interlacing **aerial roots**;
      B-grey, smooth; latex **milky**.
   Flowers: I-subglobose, axillary.
   Fruits: figs ripening golden yellow.
   Site: forest subcanopy; DL, IN, W.

1 cm

16. *Ficus tsjahela*, kiri pella (S), (DF III:237), N, 20, tree.
   Leaves: **spiral**, oblong-elliptic, base rounded, shortly
      acuminate, lateral veins **8-10** pairs; deciduous.
   Trunk: fluted stem; B-cracks like dry mud.
   Flowers: I-depressed globose, sessile, axillary,
      clusters on twigs behind leaves.
   Fruits: dull grey-purple, very small.
   Site: intermediate and rain forest subcanopy
      and canopy; parks, roadsides, gardens; IN, W.
   Uses: ornamental, shade tree.

2 cm

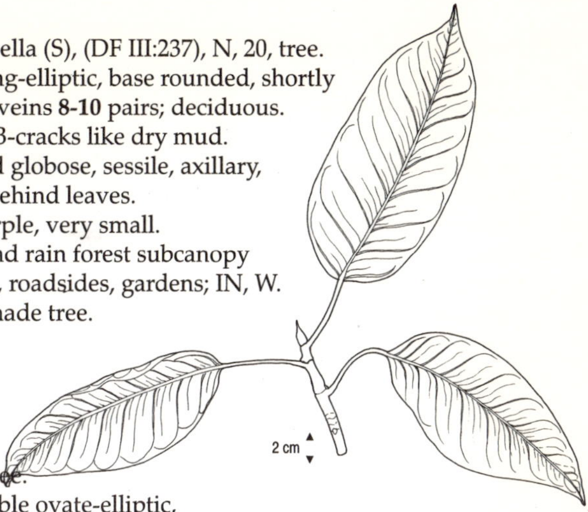

17. *Ficus virens*,
   (DF III:238), N, 20, tree.
   Leaves: **spiral**, variable ovate-elliptic,
      base rounded to subcordate,
      margins often **wavy**, lateral
      veins **7-12** pairs with
      1-2 pairs at base; deciduous.
   Trunk: tufts of **aerial roots**
      closely pressed to stem;
      B-greenish, smooth
      with faint vertical
      patterns of lenticels.
   Flowers: I-globose, paired, axillary.
   Fruits: figs ripening whitish
      pink to black.
   Site: monsoon forest canopy,
      gardens, roadsides; DL.
   Uses: ornamental, shade tree.

7 cm

2.9 cm

6

MORACEAE

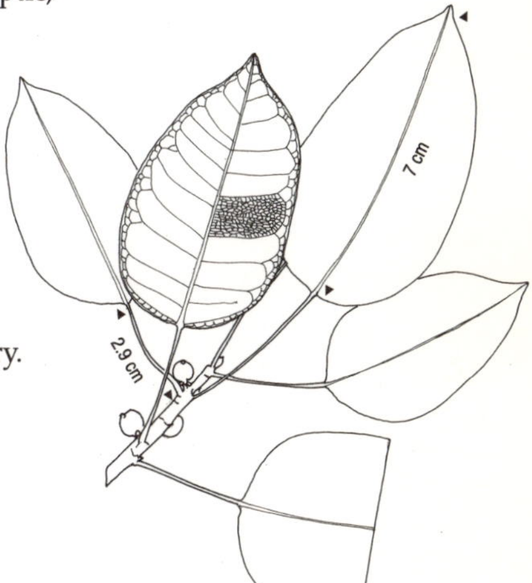

18. *Morus alba*, mulberry (E,S), I (China), 3, shrub.
   Leaves: 2-ranked, ovate, unequal-sided, round to slightly
      cordate base, long-pointed apex, margins **serrate**;
      petiole long; young shoots with **3-5-lobed** leaves;
      older branches with
      entire leaves.
   Trunk: crooked,
      sparingly branched.
   Flowers: I-axillary, distinctly
      stalked, short spikes.
   Fruits: white turning red to
      blackish purple, as long as leaf stalk.
   Site: home gardens, plantations; IN, W.
   Uses: leaves-food for silkworms;
      fruits-edible.

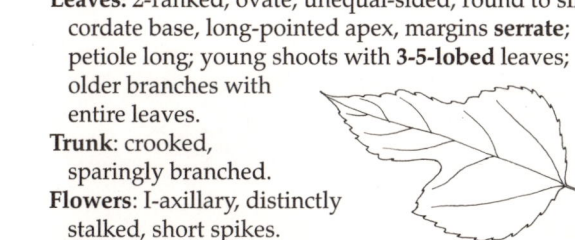

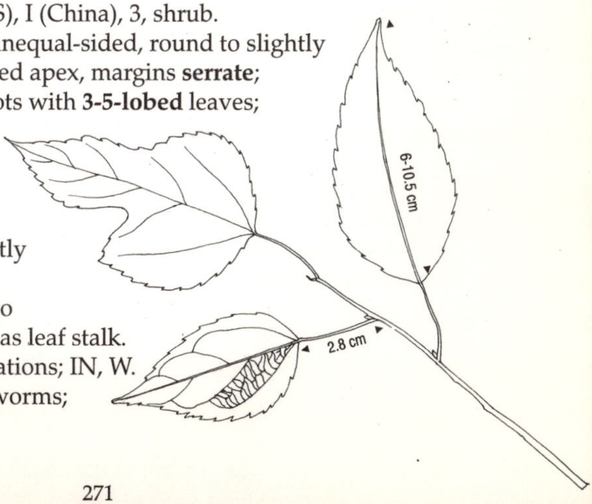

6-10.5 cm

2.8 cm

19. *Streblus asper*, geta netul (S)/pirasu (T),
    (DF III:281), N, 7, shrub to tree.
    Leaves: elliptic to obovate,
      unequal base, serrate to
      scalloped margin, lateral
      veins 4-8 pairs, **rough**.
    Trunk: B-grey; IB-whitish;
      branches often drooping
      and prostrate when young,
      densely **twiggy**, hairy;
      latex copious, **milky**.
    Flowers: yellow; I-heads,
      puberulous.
    Fruits: yellow to orange,
      globose drupe.
    Site: dry, intermediate and
      rain forest understory;
      disturbed forest areas,
      open places; DL, IN, W.
    Uses: leaves-fodder; fruit-edible;
      whole plant-medicinal; W-brushing teeth.

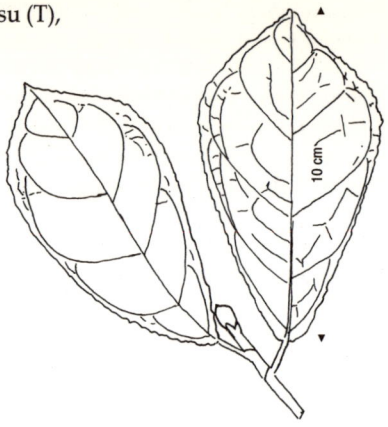

20. *Streblus taxoides*, fig-lime (E)/gongotu (S),
    (DF III:283), N, 5, small tree.
    Leaves: Variable size, elliptic to obovate, **unequal**
      wedge-shaped base, apex long-pointed,
      often **tridentate**, lateral veins
      8-12 pairs; hairy beneath;
      stipules triangular,
      deciduous.
    Trunk: B-grey; branches
      twiggy; twigs **spiny**.
    Flowers: male whitish;
      female green, solitary;
      I-male subglobose clusters.
    Fruits: unequal-sided, ellipsoid
      drupe with a fleshy base.
    Site: rocky, dry places,
      monsoon and
      intermediate forest
      understory; DL, IN.

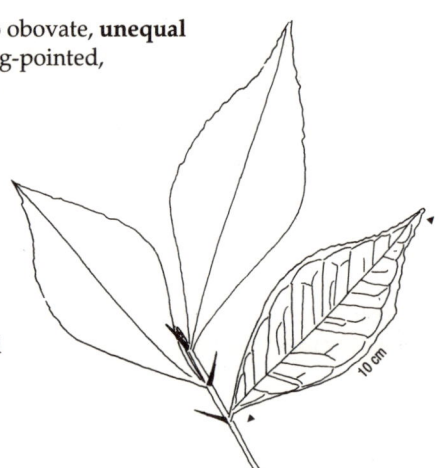

62. MORINGACEAE

FAMILY DESCRIPTION - Habit: trees. Leaves: bi-tri-pinnate with opposite leaflets. Glands at base of petiole and leaflets. **Stipules:** absent. **Flowers**: bisexual, zygomorphic, in axillary panicles or clusters. Hypanthium lined with nectary disc. **Fruits**: woody, elongate, 3-6-angled, with replum, explosively dehiscent. Seeds 3-winged or less often wingless.

FLOWER PARTS - CALYX: 5, united, unequal, spreading or reflexed, imbricate, borne on hypanthium. COROLLA: 5, free, upper two smaller and reflexed, laterals ascending and the lowest the largest, imbricate. ANDROECIUM: 5 stamens alternating with 5 staminodes. Filaments free, anthers dorsifixed, unilocular, opening lengthwise. GYNOECIUM: superior, 3 carpels, 1-locular, with 3 parietal placenta, each placentum with 2 rows of ovules.

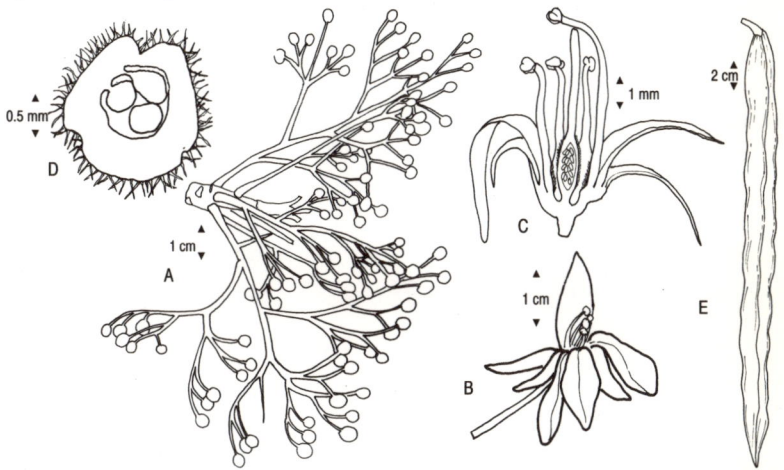

Key: inflorescence (A), full flower (B), half flower (C), transverse section of ovary (D) and fruit (E) of *Moringa oleifera.*

1. *Moringa oleifera,* horse radish tree (E)/murunga (S)/
   marunkai (T), (T I:327), I (India), 7, tree.
   Leaves: fern-like, **tri-pinnate**; leaflets
      thin, elliptic, paired with terminal.
   Trunk: B-smooth to corky
      warty; IB-greenish.
   Flowers: white,
      showy; I-axillary clusters.
   Fruits: **brown, long,** 3-angled **capsule.**
   Site: gardens; DL, IN, W.
   Uses: fruit, leaves-vegetable; live fences; roots,
      bark-condiment; seeds-ben oil, lubricant,
      perfume; whole plant-medicinal.

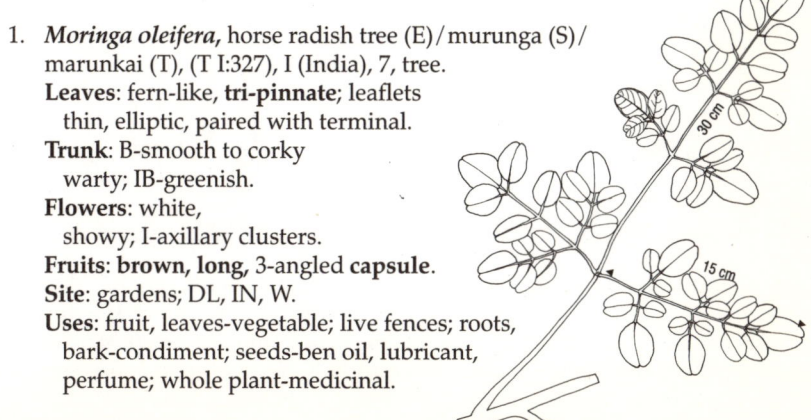

63. MYRISTICACEAE

FAMILY DESCRIPTION - Habit: trees or shrubs. Usually aromatic wood and foliage. Exudate red. **Leaves**: alternate, simple, entire, often gland-dotted. **Stipules**: absent. **Flowers**: unisexual, (plants dioecious or monoecious), actinomorphic, cymes or racemes. **Fruits**: fleshy, usually dehiscing along 2 sutures. Seed partially or completely enveloped by an often brightly coloured laciniate or subentire, conspicuous, fleshy aril.

FLOWER PARTS - CALYX: 3, rarely 2 or 5, united and saucer-or funnel-shaped with valvate lobes. COROLLA: absent. ANDROECIUM: in male flowers stamens 2 to numerous. Filaments united into a column. Anthers 2-locular, free or laterally united, opening longitudinally. GYNOECIUM: in female flowers ovary superior, 1-carpelled, 1-locular, with a solitary ovule. Placentation seemingly basal. Style short or absent.

Key: inflorescence bearing male flowers (A), a male flower in full (B), and half flower showing the fusion of the stamens into one column (C) of *Horsfieldia irya*. A female flower in full (D) showing the gynoecium (E); a male half flower showing the androecium (F) and a dehiscing fruit (G) of *Myristica fragrans.*

1. *Horsfieldia irya*, iriya (S), (T III:435 & VI:247), N, 25, tree.
   Leaves: **2-ranked**, numerous, oblong to
   linear -lanceolate, acute ends, **veins hairy**.
   Trunk: B-smooth, purplish grey, marked with leaf
   scars; W-yellowish, moderately heavy;
   young parts **tomentose**.
   Flowers: pale yellow, fragrant; male
   minute, numerous, clustered; female
   larger, fewer; I-rusty pubescent
   panicles at leafless nodes.
   Fruits: globose, rusty tomentose,
   bright pink within; seed
   with a scarlet aril.
   Site: rain forest
   subcanopy; LW.
   Uses: flowers-temple
   offerings; whole
   plant-medicinal.

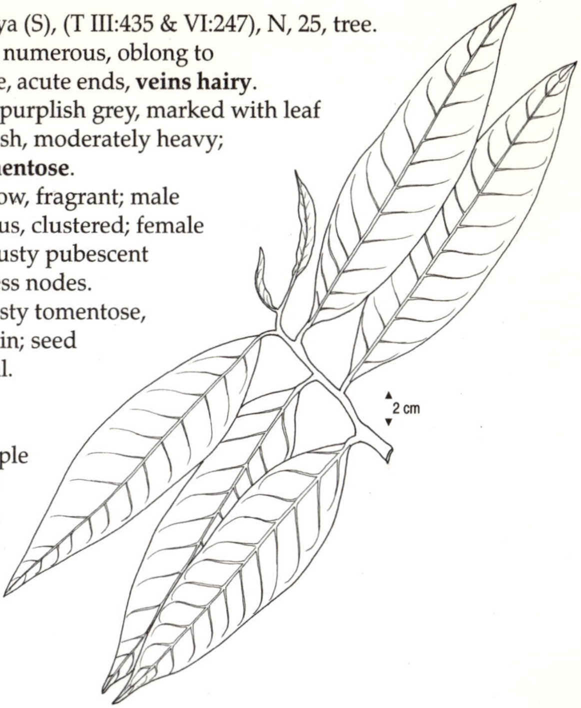

2 cm

63

2. *Horsfieldia iryaghedhi*, ruk (S),
   (T III:435 & VI:247), E, 20, tree.
   Leaves: **large**, oblong-lanceolate, acute ends,
   acuminate, beneath orange **stellate-tomentose**.
   Trunk: B-thin, slightly cracked, brownish grey;
   branches **drooping**; branchlets leaf-scarred;
   young parts **orange-tomentose**.
   Flowers: orange-yellow, fragrant; male
   numerous, very small, sessile; female
   fewer, larger; I-male dense, globose, heads of
   orange, tomentose panicles at leafless nodes;
   female short
   axillary panicles.
   Fruits: globose,
   rusty tomentose; aril
   orange, completely
   covers seed.
   Site: rain forest
   subcanopy; LW.
   Uses: flowers, B-medicinal;
   W-boat and box making;
   flowers-temple offerings.

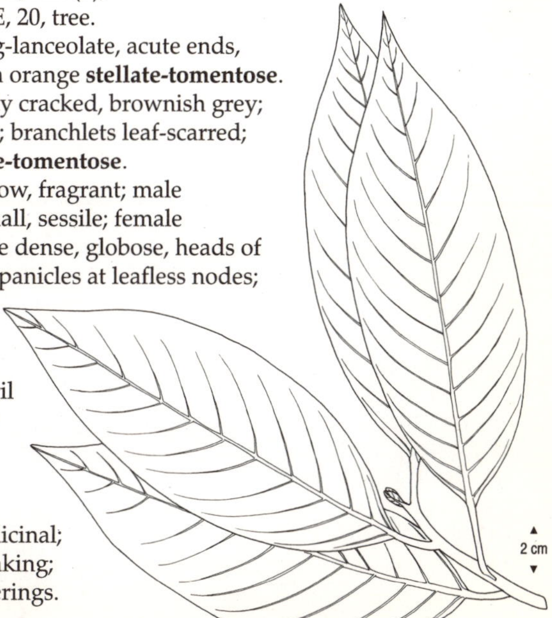

MYRISTICACEAE

2 cm

3. ***Myristica dactyloides***, malaboda (S)/palmanikam (T),
   (T III:434 & VI:247), E, 20, tree.
   Leaves: **large**, oval-lanceolate, slightly rounded
   base, leathery, lateral veins **parallel, numerous**.
   Trunk: B-smooth, orange-grey; W-light,
   pale yellow with astringent
   orange gum; young parts
   yellow-brown **puberulous**;
   bud protected by channelled
   petiole base.
   Flowers: orange-yellow;
   I- female nearly sessile
   clusters at leafless
   nodes; male more
   numerous; shortly
   racemose.
   Fruits: fleshy, leathery,
   ovoid with a scurfy
   orange pubescence; seed
   shiny brown; aril orange,
   lacerate linear segments.
   Site: intermediate and rain forest
   canopy and subcanopy; LM, W, IN.
   Uses: b, leaves-medicinal; W-light construction.

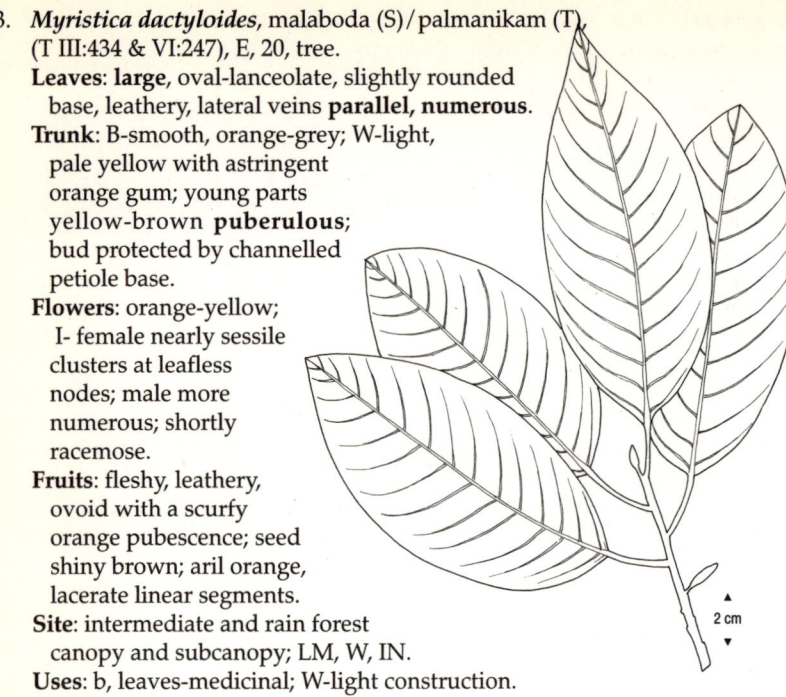

2 cm

4. ***Myristica fragrans***, nutmeg (E)/sadikka (S)/ sadikkai (T),
   I (Molluccas Is.), 20, tree.
   Leaves: ovate, shortly pointed, dark green
   above, light green below, **veins brown**.
   Trunk: spreading branches near base; B-dark grey.
   Flowers: pale yellow, bell-shaped, pendulous,
   I (female)-solitary, I (male)-
   cymose clusters, axillary.
   Fruits: pale yellow, globose,
   smooth, splitting;
   seed dark brown,
   enclosed in
   scarlet aril.
   Site: home
   gardens; IN, W.
   Uses: fruit-
   medicinal; aril,
   seed-spice.

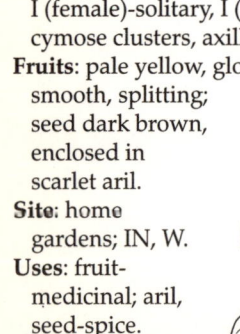
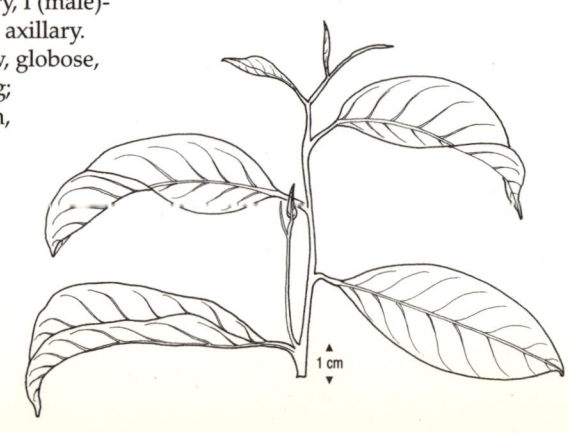

1 cm

a. *Rhizophora mucronata* (kadol) on the water's edge, a short distance inland from the mouth of the Bentota river.

b. Buds and flowers of *Rhizophora mucronata,* with a developing seedling showing viviparous germination.

c. Twigs of *Bruguiera gymnorhiza* and *Bruguiera sexangula* bearing pink and yellow flowers respectively. Germinating fruit of *B. sexangula,* always much shorter than that of *R. mucronata.*

Plate 2.

a. *Avicennia marina* in a salt water tidal mud flat in Mandathivu, with their negatively geotropic roots (pneumatophores) radiating from the trees. The species is used for fire wood, boat construction and its branches for brush pile fishing. Its roots are also medicinal.

b. Fruits of *Sonneratia caseolaris* (kirala) strung up for sale. They can be made into a delicious fruit drink.

c. Flowering twigs of *Lumnitzera racemosa*.

Plate 3.

a. Panoramic view of part of the canopy of Sinharaja rain forest, showing the crowns in flower (light cream/white) and young fruit (pinkish brown) of *Shorea trapezifolia* (thiniya dun). Note the gregarious nature of this species and the inverted bowl-shaped flowers.

b. Spreading crowns of *Shorea trapezifolia*.

Plate 4.

a. Grove of *Areca catechu* in a village on the outskirts of Sinharaja World Heritage Site.

b. Flowers of *Shorea megistophylla* (Beraliya), a canopy dominant species in Sinharaja rain forest. It is pollinated by the larger bodied rock bee (bambara) (inset) and the honey bee (mee massa). Other small stingless social bees (kana massa), beetles and flies also visit these flowers, attracted by their strong fragrance and pollen. The cotyledons of the seeds are used to make beraliya pittu, a delicious curry.

Photograph by B.M.P. Sinhakumara

Photograph by B.M.P. Sinhakumara

a. Crown of *Shorea worthingtonii* standing out conspicuously in the forest because of its bright purplish pink young leaves.

b. Fruiting twig of *Cullenia rosayroana* (kata boda).

c. Fruiting twig of *Ptycopyxis thwaitesii* (wal rambuttan).

d. Flame coloured young leaves of *Agrostistachys coriacea* (beru) growing in the understory of the Sinharaja rain forest. Its leathery leaves are used by villagers for roof thatching.

Plate 6.

Stand of *Pinus caribaea* (a) and its stems tapped for oleo resin (b).

Photograph by late Prof. S. Balasubramaniam

a. Twin trees of *Vitex leucoxylon* (nebada), with twig bearing fruit.

b. Twig of *Limonia acidissima* (diwul) bearing a mature (white) and a shriv-elled undeveloped (black) fruit. Note its compound leaves. The ripe fruits make a delicious jam and beverage, and the unripe ones an equally good 'achcharu'.

c. Flowers and fruits of *Butea monosperma* (gas kela).

d. *Drypetes sepiaria* (weera) with its pale coloured, fluting trunk. This tree is one left behind in System C of the Mahaweli Project.

Plate 8.

a.   Semi-deciduous forest of the dry zone, around Kandalama peak. Part of crown of *Azadirachta indica* (kohomba) on top left. It has many uses.

b.   *Manilkara hexandra* (palu) in the Victoria-Randenigala-Rantambe Sanctuary with details of its stem characteristics. It is a fast disappearing canopy-dominant species in the dry zone forests.

c.   *Erythrina fusca* (yak erabadu) showing pyramidal corky spines on its trunk. The species grows near water courses in the lowland wet, intermediate and dry zones.

a. Montane forests at Adam's Peak with *Camellia sinensis* (thé) in the fore-
    ground.

b. Tree of *Gordonia speciosa* (ratu mihiriya).

c. Flowering twig of *Gordonia speciosa*.

Plate 10.

Photograph by B.M.P. Sinhakumara

a. Part of the montane forest in Horton Plains, showing the short-statured trees with gnarled and twisted branches, so characteristic of this ecosystem.

b. Tree of *Rhododendron arboreum* var. *zeylanicum* (asoka) in flower.

c. An inflorescence with open flowers, and unfolding young leaves of *Rhododendron arboreum* var. *zeylanicum.*

Photograph by B.M.P. Sinhakumara

a. *Calophyllum walkeri* (keena), a canopy-dominant species in montane forests, with Kirigalpotha (Sri Lanka's second-highest mountain) in the distance.

b. Flowering twig of *Calophyllum walkeri.*

c. Twig bearing brightly-coloured young leaves of *Calophyllum walkeri,* which are responsible for its very attractive and conspicuous crown during the flushing season of this species.

Plate 12.

a. Montane forests on the ascent to Horton Plains from Pattipola and a view of a tree fern from above.

b. A tree fern *Cyathea* seen in a canopy gap of the montane forest in Horton Plains. Note the very large size of its leaves.

c. Flowers of the dominant understory *Strobilanthes* (nelu) species in Horton Plains. Its flowers are pollinated by honey bees.

Photograph by late Prof. S. Balasubramaniam

a. Sub-montane forests at Dethanagala, seen from the road from Balangoda to Bogawanthalawa. The tree in the foreground is *Shorea gardneri* (rath dun), one of the dominant species in this forest.

b. A patch of sub-montane forests in the Knuckles Range showing forest die-back.

c. A flower of *Adinandra lasiopetala* (rathu mihiriya) of Theaceae, the tea family.

Plate 14.

a. Scattered trees in a savanna at Samanalawewa.

b. *Anogeissus latifolius* (dawu), a tree frequently found in the savannas and intermediate zone forests, with its characteristic bark.

c. Young developing fruits of *Anogeissus latifolius.*

a. Intermediate zone forests, with some crowns in young flush, in the Samanalawewa area.

b. *Mangifera zeylanica* (etamba) in flower, a tree frequently seen in forests of the lowland wet and intermediate zones of the country with a twig bearing fruits.

c. *Syzygium zeylanicum* (yakada maran) in flower, a shrub along roadsides and scrub vegetation in the Samanalawewa area. Its stems have been used in the past as a source of carbon for the manufacture of steel.

Plate 16.

a. A villager's choice of species *(Swietenia macrophylla* (mahogany), *Cocos nucifera, Alstonia macrophylla, Artocarpus heterophyllus* (kos), *Areca catechu* (puwak), *Ceiba pentandra* (kotta pulun), *Gliricidia sepium, Tithonia diversifolia* (wal suriya kantha), *Musa* sp. (kesel), *Bambusa vulgaris)* in a home garden near Gampola in the mid-country.

b. Some produce from a home garden: fruits of *Durio zibethinus* (durian), *Spondias dulcis* (ambarella) and *Aegle marmelos* (beli).

c. Defoliating stand of young *Hevea brasiliensis* (rubber).

64. MYRSINACEAE

FAMILY DESCRIPTION - Habit: trees, shrubs, lianas or rarely herbs. **Leaves**: spiral, simple, usually entire with gland dots or glandular hairs. **Stipules**: absent. **Flowers**: bisexual, actinomorphic; racemes or panicles. **Fruits**: berry or drupe.

FLOWER PARTS - CALYX: 3-6, free or basally united, lobes imbricate, convolute or valvate. COROLLA: 3-6, united, rarely free, lobes imbricate, convolute or valvate. ANDROECIUM: stamens 3-6, usually epipetalous. Filaments sometimes basally united. Anthers 2-lobed, opening by longitudinal slits or apical pores. GYNOECIUM: superior or half-inferior, carpels 3-6, 1-locular. Style simple. Placentation axile, free central or basal.

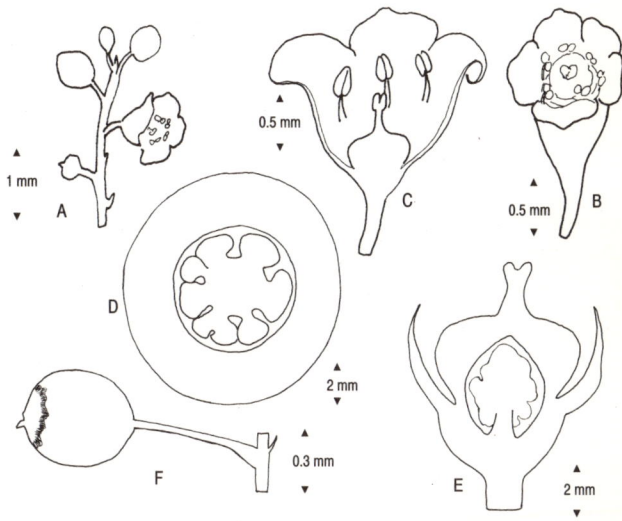

Key: inflorescence (A), full flower (B), half flower (C), transverse (D) and longitudinal (E) sections of ovary showing basal placentation, and fruit (F) of *Maesa perrottetiana*.

1. *Ardisia gardneri*, (T III:72), E, 3, shrub.
    Leaves: oval, tapering ends, subacute,
       lateral veins numerous, obscure,
       many wart-like glands near margin.
    Trunk: much branched;
       twigs slender, often zig-zag.
    Flowers: pale pink-white, small,
       on long slender pedicels;
       I-lax, terminal panicles.
    Fruits: crimson berry.
    Site: montane forest
       understory; LM, M.

2. *Ardisia paniculata*, (T III:71 & VI:178), N, 2, shrub.
   Leaves: terminally crowded, narrowly
   oblong-lanceolate, tapering ends, margins
   faintly scalloped; petiole short.
   Trunk: branchlets stout, glabrous.
   Flowers: pale pink, on long
   spreading pedicel;
   I-terminal pyramid-shaped
   panicle.
   Fruits: red berry.
   Site: montane forest
   understory; LM.

3. *Ardisia pauciflora*, (T III:73), N, 2, shrub.
   Leaves: narrowly lanceolate-oblong,
   tapering base, pointed to obtuse apex,
   margins often wavy, lateral veins
   numerous, horizontal, obscure.
   Trunk: branches slender, straggling;
   young parts reddish scaly.
   Flowers: white, small, few, usually in
   3-5's; I-axillary pedunculate racemes.
   Fruits: red berry.
   Site: montane and rain forest
   understory; LM, M, UW.

4. *Ardisia willisii*, balu dan (S),
   (T III:72 & VI:178), E, 4, shrub.
   Leaves: large, oval to obovate,
   tapering base, obtuse somewhat
   twisted apex, margins
   obscurely scalloped;
   petiole short, stout.
   Trunk: branchlets
   stout, glabrous.
   Flowers: bright pink,
   on stout long pedicels;
   I-short terminal panicles.
   Fruits: bright scarlet,
   large, pulpy berry.
   Site: coastal areas; MC.
   Uses: fruit-edible; whole
   plant-medicinal; ornamental.

5. ***Maesa perrottetiana***, ma thambiya (S),
   (T III:67 & VI:176), N, 3, shrub.
   Leaves: rather large, ovate-lanceolate,
   acute base, pointed apex, margins
   coarsely dentate, undulate,
   lateral veins conspicuous.
   Trunk: many lenticels; twigs slender.
   Flowers: white, very small;
   I-lax axillary panicles.
   Fruits: pale orange-cream berry;
   small, smooth, globose,
   bluntly ribbed.
   Site: montane
   forest
   understory; M.

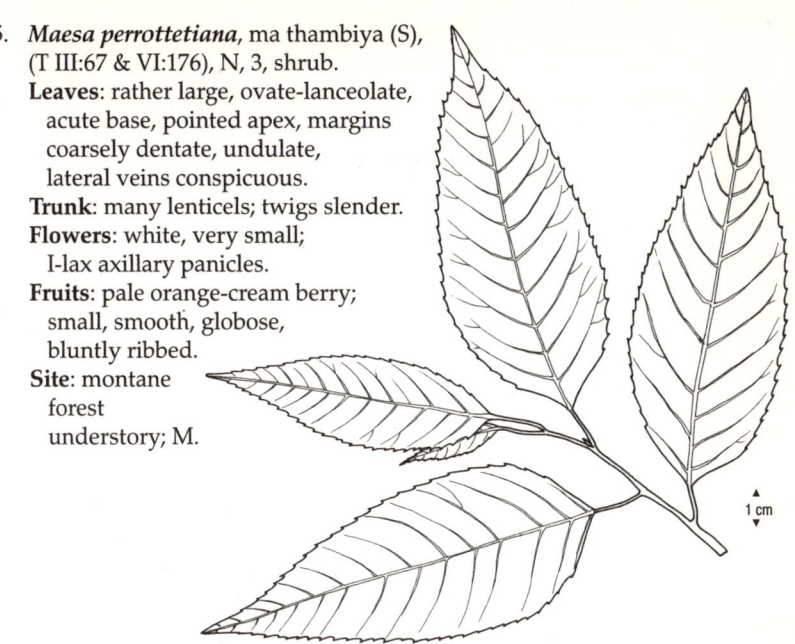

1 cm

64

MYRSINACEAE

6. ***Myrsine robusta***, (T III:68 & VI:177), N, 9, shrub to tree.
   Leaves: closely placed, oval to oblong-lanceolate,
   tapering base, obtuse, often
   notched apex; sessile.
   Trunk: branchlets thick, dark
   brown, leaf-scarred, glabrous.
   Flowers: pale yellow-pink,
   numerous, in small clusters
   on nodules at leafless nodes.
   Fruits: purple, shiny berry.
   Site: montane forest
   understory; LM, M.

6.5 cm

0.5 cm

65. MYRTACEAE

FAMILY DESCRIPTION - Habit: trees or shrubs. **Leaves**: opposite, rarely spiral, simple, mostly entire. **Stipules**: absent or rare, very small. **Flowers**: bisexual or polygamous by abortion, actinomorphic; cymes, racemes or panicles. Hypanthium present. **Fruits**: 1-to many-seeded berry, loculicidal capsule, drupe or nut.

FLOWER PARTS - CALYX: 4-5, mostly inconspicuous or absent in *Eucalyptus,* free or basally united. COROLLA: 4 or 5, free or united forming an operculum or calyptra in *Eucalyptus,* imbricate. ANDROECIUM: stamens numerous, borne on rim of hypanthium. Filaments free or basally united into 4-5, less often 2 groups. Anthers 2-locular, opening lengthwise or rarely by terminal pores. GYNOECIUM: inferior, carpels usually 2-5, with 1 to many loculi. Placentation mostly axile, rarely parietal. Ovules 2 to many per loculus.

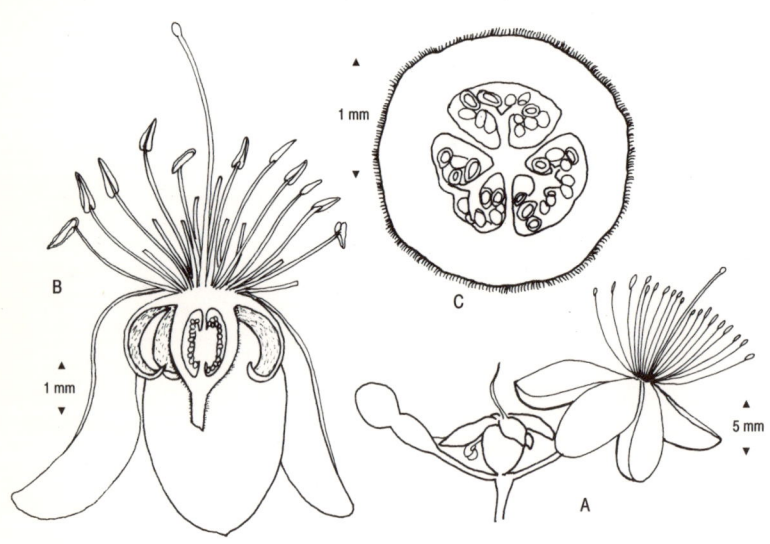

Key: inflorescence (A), half flower (B) showing gynoecium, androecium and other floral parts, and transverse section of ovary (C) of *Psidium guajava*.

1. *Eucalyptus camaldulensis*, Murray red gum (E),
   (DF II:462), I (Australia), 30, tree.
   Leaves: lanceolate-acuminate;
      flush **slightly whitish**.
   Trunk: B-**smooth, greyish**
      white, sometimes fibrous.
   Flowers: 5-10 in a cluster;
      I-axillary umbels.
   Fruits: hemispherical to
      broadly conical capsule.
   Site: forest plantations; DL, M.
   Uses: stem exudate-medicinal;
      W-fuelwood.

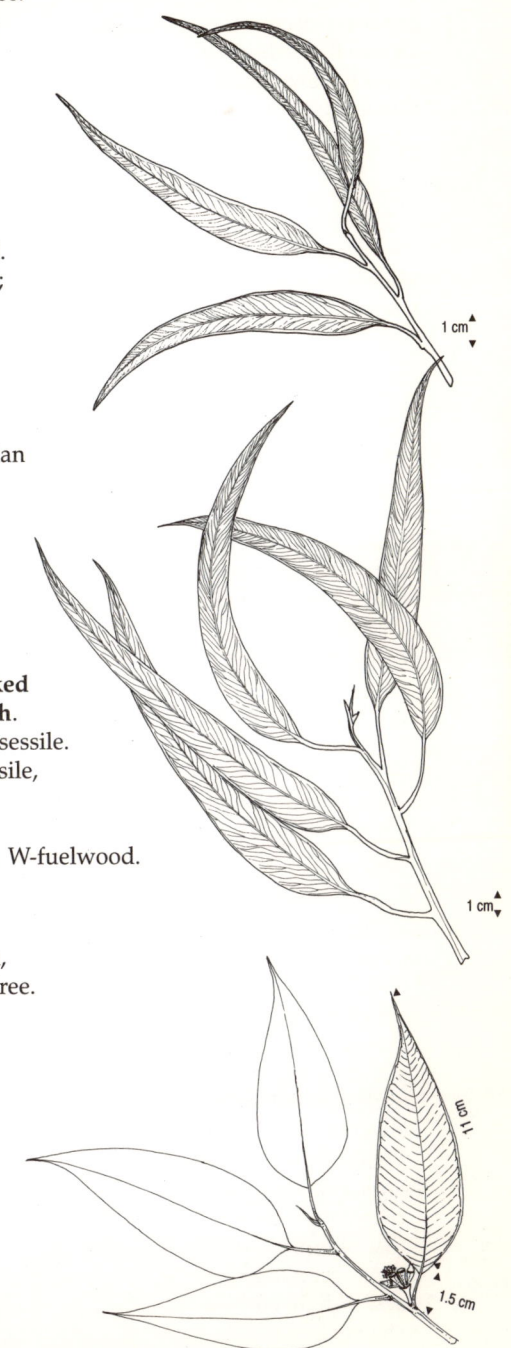

2. *Eucalyptus globulus*, Tasmanian
   blue gum (E), (DF II:462),
   I (Australia), 15, tree.
   Leaves: dark green,
      lanceolate, more or
      less **sickle-shaped**,
      pointed apex; flush
      prominently whitish.
   Trunk: B-**smooth, scroll-marked**
      with flakes persisting, **bluish**.
   Flowers: solitary, axillary, subsessile.
   Fruits: depressed globose, sessile,
      4-ribbed capsule.
   Site: tea estates; M.
   Uses: ornamental; shade tree; W-fuelwood.

3. *Eucalyptus grandis*, rose gum,
   (DF II:459), I (Australia), 30, tree.
   Leaves: narrowly lanceolate,
      pointed apex, margins
      wavy, veins slender.
   Trunk: B-**smooth**,
      white to pale,
      whitish.
   Flowers: 3-10 in a cluster;
      I-axillary umbels.
   Fruits: whitish,
      pear-shaped capsule.
   Site: forest plantations; M.
   Uses: W-heavy construction.

4. **Eucalyptus microcorys**, tallow wood (E),
   (DF II:469), I (Australia), 30, tree.
   Leaves: lanceolate, pointed apex.
   Trunk: B-**persistently
   flaky, corrugated**.
   Flowers: 4-8 in a cluster;
   I-axillary or terminal umbels.
   Fruits: pear-shaped to
   cylindrical-to
   club-shaped,
   truncate capsule; disc obscure.
   Site: forest plantations; DL.
   Uses: W-fuelwood, poles.

5. **Eucalyptus pilularis**, black butt (E),
   (DF II:468), I (Australia), 25, tree.
   Leaves: lanceolate, shiny.
   Trunk: B-**rough, above
   smooth, greenish white**.
   Flowers: 6-12 in a cluster;
   I-axillary umbels.
   Fruits: ovoid to
   subglobose
   capsule; disc small.
   Site: tea estates; M.
   Uses: ornamental, shade
   tree; W-poles, flooring,
   railway sleepers.

6. **Eucalyptus robusta**, swamp
   mahogany (E), (DF II:461),
   I (Australia), 25, tree.
   Leaves: broadly lanceolate,
   pointed apex, veins
   slender, almost parallel.
   Trunk: B-**rough, fibrous**.
   Flowers: 5-10 in a cluster;
   I-axillary to subterminal.
   Fruits: cylindrical to
   urceolate capsule;
   disc oblique.
   Site: forest plantations; M.
   Uses: W-light construction,
   charcoal.

7. **Eugenia bracteata**, thambiliya (S),
   (DF II:416), N, 10, shrub to tree.
   Leaves: elliptic to
     subovate-oblong,
     broadly obtuse to
     pointed apex, acute
     base, lateral veins
     10 pairs with **1**
     short intermediary
     in between;
     aromatic
     clove-like smell.
   Trunk: B-smooth,
     sometimes
     peeling, pale grey-brown.
   Flowers: cream, showy;
     I-axillary or terminal cymes.
   Fruits: green to yellow to red,
     subglobose, berry; persistent calyx.
   Site: secondary monsoon and intermediate
     forest, especially near coast; DC, DL, IN.
   Uses: whole plant-medicinal.

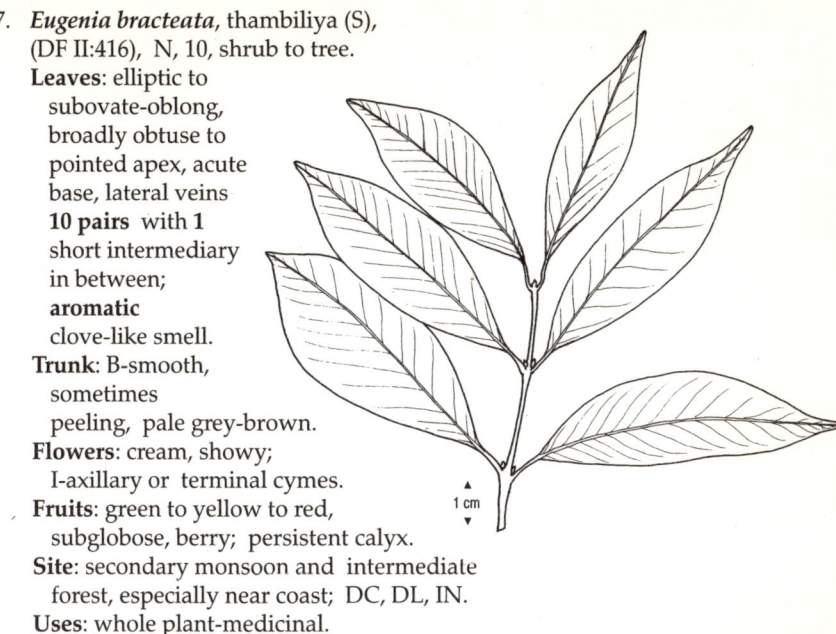

1 cm

65

MYRTACEAE

8. **Eugenia mabaeoides**, (DF II:413),
   E, 5, shrub to small tree.
   Leaves: variable, **spatulate to elliptic**,
     tapered wedge-shaped base,
     rounded to obtuse apex,
     lateral veins **5-7** pairs,
     ascending with an
     intramarginal vein,
     leathery, distinctly
     dotted beneath.
   Trunk: greenish,
     short; young
     shoots pubescent.
   Flowers: pale green,
     1-3 axillary, cup-shaped.
   Fruits: crimson, succulent,
     ellipsoid berry; prominent
     persistent calyx.
   Site: montane forest understory; M.

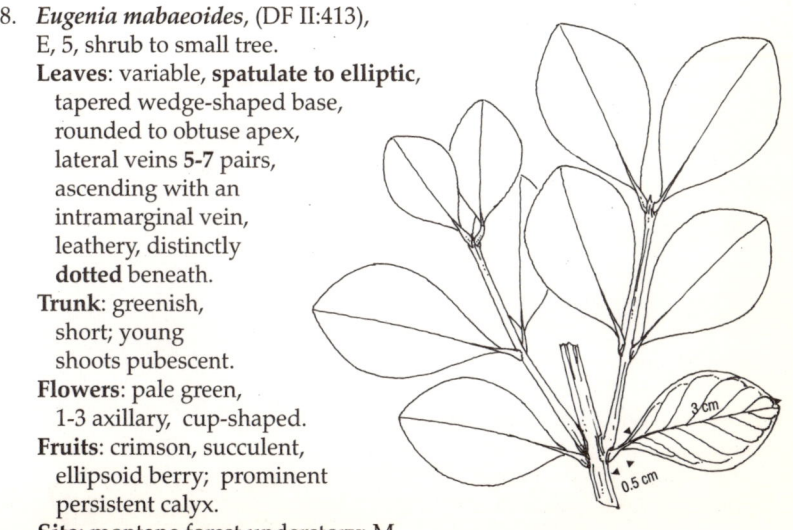

3 cm

0.5 cm

9. *Eugenia thwaitesii*, (DF II:415), N, 5, shrub to small tree.
   Leaves: variable, ovate-lanceolate to
      elliptic, wedge-shaped base, pointed
      to obtuse apex, lateral veins **8-15**
      pairs, elevated beneath with
      shorter, less distinct intermediaries.
   Trunk: twigs becoming conspicuously
      cream-coloured.
   Flowers: white, small, sessile;
      I-terminal or axillary clusters.
   Fruits: yellowish crimson, depressed
      globose, pendulous, berry,
      prominent persistent calyx.
   Site: dry patana grasslands,
      rocky or shallow soils; IN, W.

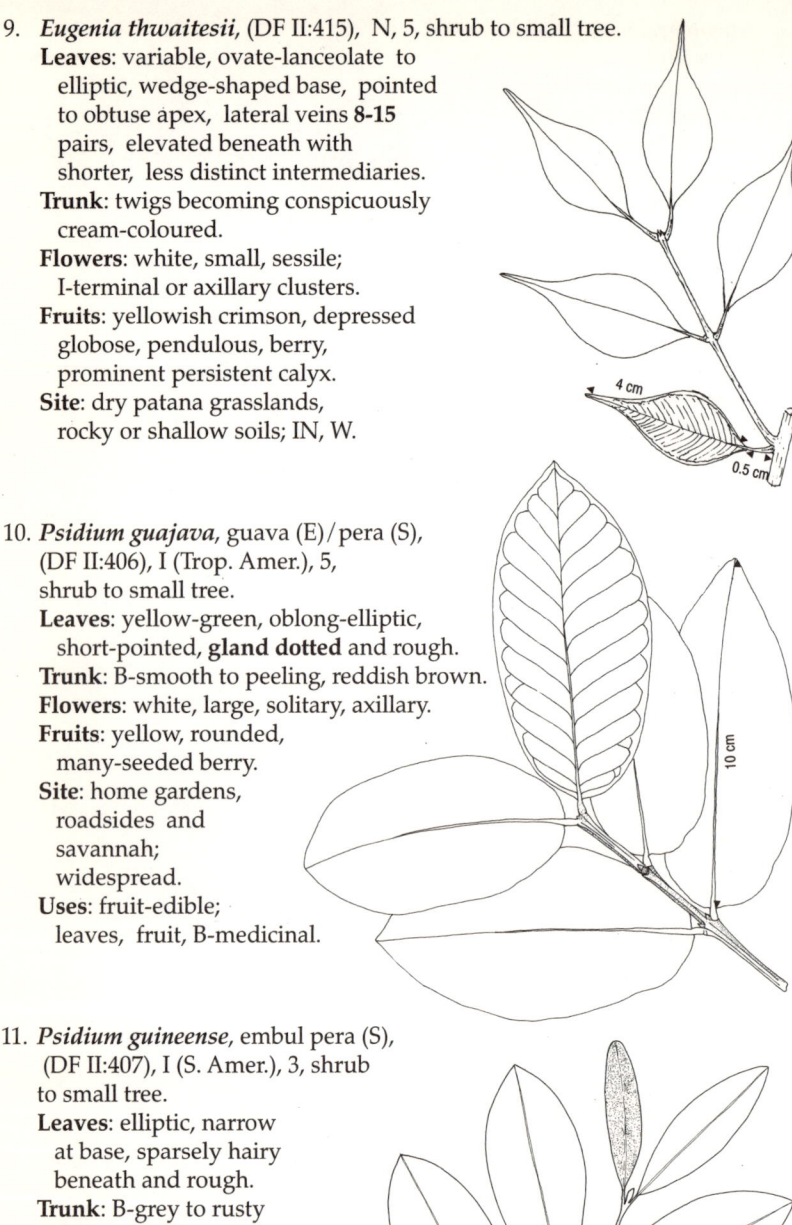

10. *Psidium guajava*, guava (E)/pera (S),
      (DF II:406), I (Trop. Amer.), 5,
      shrub to small tree.
   Leaves: yellow-green, oblong-elliptic,
      short-pointed, **gland dotted** and rough.
   Trunk: B-smooth to peeling, reddish brown.
   Flowers: white, large, solitary, axillary.
   Fruits: yellow, rounded,
      many-seeded berry.
   Site: home gardens,
      roadsides and
      savannah;
      widespread.
   Uses: fruit-edible;
      leaves, fruit, B-medicinal.

11. *Psidium guineense*, embul pera (S),
      (DF II:407), I (S. Amer.), 3, shrub
      to small tree.
   Leaves: elliptic, narrow
      at base, sparsely hairy
      beneath and rough.
   Trunk: B-grey to rusty
      brown, scroll-marked.
   Flowers: white, large; I-axillary.
   Fruits: pale yellowish green,
      pear-shaped, fleshy berry.
   Site: dry patana grasslands,
      scrub; LM, IN, UW.
   Uses: fruit-edible.

12. *Psidium littorale*, cheena pera (S),
    I (Trop. Amer.), 3, shrub to small tree.
    Leaves: broadly obovate, narrowly
       pointed base, rounded to
       shortly pointed apex,
       glabrous, intramarginal vein.
    Trunk: B-grey to pinkish
       brown, scroll-marked,
       smooth, twigs
       4-angled.
    Flowers: white; I-axillary.
    Fruits: red, fleshy, globose berry.
    Site: home gardens,
       scrub; LM, IN, W.
    Uses: fruit-edible; ornamental.

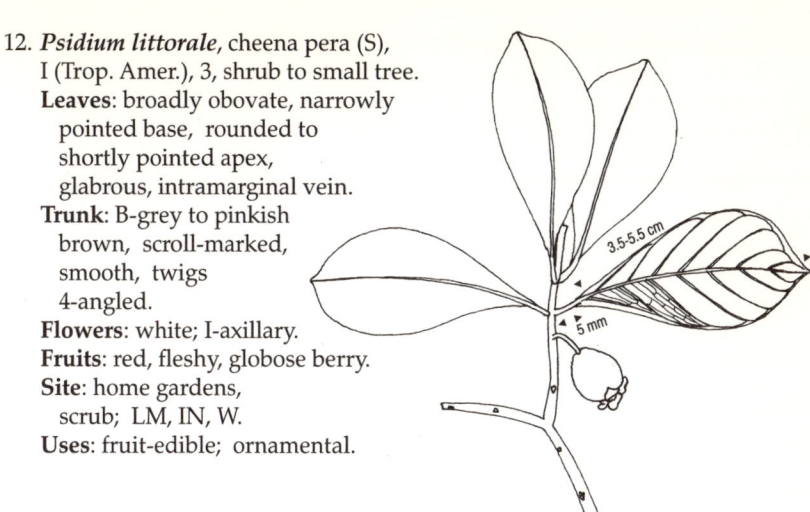

65

13. *Rhodomyrtus tomentosa*, seeta pera
    (DF II:405), N, 3, shrub to small tree.
    Leaves: elliptic, 3-veined, hairy, veins
       beneath **dark grey, prominently
       elevated**, **depressed** above, leathery.
    Trunk: much branched; B-thinly flaky,
       orange-brown; twigs stout; young
       parts dense **greyish velvety** hairs.
    Flowers: pink, large; I-axillary,
       long pedunculate cymes.
    Fruits: globose berry;
       persistent calyx; purple pulp.
    Site: montane forest openings
       and fringes; M.
    Uses: fruits-edible.

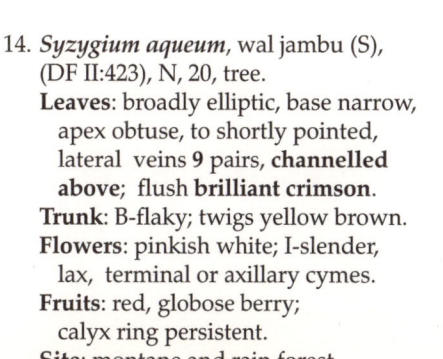

MYRTACEAE

14. *Syzygium aqueum*, wal jambu (S),
    (DF II:423), N, 20, tree.
    Leaves: broadly elliptic, base narrow,
       apex obtuse, to shortly pointed,
       lateral veins **9** pairs, **channelled
       above**; flush **brilliant crimson**.
    Trunk: B-flaky; twigs yellow brown.
    Flowers: pinkish white; I-slender,
       lax, terminal or axillary cymes.
    Fruits: red, globose berry;
       calyx ring persistent.
    Site: montane and rain forest
       canopy, secondary forest;
       M, W.

15. **Syzygium aromaticum**, clove (E)/
   karabu (S)/karabu (T), (DF II:434),
   I (Molluccas Is.),10, tree.
   Leaves: oblanceolate,
      lateral veins many,
      obscure, parallel, beneath
      densely dotted; aromatic.
   Trunk: B-smooth,
      pale brown.
   Flowers: pale
      green; I-cymes.
   Fruits: red, oblong berry
      with persistent calyx.
   Site: home gardens; IN, W.
   Uses: flower buds-spice, medicinal.

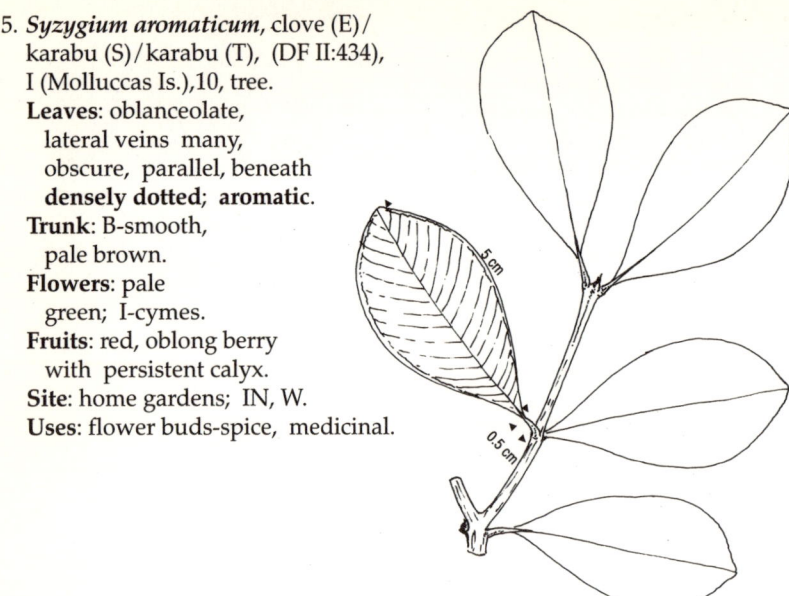

16. **Syzygium assimile**, damba (S), (DF II:451), N, 15, tree.
   Leaves: elliptic obovate, base narrow, apex abruptly down-curved,
      lateral veins **12** pairs with many intermediaries,
      marginal vein present: flush **brillant crimson**.
   Trunk: fluted; B-roughly irregularly flaky,
      red-brown to grey.
   Flowers: cream, many;
      I-terminal and axillary cymes.
   Fruits: green, subglobose,
      berry; persistent calyx.
   Site: montane and
      rain forest
      subcanopy;
      LM, UW.

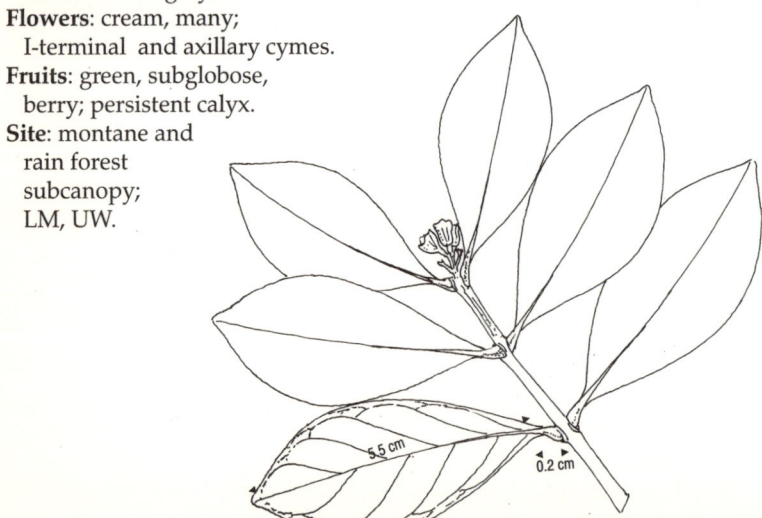

17. *Syzygium caryophyllatum*, dan (S), (DF II:450), N, 10, tree.
    Leaves: frequently **subopposite**,
    obovate to oblanceolate, base
    wedge-shaped, short pointed
    apex, midrib **channelled
    above**, lateral veins many,
    intermediaries and tertiaries,
    intramarginal vein present,
    densely dotted beneath;
    flush **brownish red**.
    Trunk: B-smooth to minutely
    cracked, pale grey-brown.
    Flowers: white; I-much branched
    axillary or terminal long dense cymes.
    Fruits: dark purple, succulent,
    subglobose berry with
    persistent calyx.
    Site: secondary forest
    on sandy soils; IN, W.
    Uses: fruit-edible; leaves, B-medicinal.

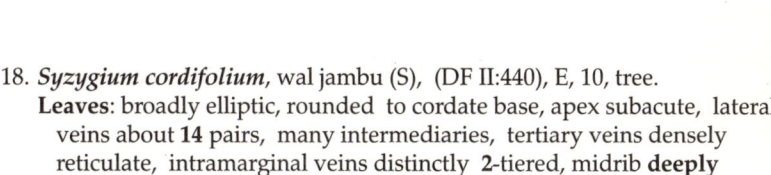

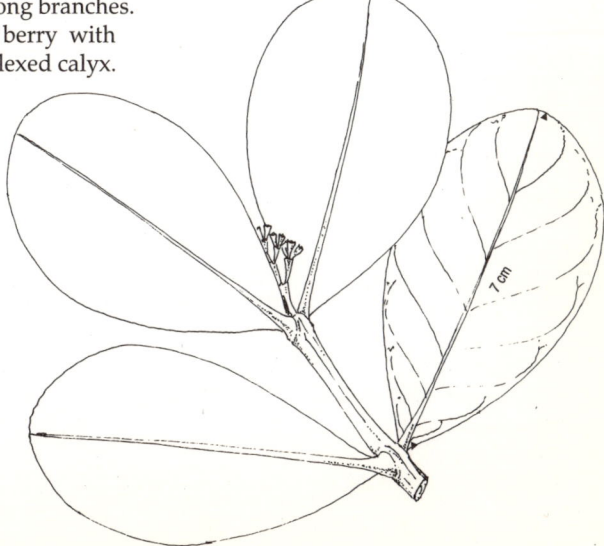

18. *Syzygium cordifolium*, wal jambu (S), (DF II:440), E, 10, tree.
    Leaves: broadly elliptic, rounded to cordate base, apex subacute, lateral
    veins about **14** pairs, many intermediaries, tertiary veins densely
    reticulate, intramarginal veins distinctly **2-tiered**, midrib **deeply
    channelled**; leathery; flush **purplish red**.
    Trunk: B-finely ridged, fissured, pale brown-grey.
    Flowers: white; I-terminal or axillary
    cymes with long branches.
    Fruits: globose berry with
    persistent reflexed calyx.
    Site: montane
    and rain
    forest
    understory,
    often along
    streams in
    moist forest;
    LM, W.

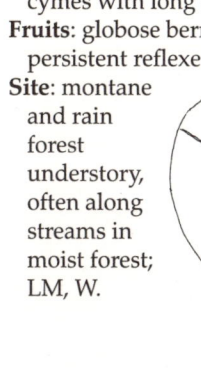

19. *Syzygium cumini*, madan (S)/nava (T),
    (DF II:443), N, 20, shrub to tree.
    Leaves: yellowish green, elliptic
      to ovate-lanceolate, prominent
      slender apex, margin **wavy**,
      lateral veins **20-25** pairs,
      densely reticulate tertiaries,
      intramarginal vein present, pitted
      above, **minutely dotted** beneath.
    Trunk: B-thinly flaky,
      pale yellow-brown;
      IB-fibrous, pale brown.
    Flowers: white; I-axillary
      or terminal cymes.
    Fruits: purple, broadly
      ellipsoid berry; prominent calyx.
    Site: monsoon forest canopy along
      waterways, reservoirs, and near
      coastal sand dunes; DC, DL.
    Uses: W-heavy construction;
      fruit-edible; fruit, B-medicinal.

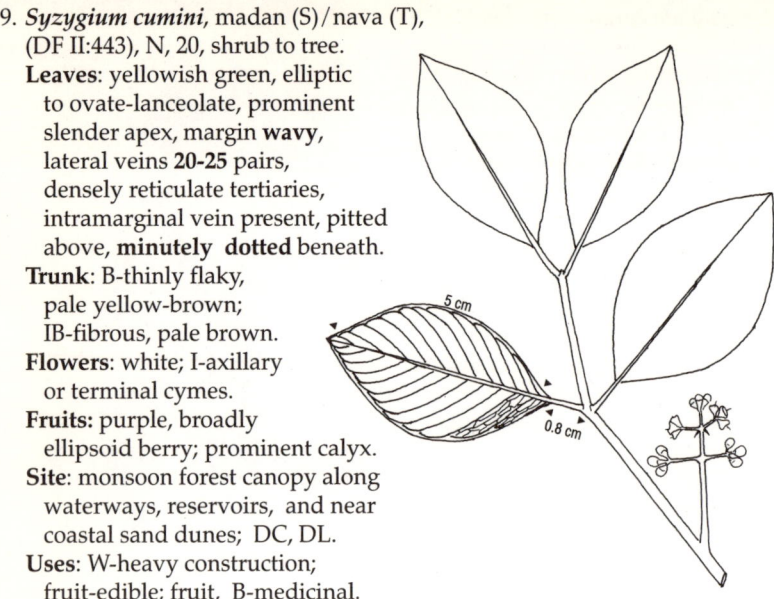

20. *Syzygium gardneri*, damba (S)/nava (T), (DF II:447), N, 20, tree.
    Leaves: ovate-lanceolate, base narrow, apex long-pointed,
      margin wavy, lateral veins more than **25** pairs,
      densely reticulate tertiaries, intramarginal
      vein present; old leaves yellow; flush **red**.
    Trunk: B-smooth, pale grey to cream-brown;
      frequently suckering at base; twigs
      conspicuous, **pale green**.
    Flowers: white, funnel-shaped;
      I-much branched
      terminal or
      axillary
      cymes.
    Fruits: green,
      ellipsoid-
      obovoid, no
      collar, deep
      cavity at apex.
    Site: intermediate and
      rain forest; along
      waterways; IN, W.
    Uses: W-construction.

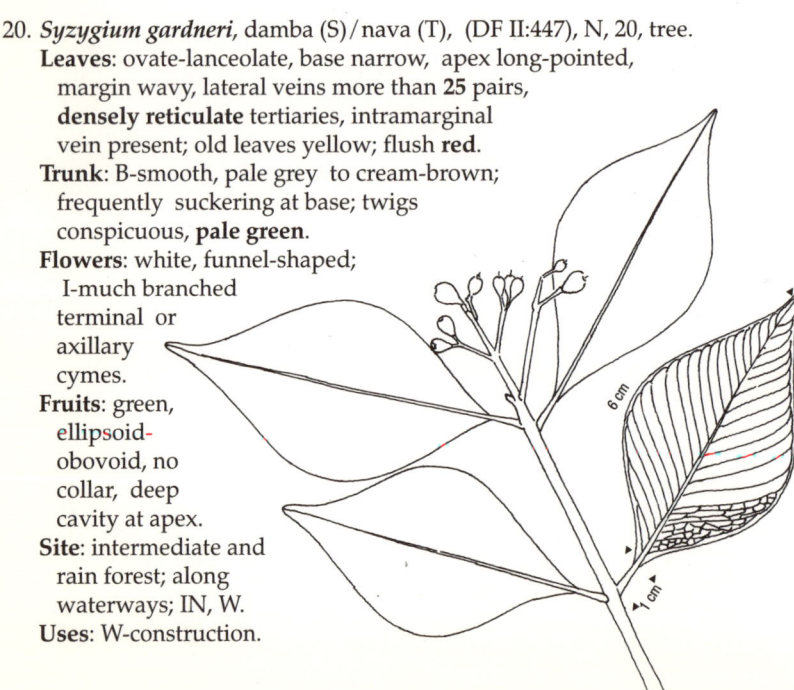

21. *Syzygium jambos*, rose apple (E)/veli jambu (S),
    (DF II:427), I (Malesia), 10, tree.
    Leaves: **dark green**, lanceolate, long
        pointed ends, beneath **gland dotted**.
    Trunk: B-smooth to fissured, brown.
    Flowers: pale yellow-white, large: I-terminal.
    Fruits: pale yellow tinged pink, rounded,
        large, subglobose berry.
    Site: home gardens; IN, W.
    Uses: fruit-edible; ornamental,
        W-basket work.

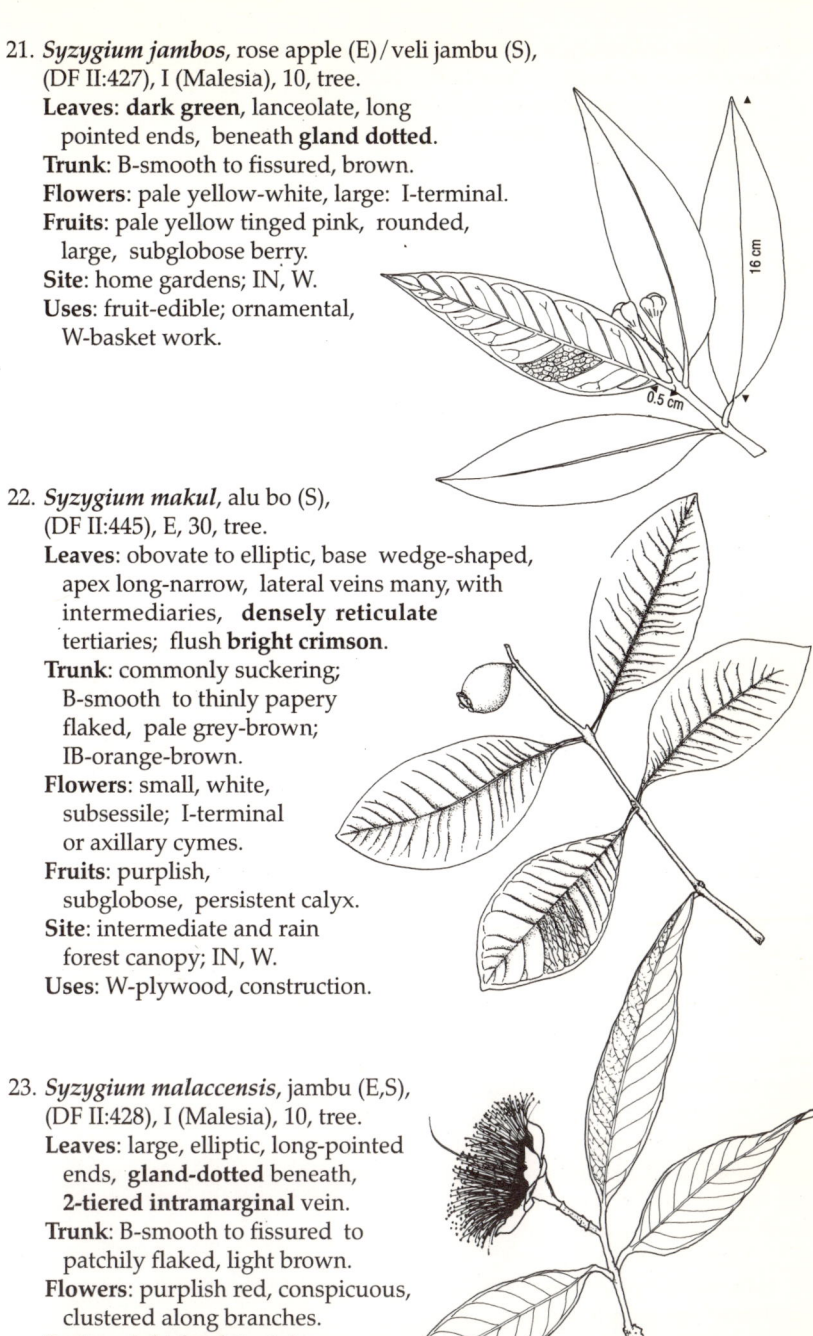

22. *Syzygium makul*, alu bo (S),
    (DF II:445), E, 30, tree.
    Leaves: obovate to elliptic, base wedge-shaped,
        apex long-narrow, lateral veins many, with
        intermediaries, **densely reticulate**
        tertiaries; flush **bright crimson**.
    Trunk: commonly suckering;
        B-smooth to thinly papery
        flaked, pale grey-brown;
        IB-orange-brown.
    Flowers: small, white,
        subsessile; I-terminal
        or axillary cymes.
    Fruits: purplish,
        subglobose, persistent calyx.
    Site: intermediate and rain
        forest canopy; IN, W.
    Uses: W-plywood, construction.

23. *Syzygium malaccensis*, jambu (E,S),
    (DF II:428), I (Malesia), 10, tree.
    Leaves: large, elliptic, long-pointed
        ends, **gland-dotted** beneath,
        2-tiered intramarginal vein.
    Trunk: B-smooth to fissured to
        patchily flaked, light brown.
    Flowers: purplish red, conspicuous,
        clustered along branches.
    Fruits: pinkish white, juicy,
        sour, pear-shaped berry.
    Site: home gardens; IN, W.
    Uses: fruit-edible; ornamental.

65

MYRTACEAE

24. *Syzygium micranthum*, (DF II:448), E, 15, tree.
    Leaves: ovate-lanceolate to elliptic, abruptly
    pointed, lateral veins **10-13** pairs, irregular
    intermediaries, **densely reticulate**; flush
    red-brown; **weeping** crown.
    Trunk: B-flaky to smooth,
    orange-brown; branchlets
    cylindrical; twigs flattened.
    Flowers: pale green to white,
    sessile, pendant; I-terminal
    or axillary cymes.
    Fruits: red, globose
    berry; persistent calyx.
    Site: rain forest subcanopy
    and understory; LM, UW, W.

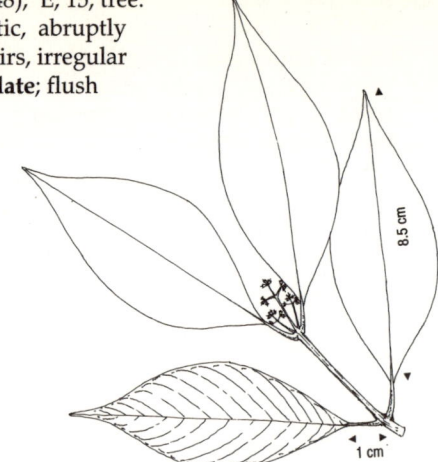

8.5 cm

1 cm

25. *Syzygium neesianum*, panu kera (S),
    (DF II:442), E, 25, tree.
    Leaves: oblong to elliptic, base rounded
    to subcordate, lateral veins **20** pairs,
    many short intermediaries, **densely
    reticulate** tertiaries, **2-tiered**
    intramarginal vein; flush **purplish-red**.
    Trunk: B-papery, shallowly flaky, pale
    brown; branches pendulous.
    Flowers: white; I-spreading,
    axillary or terminal cymes.
    Fruits: globose, woody berry;
    calyx persistent, unlobed.
    Site: rain forest subcanopy; W.
    Uses: W-heavy construction.

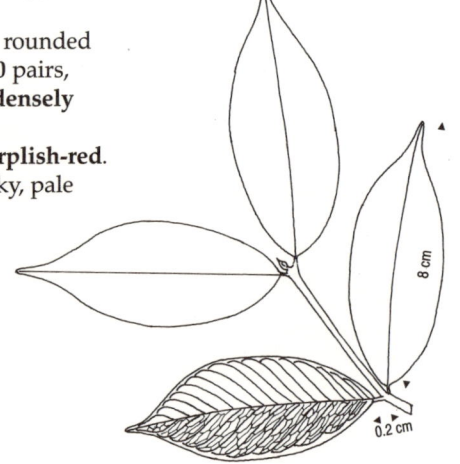

8 cm

0.2 cm

26. *Syzygium oliganthum*, (DF II:437), E, 10, tree.
    Leaves: obovate to oblong, base tapered,
    inconspicuous lateral veins, about 8 pairs,
    prominent intermediaries, stiff and leathery.
    Trunk: crooked; B-smooth, reddish brown;
    twigs **quadrangular to winged**.
    Flowers: pinkish to white, small,
    sessile, densely clustered;
    I-terminal or axillary cymes.
    Fruits: ripening red to black, globose
    berry; persistent calyx, unlobed.
    Site: montane forest; M, LM.

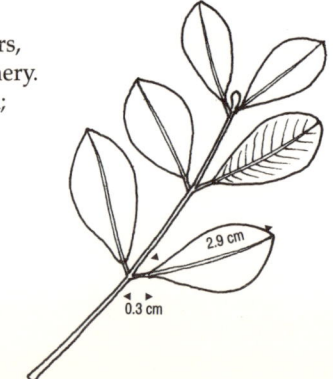

2.9 cm

0.3 cm

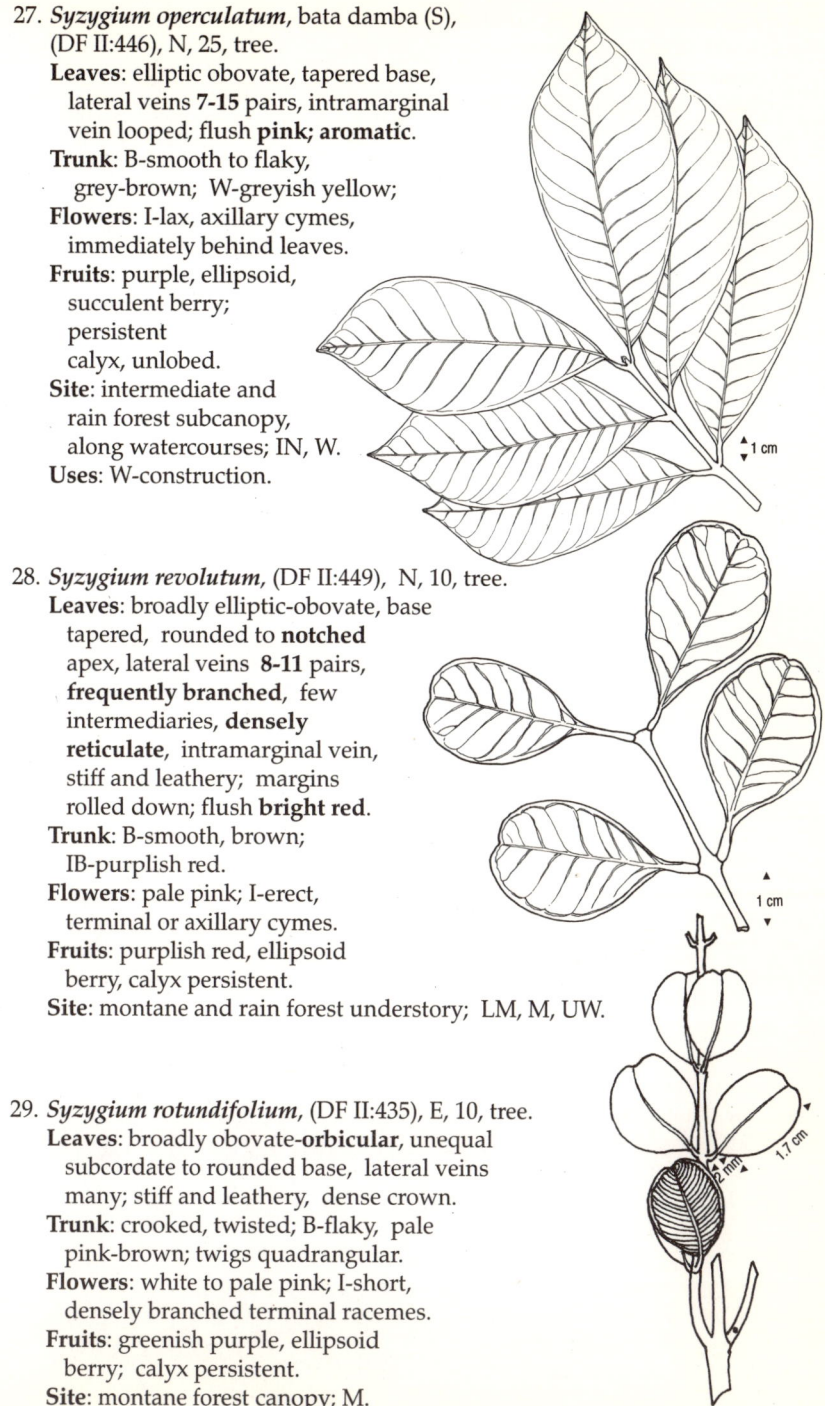

27. *Syzygium operculatum*, bata damba (S),
(DF II:446), N, 25, tree.
 Leaves: elliptic obovate, tapered base,
 lateral veins **7-15** pairs, intramarginal
 vein looped; flush **pink; aromatic**.
 Trunk: B-smooth to flaky,
 grey-brown; W-greyish yellow;
 Flowers: I-lax, axillary cymes,
 immediately behind leaves.
 Fruits: purple, ellipsoid,
 succulent berry;
 persistent
 calyx, unlobed.
 Site: intermediate and
 rain forest subcanopy,
 along watercourses; IN, W.
 Uses: W-construction.

1 cm

65

28. *Syzygium revolutum*, (DF II:449), N, 10, tree.
 Leaves: broadly elliptic-obovate, base
 tapered, rounded to **notched**
 apex, lateral veins **8-11** pairs,
 frequently branched, few
 intermediaries, **densely
 reticulate**, intramarginal vein,
 stiff and leathery; margins
 rolled down; flush **bright red**.
 Trunk: B-smooth, brown;
 IB-purplish red.
 Flowers: pale pink; I-erect,
 terminal or axillary cymes.
 Fruits: purplish red, ellipsoid
 berry, calyx persistent.
 Site: montane and rain forest understory; LM, M, UW.

1 cm

MYRTACEAE

29. *Syzygium rotundifolium*, (DF II:435), E, 10, tree.
 Leaves: broadly obovate-**orbicular**, unequal
 subcordate to rounded base, lateral veins
 many; stiff and leathery, dense crown.
 Trunk: crooked, twisted; B-flaky, pale
 pink-brown; twigs quadrangular.
 Flowers: white to pale pink; I-short,
 densely branched terminal racemes.
 Fruits: greenish purple, ellipsoid
 berry; calyx persistent.
 Site: montane forest canopy; M.

2 mm   1.7 cm

30. *Syzygium rubicundum*, kuretiya, kurumbatiya
    (S), (DF II:437), N, 25, tree.
    Leaves: narrowly oblong-elliptic,
      lateral veins more than **25** pairs,
      intramarginal vein; flush
      rose-pink turning
      olive-green.
    Trunk: small narrow
      buttresses; B-smooth,
      oblong-flaky, pale
      orange-brown;
      twigs quadrangular.
    Flowers: pale pink,
      many; I-long, slender,
      subterminal or axillary cymes.
    Fruits: red-purple, small, juicy,
      globose berry, calyx persistent.
    Site: rain forest subcanopy; W.
    Uses: W-construction.

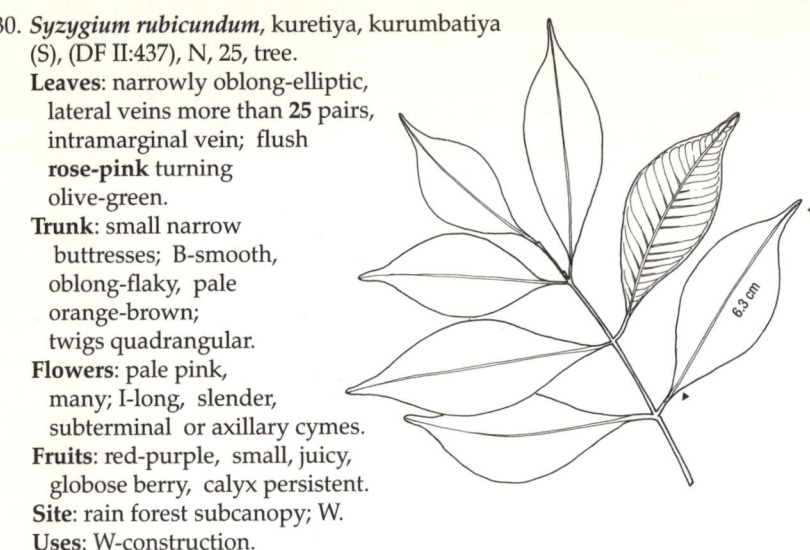

31. *Syzygium spathulatum*, (DF II:438), E, 10, tree.
    Leaves: obovate to oblanceolate, round-pointed apex,
      lateral veins **6-12** pairs, rigidly leathery, **reddish
      brown to golden** hairy beneath;
      flush **bright copper**-red.
    Trunk: B-rough, cracked, flaky, orange-pink;
      branches twisted; twigs quadrangular.
    Flowers: pale pink, small; I-terminal
      or axillary cymes.
    Fruits: purple, globose berry.
    Site: intermediate,
      montane and rain
      forest understory,
      scrub, and
      grassland;
      IN, LM, W.

292

32. *Syzygium umbrosum*, weli damba (S)/ naval (T),
(DF II:439), E, 10, tree.
Leaves: **spatulate**, base tapered,
lateral veins **5-10** pairs,
obscure, reticulation
obscure; flush
purplish crimson.
Trunk: B-smooth
to flaky, pale
pink-brown;
twigs quadrangular.
Flowers: white petals,
crimson calyx;
I-terminal or
axillary cymes.
Fruits: subglobose
berry, calyx persistent.
Site: montane and rain
forest understory;
M, LM, UW.

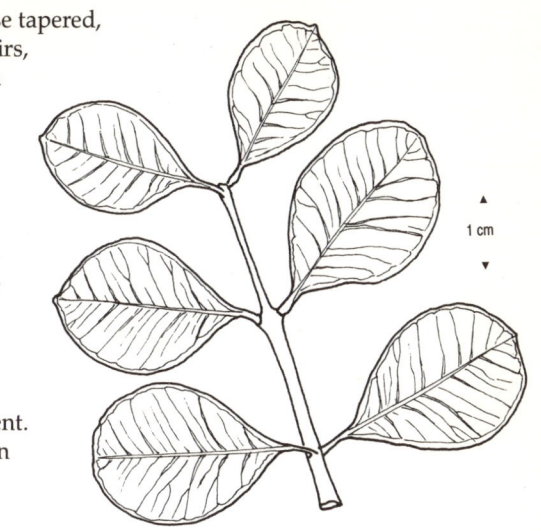

1 cm

65

33. *Syzygium zeylanicum*, yakada maran (S)/
maranda (T), (DF II:431), N, 10, tree.
Leaves: variable, lanceolate, apex
long-pointed, lower surface **minutely
pitted**, lateral veins **12-15** pairs,
obscure, with intermediaries,
strongly aromatic; intramarginals;
flush **pale pink**.
Trunk: crooked; b-smooth or irregularly
cracked, flaky, pale grey-brown to
green; twigs slightly quadrangular.
Flowers: white,
densely
clustered;
terminal or
axillary racemes.
Fruits: white, broadly ellipsoid
to subglobose berry;
calyx persistent.
Site: rocky
summits, sandy
coastal plains;
DL, IN, LM, WC.
Uses: W-fuelwood;
whole plant medicinal.

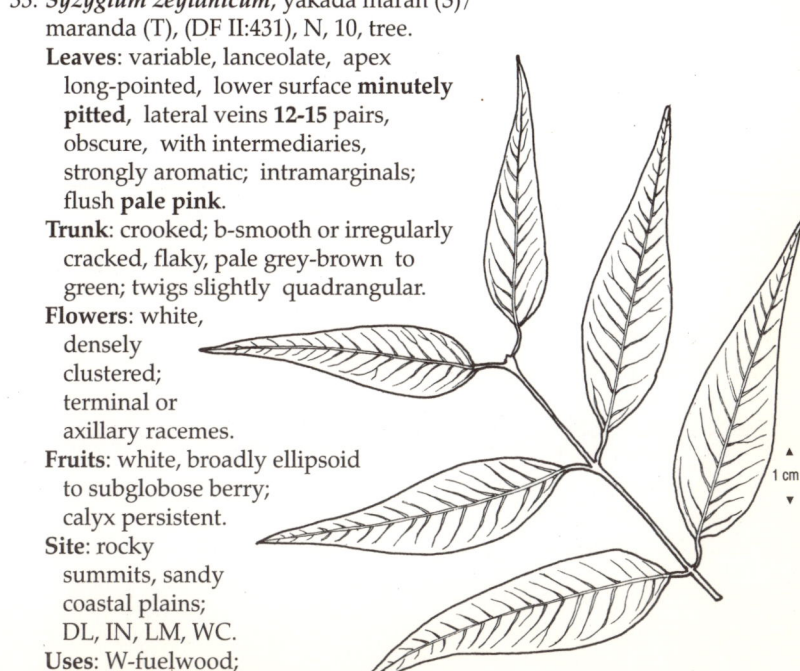

1 cm

MYRTACEAE

66. NYCTAGINACEAE

FAMILY DESCRIPTION - Habit: trees, shrubs or herbs. **Leaves**: usually opposite, simple, entire. **Stipules**: absent. **Flowers**: bisexual, rarely unisexual on monoecious plants, actinomorphic, in cymose inflorescences. Sometimes surrounded by brightly coloured bracts. **Fruits**: achene or nut, often enclosed in persistent base of calyx tube.

FLOWER PARTS - CALYX: 4-5, united and tubular, often petaloid, lobes valvate. COROLLA: absent. ANDROECIUM: stamens 1 to many, free or basally united into a tube. Anthers 2-locular, opening lengthwise. Annular nectary disk often around ovary. GYNOECIUM: superior, 1-carpelled, 1-locular, with a solitary ovule. Placentation basal.

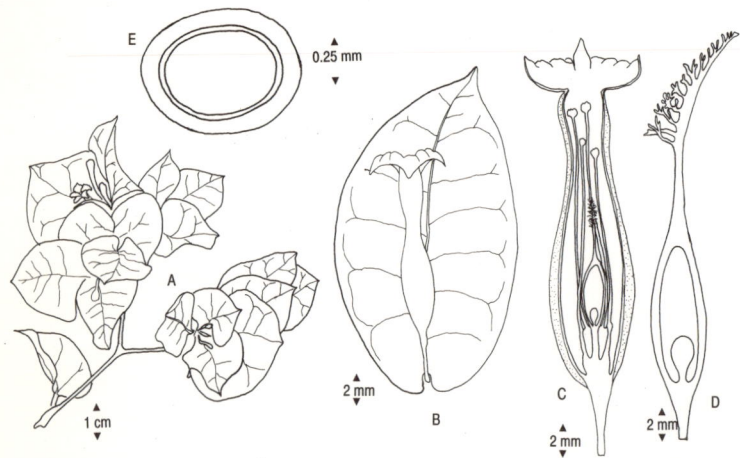

Key: inflorescence (A), a single flower with the bract (B), half flower (C), longitudinal section of gynoecium (D), and transverse section of ovary (E) of *Bougainvillea spectabilis*.

1. *Bougainvillea spectabilis*, bougainvillea (E)/ boganvilla (S), I (S. Amer.), 2-10, scandent shrub.
    Leaves: opposite to subopposite or alternate, variable in shape, entire.
    Trunk: branches often climbing, thorny.
    Flowers: **crimson, white, orange, pink to purple**, conspicuous bracts surrounding 3 small flowers.
    Fruits: membranous, enclosed in persistent perianth.
    Site: gardens, roadsides; widespread.
    Uses: ornamental, hedges.

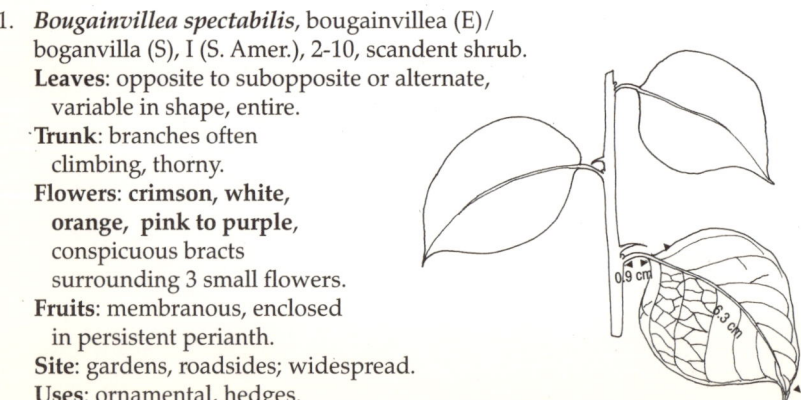

2. **Pisonia grandis**, lettuce tree (E), wata banga (S),
   lechchai kottai (T), I (Malaysia), 10, tree.
   Leaves: opposite or alternate, ovate to oblong,
      base rounded, apex pointed; flush pale
      yellow, mature leaf yellowish green.
   Trunk: low branching, branches stout.
   Flowers: pale green, small;
      I-terminal corymbose panicles.
   Fruits: long-stalked with leathery
      perianth, 5-angled
      with 1 row of prickles.
   Site: home gardens; IN, W.
   Uses: ornamental; leaves-medicinal,
      vegetable.

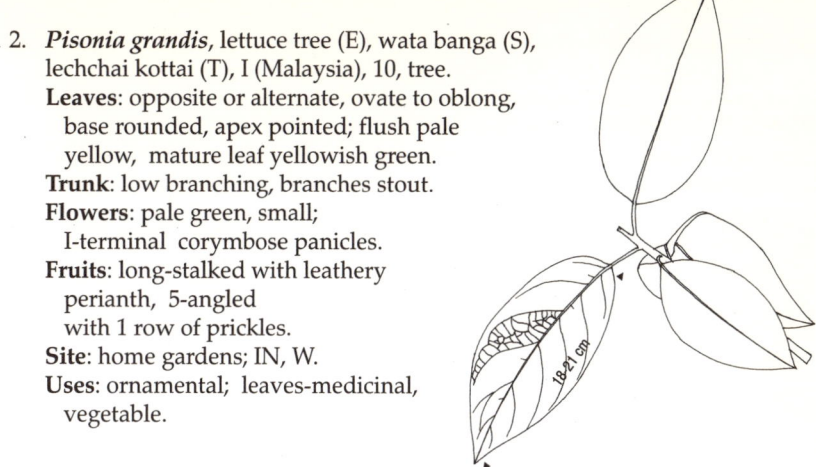

67. NYCTANTHACEAE

FAMILY DESCRIPTION - Habit: small trees. **Leaves:** simple, decussate, often lobed or dentate. **Stipules:** absent. **Flowers:** bisexual, salver-shaped, actinomorphic, terminal trichotomous cymes. **Fruits:** flattened, capsule.

FLOWER PARTS - CALYX: 4-8, united, tubular, rim denticulate or truncate. COROLLA: 4-8, basally tubular, mostly yellow to orange, contorted. ANDROECIUM: stamens 2, epipetalous, near apex of corolla tube. GYNOECIUM: stigma shortly bifid; style cylindrical; ovary superior, 2-locular, 1 erect ovule per loculus with basal placentation.

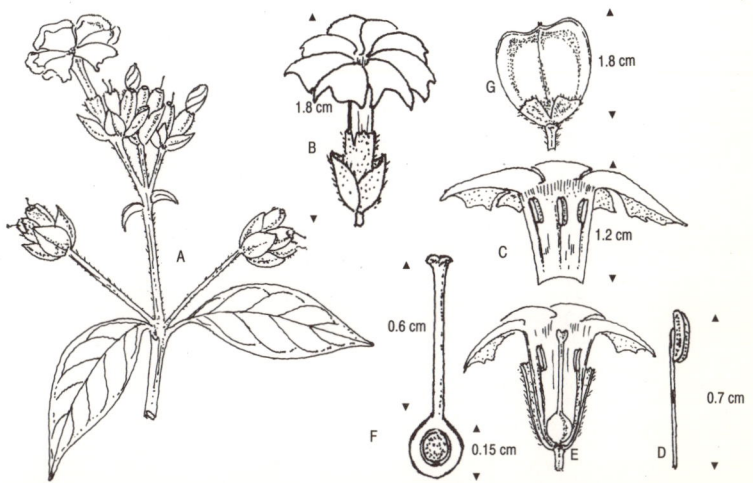

Key: inflorescence (A), single flower with bracts (B), part of half flower showing epipetalous stamens (C), a single stamen (D), half flower (E), longitudinal section of ovary (F) and fruit (G) of *Nyctanthes arbor-tristis*.

1. **Nyctanthes arbor-tristis,** tree of sadness (E),
   sepalika (S), (DF IV:179), I (India), 3, small tree.
   Leaves: decussate, ovate to oblong, rounded
       to pointed base, long-pointed apex, margin
       entire to **coarsely** and unequally
       dentate, roughly hairy.
   Trunk: much branched;
       twigs hairy, 4-sided.
   Flowers: white and orange, sessile,
       fragrant, opening at sunset; I-axillary
       or terminal panicles, each with 2-7
       flowers; peduncles four-sided and hairy.
   Fruits: flattened, rounded capsule.
   Uses: flowers-temple offerings, yellow
       dye, perfumery; leaves-sandpaper;
       W-fuelwood; ornamental.

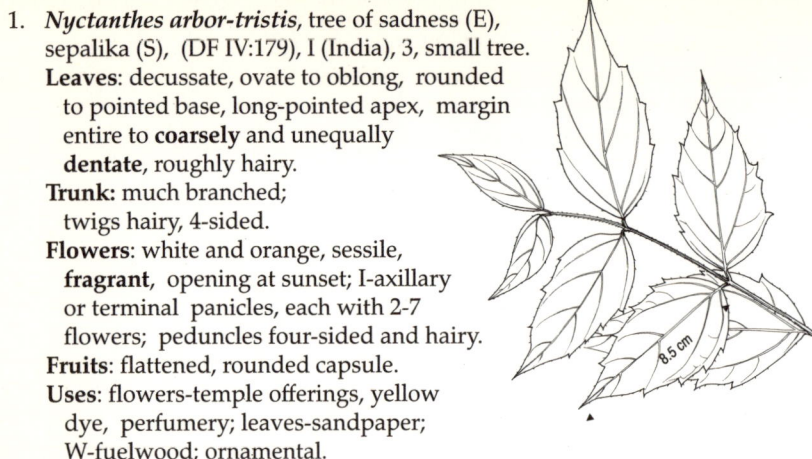

68. OCHNACEAE

FAMILY DESCRIPTION - Habit: trees, shrubs, herbs. **Leaves**: spiral, simple, rarely pinnate, with many parallel lateral veins. **Stipules**: present, sometimes deeply lobed. **Flowers**: bisexual, usually actinomorphic, racemes or cymes. **Fruits**: carpels of fruit quite separate on enlarged torus and drupaceous. Seeds 1 per carpel.

FLOWER PARTS - CALYX: 4-5, rarely 10, free, imbricate, or rarely contorted. COROLLA: 4-10, free, imbricate, or contorted. ANDROECIUM: stamens 5, 10 or many, sometimes in 3-5 whorls, sometimes borne on an androgynophore. Anthers linear, basifixed, opening lengthwise or by a terminal pore. GYNOECIUM: superior, 2-5-carpelled and 2-5-locular, rarely 10-15 locular. Carpels entire or lobed and united by common style. Ovules 1 to many per loculus. Placentation axile.

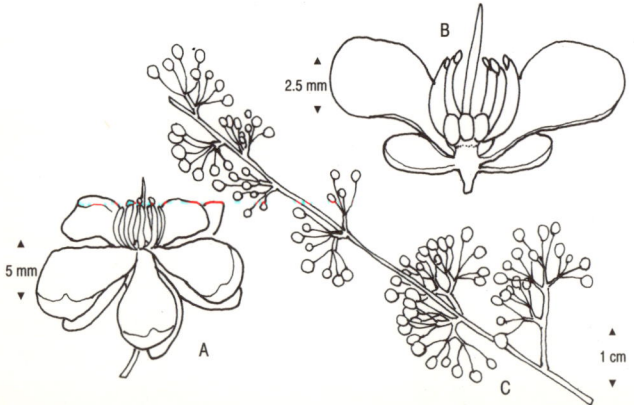

Key: full flower (A), half flower (B) and inflorescence (C) of *Gomphia serrata*.

1. ***Ochna lanceolata***, bo kera (S)/katkarai (T),
    (DF VI:247), N, 4, shrub to small tree.
    Leaves: lanceolate to rhomboid, acute to
      rounded base, acute apex, margins
      finely serrate, pale beneath.
    Trunk: much branched; W-light, soft,
      pale yellow; young parts glabrous.
    Flowers: bright yellow,
      sepals red, solitary to 2-3.
    Fruits: smooth,
      purplish black
      carpels.
    Site: monsoon,
      intermediate and rain
      forest understory;
      DL, IN, W.
    Uses: whole plant
      medicinal, W-light
      construction.

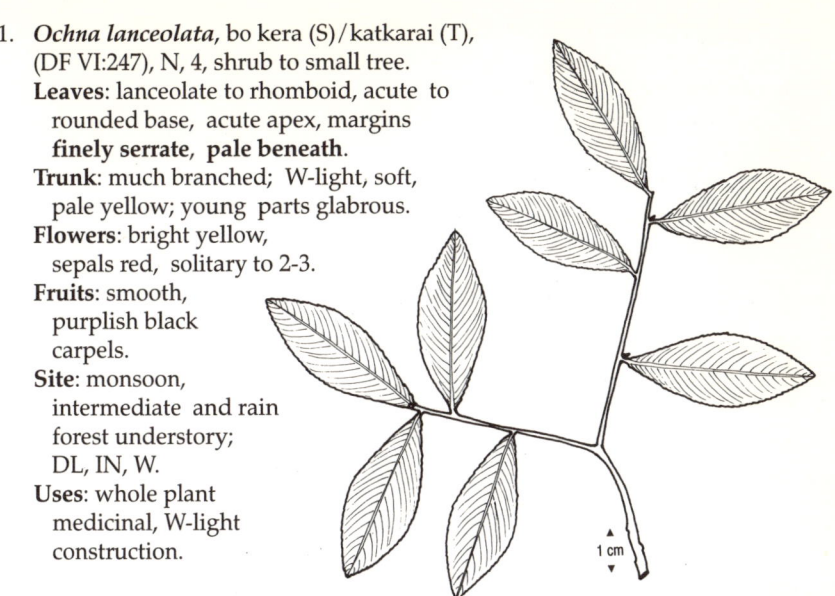

1 cm

2   ***Ouratea serrata***, bo kera (S), (DF VI:252), N, 15, tree.
    Leaves: **2-ranked**, lanceolate, acute ends, margins **finely
      serrate**, lateral veins numerous, very close, 2 marginals.
    Trunk: much-branched; W-hard, pale
      brownish; young parts glabrous.
    Flowers: yellow, numerous, sepals
      red; I-large axillary or terminal
      pyramidal panicles.
    Fruits: purple-black,
      ovoid-reniform, shiny,
      surrounded by sepals,
      with 5 or less carpels
      attached to central stalk.
    Site: intermediate, montane
      forest canopy and rain
      forest understory;
      LM, IN, W.
    Uses: W-light construction.

8 cm

3 cm

69. OLACACEAE

FAMILY DESCRIPTION - Habit: trees, shrubs or lianas. **Leaves**: alternate, simple, entire. **Stipules**: absent. **Flowers**: bisexual or polygamodioecious, actinomorphic, in axillary panicles, racemes or heads. **Fruits**: 1-seeded drupe or nut often enclosed by accrescent calyx.

FLOWER PARTS - Calyx: 4-6, united and cupular, imbricate or open in bud. Corolla: 4-6, free or variously united, valvate. Nectary disc present, often annular. Androecium: stamens 4-12, free or rarely monadelphous, some occasionally antherless. Anthers 2-locular, opening lengthwise or by pore-like apical slits. Gynoecium: superior, sometimes seemingly inferior by adnation to surrounding disc. Carpels 2-5, basally 2-5-locular and apically 1 locular by incomplete septation. Ovule solitary in each loculus. Placentation free central or typically axile.

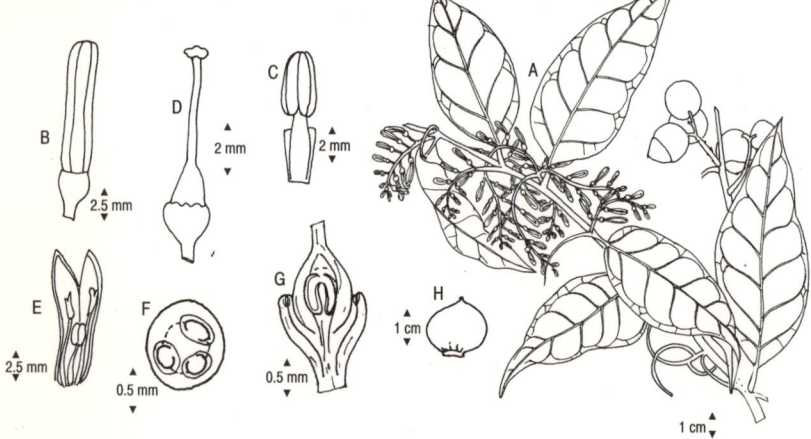

Key: inflorescence and fruit-bearing twig (A), full flower (B), single stamen (C), gynoecium (D), half flower (E), ovary in transverse (F), and longitudinal sections (G), and a single fruit (H) of *Olax zeylanica*..

1. *Olax zeylanica*, mella (S),
    (T I:257), E, 3, small tree.
    Leaves: ovate-oblong, acute base, acute apex.
    Trunk. young branches **yellow**, acutely
        angled, **finely ridged**, glabrous.
    Flowers: few, very small;
        I-axillary racemes.
    Fruits: scarlet, broadly ovoid,
        strongly apiculate, smooth
        drupe, base with cup-shaped calyx.
    Site: intermediate forest subcanopy; IN.
    Uses: leaves-vegetable, medicinal.

2. **Strombosia nana,** hora kaha (S), E, 10, tree.
    Leaves: spirally arranged, subovate to
        elliptic, rounded to pointed apex,
        inconspicuous lateral veins.
    Trunk: B-smooth, finely fissured,
        dark to light brown; IB-white.
    Flowers: I-axillary clusters.
    Fruits: solitary, club-shaped.
    Site: rain forest understory; W.

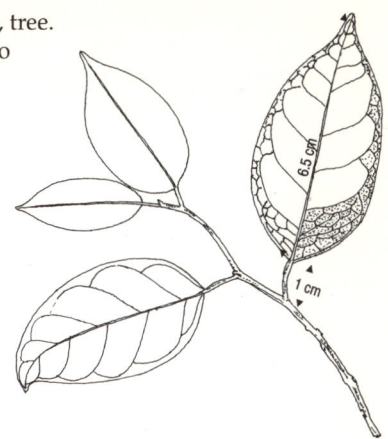

70. OLEACEAE

70

FAMILY DESCRIPTION - Habit: trees, shrubs or sometimes lianas. **Leaves**: opposite, (spiral in *Jasminum*), simple or pinnate. **Stipules**: absent. **Flowers**: bisexual or rarely unisexual, actinomorphic, cymose inflorescences or solitary. **Fruits**: loculicidally dehiscent capsule, berry, drupe or samara.

FLOWER PARTS - CALYX: typically 4, rarely 4-15, united, valvate. Rarely absent. COROLLA: typically 4, rarely 4-12, free or united, imbricate, valvate or convolute. Sometimes absent. ANDROECIUM: typically 2, rarely 4, epipetalous. Filaments free. Anthers 2-locular, loculi back to back, opening lengthwise. GYNOECIUM: superior, 2 carpels, 2 locular. Ovules usually 2 per loculus. Placentation axile. Style simple with capitate or bifid stigma.

OLEACEAE

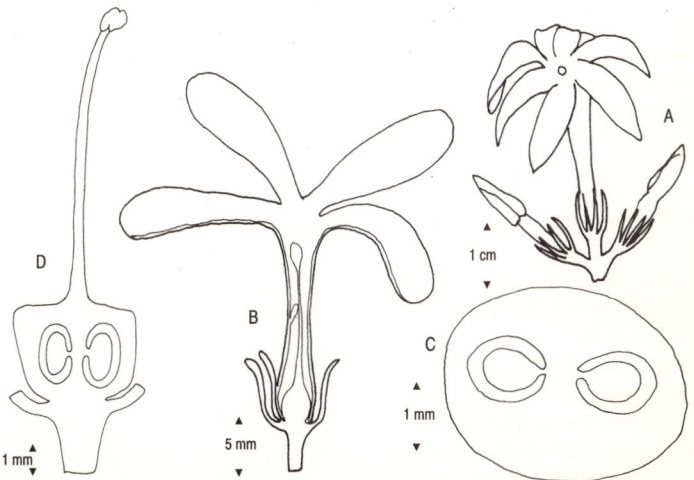

Key: part of the inflorescence (A), half flower (B), transverse section of ovary (C), and longitudinal section of gynoecium (D) of *Jasminum pubescens*.

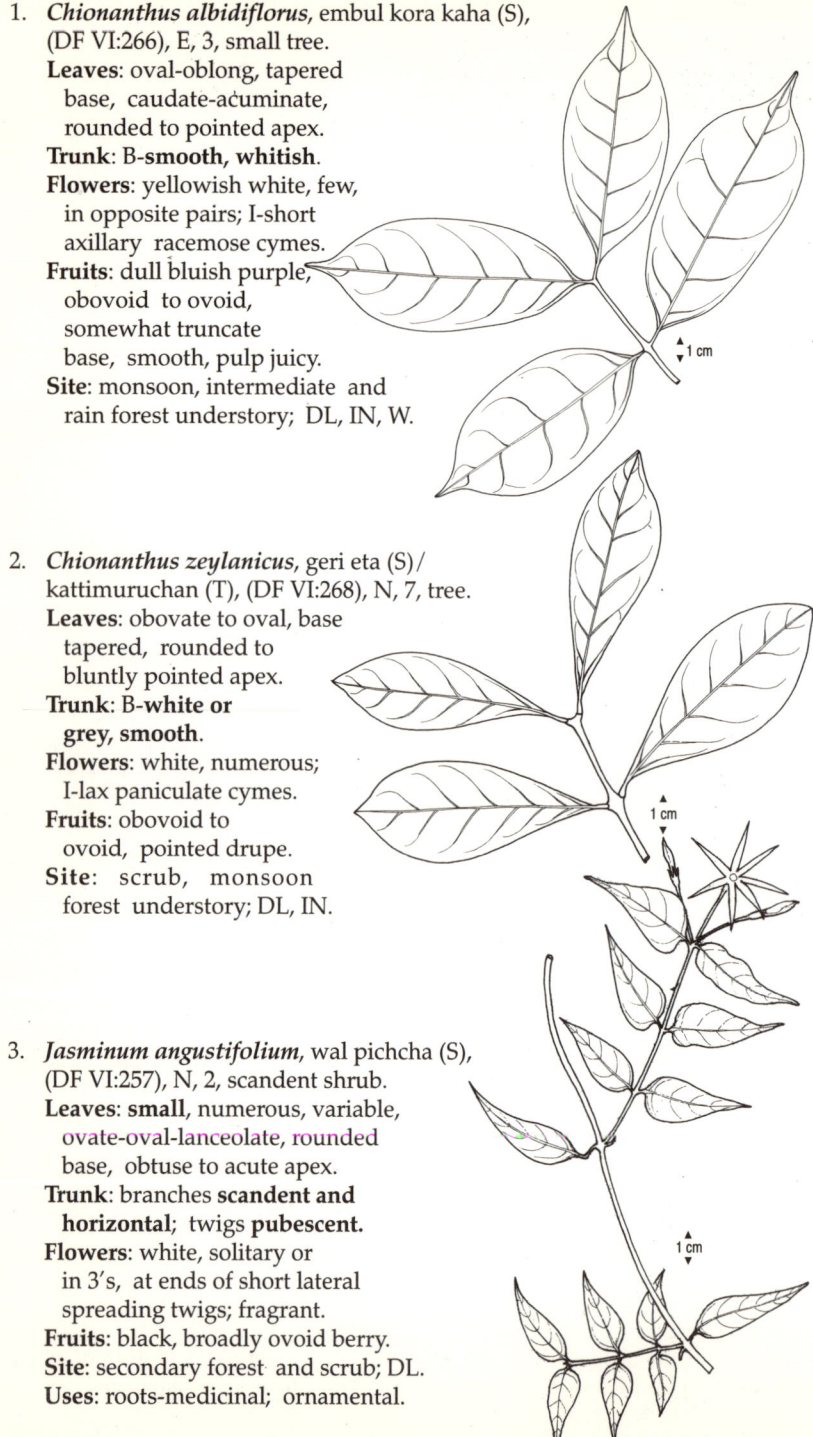

1. *Chionanthus albidiflorus*, embul kora kaha (S), (DF VI:266), E, 3, small tree.
   Leaves: oval-oblong, tapered base, caudate-acuminate, rounded to pointed apex.
   Trunk: B-**smooth, whitish**.
   Flowers: yellowish white, few, in opposite pairs; I-short axillary racemose cymes.
   Fruits: dull bluish purple, obovoid to ovoid, somewhat truncate base, smooth, pulp juicy.
   Site: monsoon, intermediate and rain forest understory; DL, IN, W.

2. *Chionanthus zeylanicus*, geri eta (S)/ kattimuruchan (T), (DF VI:268), N, 7, tree.
   Leaves: obovate to oval, base tapered, rounded to bluntly pointed apex.
   Trunk: B-**white or grey, smooth**.
   Flowers: white, numerous; I-lax paniculate cymes.
   Fruits: obovoid to ovoid, pointed drupe.
   Site: scrub, monsoon forest understory; DL, IN.

3. *Jasminum angustifolium*, wal pichcha (S), (DF VI:257), N, 2, scandent shrub.
   Leaves: **small**, numerous, variable, ovate-oval-lanceolate, rounded base, obtuse to acute apex.
   Trunk: branches **scandent and horizontal**; twigs **pubescent**.
   Flowers: white, solitary or in 3's, at ends of short lateral spreading twigs; fragrant.
   Fruits: black, broadly ovoid berry.
   Site: secondary forest and scrub; DL.
   Uses: roots-medicinal; ornamental.

4. *Jasminum auriculatum*, (DF VI:260), N, 2, shrub.
   Leaves: pinnate; **3** leaflets, laterals **very small**,
      often absent, central one broadly ovate to oval,
      mucronate apex, glabrous or pubescent.
   Trunk: branches more or less pubescent.
   Flowers: white, small, numerous, fragrant;
      I-pubescent, lax corymbose cymes.
   Fruits: purple-black, globose berry.
   Site: monsoon forest understory, scrub; DL.

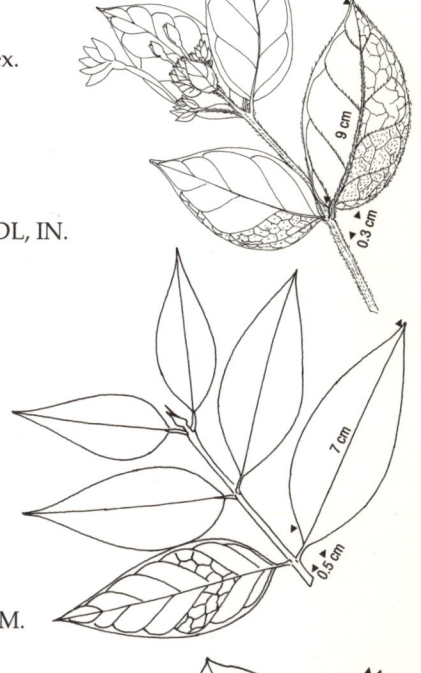

5. *Jasminum rottlerianum*, (DF VI:259),
   N, 2, scandent shrub.
   Leaves: oval, rounded to
      subcordate base, **mucronate** apex.
   Trunk: branches **horizontal**
      and **scandent**.
   Flowers: white, in 3's, dense;
      I-small corymbose cymes.
   Fruits: black globose berry.
   Site: secondary forest and scrub; DL, IN.

6. *Ligustrum robustum*, bora (S),
   (DF VI:270), N, 8, tree.
   Leaves: oval to lanceolate, acute
      base, tapering pointed apex.
   Trunk: branchlets with many
      white lenticels.
   Flowers: white, numerous,
      fragrant; I-large, erect,
      terminal pyramidal panicles.
   Fruits: purple, ovoid drupe.
   Site: montane forest understory; LM.

7. *Olea polygama*, (DF VI:264), N, 10, tree.
   Leaves: **terminally crowded, small**, oval,
      acute ends, margins **rolled down**,
      faintly serrate or entire, lateral veins
      few, prominent with intramarginal.
   Trunk: B-whitish brown.
   Flowers: white, clusters of 2-7;
      I-small, racemose, on long-
      stalked axillary panicles.
   Fruits: ovoid drupe.
   Site: montane forest subcanopy; M.

71. PITTOSPORACEAE

FAMILY DESCRIPTION - Habit: trees, shrubs or lianas. Sometimes spiny. **Leaves**: spiral, simple. **Stipules**: absent. **Flowers**: usually bisexual, actinomorphic, solitary or in corymbs or compact clusters. **Fruits**: loculicidal capsule or berry. Seeds often in sticky pulp.

FLOWER PARTS - CALYX: 5, free or sometimes basally united forming a tube, lobes imbricate. COROLLA: 5, usually basally united forming a tube, lobes imbricate. ANDROECIUM: stamens 5, free or united. Anthers 2-locular, opening lengthwise or by pores. GYNOECIUM: superior, 2-5-carpelled and locular, with axile placentation, or 2-carpelled and 1-locular with parietal placentation. Style simple.

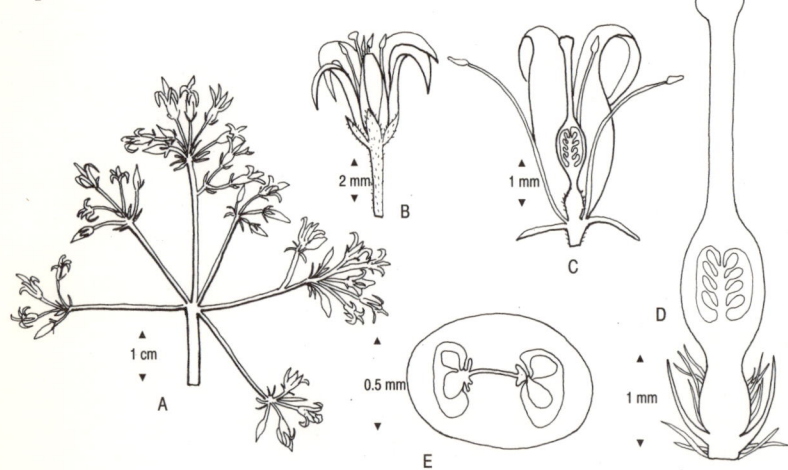

Key: inflorescence (A), full flower (B), half flower (C), longitudinal section of gynoecium (D), and transverse section of ovary (E) of *Pittosporum ceylanicum*.

1. *Pittosporum ceylanicum,* ketiya (S), (DF II:474), E, 8, tree.
   Leaves: terminally clustered, spiral, obovate to oblanceolate, tapered base, veins reticulate; young shoots and petioles **glandular**.
   Trunk: B-greyish white, warty, lenticellate.
   Flowers: yellowish white; I-pseudo-terminal umbels, paniculate.
   Fruits: yellowish, subglobose capsules, valves woody, hard, mango smell; seeds angular; pulp orange-red, sticky.
   Site: secondary forest, rocky outcrops; M, LM, UW.

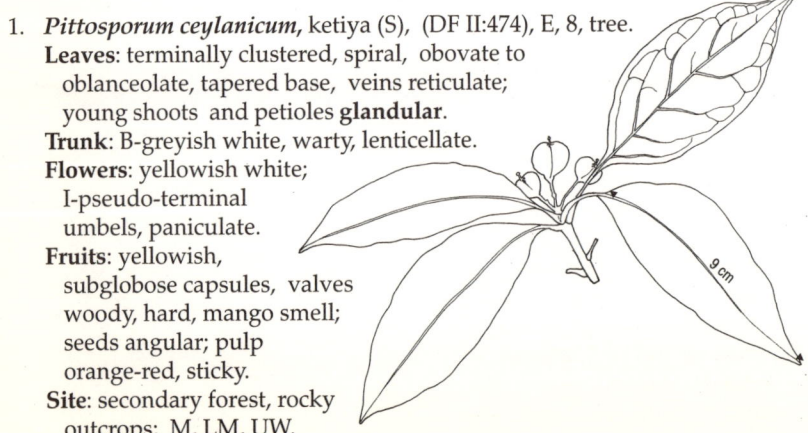

2. **Pittosporum ferrugineum**, kaputu gas (S), (DF II:476), I (SE. Asia), 10, tree.

Leaves: spiral, terminally crowded, obovate to oval-lanceolate, pointed ends, margins wavy; flush **rusty pubescent**.

Trunk: B-warty, prominent light brown lenticels; young twigs rusty pubescent.

Flowers: white; I-axillary and terminal, fine brown hairs.

Fruits: orange, flattened globose capsule, splitting in 2; seeds bright scarlet.

Uses: gardens, roadsides; W.

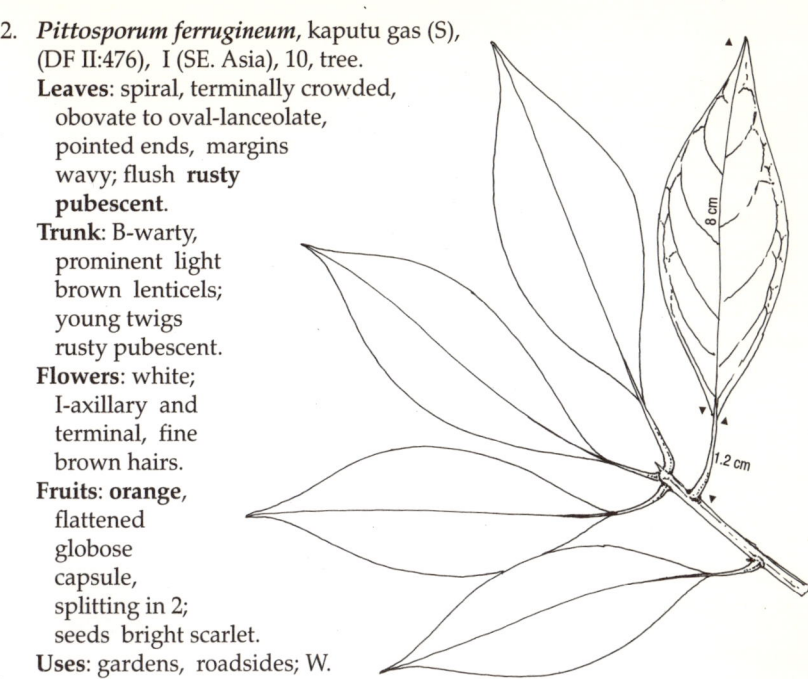

3. **Pittosporum tetraspermum**, (DF II:475), N, 3, shrub.

Leaves: terminally clustered, oval to oblong to elliptic, acute base, subacute apex, veins reticulate, translucent, distinct; flush hairy.

Trunk: branching **bifurcate or whorled** at top; B-greyish brown, lenticellate; IB-fibrous, white; young branches puberulous; **gummy** exudate.

Flowers: yellowish green, fragrant; I-pseudo-terminal, umbellate cymes.

Fruits: capsule subglobose, tipped by withered style; seeds darkish red with orange red aril.

Site: montane forest subcanopy, waterways; M.

72. PROTEACEAE

FAMILY DESCRIPTION - Habit: trees or shrubs, rarely herbs. **Leaves**: usually spiral, simple to pinnate or bipinnate. Often xeromorphic. **Stipules**: absent. **Flowers**: usually bisexual, actinomorphic or zygomorphic, in bracteate heads, spikes or racemes. **Fruits**: follicle, nut, achene or drupe. Seeds often winged.

FLOWER PARTS - CALYX: 4, corolla-like, coloured, united and valvate in bud. Variously split when open. COROLLA: apparently an annular, usually 4-lobed nectary disc, or represented by 2-4 scales. ANDROECIUM: stamens 4. Filaments usually adnate to calyx tube or lobes, rarely free. Anthers free, 2-locular, opening lengthwise. GYNOECIUM: superior. 1-carpelled, 1-locular. Ovules 1 or 2, parietal when more than 1. Style slender, elongated. Stigma often bulbous.

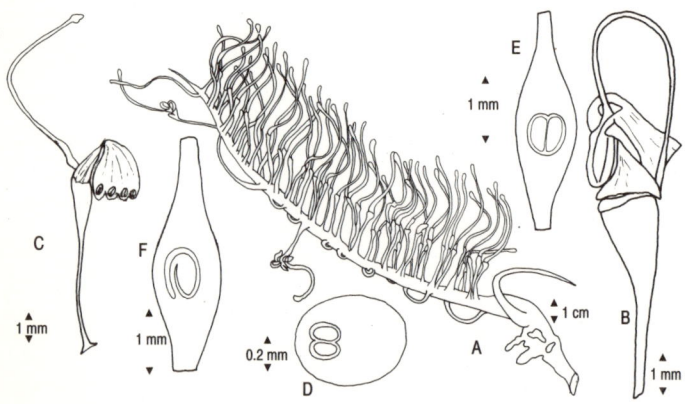

Key: inflorescence (A), full flower with style bent over (B), and style distended (C), ovary in transverse section (D), and two longitudinal sections at right angles to each other to show the two ovules within one carpel and placentation (E, F) of *Grevillea robusta*.

1. *Grevillea robusta*, silky oak (E)/sabukku (S,T),
    (DF II:485), I (Australia), 15, tree.
    Leaves: pinnate; leaflets opposite to subopposite, **deeply incised** acute lobes, upper surface glabrous, lower surface **tomentose**.
    Trunk: B-greyish brown, furrowed; rusty brown tomentose twigs.
    Flowers: orange to golden brown; many-flowered racemes.
    Fruits: blackish purple, compressed
    ovoid, with curved stalk;
    seed flat, winged.
    Site: tea estates; LM, M.
    Uses: shade tree, windbreaks;
    W-fuelwood, furniture;
    ornamental, support
    tree for pepper vines.

73. PUNICACEAE

FAMILY DESCRIPTION - Habit: trees or shrubs, often spiny. **Leaves**: mostly opposite or subopposite or fascicled, simple. **Stipules**: rudimentary or absent. **Flowers**: bisexual, actinomorphic, solitary or clustered at branch apices, with tubular or urceolate hypanthium. **Fruits**: berry with a leathery pericarp and persistent calyx lobes. Seeds with fleshy testa that forms the pulp within the fruit.

FLOWER PARTS - CALYX: 5-8 on hypanthium, fleshy, valvate. COROLLA: 5-8, free, crumpled in bud, imbricate. ANDROECIUM: numerous stamens. Filaments slender, free. Anthers 2-locular, dorsifixed, opening lengthwise. GYNOECIUM: inferior, carpels and loculi 7-15, with axile placentation. During development an outer whorl of carpels becomes superimposed on the inner whorl, making the placentation of carpels of the upper whorl appear parietal. Ovules numerous per loculus. Style slender and simple.

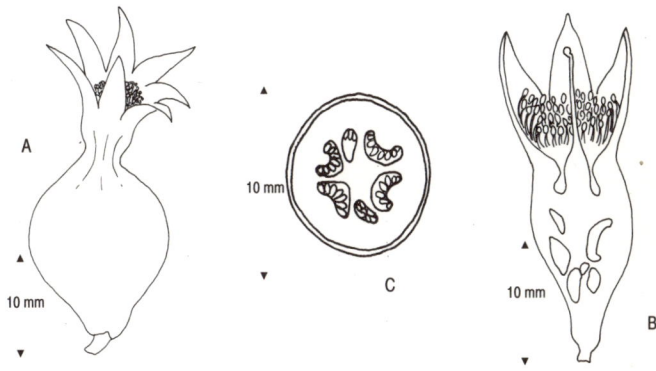

Key: developing fruit (A), half flower without the corolla (B), and transverse section of the ovary (C) of *Punica granatum*.

1. *Punica granatum*, pomegranate (E)/ delun (S)/madalai (T), (DF VI:318), I (India), 3, shrub or tree.
   Leaves: opposite, often **crowded** on short shoots, variable.
   Trunk: twigs frequently ending in **spines**.
   Flowers: red, **showy**, solitary or in terminal cymes.
   Fruits: yellowish red berry, large, conspicuous, divided into fleshy chambers containing many seeds surrounded by pulp.
   Site: home gardens; widespread.
   Uses: fruit-edible; fruit, leaves, roots-medicinal.

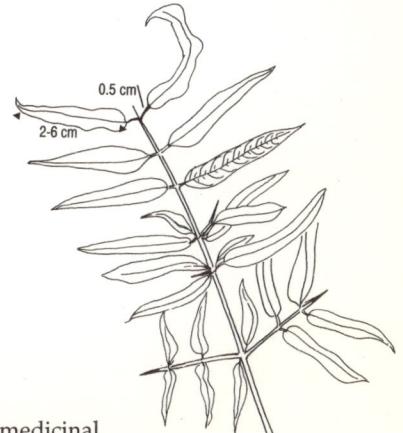

74. RHAMNACEAE

FAMILY DESCRIPTION - **Habit**: trees, shrubs or lianas, rarely herbs. Hooks, tendrils or twining stems used for climbing. **Leaves**: spiral or opposite, simple, pinnately veined or with several main veins from the base. **Stipules**: small and spiny or absent. **Flowers**: bisexual, rarely unisexual on monoecious plants, actinomorphic, in axillary corymbose or cymose inflorescences. **Fruits**: variable, drupe, or dry dehiscent capsule, or rarely a samara.

FLOWER PARTS - CALYX: 5, rarely 4. United and tubular, valvate. COROLLA: 5, rarely 4. Often hooded and concave holding anthers, rarely absent. ANDROECIUM: stamens 4 or 5, arising from a disc and enclosed by the petals. Anthers 2- locular, opening lengthwise. GYNOECIUM: superior or seemingly inferior, 2-4 carpels and 2-4 loculi. Ovules 1, rarely 2, per loculus. Placentation basal. Style shortly lobed.

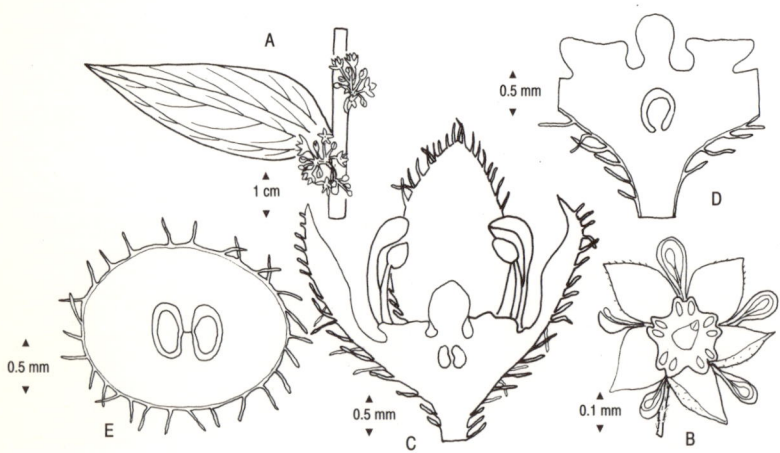

Key: inflorescence (A), full flower (B), half flower showing androecium with disc and gynoecium (C), longitudinal section of gynoecium with disc (D), and transverse section of ovary (E) of *Zizyphus oenoplia*.

1. *Colubrina asiatica,* telhiriya (S)/ mayir
   manikkam (T), (T I:285), N, 2, shrub.
   Leaves: ovate, rounded to
      subcordate base, rounded to
      pointed apex, margin **serrate**,
      somewhat **3**-veined at base.
   Trunk: much-branched.
   Flowers: greenish yellow; I-small,
      axillary, nearly sessile cymes.
   Fruits: smooth, globose capsule.
   Site: secondary forest and scrub; DL.
   Uses: B-medicinal.

2. **Rhamnus wightii**, (T I:283), N, 3, large shrub.
   Leaves: ovate, rounded base, rounded to pointed
   apex, margins finely **serrate**, glandular.
   Trunk: young parts puberulous.
   Flowers: yellowish green;
   I-axillary clusters of 1-5.
   Fruits: reddish purple, globose,
   smooth berry, tipped with
   persistent style and supported
   by flat persistent calyx.
   Site: montane forest understory; M.
   Uses: B-medicinal.

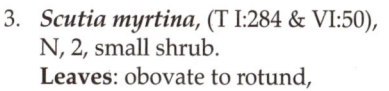
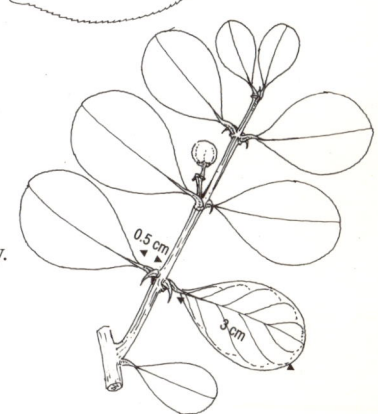

3. **Scutia myrtina**, (T I:284 & VI:50),
   N, 2, small shrub.
   Leaves: obovate to rotund,
   acute base, rounded or
   truncate apex, often notched.
   Trunk: straggling, spreading branches
   with small sharp, hooked axillary
   spines; young parts puberulous.
   Flowers: pale green; I-short,
   pedunculate, axillary umbels.
   Fruits: globular, pointed, smooth berry.
   Site: secondary forest and scrub; DL.

4. **Zizyphus jujuba**, masan (S)/ilantai (T),
   (T I:280), N, 4, small tree.
   Leaves: broadly oblong to oval to rotund,
   rounded ends, margins irregularly
   denticulate, beneath densely
   hairy; stipular **prickles**,
   one **curved**, one s**traight**.
   Trunk: much-branched; B-dark grey,
   deep longitudinal fissures;
   W-pale reddish, heavy, hard;
   branchlets elongated, **woolly**.
   Flowers: greenish white; I-small
   axillary clusters of paniculate cymes.
   Fruits: yellow, globose, fleshy, smooth drupe.
   Site: monsoon forest understory; DL.
   Uses: fruit-edible, medicinal; W-turnery;
   seeds-beads.

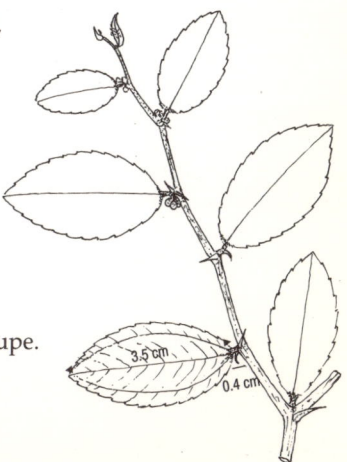

5. ***Zizyphus oenoplia***, hin eraminiya (S)/churai (T),
   (T I:281), N, 2, scrambling shrub.
   Leaves: 2-ranked, numerous, lanceolate
   to oval, unequal-sided, obliquely
   subacute base, acute apex, entire
   to minutely denticulate, silky
   pubescent beneath;
   prickles hooked.
   Trunk: B-rough;
   branchlets **red**
   pubescent.
   Flowers: pale greenish
   yellow, small, hairy;
   I-axillary cymes crowded
   on contracted branches.
   Fruits: dark purple to
   black, small, ovoid,
   shining, pointed drupes.
   Site: monsoon forest
   understory and scrub; DL.
   Uses: roots, leaves, B-medicinal;
   fruit-edible.

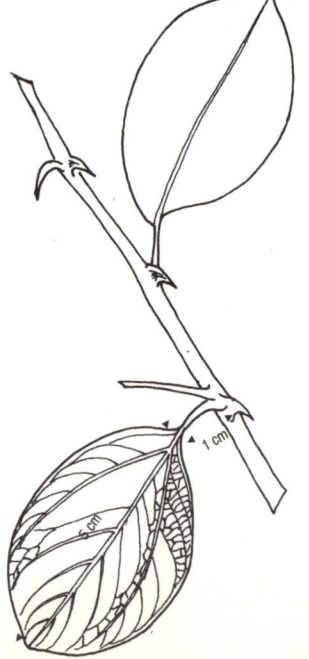

6. ***Zizyphus rugosa***, maha eraminiya
   (S)/ churai (T), (T I:282), N, 2, shrub.
   Leaves: broadly oval,
   unequal-sided, oblique base,
   shortly acuminate, **denticulate**
   margin, **densely yellow-brown**
   tomentose when young.
   Trunk: long twiggy branches;
   tomentose when young.
   Flowers: greenish, very small,
   numerous; I-tomentose cymes.
   Fruits: small, globose, smooth,
   tipped with a point.
   Site: secondary forest understory; W.
   Uses: whole plant-medicinal.

75. RHIZOPHORACEAE

FAMILY DESCRIPTION - Habit: trees or shrubs, often in mangroves. Branchlets swollen at nodes. **Leaves**: opposite, rarely spiral, simple. **Stipules**: interpetiolar, caducous and sheathing terminal bud, or rarely absent. **Flowers**: usually bisexual, rarely unisexual (*Anisophyllea*), actinomorphic, axillary inflorescences or solitary. **Fruits**: usually a berry with persistent calyx, rarely a capsule. Embryo viviparous in mangroves.

FLOWER PARTS - CALYX: 3-16, more or less basally united and adnate to the ovary below or free. COROLLA: equal in number to calyx, usually small, often notched, bifid or lacerate, free, convolute or inflexed in bud. ANDROECIUM: 2-4 times as many as sepals, often in pairs opposite the petals. Filaments short, often basally united, attached around or on a disc, sometimes without a disc. Anthers 4-to many-locular, usually dehiscing longitudinally. GYNOECIUM: superior, half-inferior or inferior. Carpels and loculi 2-4, rarely unilocular by suppression of septa. Ovules 2, rarely more, per loculus. Placentation pendulous. Style with simple or lobed stigma.

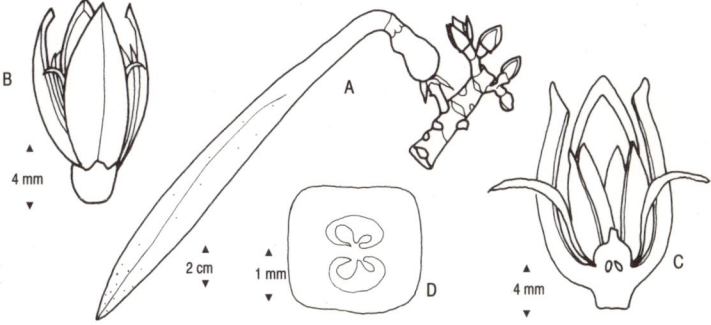

Key: inflorescence and developing fruit (A), full flower (B), half flower (C), and transverse section of ovary (D) of *Rhizophora apiculata*.

1. *Anisophyllea cinnamomoides*, weli piyanna (S), (DF II:500), E, 25, tree.
   Leaves: **dimorphic**; ordinary leaves **2-ranked**, oblong to ovate, tapering to unequal base; other leaves small, deciduous, between the ordinary, persistent larger leaves.
   Trunk: B-pale grey-brown; horizontal or drooping branches.
   Flowers: I-greenish white, 2-3 short supra-axillary racemes.
   Fruits: greenish white, ellipsoid, one seeded berry.
   Site: rain forest canopy; W.
   Uses: W-light construction.

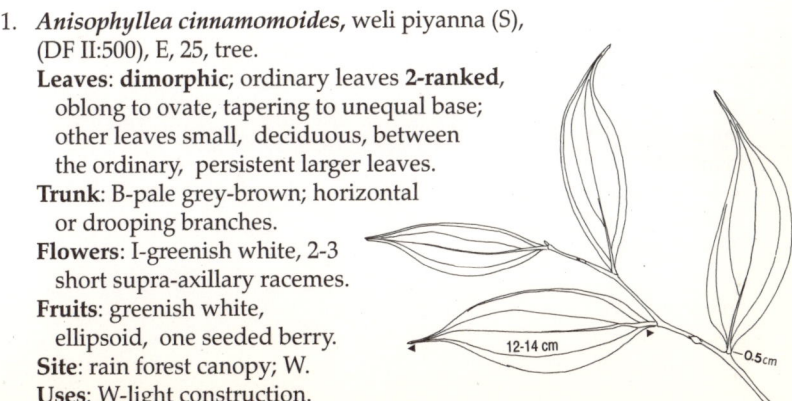

2. **Bruguiera cylindrica**, (DF II:492), N, 7, small tree.
   Leaves: lanceolate, tapering base, **black-dotted** beneath; long stipules.
   Trunk: knee roots; B-fissured or smooth when mature.
   Flowers: greenish, axillary.
   Fruits: green, oblong to ovoid, smooth, slender, cylindrical; **viviparous**.
   Site: mangroves; DC, MC.

3. **Bruguiera sexangula**,
   (DF II:493), N, 15, tree.
   Leaves: elliptic to oblong, acute ends, **black-dotted** beneath; stipules green or yellowish.
   Trunk: B-smooth, greyish with few large lenticels; knee roots.
   Flowers: yellow-brown, nodding; petals fringed with hairs.
   Fruits: oblong to ovoid; **viviparous**.
   Site: mangroves; MC.

4. **Carallia brachiata**, dawata (S),
   (DF II:496), N, 25, tree.
   Leaves: broadly elliptic, tapering base, paler underneath with scattered **black dots**.
   Trunk: B-smooth, grey to dark brown; W-hard, orange; disinctive branching.
   Flowers: creamy-white, small; I-small heads in axillary cymes.
   Fruits: red, smooth, small, berry-like; seeds with thick orange testa.
   Site: rain forest subcanopy; W.
   Uses: ornamental, W-construction, furniture.

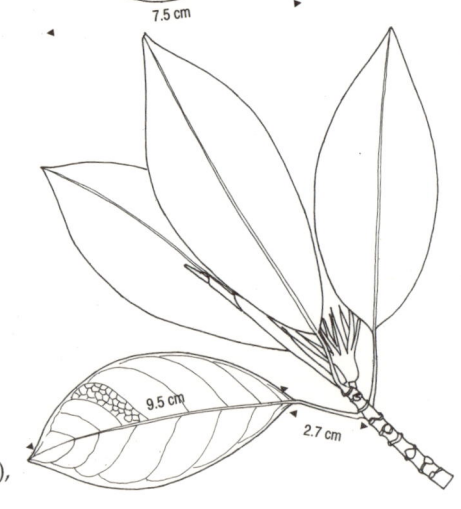

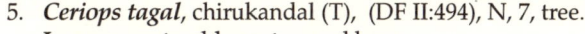

5. *Ceriops tagal*, chirukandal (T), (DF II:494), N, 7, tree.
   Leaves: ovate-oblong, tapered base,
      rounded apex.
   Trunk: twigs marked by leaf and
      stipule-scars; prop roots.
   Flowers: white, small.
   Fruits: reddish brown,
      ribbed, pendulous;
      viviparous.
   Site: mangroves; DC, MC.

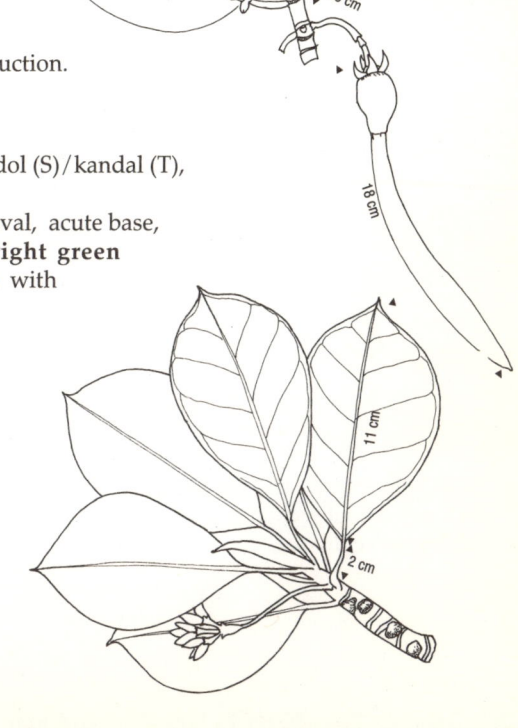

6. *Rhizophora apiculata*, kadol (S)/
   kandal (T), (DF II:490), N, 10, tree.
   Leaves: broadly ovate, acute base,
      tapered apex, **dark green** above,
      pale green with **brownish** dots
      beneath; spreading crown.
   Trunk: **prop roots**.
   Flowers: creamy-white,
      axillary, sessile, in pairs.
   Fruits: brown, conical to
      ovoid; pendulous,
      viviparous.
   Site: mangroves; DC, MC.
   Uses: B-tanning fish nets;
      W-fuelwood, light construction.

7. *Rhizophora mucronata*, kadol (S)/kandal (T),
   (DF II:491), N, 10, tree.
   Leaves: broadly elliptic to oval, acute base,
      bluntly acute apex, **bright green**
      above, **pale green**-dotted with
      tiny **red** spots beneath.
   Trunk: **prop roots**.
   Flowers: white, sepals
      pale yellow, in fours;
      I-axillary cymes.
   Fruits: dark brown,
      ovoid to conical, rough,
      pendulous; **viviparous**.
   Site: mangroves; DC, MC.
   Uses: W-fuelwood, light
      constuction; B-tanning
      fish nets.

75

RHIZOPHORACEAE

76. ROSACEAE

FAMILY DESCRIPTION - Habit: trees, shrubs or herbs, often thorny, sometimes climbing. **Leaves**: spiral or rarely opposite, simple or compound. **Stipules**: mostly present and paired, sometimes adnate to petiole. **Flowers**: bisexual, actinomorphic, solitary or in cymes. **Fruits**: head of follicles, achenes or druplets, or pome.

FLOWER PARTS - Calyx: typically 5, basally united, imbricate, often appearing as lobes of hypanthium. Corolla: typically 5, same number as calyx, rarely absent, free, imbricate. Androecium: stamens numerous, in whorls of 5. Filaments usually free and attached to hypanthium. Anthers 2-locular, opening lengthwise. Inner surface of hypanthium nectariferous. Gynoecium: superior or inferior, carpels and loculi 1 to many, free or variously united, often adnate to hypanthium, or situated within it. 2 or more ovules per loculus. Placentation axile, when carpels are united. Styles free or united.

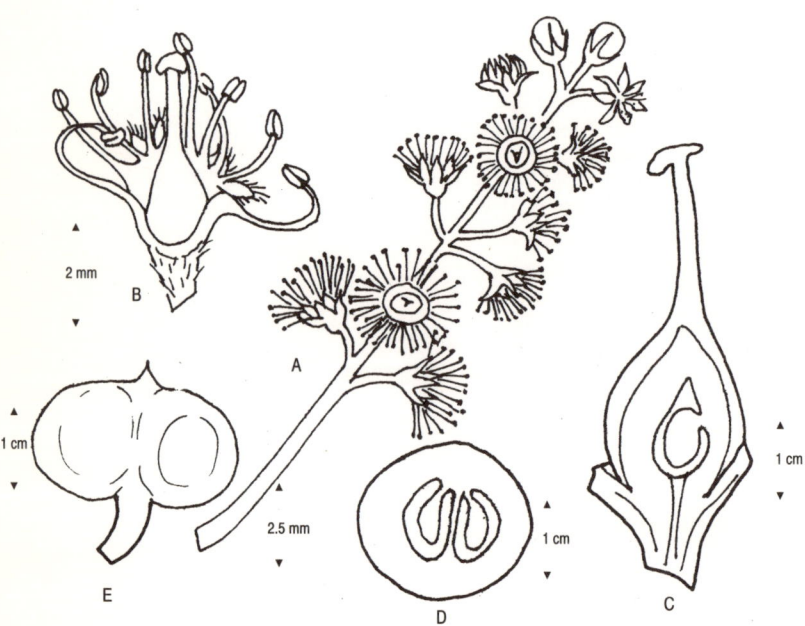

Key: inflorescence (A), single flower with some petals removed (B), gynoecium in longitudinal section (C), transverse section of ovary (D), and whole fruit (E) of *Prunus ceylanica*.

1. *Photinia integrifolia,* lunu warala (S),
    (DF III:333), N, 5, shrub to small tree.
    Leaves: oblanceolate to
       oval-oblong, rounded base,
       subacute apex; flush red.
    Trunk: B-wrinkled; young
       parts glabrous.
    Flowers: white to pink;
       I-terminal corymbose
       panicles.
    Fruits: globose, berry-like
       pome, calyx lobes at top,
       persistent bifid style.
    Site: montane forest
       understory, gaps
       and fringes; M.

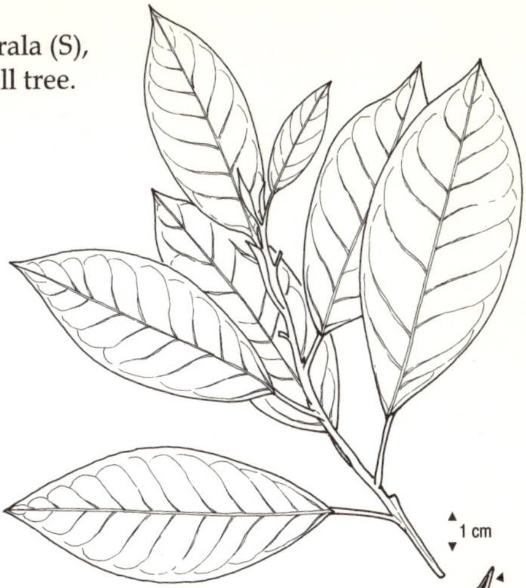

2. *Prunus cerasoides,* Indian cherry (E)/
    padma kashta (S), (DF III:373), I (India), 13, tree.
    Leaves: ovate to oblong-lanceolate, cuneate base,
       sharply **serrate**; **circular gland** at apex of petiole;
       stipules deeply lobed, glandular, deciduous.
    Trunk: B-smooth, reddish brown, with **lenticels**.
    Flowers: white to pink; I-3-flowered clusters or umbels.
    Fruits: yellowish red drupe, ellipsoid,
       glabrous; seeds furrowed.
    Site: home gardens; M.
    Uses: W-medicinal;
       fruit-medicinal.

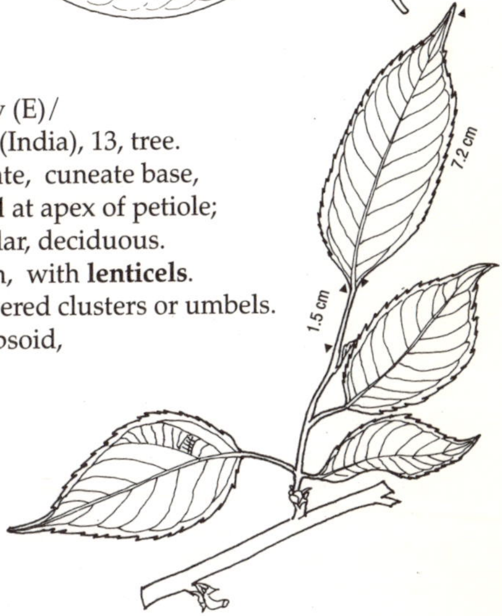

3. *Prunus ceylanica,* (DF III:376), N, 15, tree.
    Leaves: oval to oval-lanceolate,
       rounded base, acute apex;
       2 basal glands at leaf base.
    Trunk: iB-**darkening on
       exposure, almond-
       scented**; W-white, soft.
    Flowers: white, sweet-
       scented; I-racemes.
    Fruits: green to purplish
       black, ellipsoid drupe,
       bitter-almond smell.
    Site: montane forest canopy; M.

ROSACEAE

76

▲
1 cm
▼

7.2 cm

1.5 cm

4. *Prunus walkeri*, gulu mora (S), (DF III:375), E, 25, tree.
   Leaves: ovate to oblong-lanceolate,
       rounded base, slightly pointed apex,
       pubescent beneath; young leaves
       densely **golden tomentose**;
       2 **circular glands** at leaf base.
   Trunk: B-smooth, reddish,
       horizontal lenticels;
       W-hard, close-grained,
       heavy, yellowish white,
       almond-scented.
   Flowers: creamy white,
       solitary or I-axillary racemes.
   Fruits: pubescent, globose, drupe.
   Site: rain forest subcanopy,
       secondary forest
       gaps and fringes; W.
   Uses: W-construction, fuelwood.

5. *Pyrus communis*, common pear (E)/pears (S),
   (DF III:337), I (Europe), 7, tree.
   Leaves: broadly ovate to orbicular, rounded
       base, short pointed apex, finely **serrate**.
   Trunk: branches slightly pubescent.
   Flowers: white or pink; I-corymbose
       or umbellate racemes.
   Fruits: **green, pyriform**,
       very warty with
       persistent
       calyx lobes at top.
   Site: home gardens; M.
   Uses: fruit-edible.

6.5 cm   4.5 cm

77. RUBIACEAE

FAMILY DESCRIPTION - Habit: trees, shrubs, lianas or herbs. **Leaves**: opposite or whorled, rarely spiral, simple, usually entire. **Stipules**: present, inter-or intra-petiolar, free or united, sometimes leafy and indistinguishable from leaves. **Flowers**: mostly bisexual, actinomorphic, rarely zygomorphic, in cymes or rarely solitary. **Fruits**: capsule, berry, drupe or schizocarp.

FLOWER PARTS - CALYX: 4-5, united, sometimes adnate to ovary. COROLLA: 4-10, united, tubular, contorted, imbricate or valvate. ANDROECIUM: stamens as many as petals, epipetalous. Filaments free. Anthers 2-locular, opening lengthwise. GYNOECIUM: ovary inferior, 2 or more carpels and loculi, with axile placentation, rarely 1-locular with parietal placentation. Nectary disc often present at the top. Style slender, variously lobed.

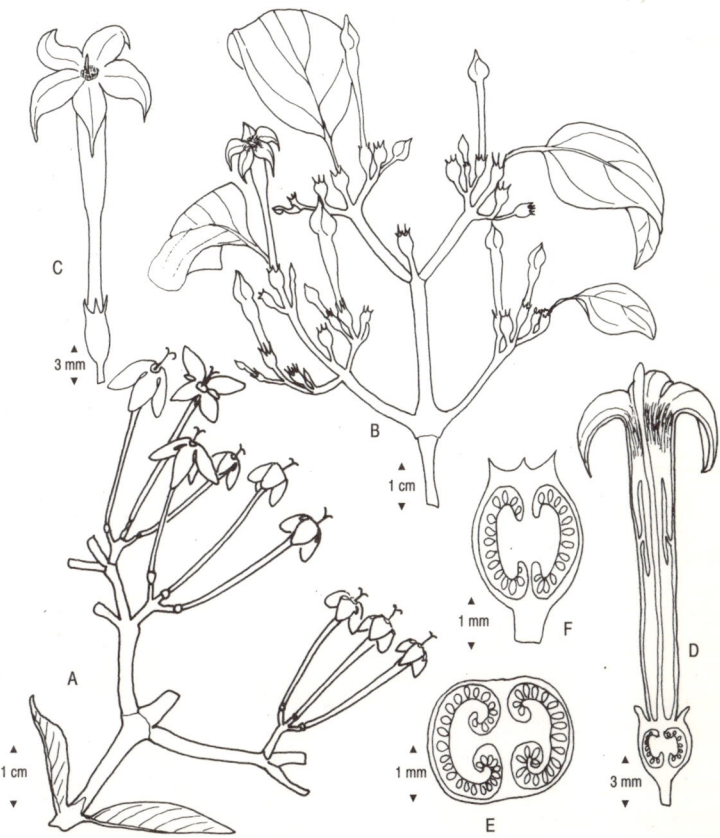

Key: part of inflorescence (A), of *Ixora macrothyrsa*. Inflorescence (B), full flower (C), half flower (D), and ovary in transverse (E) and longitudinal sections (F) of *Mussaenda frondosa*.

1. *Adina cordifolia*, kolon (S)/manchal kadampa (T),
   (T II:293), N, 20, tree.
   Leaves: closely spaced, very broadly ovate,
     cordate base, rounded to obtuse apex,
     veins minutely reticulate; young
     parts minutely stellate-pubescent.
   Trunk: branches **horizontal**;
     B-brownish grey, thick,
     soft, furrowed; W-pale
     yellow, durable, hard, heavy.
   Flowers: yellow, in 3's, axillary.
   Fruits: very hairy,
     conical capsule.
   Site: monsoon and
     intermediate forest
     canopy; DL, IN.
   Uses: B-medicinal;
     W-construction, carving.

2. *Allaeophania decipiens*, (T II:301), E, 2, shrub.
   Leaves: lanceolate, acute base, pointed
     apex, pubescent on veins.
   Trunk: numerous stems from base.
   Flowers: bluish white with red anthers,
     dimorphic, sessile; I-crowded,
     dense axillary whorls.
   Fruits: yellow, smooth, very small
     drupe; crowned with calyx limb.
   Site: montane and scrub
     forest understory; M.

3. *Benkara malabarica*, maha geta
   kulu (S)/ pudan (T),
   (T II:331), N, 3, shrub.
   Leaves: lanceolate to oblong, tapered
     base, pointed apex; short petiole.
   Trunk: erect, much branched; branchlets
     reduced to short, sharp, **axillary spines**.
   Flowers: lemon-yellow;
     I-few flowered cymes.
   Fruits: red, globose
     berry, tipped with
     ring scar of calyx.
   Site: monsoon and
     intemediate forest
     understory, scrub; DL, IN.

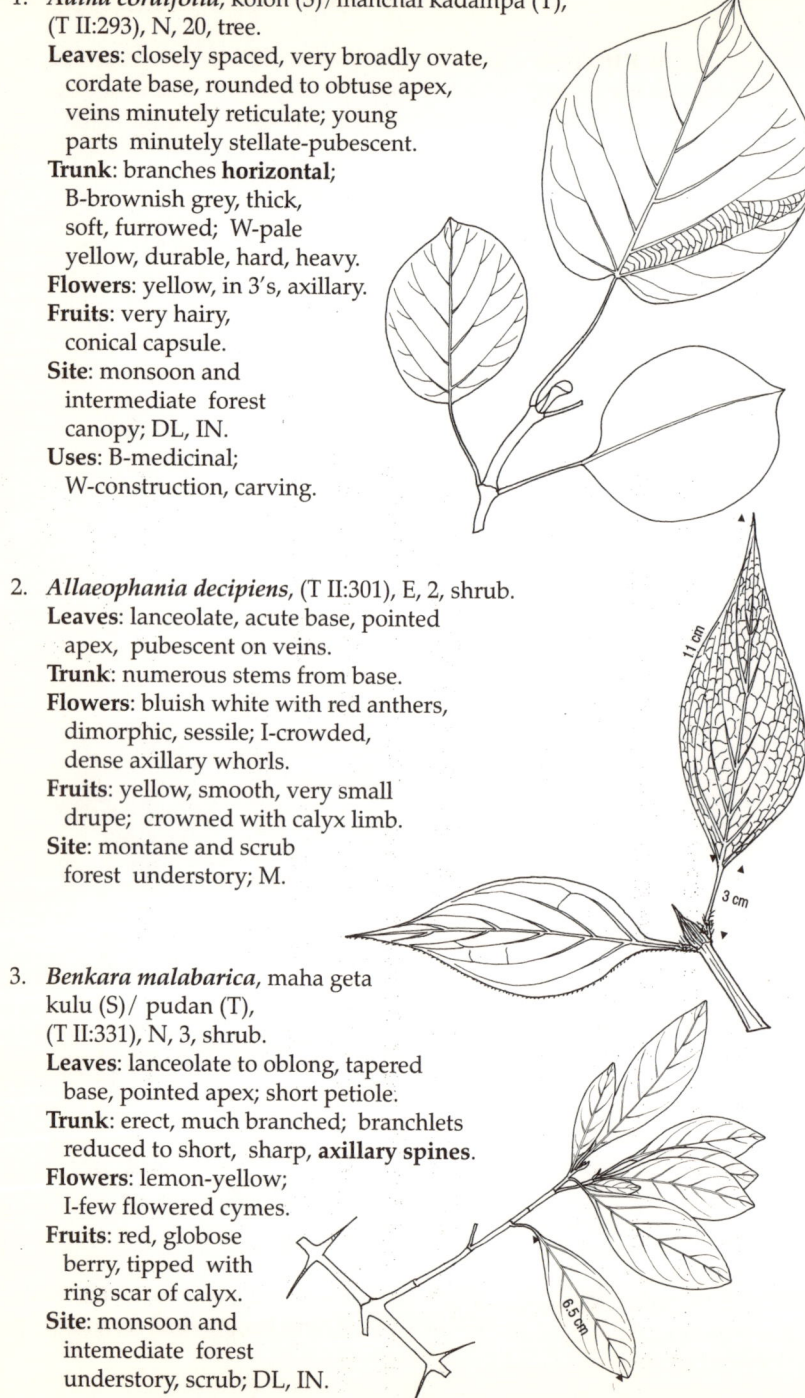

4. *Canthium coromandelicum*, kara (S)/karai (T),
   (T II:346 & VI:152), N, 3, shrub.
   Leaves: often clustered, **variable** in
   size, in moist regions larger, rotund
   to oval, elongate to obtuse ends,
   whitish beneath, veins reticulate.
   Trunk: W-very hard,
   coarse-grained;
   branches stout,
   rigid; supra
   axillary
   spines
   numerous, in
   moist regions
   often without
   spines.
   Flowers:
   yellow, very
   small, on slender
   pedicels; I-cymes.
   Fruits: yellow, ovoid,
   compressed drupe;
   Site: monsoon forest understory; DL.
   Uses: fruit-edible; leaves-vegetable.

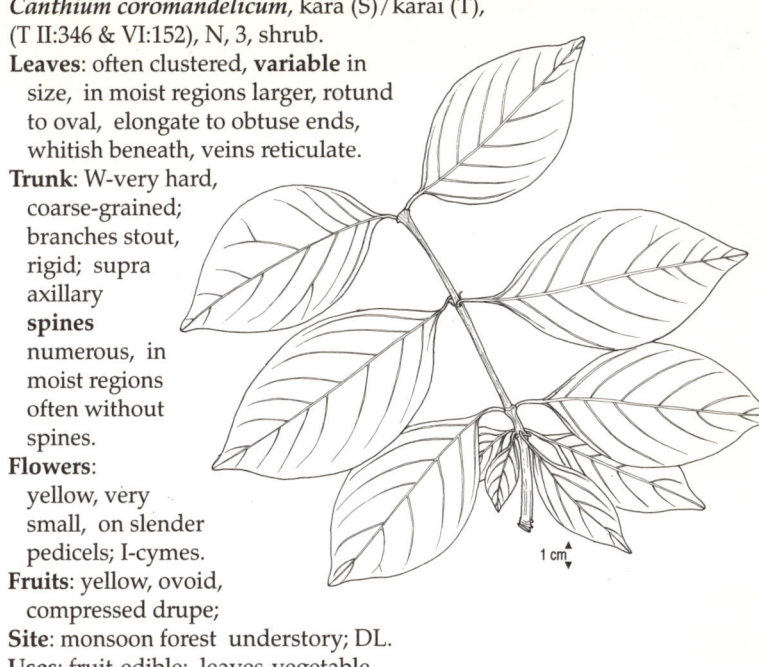

1 cm

5. *Canthium dicoccum*, panderu (S)/vatchikuran (T),
   (T II:343 & VI:151), E, 15, tree.
   Leaves: lanceolate to oblong, pointed
   to obtuse apex, lateral veins few,
   oblique; hard, beaked fruit-like
   galls often on young branches.
   Trunk: B-smooth to vertically
   furrowed, whitish; W-greyish,
   hard, heavy; branchlets
   drooping, thickened at
   nodes, without **spines**.
   Flowers: white, small,
   numerous; I-lax,
   corymbose cymes.
   Fruits: purple, globose
   drupe; black, kidney
   shaped, warty pyrenes.
   Site: monsoon and
   intermediate forest
   subcanopy and rain forest
   understory; DL, IN, W.
   Uses: W-light construction.

1 cm

6. *Canthium montanum*, wal buruta (S), (T II:343), E, 10, tree.
   Leaves: rotund, rounded ends,
      margins rolled down, stiff,
      veins few, prominent,
      reticulate beneath.
   Trunk: B-blackish, flaking;
      W-hard, heavy, fine-grained;
      branches horizontal;
      branchlets numerous,
      stout, thickened at
      nodes; twigs
      quadrangular.
   Flowers: green;
      I-umbellate cymes.
   Fruits: ovoid drupe.
   Site: montane forest canopy; M.
   Uses: W-heavy construction.

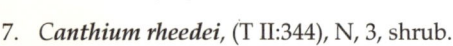

7. *Canthium rheedei*, (T II:344), N, 3, shrub.
   Leaves: ovate, cordate to rounded base, obtuse apex,
      lateral veins prominent, glands beneath in
      vein axils, few bristly hairs beneath.
   Trunk: much-branched; branchlets
      spreading, roughly pubescent,
      sharp supra-axillary **spines**.
   Flowers: pale greenish yellow, small;
      I-solitary or few in small sessile cymes.
   Fruits: broader than long drupe;
      strongly warted pyrenes.
   Site: rain forest understory; W.

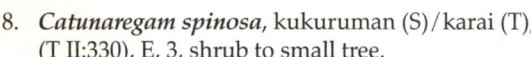

8. *Catunaregam spinosa*, kukuruman (S)/karai (T),
   (T II:330), E, 3, shrub to small tree.
   Leaves: clustered, obovate-oval to
      spathulate, base tapered, obtuse to
      pointed apex, veins reticulate; **sessile**.
   Trunk: branches **horizontal**, rigid, **paired**
      spines sharp, woody; lateral
      branchlets short, suppressed.
   Flowers: yellowish white, 1-3 at
      ends of suppressed branchlets.
   Fruits: yellow, globose berry, with large calyx.
   Site: monsoon forest understory, scrub,
      especially on coastal sands; DC, DL.
   Uses: fruit pulp-fish poison, emetic;
      root, fruit, B-medicinal.

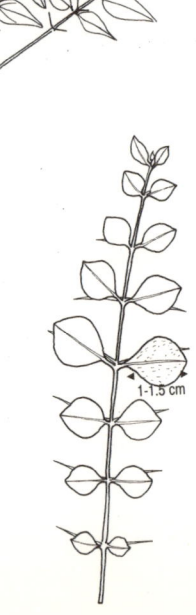

9. *Chassalia ambigua*, (T II:362 & VI:152), N, 2, shrub.
   Leaves: oval to linear to lanceolate, tapered base,
     pointed to obtuse apex, lateral veins
     curved merging within margin.
   Trunk: twigs slender, compressed.
   Flowers: pink to white, sessile;
     calyx purple; I-small,
     trichotomous, terminal cymes.
   Fruits: depressed-globose drupe,
     tipped with red calyx;
     pulp inky purple.
   Site: rain forest
     understory; W.

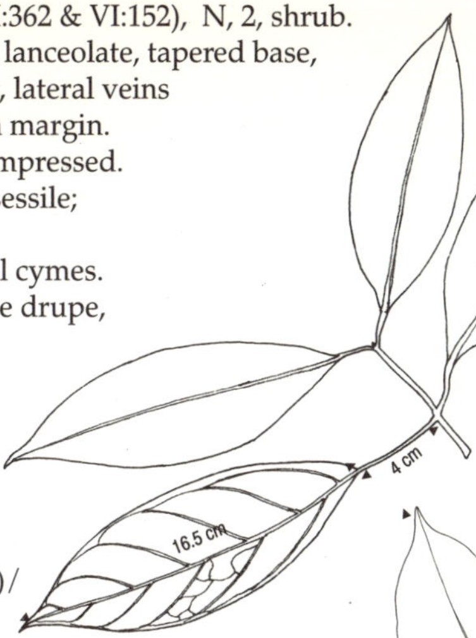

10. *Coffea arabica*, coffee (E)/
    kopi (S), I (Ethiopia), 3,
    shrub to small tree.
   Leaves: dark green, **2-ranked**, elliptic,
     pointed ends, slightly **wavy** margin; **drooping**.
   Trunk: much-branched.
   Flowers: white, sessile,
     fragrant, several
     together at leaf bases.
   Fruits: red to brown, elliptic
     drupe; 2 pyrenes.
   Site: home gardens; W.
   Uses: seed-beverage.

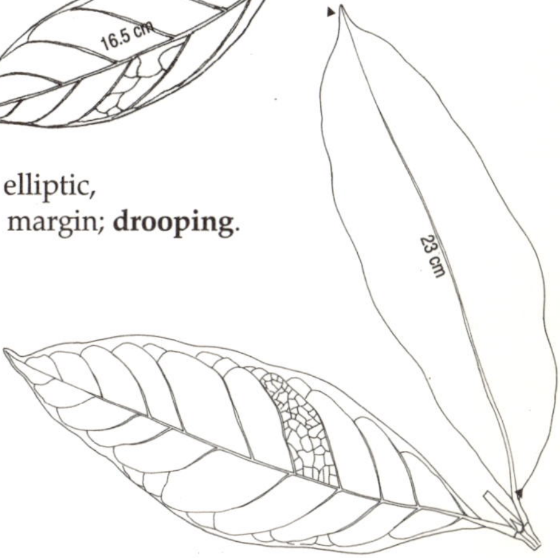

11. *Dichilanthe zeylanica*, gal ehatu (S),
    (T II:339) E, 20, tree.
   Leaves: lanceolate, acute base, short-
     pointed apex; **persistent ring**-like
     stipule becomes coated with **resin**.
   Trunk: B-rough greyish
     black; branchlets stout,
     swollen at nodes.
   Flowers: scarlet or yellow,
     6-8 together; I-shortly-
     stalked, terminal head.
   Fruits: cluster of small
     drupes falling together
     with subtending leaves
     giving it a false winged
     appearance.
   Site: rain forest subcanopy; W.

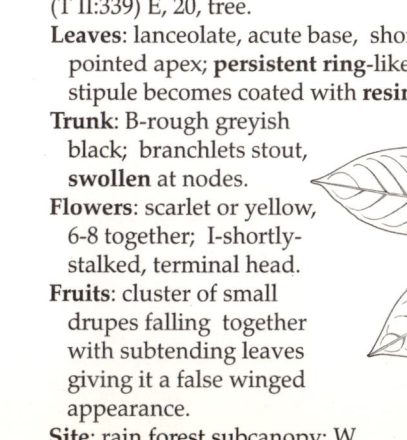

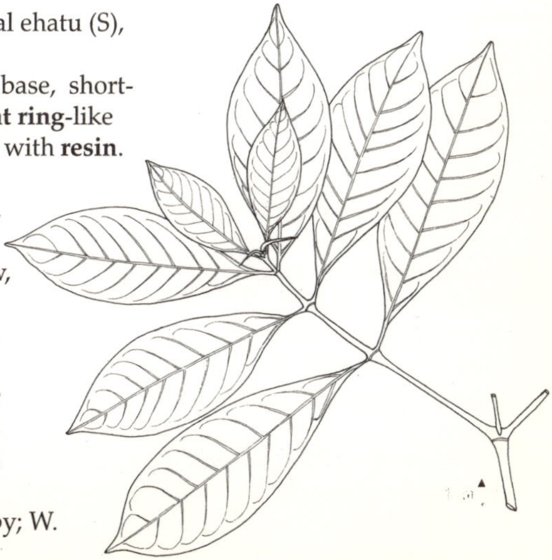

12. *Gaertnera vaginans*, pera tambala (S),
    (T III:177 & VI:197)), N, 2, shrub.
    Leaves: **large**, oblong-lanceolate, tapering
       base, **acute** apex, lateral veins arched;
       stipular sheath bifid with acute tips.
    Trunk: branches stout,
       compressed, nodes thickened.
    Flowers: white; I-large, lax, pyramidal cymes.
    Fruits: smooth berry
       on enlarged
       persistent
       calyx.
    Site: rain forest
       understory; W.
    Uses: ornamental.

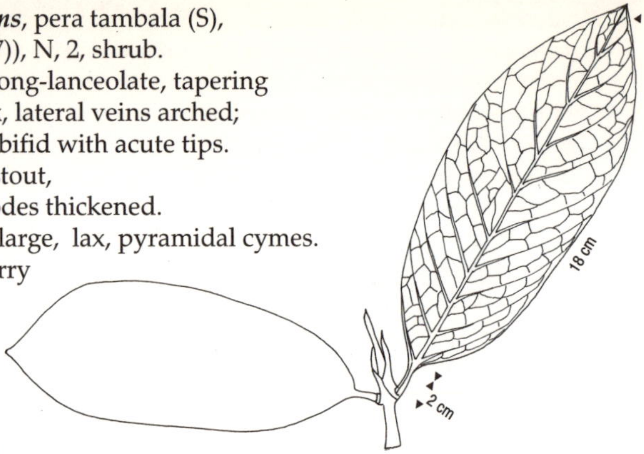

13. *Gaertnera walkeri*,
    (T III:178 & VI:197), E, 2, shrub.
    Leaves: oval to linear-
       lanceolate, tapering base,
       caudate to acute apex;
       stipular sheath
       nearly truncate.
    Trunk: branches nodally thickened.
    Flowers: white; I-large cymes.
    Fruits: dark blue, depressed
       globose berry.
    Site: rain forest
       understory; W.

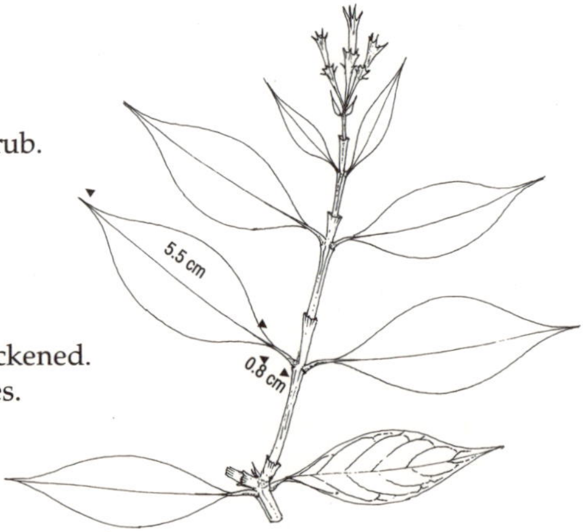

14. *Guettarda speciosa*, nil pichcha (S)/
    panir (T), (T II:338), N, 4, small tree.
    Leaves: terminally **crowded**,
       obovate to rotund, cordate
       base, usually rounded apex,
       pubescent; flush pink.
    Trunk: branchlets stout,
       pubescent, **leaf-scarred**.
    Flowers: white, sweet-scented,
       jasmine-like, nearly
       sessile; I-cyme.
    Fruits: hard,
       depressed-globose drupe.
    Site: home gardens; MC.

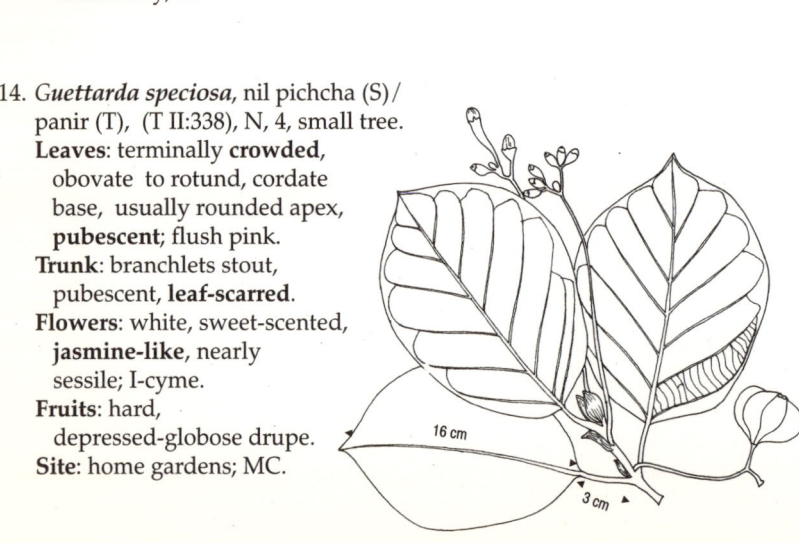

15. *Hedyotis fruticosa*, weraniya (S), (T II:304), N, 3, shrub.
   Leaves: narrowly lanceolate, base tapered, long-pointed
   apex, **dark green** above, **pale whitish** green beneath.
   Trunk: B-white, smooth; twigs **quadrangular**,
   glabrous, **whitish**.
   Flowers: white, sessile; I-numerous
   cymes in terminal and axillary panicles.
   Fruits: oblong to oval, pubescent capsule.
   Site: secondary forest and scrub; W.
   Uses: whole plant-medicinal.

16. *Hedyotis lawsoniae*,
   (T II:310), E, 2, shrub.
   Leaves: varying in size,
   oval to lanceolate,acute ends.
   Trunk: branches supra-axillary;
   twigs bluntly 4-angled,
   glabrous, often **purplish**.
   Flowers: white, dimorphic;
   I-numerous, large compound
   terminal and axillary cymes.
   Fruits: ovoid to globose capsule
   that projects beyond persistent
   calyx, splitting in two.
   Site: montane forest
   understory, gregarious; M.
   Uses: ornamental.

17. *Hedyotis lessertiana*,
   (T II:309) E, 3, shrub.
   Leaves: varying in size, lanceolate,
   tapered base, acute apex, stiff,
   corrugated-like, lateral veins
   numerous, prominent.
   Trunk: branchlets
   stout, **cylindrical**.
   Flowers: white, dimorphic;
   I-numerous cymes in
   large spreading
   panicles, axillary.
   Fruits: broadly oblong
   to ovoid capsule,
   splitting in two.
   Site: montane forest
   understory; M.
   Uses: ornamental.

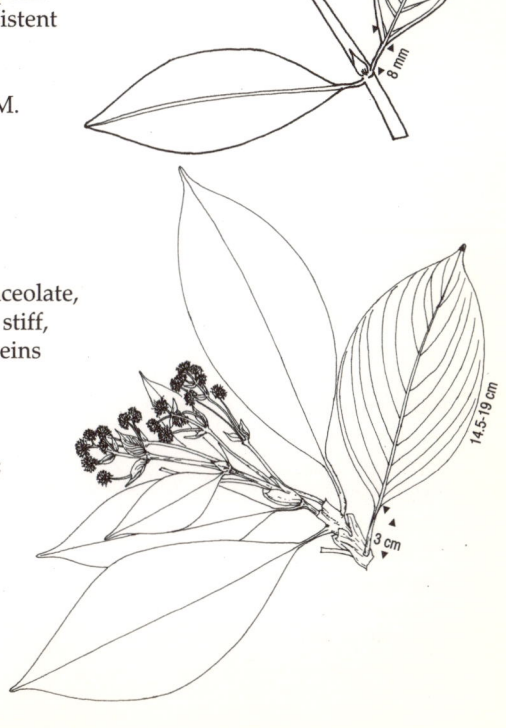

18. *Ixora arborea*, maha ratambala (S)/ karankutti (T),
    (T II:348), N, 4, small tree.
    Leaves: oval-oblong to obovate, rounded base,
        pointed apex, **dark green** above, **pale** below.
    Trunk: B-thick, reddish brown, peeling in
        irregular pieces; W-hard, heavy, brownish
    yellow; branchlets somewhat compressed.
    Flowers: white to pinkish, sweet-scented,
        very small, numerous; I-stalked, lax,
        terminal, trichotomous cymes.
    Fruits: small, globose berry.
    Site: monsoon forest understory; DL.

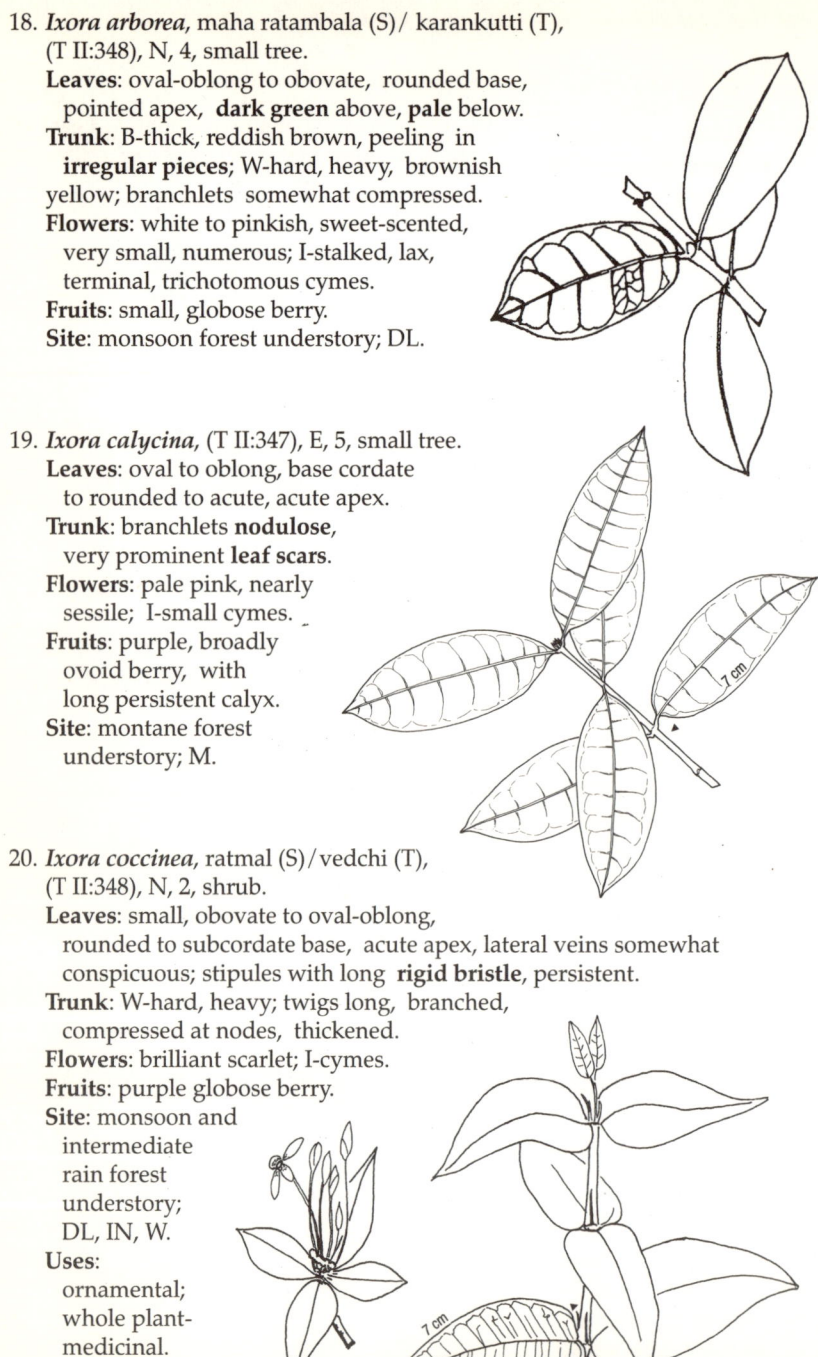

19. *Ixora calycina*, (T II:347), E, 5, small tree.
    Leaves: oval to oblong, base cordate
        to rounded to acute, acute apex.
    Trunk: branchlets **nodulose**,
        very prominent **leaf scars**.
    Flowers: pale pink, nearly
        sessile; I-small cymes.
    Fruits: purple, broadly
        ovoid berry, with
        long persistent calyx.
    Site: montane forest
        understory; M.

20. *Ixora coccinea*, ratmal (S)/vedchi (T),
    (T II:348), N, 2, shrub.
    Leaves: small, obovate to oval-oblong,
        rounded to subcordate base, acute apex, lateral veins somewhat
        conspicuous; stipules with long **rigid bristle**, persistent.
    Trunk: W-hard, heavy; twigs long, branched,
        compressed at nodes, thickened.
    Flowers: brilliant scarlet; I-cymes.
    Fruits: purple globose berry.
    Site: monsoon and
        intermediate
        rain forest
        understory;
        DL, IN, W.
    Uses:
        ornamental;
        whole plant-
        medicinal.

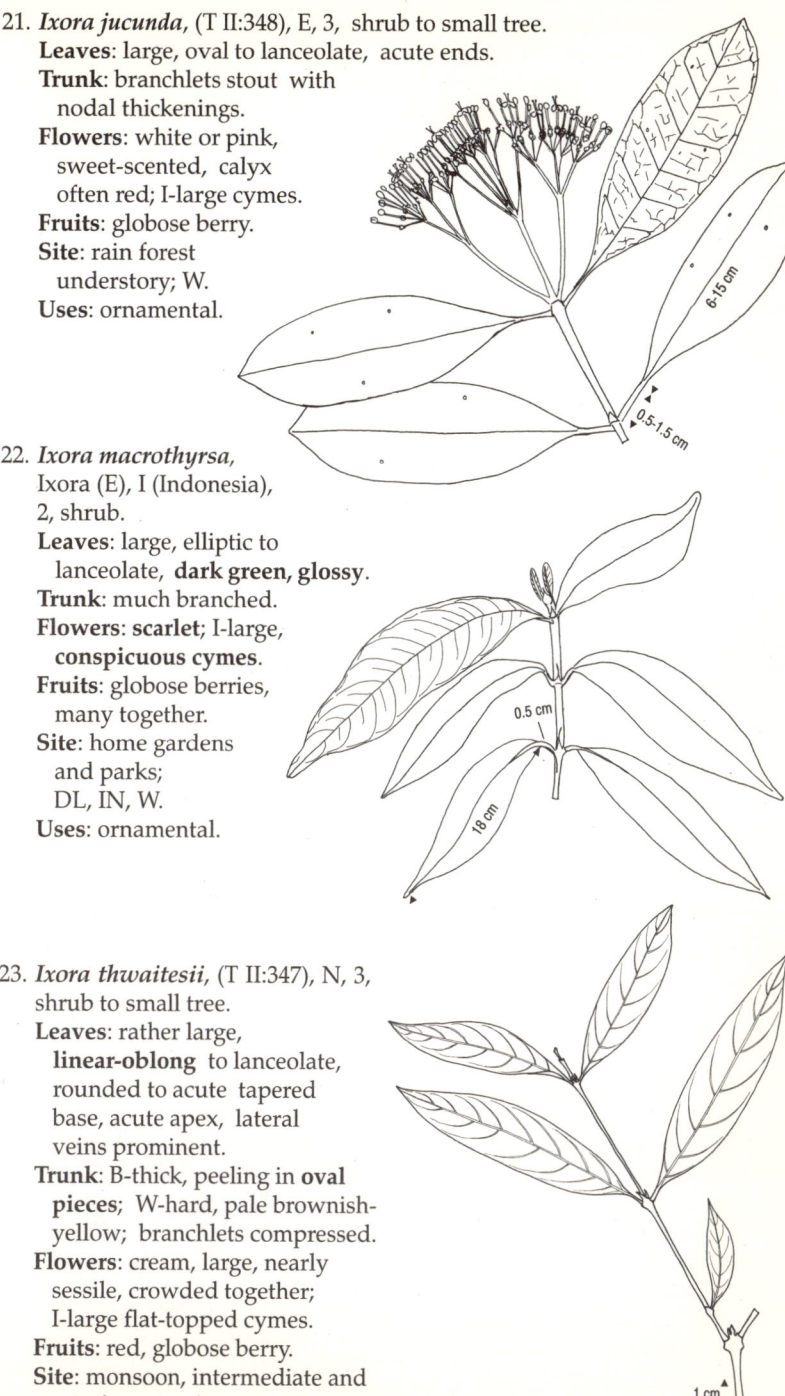

21. *Ixora jucunda*, (T II:348), E, 3, shrub to small tree.
    Leaves: large, oval to lanceolate, acute ends.
    Trunk: branchlets stout with
       nodal thickenings.
    Flowers: white or pink,
       sweet-scented, calyx
       often red; I-large cymes.
    Fruits: globose berry.
    Site: rain forest
       understory; W.
    Uses: ornamental.

    6-15 cm
    0.5-1.5 cm

22. *Ixora macrothyrsa*,
    Ixora (E), I (Indonesia),
    2, shrub.
    Leaves: large, elliptic to
       lanceolate, **dark green, glossy**.
    Trunk: much branched.
    Flowers: **scarlet**; I-large,
       conspicuous cymes.
    Fruits: globose berries,
       many together.
    Site: home gardens
       and parks;
       DL, IN, W.
    Uses: ornamental.

    0.5 cm
    18 cm

23. *Ixora thwaitesii*, (T II:347), N, 3,
    shrub to small tree.
    Leaves: rather large,
       linear-oblong to lanceolate,
       rounded to acute tapered
       base, acute apex, lateral
       veins prominent.
    Trunk: B-thick, peeling in **oval
       pieces**; W-hard, pale brownish-
       yellow; branchlets compressed.
    Flowers: cream, large, nearly
       sessile, crowded together;
       I-large flat-topped cymes.
    Fruits: red, globose berry.
    Site: monsoon, intermediate and
       rain forest understory; DL, IN, W.

    1 cm

24. *Lasianthus oliganthus*, (T II:366), E, 2, shrub.
  Leaves: acute to obtuse base,
    pointed apex, **4-5** lateral veins.
  Trunk: branches slender, slightly
    rough, adpressed hairs.
  Flowers: very small,
    sessile, solitary,
    or 2 or 3.
  Fruits: very small,
    depressed berry,
    truncate at top.
  Site: rain forest
    understory; W.

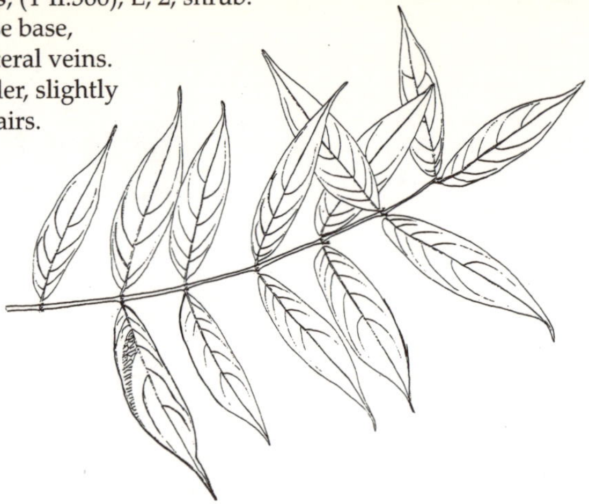

25. *Lasianthus strigosus*, wal kopi (S), (T II:367), E, 2, shrub.
  Leaves: oblong-lanceolate, pointed ends, **mucronate**,
    lateral veins **7-9**, prominent, pubescent beneath.
  Trunk: stems rough, adpressed hairs,
    stout, cylindrical.
  Flowers: white,
    sessile, crowded.
  Fruits: nearly
    globose berry,
    with tubular calyx.
  Site: montane and
    rain forest
    understory;
    LM, W.

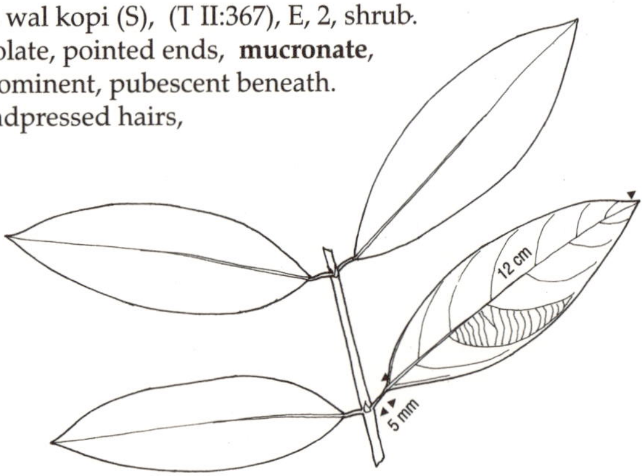

26. *Lasianthus varians*, (T II:368), E, 2, shrub.
  Leaves: usually small, oval to lanceolate,
    base acute, margins often rolled
    down, lateral veins **slender**.
  Trunk: stems much-branched,
    compressed when young,
    thickened nodes.
  Flowers: white, sessile, small
    clusters, on short peduncles.
  Fruits: black, broadly ovoid
    berry, with erect
    tooth-like calyx.
  Site: montane forest
    understory; M.

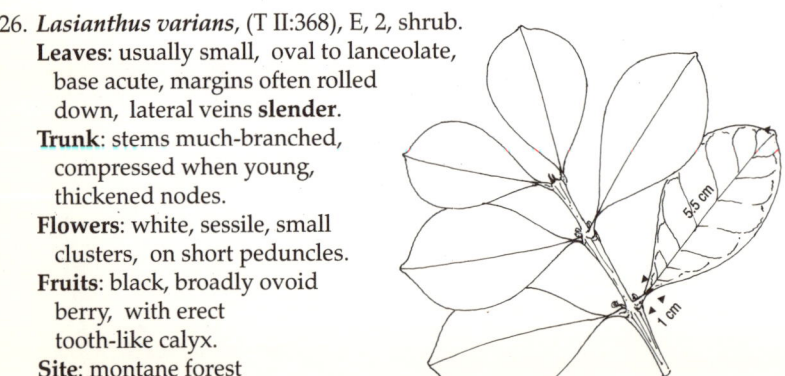

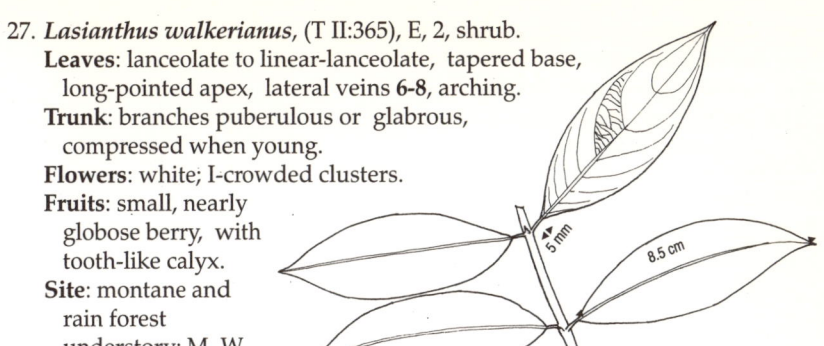

27. *Lasianthus walkerianus*, (T II:365), E, 2, shrub.
Leaves: lanceolate to linear-lanceolate, tapered base,
long-pointed apex, lateral veins **6-8**, arching.
Trunk: branches puberulous or glabrous,
compressed when young.
Flowers: white; I-crowded clusters.
Fruits: small, nearly
globose berry, with
tooth-like calyx.
Site: montane and
rain forest
understory; M, W.

28. *Mitragyna parvifolia*, helamba (S)/chelampai (T),
(T II:294 & VI:140),N, 15, tree.
Leaves: rotund to broadly oval, **rounded** ends,
small tufts of hair in vein axils beneath.
Trunk: B-rather smooth, whitish grey, flaking;
W-hard, heavy, smooth, pinkish.
Flowers: greenish yellow,
scented; I-heads.
Fruits: crowded,
separate capsules.
Site: monsoon and
intermediate forest
and scrub; often near
water courses; DL, IN.
Uses: roots and stems-medicinal;
W-furniture, construction.

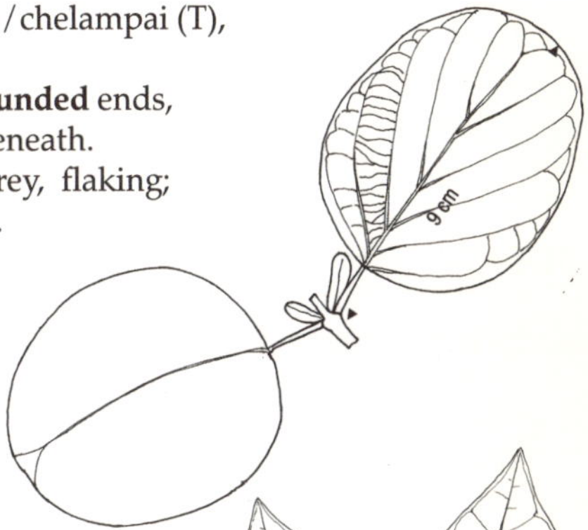

29. *Morinda citrifolia*, ahu (S)/nuna (T),
(T II:354 & VI:152), N, 4, small tree.
Leaves: closely placed, large, **shiny**,
oval, tapered base, short-pointed
apex, lateral veins conspicuous,
pale green, glandular pits in
vein axils beneath; stipules persistent.
Trunk: B-yellowish white; branchlets
stout with **leaf scars**.
Flowers: white, rather large;
I-heads, solitary or 2 or 3 together.
Fruits: white-green, globose, very fleshy
drupe; pyrenes ovoid with winged edge.
Site: home gardens, scrub; DC, MC.
Uses: root B-red dye; fruit-edible;
ornamental; whole plant-medicinal.

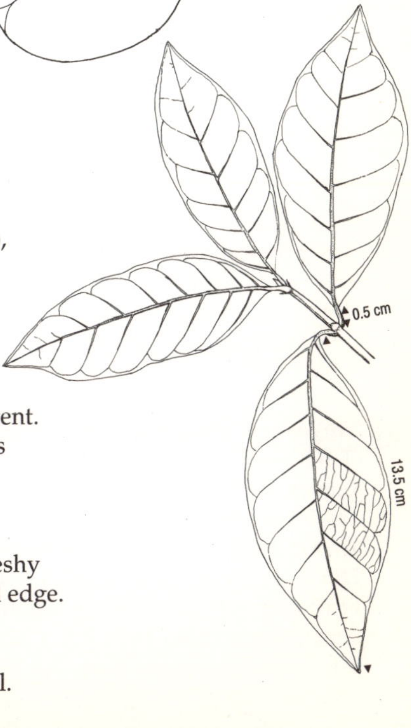

30. **Morinda tinctoria**, ahu (S)/manchavanna (T), (T II:354), N, 4, small tree.
   Leaves: **dull greenish**, lanceolate, base
      tapering, pointed apex, **tufts of
      hair** in vein axils beneath;
      stipules deciduous.
   Trunk: B-thick, spongy, deep
      longitudinal furrows,
      branchlets thickened at
      nodes, **leaf scars** prominent.
   Flowers: white-green, sweet-
      scented; I-globose heads,
      solitary or 2 or 3 together.
   Fruits: white-green, globose, fleshy
      drupe; pyrenes usually 4, oblong.
   Site: monsoon forest understory; DL.
   Uses: root B-red dye; leaves-
      medicinal; W-light
      construction.

31. **Mussaenda frondosa**, mussenda (E,S), (T II:323),
      N, 3, scandent shrub.
   Leaves: ovate to lanceolate, rounded base,
      shortly acuminate, veins prominent
      beneath, finely **velvety**.
   Trunk: branches **climbing**, long,
      spreading, velvety, cylindrical.
   Flowers: brillant orange, rather
      small; **one sepal conspicu-
      ously enlarged** resembling a
      leaf but **creamy-white to
      pink**; I-terminal cymes.
   Fruits: globose, faintly 2-lobed,
      slightly scabrous berry.
   Site: scrub, roadsides; W.
   Uses: whole plant-medicinal;
      enlarged sepal-vegetable.

32. **Nargedia macrocarpa**, wal kopi (S),
      (T II:334), E, 3, shrub.
   Leaves: large, oblong to lanceolate,
      pointed ends.
   Trunk: B-smooth, **polished yellow**.
   Flowers: cream, rather small, 1-3 in axils.
   Fruits: broadly ovoid berry, slightly
      tapering at ends; seeds 2-3.
   Site: rain forest understory; W.

33. *Nauclea orientalis*, bakmi (S)/vammi (T), (T II:292 & VI:140), N, 15, tree.
   Leaves: broadly ovate to oval, slightly cordate base, rounded apex, sometimes pubescent beneath; young parts minutely stellate pubescent.
   Trunk: B-**silvery grey**, smooth; W-yellow, light, soft; branches prominently **leaf scarred**.
   Flowers: pale yellow; I-heads on stout peduncles, often in 3's.
   Fruits: globose, fleshy, multiple berry; seeds black.
   Site: monsoon and intermediate forest subcanopy, near water; DL, IN.
   Uses: leaves, B-medicinal; fruit-edible.

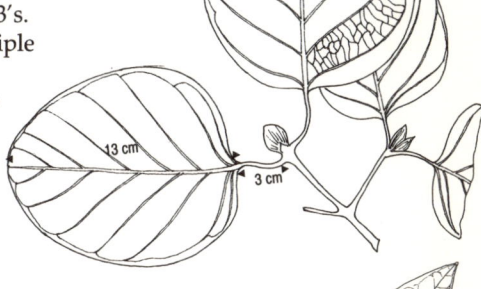

34. *Pavetta gleniei*, vetpavaddai (T), (T II:351), E, 3, shrub.
   Leaves: oblong-oval to lanceolate, ends pointed, pubescent, scattered warts.
   Trunk: B-**pale cinnamon-brown, smooth**; twigs compressed, **pubescent**.
   Flowers: white, small; I-lax, trichotomous, hairy cymes shorter than leaf.
   Fruits: globose, hairy segmented berry, crowned with erect calyx.
   Site: monsoon forest understory; DL.

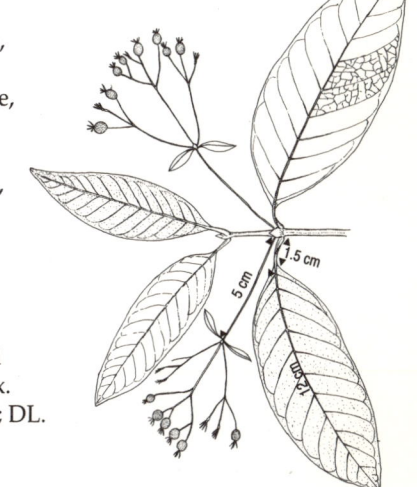

35. *Pavetta indica*, pawatta (S)/pavaddai (T), (T II:349), N, 3, shrub to small tree.
   Leaves: dark green, lanceolate, tapered base, pointed apex, scattered **hard warts**.
   Trunk: B-**smooth, yellowish white**; twigs cylindrical, glabrous.
   Flowers: greenish white, numerous; I-lax, corymbose, terminal cymes.
   Fruits: black, polished, globose, berry.
   Site: monsoon, intermediate and rain forest understory; DL, IN, W.
   Uses: whole plant-medicinal.

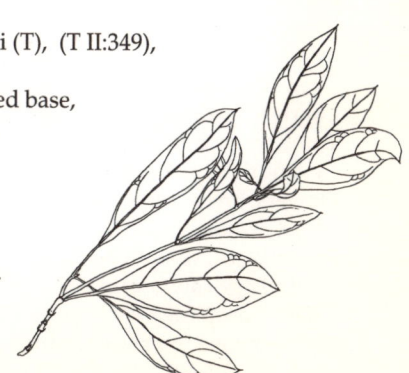

36. *Psychotria nigra*, (DF VI:345), N, 2, shrub.
    Leaves: rather large, oblong to lanceolate,
    pointed ends, lateral veins distinct.
    Trunk: branchlets stout.
    Flowers: greenish, nearly
    sessile with brown hair at
    base; I-long-stalked,
    pyramidal, whorled cymes.
    Fruits: black, broadly ovoid berry;
    pyrenes without furrows.
    Site: montane and rain forest
    understory; damp areas; M, W.

37. *Psychotria sordida*, (DF VI:352), E, 2, shrub.
    Leaves: small, lanceolate-
    oblong, base tapered,
    long-pointed apex, lateral
    veins inconspicuous.
    Trunk: branchlets slender.
    Flowers: greenish-white,
    sessile, clusters of 3's;
    I-3-branched, short,
    spreading, sessile cymes.
    Fruits: black, oblong berry,
    capped by small calyx;
    pyrenes with 2
    deep furrows.
    Site: montane forest
    understory; M.

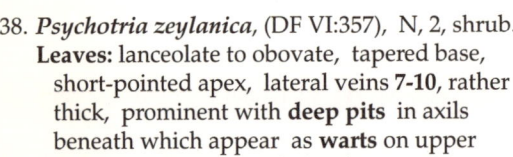

38. *Psychotria zeylanica*, (DF VI:357), N, 2, shrub.
    Leaves: lanceolate to obovate, tapered base,
    short-pointed apex, lateral veins **7-10**, rather
    thick, prominent with **deep pits** in axils
    beneath which appear as **warts** on upper
    surface, paler silvery beneath.
    Trunk: branches shiny, stout,
    cylindrical, glabrous.
    Flowers: green, rather large;
    I-lax cymes.
    Fruits: black, smooth berry,
    with large calyx; pyrenes
    with 2 deep furrows.
    Site: montane forest understory; M.

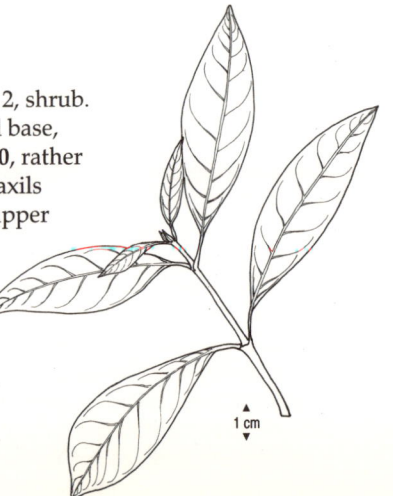

39. *Randia gardneri*, (T II:331), E, 20, tree.
    Leaves: narrowly oblong-lanceolate, pointed ends, attenuate to caudate,
      somewhat sickle-shaped, shining above,
      paler beneath; short petiole.
    Trunk: branches slender, **drooping**,
      cylindrical, **without spines**.
    Flowers: pale yellow or white;
      I-lax, axillary, spreading cymes.
    Fruits: pear-shaped to globose,
      broadly ovoid berry.
    Site: intermediate and rain
      forest subcanopy; IN, W.

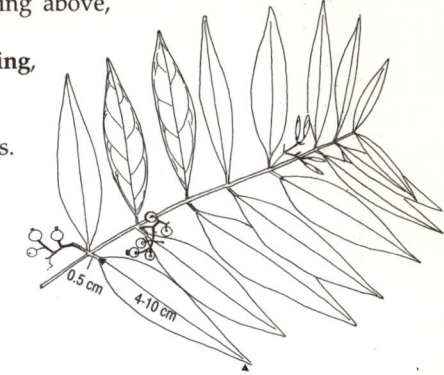

40. *Saprosma foetens*, (T II:369 & VI:153), N, 3, shrub.
    Leaves: lanceolate to oval, pointed ends, lateral veins
      curved with **glandular pits** in
      axils, **fetid odour** when crushed.
    Trunk: B-pale yellowish brown,
      smooth; branches slender, erect;
      branchlets brittle, flattened.
    Flowers: pale sulphur-yellow,
      large, axillary or terminal.
    Fruits: bright blue berry,
      tipped with small calyx.
    Site: montane forest
      understory; M, LM.

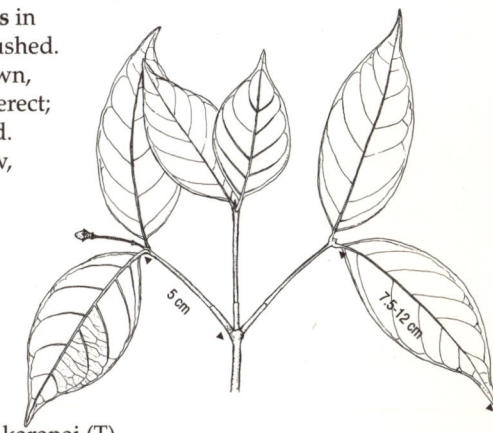

41. *Tarenna asiatica*, tarana (S)/karanai (T),
    (T II:328 & VI:150), N, 4, shrub to small tree.
    Leaves: oblong to lanceolate, pointed ends, lateral veins prominent
      beneath, with small **pits** in axils, rather thick; buds exude
      shiny waxy secretion which dries bright green-yellow.
    Trunk: B-smooth, pale grey; W-hard, heavy, yellowish, durable; twigs
      compressed, glabrous.
    Flowers: white; I-trichoto-
      mous cymes.
    Fruits: dull green to black,
      globose, glabrous berry.
    Site: monsoon and rain forest
      understory; DL, W.
    Uses: fruit-medicinal;
      sticks-buffalo goads;
      W-constructing
      granaries.

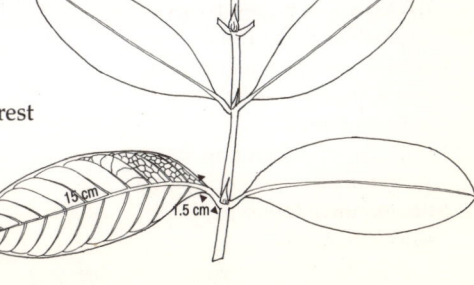

42. *Timonius jambosella*, angana (S), (T II:338), E, 3, shrub to small tree.
   Leaves: closely placed, oval to lanceolate, pointed ends,
      tufts of hair in vein axils beneath.
   Trunk: branchlets slender, cylindrical,
      glabrous, leaf scars.
   Flowers: yellow, sessile; I-female
      solitary or 3 together on
      peduncle, male, short
      spreading cymes.
   Fruits: dark green,
      globose, lobed to
      ribbed, smooth berry
      tipped with calyx.
   Site: montane and rain
      forest understory; LM, W.

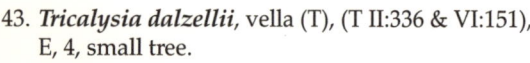

43. *Tricalysia dalzellii*, vella (T), (T II:336 & VI:151),
   E, 4, small tree.
   Leaves: lanceolate or oval, pointed ends.
   Trunk: B-very smooth, white to
      yellowish; W-white, hard, heavy;
      twigs compressed, thickened at nodes.
   Flowers: greenish white, small,
      nearly sessile; I-dense cymes.
   Fruits: dark green, globose,
      smooth, berry.
   Site: monsoon and intermediate
      forest understory, scrub; DL, IN.

44. *Tricalysia erythrospora*, (T II:336 & VI:151), E, 4, shrub to small tree.
   Leaves: oval, base tapered, apex rounded to abruptly
      pointed, margin rolled down, glandular
      pits in vein axils beneath;
      petiole very short.
   Trunk: branchlets stout,
      thickened at nodes; twigs
      compressed, glabrous.
   Flowers: pale greenish yellow,
      small, nearly sessile;
      I-dense cymes.
   Fruits: ovoid-subglobose berry,
      pulp purple; seeds bright red.
   Site: montane forest
      understory; M.

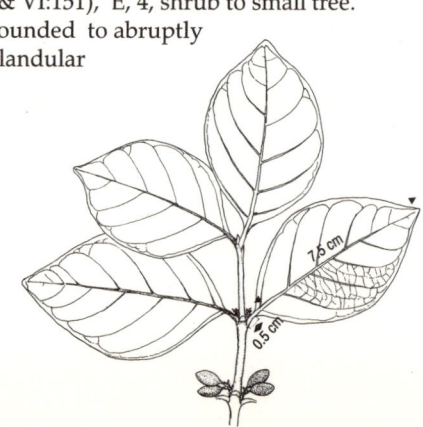

45. *Urophyllum ellipticum*, (T II:326), E, 3, shrub.
   Leaves: large, lanceolate, acuminate, ends
      pointed, veins prominent beneath.
   Trunk: branches cylindrical, stiff,
      glabrous; young branches
      4-angled, pubescent.
   Flowers: white, crowded on
      short pedicels; I-cymes
      small, nearly sessile.
   Fruits: dull orange-yellow,
      small berry, crowned by
      calyx rim.
   Site: montane and rain forest
      understory; M, W.

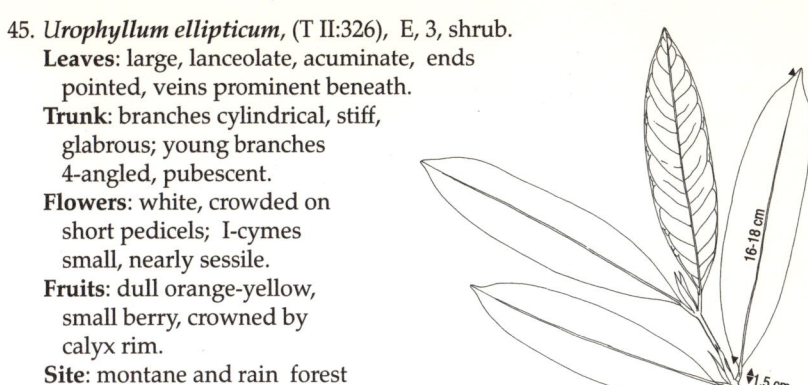

46. *Urophyllum zeylanicum*, wal handun (S),
   (TII:326 & VI:150), N, 3, shrub.
   Leaves: lanceolate, pointed
      ends, viens reticulate.
   Trunk: branches spreading,
      cylindrical, glabrous.
   Flowers: pale yellow;
      I-irregularly umbellate,
      pedunculate cymes.
   Fruits: orange-yellow, waxy,
      smooth, depressed globose
      berry, crowned by calyx rim.
   Site: montane forest
      understory; M.

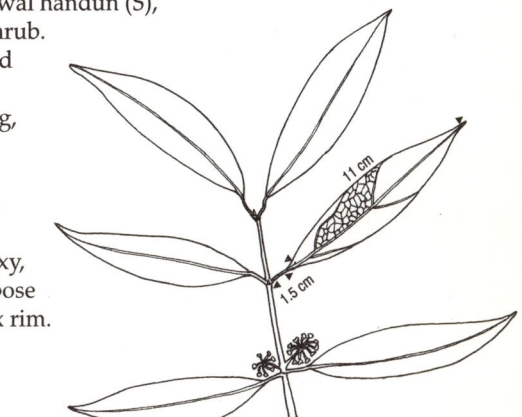

47. *Wendlandia bicuspidata*,
   rawan idala, wana idala (S), (T II:297) N, 5, small tree.
   Leaves: 3 at a node, lanceolate to oval, tapering
      base, pointed apex, margins **wavy**,
      veins beneath **prominent pink**
      and pubescent; flush **pink**.
   Trunk: B-**orange-red**, peeling in **fibrous**
      strips; W-heavy, hard, brownish-red;
      branchlets densely **pubescent**.
   Flowers: white, sessile, numerous;
      I-dense, pyramidal, terminal panicles.
   Fruits: small capsules; seeds
      numerous, minute.
   Site: montane and
      rain forest gaps
      and fringes; M, W.
   Uses: W-turnery, light
      construction; whole
      plant-medicinal.

78. RUTACEAE

FAMILY DESCRIPTION - Habit: trees or shrubs, sometimes lianas or herbs. Sometimes thorny. **Leaves**: usually spiral, rarely opposite, simple or compound, sometimes reduced to spines, undersurface translucent gland-dotted. Citrus-like aromatic odour. **Stipules**: absent. **Flowers**: usually bisexual, rarely unisexual (plants dioecious), actinomorphic, rarely zygomorphic, cymes or racemes, rarely solitary. **Fruits**: berry or drupe or hesperidium, rarely a capsule.

FLOWER PARTS - CALYX: 4-5, rarely 2-3, free or sometimes basally united, often imbricate. COROLLA: petals twice as many as sepals, free or rarely basally united to form a tube, imbricate or valvate, rarely absent. ANDROECIUM: stamens 4-10 or more, free or rarely united to form a staminal tube, equal or alternating long and short, all or the longer ones fertile, shorter ones staminodal. Anthers 2-locular, with 1 to several glands. Nectary disc annular, sometimes one-sided or accrescent as a gynophore in fruit. GYNOECIUM: superior, carpels usually 4-5, rarely 2, 3 or more, united, sometimes basally free but united by their styles, 1, 2 or more ovules per loculus. Placentation axile.

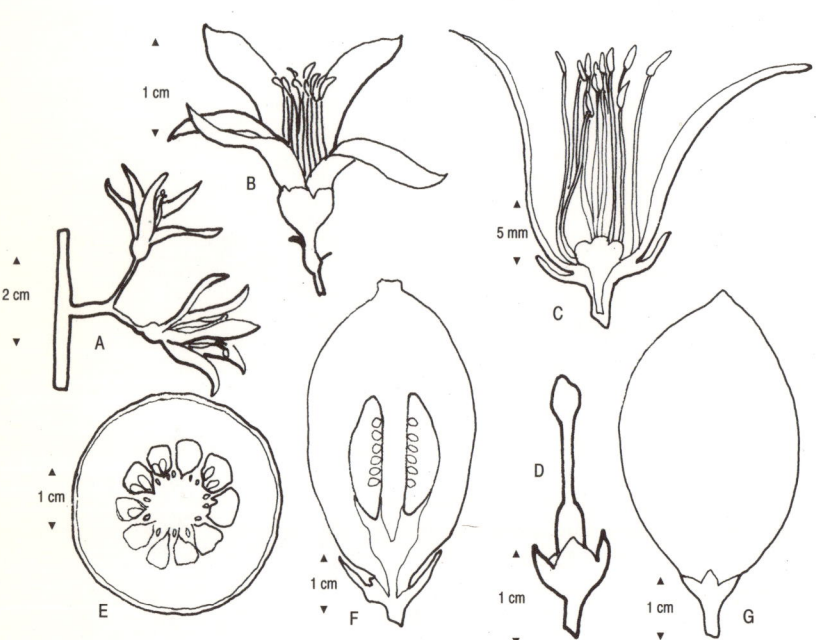

Key: inflorescence (A), full flower (B) and longitudinal section (C), gynoecium (D) with other floral parts removed, developing fruit in transverse (E) and longitudinal (F) sections to show the axile placentation, and a developing fruit (G) of *Citrus* x *limon.*

1. *Acronychia pedunculata*, ankenda (S), (DF V:412), N, 5, small tree.
   Leaves: **simple**, opposite; elliptic to suboblong, base usually tapered,
   apex obtusely pointed, lateral veins
   3-9 pairs; aromatic when crushed.
   Trunk: B-pale, smooth.
   Flowers: greenish white;
   I-axillary, corymbose
   panicles.
   Fruits: cream to brownish
   yellow, subglobose,
   apex short pointed.
   Site: montane and rain
   forest understory, gaps
   and fringes; M, W.
   Uses: B-medicinal.

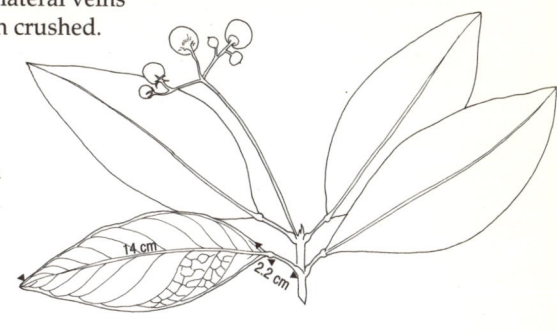

14 cm · 2.2 cm

2. *Aegle marmelos*, beli (S)/vilvam (T), (DF V:414), N, 5, small tree.
   Leaves: alternate, **trifoliolate**; leaflets ovate to elliptic,
   long-pointed, **scalloped**; aromatic when crushed.
   Trunk: branchlets green; **spines** axillary, paired or single.
   Flowers: white; I-few flowered, axillary,
   paniculate racemes.
   Fruits: pale green, globose,
   hard, woody shell.
   Site: home gardens;
   DL, IN, W.
   Uses: fruit-edible,
   unripe pulp as sealer
   for winnowing fans;
   leaves-Hindu rituals;
   whole plant-medicinal.

9 mm

3. *Atalantia ceylanica*, yakinaran (S)/peykuruntu (T),
   (DF V:416), N, 3, shrub.
   Leaves: **simple**; ovate to elliptic to lanceolate, pale yellowish green
   beneath, lateral veins **5-10** pairs; aromatic when crushed.
   Trunk: much-branched; B-grey-
   brown, lenticellate; **spines**
   slender, solitary, axillary.
   Flowers: white; I-axillary,
   cymose to racemose,
   sometimes clustered.
   Fruits: subglobose.
   Site: monsoon, intermediate
   and rain forest understory,
   scrub; DL, IN, W.
   Uses: leaves, roots-
   medicinal.

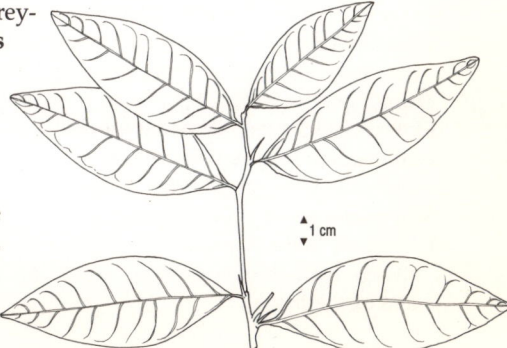

1 cm

4. *Atalantia monophylla*, perunkuruntu (T), (DF V:418), N, 3, shrub.
    Leaves: **simple**; ovate to elliptic, broadly tapered
        base, obtuse apex, clearly **notched**, margins
        entire to wavy, lateral veins **8-14** pairs;
        aromatic when crushed.
    Trunk: spiny, glabrous.
    Flowers: white; I-axillary,
        clustered racemes.
    Fruits: yellowish green,
        globose, densely
        gland-dotted.
    Site: monsoon forest
        understory; rocky coasts,
        sandy hills; DC, DL.

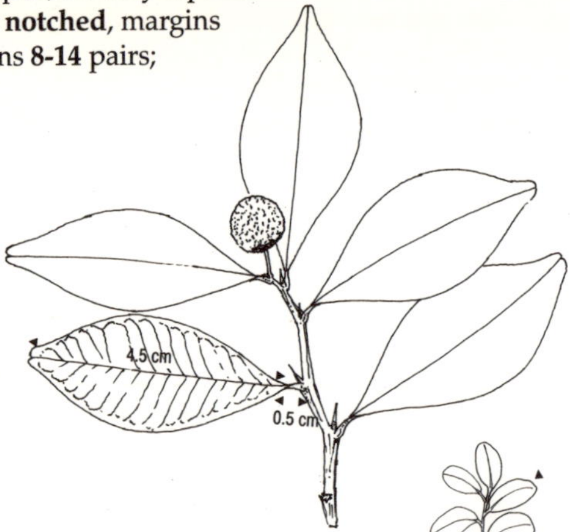

5. *Chloroxylon swietenia*, satin wood (E) burutha (S)/
    mutirai (T), (DF V:421), N, 30, tree.
    Leaves: **pinnate**; leaflets small, oblique,
        margins entire; aromatic when crushed.
    Trunk: W-hard, heavy, lustrous, satiny;
        young parts greyish, pubescent.
    Flowers: white; I-pyramidal
        racemose panicles.
    Fruits: black, 3-valved,
        oblong-ovoid capsule.
    Site: monsoon and intermediate
        forest canopy; DL, IN.
    Uses: W-furniture, cabinet wood.

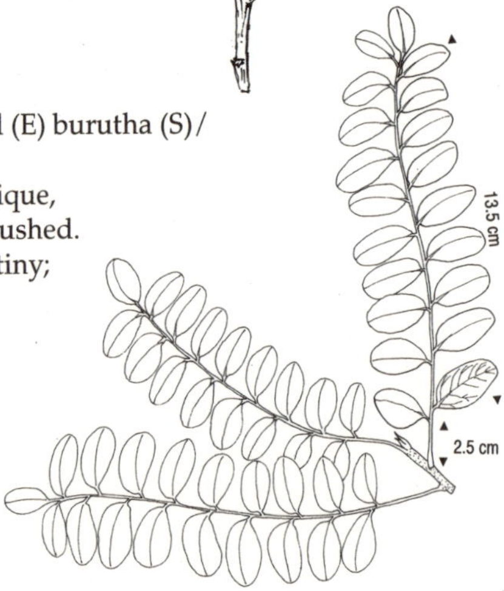

6. *Citrus* x *aurantiifolia*, lime (E)/dehi (S)/desi kai (T),
    (DF V:424), I (China), 2, shrub.
    Leaves: simple; elliptic to oblong-ovate, acute apex,
        rounded base, margins scalloped; petioles
        narrowly winged; aromatic when crushed.
    Trunk: spines short, stiff.
    Flowers: white, small, axillary in 2-7's.
    Fruits: **green to greenish-yellow**,
        globose-ovoid, sour.
    Site: home gardens; widespread.
    Uses: fruit-edible, used in
        rituals; fruit, root,
        leaves-medicinal.

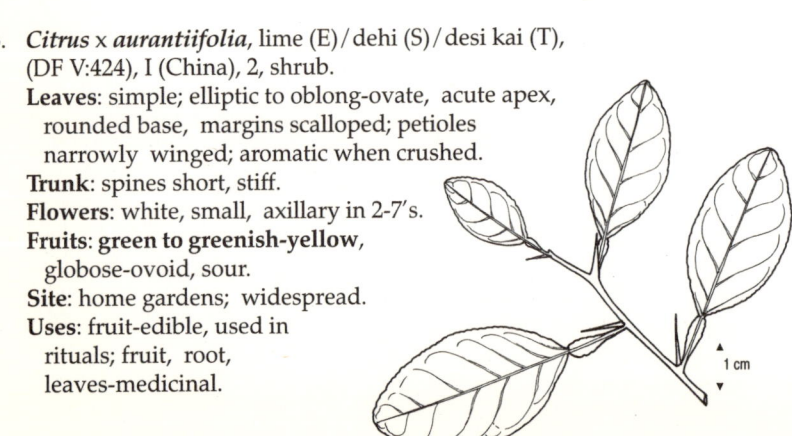

7. **Citrus grandis**, pomelo (E, T)/jambola (S), (DF V:426),
   I (Malesia), 4, small tree.
   Leaves: simple; large, ovate to elliptic, rounded to
   subcordate base, pointed apex, margin
   shallowly scalloped; petiole **broadly
   winged**; aromatic when crushed.
   Trunk: branches angular, can be very
   spiny; young parts pubescent.
   Flowers: white, single or
   clustered, large.
   Fruits: pale green to yellow,
   very large, subglobose to
   pear shaped, gland-pitted.
   Site: home gardens; widespread.
   Uses: fruit-edible.

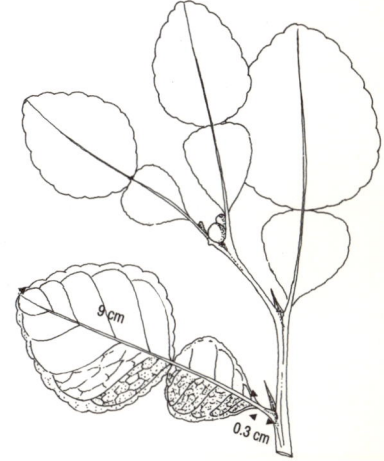

8. **Citrus hystrix**, leech lime (E)/kudalu
   dehi (S), (DF V:432), I (Malesia), 4,
   shrub to small tree.
   Leaves: simple; lanceolate, margins
   scalloped; petiole **broadly winged**,
   obcordate; aromatic when crushed.
   Trunk: branchlets compressed when
   young, stiff, axillary **spines**.
   Flowers: white to pinkish, fragrant;
   I-axillary, 1-5 flowers.
   Fruits: **green to yellow**, sub-globose,
   warty, glandular, slightly bitter.
   Site: home gardens; widespread.
   Uses: fruit-edible, shampoo, leech
   repellent, medicinal; leaves-spice.

9. **Citrus x limon**, lemon (E)/lemon (S), (DF V:428),
   I (Malesia), 2, shrub.
   Leaves: simple; ovate, apex acute,
   margins **serrate**; petiole **narrowly
   winged**; aromatic when crushed.
   Trunk: branches **spiny**.
   Flowers: white; I-axillary.
   Fruits: light to deep **yellow**,
   ovoid, glandular.
   Site: home gardens;
   DL, IN, W.
   Uses: fruit-edible.

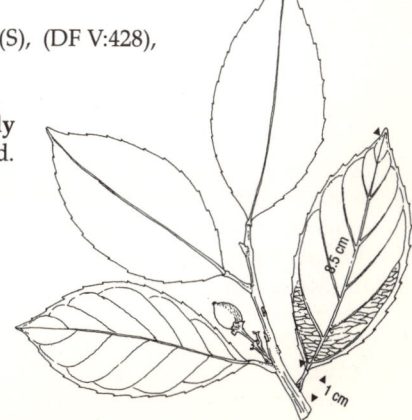

10. ***Citrus medica***, citron (E)/siderang (S), (DF V:430),
I (India), 4, shrub to small tree.
  Leaves: simple; elliptic to ovate, rounded to
    tapered base, rounded apex, margins
    serrate; petiole short, **wingless** or
    nearly so; aromatic when crushed.
  Trunk: branchlets angular with
    axillary **spines**, often
    purplish when young.
  Flowers: pink to purplish;
    I-few flowered, short
    axillary racemes.
  Fruits: **yellow**, globose or
    ovoid, slightly to
    considerably rough, warty.
  Site: home gardens; widespread.
  Uses: fruit-edible, medicinal;
    leaves-medicinal.

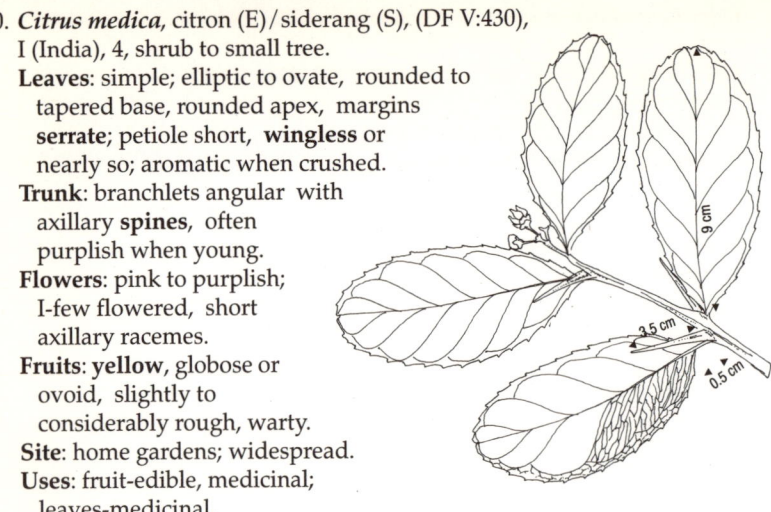

11. ***Citrus reticulata***, mandarin (E)/naran (S), (DF V:430),
I (Malesia), 3, shrub to small tree.
  Leaves: simple; lanceolate to broadly
    lanceolate, margins irregularly
    scalloped; petiole short,
    wingless or slightly winged;
    aromatic when crushed.
  Trunk: branchlets slender
    with axillary **spines**.
  Flowers: whitish,
    solitary or in clusters.
  Fruits: **orange to greenish
    orange**, subglobose,
    pericarp easily separable, sweet.
  Site: home gardens; IN, M, W.
  Uses: fruit-edible.

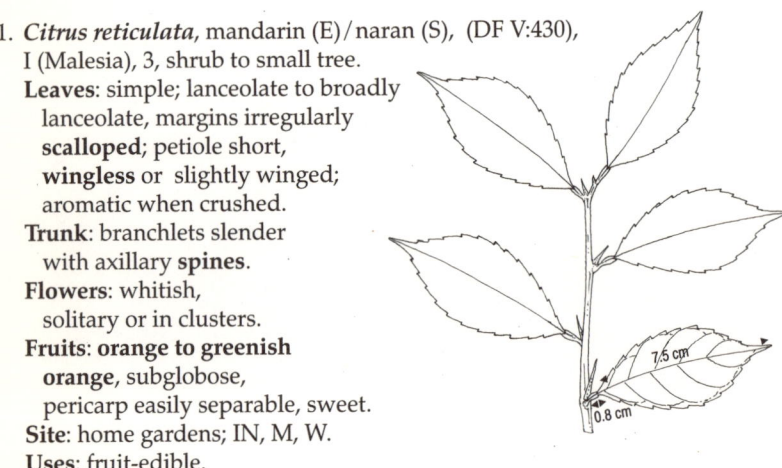

12. ***Clausena indica***, mi gon karapincha (S)/pannai (T),
(DF V:436), N, 5, shrub to small tree.
  Leaves: **pinnate**; leaflets **alternate**, basal smaller,
    asymmetric, ovate to lanceolate, lateral veins **4-5**
    pairs, margin subentire; aromatic when crushed.
  Trunk: branchlets greyish, scattered glands.
  Flowers: greenish or white; I-axillary
    and terminal panicles.
  Fruits: cream to dull pink,
    subglobose, smooth,
    glandular.
  Site: dry scrub,
    woodlands; DL.

13. *Euodia lunu-ankenda*, lunu ankenda (S), (DF V:440), N, 8, tree.
  Leaves: **trifoliolate**; leaflets elliptic to oblanceolate, tapered base,
    rounded to pointed apex, lateral veins **7-10** pairs with a
    looped marginal vein; aromatic when crushed.
  Trunk: B-smooth, grey; branchlets
    compressed, greenish.
  Flowers: yellowish green, small;
    I-axillary cymes of 9-12 flowers.
  Fruits: brown,4-lobed, glandular,
    capsules; segments ventrally
    dehiscent, exposing
    yellowish endocarp.
  Site: intermediate, montane
    and rain forest, gaps and
    fringes; IN, M, W.
  Uses: whole plant-medicinal.

14. *Glycosmis mauritiana*, (DF V:446), N, 3, shrub.
  Leaves: **pinnate**; leaflets small, elliptic to
    obovate, pointed ends, margin
    entire, lateral veins **5-8** pairs;
    aromatic when crushed.
  Trunk: B-greyish to brown, finely cracked.
  Flowers: white; I-short axillary panicles.
  Fruits: pink, subglobose, pericarp thin.
  Site: monsoon forest understory, rocky
    and sandy areas, scrub; DL.

15. *Glycosmis pentaphylla*, dodam
    pana (S)/ kula pannai (T),
    (DF V:449), N, 3, shrub.
  Leaves: **pinnate**; leaflets narrowly
    elliptic, acuminate ends, lateral
    veins **6-12** pairs; finely punctate
    and whitish beneath;
    aromatic when crushed.
  Trunk: B-thin, pale brownish;
    W-pale, hard; branchlets
    greenish.
  Flowers: white; I-axillary
    panicles, fragrant.
  Fruits: white to pink,
    subglobose.
  Site: monsoon forest
    understory, scrub; DL.
  Uses: W-tool handles; twigs-tooth
    picks; leaves-medicinal.

16. *Limonia acidissima*, wood apple (E)/diwul (S)/vila (T),
    (DF V:451), N, 10, tree.
    Leaves: **pinnate**; leaflets paired with terminal, lateral
       leaflets sessile, obovate, rounded ends; rachis
       narrowly winged; aromatic when crushed.
    Trunk: B-pale grey to whitish; **spiny**.
    Flowers: white, green or reddish,
       numerous; I-axillary cymose panicles.
    Fruits: whitish, woody, globose,
       chocolate brown sticky pulp.
    Site: monsoon forest subcanopy,
       scrub, home gardens; DL.
    Uses: fruit-edible; whole plant-medicinal;
       root stock used for grafting **Citrus.**

17. *Micromelum minutum*, wal karapincha (S)/
    kakai palai (T), (DF V:454), E, 5, small tree.
    Leaves: **pinnate**; leaflets paired with terminal,
       asymmetric, ovate lanceolate, pointed ends,
       margin **wavy to faintly scalloped,**
       aromatic when crushed.
    Trunk: branchlets yellow-brown to greyish.
    Flowers: white; I-terminal, cymose panicles.
    Fruits: green-yellow, ellipsoid-oblong, scented.
    Site: monsoon forest understory, scrub; DL.

18. *Murraya koenigii*, curry leaf (E)/karapincha (S)/
    karivempu (T), (DF V:458), N, 5, small tree.
    Leaves: **pinnate**, spiral, **crowded** at ends;
       leaflets oblong-lanceolate, asymmetric,
       apex acute, margin **minutely scalloped**.
    Trunk: B-smooth; branchlets
       green to dark grey.
    Flowers: white, small, numerous;
       I-terminal corymbose panicles.
    Fruits: purplish black berry, subglobose.
    Site: monsoon forest gaps and fringes,
       scrub, home gardens; widespread.
    Uses: leaves-curry flavouring;
       whole plant-medicinal.

30 cm

19. **Murraya paniculata**, orange jasmine (E)/etteria (S),
    (DF V:459), N, 4, shrub to small tree.
    Leaves: **pinnate**; leaflets alternate, ending in a terminal
      one, ovate to elliptic, base tapered to rounded,
      long-pointed apex, lateral veins **5-8** pairs.
    Trunk: B-Whitish, thin, smooth.
    Flowers: white, few; I-axillary
      panicles; fragrant.
    Fruits: ovoid, densely hairy,
      1-2-seeded berry.
    Site: monsoon forest understory, rocky
      outcrops, limestone scrub; DL.
    Uses: ornamental.

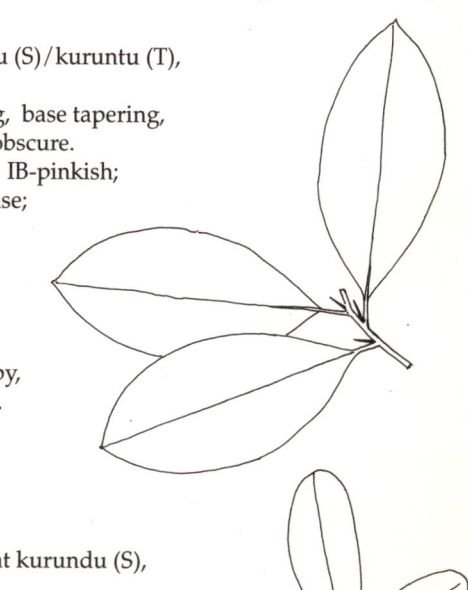

20. **Pamburus missionis**, pamburu (S)/kuruntu (T),
    (DF V:465), N, 10, tree.
    Leaves: **simple**; elliptic oblong, base tapering,
      apex pointed, lateral veins obscure.
    Trunk: B-thick, fissured, grey; IB-pinkish;
      W-straw to white, hard, dense;
      branches with **spines**.
    Flowers: white, few, fragrant;
      I-axillary racemes.
    Fruits: globose, like a small
      lime; pulp mucilaginous.
    Site: monsoon forest subcanopy,
      sandy coastal scrub; DC, DL.
    Uses: W-light construction.

21. **Pleiospermium alatum**, tumpat kurundu (S),
    (DF V:470), N, 5, small tree.
    Leaves: **trifoliolate**; lateral leaflets smaller,
      obovate to oblanceolate, base tapering, apex
      rounded; petiole **narrowly winged**.
    Trunk: W-heavy, yellowish, hard; branchlets
      often with single axillary **spines**.
    Flowers: white, small; I-clustered,
      axillary racemes.
    Fruits: globose, like a
      small lime, aromatic.
    Site: monsoon forest
      understory, scrub; DL.
    Uses: W-turnery, utensils.

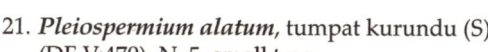

79. SABIACEAE

FAMILY DESCRIPTION - Habit: trees, shrubs or lianas. **Leaves**: spiral, simple or compound. **Stipules**: absent. **Flowers**: bisexual or plants polygamodioecious, zygomorphic in panicles or thyrses. **Fruits**: berry, sometimes dry and leathery.

**FLOWER PARTS - **Calyx: 3-5, free or basally united, imbricate. Corolla: 4-5, sometimes basally united, the two inner smaller, the outer broader, imbricate. Androecium: stamens 4-5, free or epipetalous, sometimes only 2 fertile. Anthers 2-locular with thick connective. Gynoecium: superior, 2-carpelled, 2-locular with 1-2 ovules in each. Styles 2, often united.

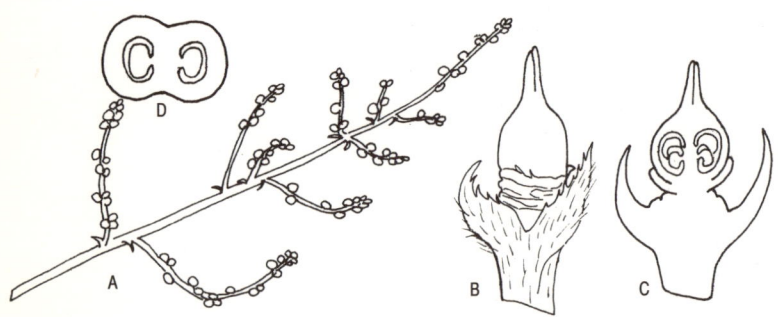

Key: inflorescence (A), gynoecium in full (B), longitudinal section of gynoecium, (C) and transverse (D) section of the ovary of *Meliosma pinnata*.

1. *Meliosma pinnata*, nika dawula (S)/kusavi (T), (DF III:383), N, 10, tree.
    Leaves: pinnate; leaflets ovate-oblong to elliptic,
        base rounded or pointed, pointed apex,
        margin entire or dentate, hairy beneath.
    Trunk: twigs rusty, young parts hairy.
    Flowers: yellowish-white,
        small, sessile; I-dense,
        spreading, pyramidal,
        reddish brown hairy, axillary
        and terminal panicles.
    Fruits: one-seeded drupe.
    Site: montane forest canopy
        and subcanopy; M, LM.
    Uses: ornamental.

2. **Meliosma simplicifolia**, elbedda (S), (DF III:380), N, 10, tree.
   Leaves: obovate to lanceolate to elliptic, **variable** size,
     margins entire to dentate, veins ascending, often
     looped near margin, **tufts of hair**
     in axils of veins beneath.
   Trunk: B-white, smooth, marked
     with large leaf scars.
   Flowers: yellowish white, sessile;
     I-densely spreading, pubescent panicles.
   Fruits: purplish, shiny, globose drupe.
   Site: montane forest canopy and
     rain forest understory;
     M, LM, W.

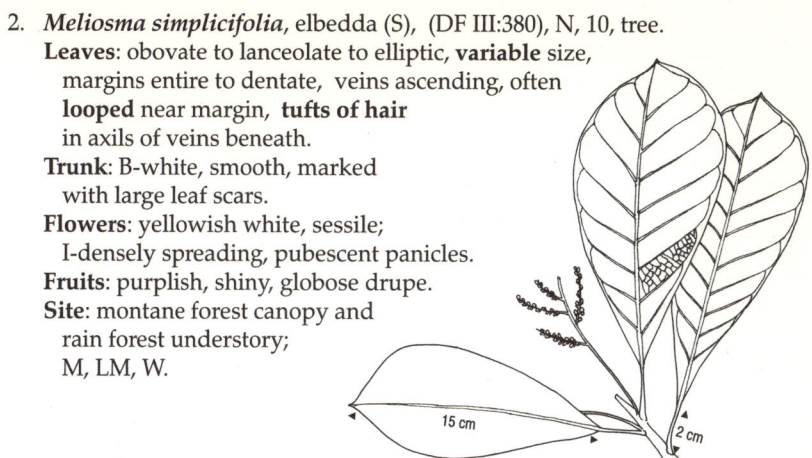

80. SALVADORACEAE

FAMILY DESCRIPTION - Habit: trees or shrubs, sometimes scrambling, unarmed or with axillary spines. **Leaves**: opposite, simple. **Stipules**: rudimentary or absent, or modified to spines. **Flowers**: bisexual or unisexual, (plants dioecious), actinomorphic, in dense axillary clusters or panicles. **Fruits**: 1-seeded berry or drupe.

FLOWER PARTS - CALYX: 2-4, rarely 5, united, imbricate. COROLLA: 4, free or partly united at base. ANDROECIUM: stamens 4, epipetalous. Filaments free or basally united. Anthers 2-locular, with loculi back to back, opening lengthwise. Disc absent or glands present between filaments. GYNOECIUM: superior, 1-2 locular, each with 1-2 basal erect ovules. Style short.

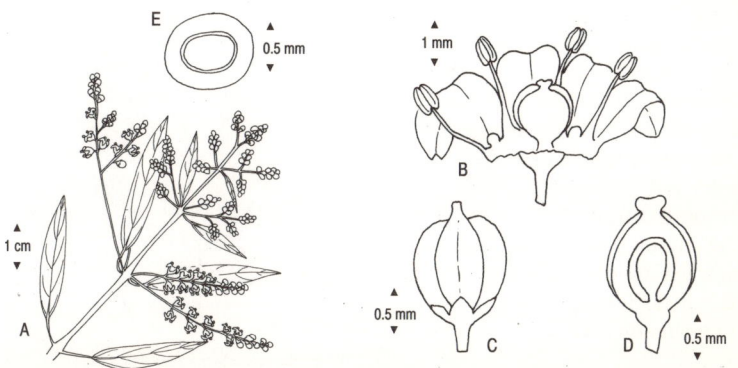

Key: inflorescence (A), an opened-out flower (B), the ovary in full (C), in longitudinal (D), and transverse (E) sections of *Salvadora persica*.

1. *Azima tetracantha*, katu niyada (S)/iyanku (T), (DF VII:397), N, 2, shrub.
   Leaves: oval, pointed ends, **mucronate** apex,
   thick, shiny; 4 short **spines** at node.
   Trunk: branches opposite, spreading;
   branchlets bluntly quadrangular;
   young twigs pubescent.
   Flowers: greenish white,
   sessile; male small
   crowded clusters;
   female solitary.
   Fruits: white, globose,
   glabrous drupe.
   Site: thorn scrub; DL.

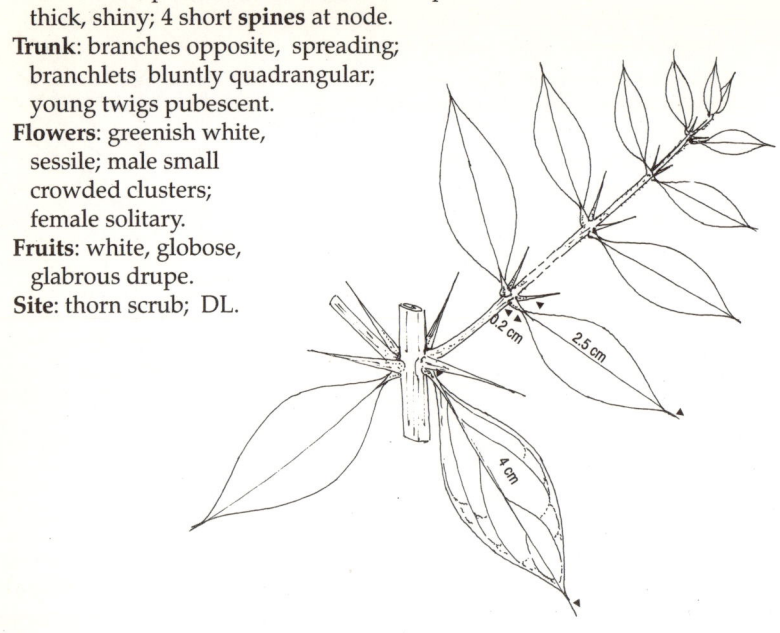

2. *Salvadora persica*, mustard tree (E)/ malittan (S)/uvay (T),
   (DF VII:398), N, 4, small tree.
   Leaves: numerous, oblong-ovate, tapered
   or rounded at base, pointed
   apex, whitish green.
   Trunk: B-rough, furrowed; Branches
   erect; branchlets long,
   drooping, slender.
   Flowers: greenish white, small,
   very numerous; I-spreading
   racemes or panicles.
   Fruits: red, globose,
   smooth drupe.
   Site: coastal thorn scrub; DC.
   Uses: leaves- for smoking
   bananas; whole
   plant-medicinal.

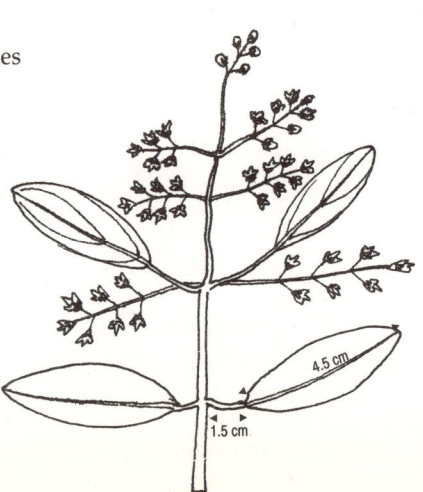

81. SANTALACEAE

FAMILY DESCRIPTION - Habit: hemiparasitic trees, shrubs or herbs. Sometimes thorny or xeromorphic. **Leaves**: usually opposite, simple, entire, sometimes scale-like. **Stipules**: absent. **Flowers**: bisexual or unisexual, actinomorphic. **Fruits**: nut or drupe.

FLOWER PARTS - PERIANTH: of one whorl, 3-8, free or basally united in to a fleshy tube. ANDROECIUM: 3-8, stamens often attached to perianth, anthers dehiscing longitudinally. Nectary disc present. GYNOECIUM: inferior to superior, carpels 2-5, ovary unilocular or basally divided. Placentation basal.

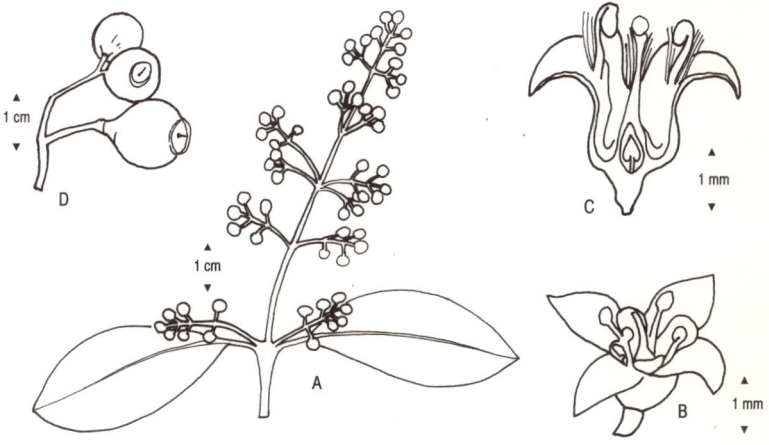

Key: inflorescence (A), full flower (B) and half flower (C) showing gynoecium with basal placentation and fruits (D) of *Santalum album*.

1. ***Santalum album,*** white sandal wood (E)/ sudu handun (S), I (India), 10, tree.
   Leaves: opposite, simple, pointed ends, leathery, margins entire and cartilaginous, veins faint, slightly depressed; shiny above and pale beneath; petiole long.
   Trunk: branchlets stiff, erect; B-smooth, greyish brown; twigs glabrous.
   Flowers: purplish red, small; I-terminal and axillary paniculate cymes.
   Fruits: black, small, round, fleshy, one-seeded drupe.
   Site: home gardens, scrub; IN, W.
   Uses: seed-medicine; W-cabinets; wood oil-cosmetics, medicine.

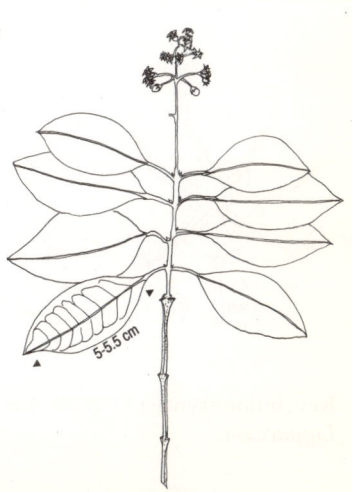

82. SAPINDACEAE

FAMILY DESCRIPTION - Habit: trees, shrubs, lianas or herbaceous climbers. **Leaves**: usually spiral, rarely opposite, compound or rarely simple. **Stipules**: absent or small. **Flowers**: bisexual or unisexual, on polygamous or polygamodioecious plants, actinomorphic or zygomorphic, in cymes or thyrses, rarely solitary. **Fruits**: variable, fleshy or dry, dehiscent or indehiscent. Seeds often arillate.

FLOWER PARTS - Calyx: 4 or 5, free or sometimes basally united, usually imbricate. Corolla: 3-5, or rarely more, often absent, free, often with scaly or hair-tufted nectaries at the base on their inner side. Annular or 1-sided disc present. Androecium: usually 8, rarely 4 or 10 or more. Filaments free, often hairy. Anthers 2-locular, opening lengthwise. Reduced ovary present in male flowers. Gynoecium: superior, usually 3-carpelled, 3-locular, with 1-2 or rarely more ovules per loculus. Placentation axile. Style mostly 1, rarely 2-4, stigma lobed.

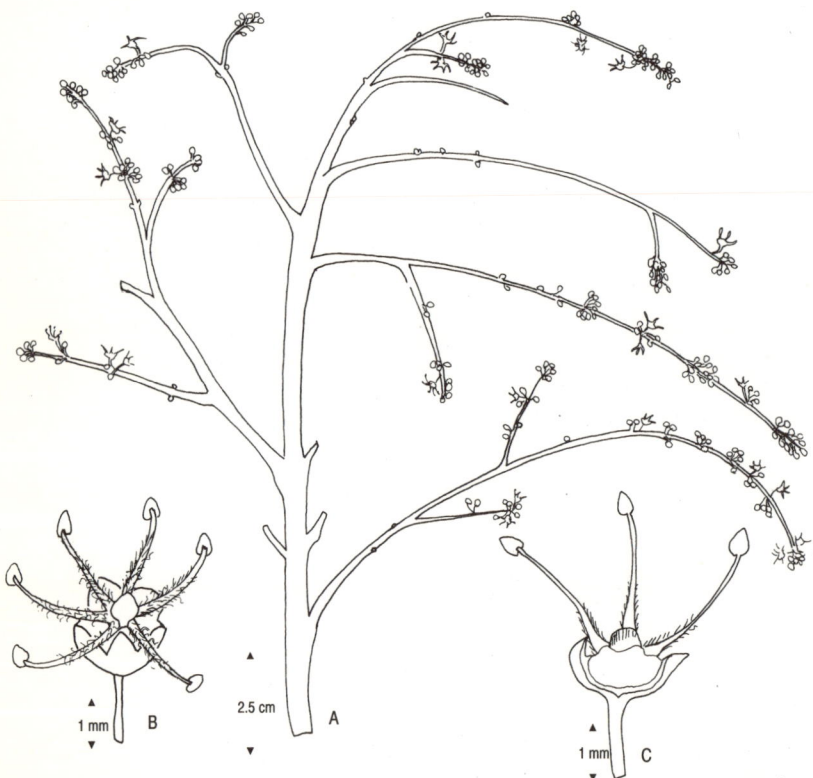

Key: inflorescence (A), male flower in full (B), and in section (C) of *Nephelium lappaceum*.

1. ***Allophylus cobbe***, kobbe (S)/amavai (T),
   (T I:303), N, 3, shrub to small tree.
   Leaves: **trifoliolate**, rachis **pubescent**;
   Leaflets with pointed base, faintly
   serrate, softly pubescent, **tufts of**
   hair in vein axils beneath.
   Trunk: much-branched, **pubescent**.
   Flowers: greenish, very small;
   I-axillary panicles.
   Fruits: red, solitary, usually
   globose-ovoid.
   Site: montane and rain
   forest understory;
   LM, W.
   Uses: W-used to
   make bows by
   'veddas'; leaves,
   B, roots-medicinal.

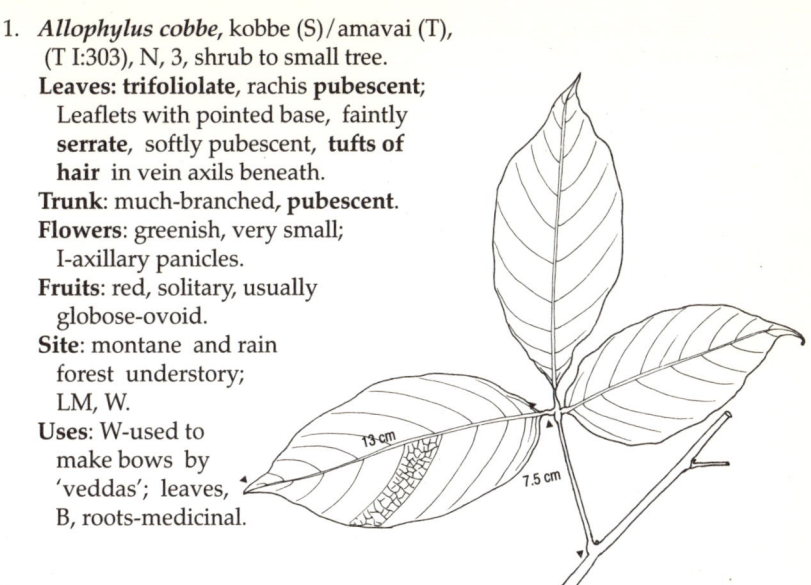

82

SAPINDACEAE

2. ***Allophylus zeylanicus***, wal kobbe (S),
   (T I:302), E, 3, shrub to small tree.
   Leaves: simple; variable, ovate to
   lanceolate, pointed ends, margins
   entire or with some **coarse**
   teeth at apex.
   Trunk: B-whitish, young parts
   glabrous; W-yellowish
   brown, soft.
   Flowers: greenish white
   or white; I-racemes.
   Fruits: red, ovoid, smooth.
   Site: montane and rain
   forest understory;
   M, LM, W.
   Uses: whole plant-
   medicinal.

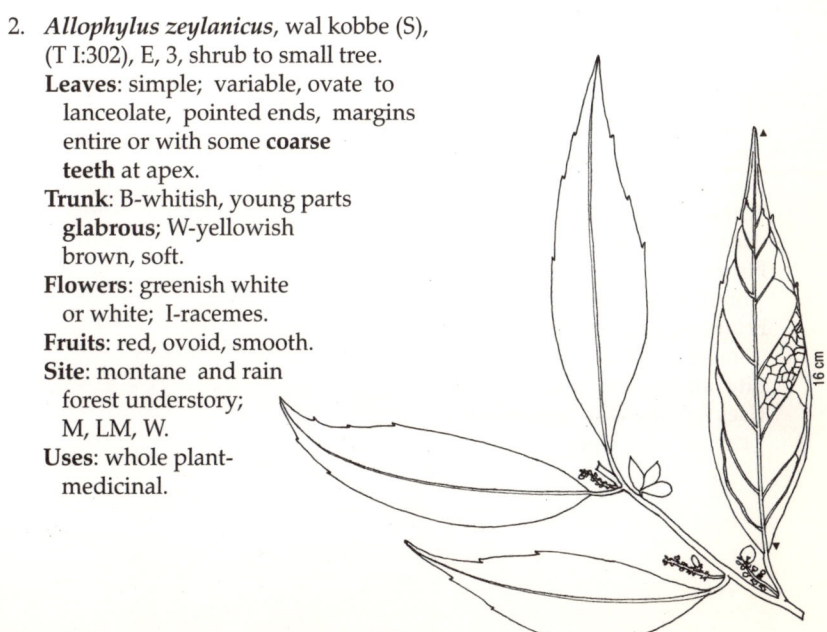

3. *Dimocarpus longan*, mora (S)/nurai (T), (T I:309), N, 10, tree.
   Leaves: **pinnate**, rachis **finely pubescent**; leaflets
   oblong-lanceolate, base pointed and unequal,
   apex pointed, **wavy** margins, often
   puberulous beneath, often with insect galls.
   Trunk: B-smooth to flaking, yellowish grey;
   W-hard, pale reddish; young parts
   fine **reddish brown pubescent**.
   Flowers: pale yellow; I-large, lax,
   pubescent, pyramidal,
   terminal panicles.
   Fruits: brown, ovoid, stellate
   hair, supported on persistent
   calyx; seeds black, covered
   with pulpy white aril.
   Site: monsoon and intermediate
   forest subcanopy and rain
   forest understory; DL, IN,W.
   Uses: fruit-edible; W-medium
   construction; whole
   plant-medicinal.

1 cm

4. *Filicium decipiens*, pihimbiya (S) / chittira vempu
   (T), (T I:240 & VI:44), N, 20, tree.
   Leaves: fern-like, **pinnate**, closely placed, spreading,
   rachis **winged**; leaflets opposite or alternate, sessile,
   unequal base, **linear-oblong**, **entire**, **wavy**
   margins, lateral veins numerous, **parallel**.
   Trunk: B-reddish grey, rough, prominent **large
   leaf-scars**; W-pale red, rather hard, heavy.
   Flowers: pinkish-white, small,
   numerous; I-axillary panicles.
   Fruits: purple, ovoid, pointed,
   smooth, shiny drupe.
   Site: intermediate forest canopy; IN.
   Uses: ornamental; W-furniture,
   cabinet wood, heavy construction.

9 cm

5 cm

6 cm

5. *Glenniea unijuga*, wal mora (S)/kuma (T), (T I:305 & VI:57), E, 10, tree.
   Leaves: **pinnate**, rachis **swollen**; leaflets opposite or alternate, oval-
   lanceolate, pointed ends, **wavy** margins, veins prominently **reticulate**.
   Trunk: B-whitish, thin; W-yellowish white, heavy.
   Flowers: greenish white,
   shortly pedunculate clusters;
   I-pilose, male terminal,
   paniculate; bisexual
   axillary, unbranched.
   Fruits: green, depressed-
   globose, 2-3-lobed, smooth.
   Site: monsoon and
   intermediate forest
   subcanopy; DL, IN.
   Uses: light construction.

6. *Harpullia arborea*, na imbul (S),
   (T I:311 & VI:59), N, 20, tree.
   Leaves: **pinnate**, **pubescent** rachis;
   leaflets opposite or alternate,
   lanceolate, base pointed and
   unequal, pointed to rounded apex.
   Trunk: fluted; B-pale, smooth; young
   parts finely **yellowish-pubescent**.
   Flowers: sulphur-yellow to greenish;
   I-long, pendulous axillary panicles.
   Fruits: bright orange, pendulous,
   inflated capsule, tipped with
   persistent style; seeds black.
   Site: intermediate forest canopy; IN.
   Uses: ornamental; fruit-soap.

7. *Lepisanthes senegalensis*, gal kuma (S), (T I:307 & VI:59), N, 10, tree.
   Leaves: **pinnate with one pair of leaflets**,
   rachis **glabrous**; leaflets narrow-
   lanceolate, pointed base, apex
   tapering, **entire, wavy** margins.
   Trunk: young parts **pubescent**.
   Flowers: white, numerous, in clusters;
   I-small, terminal and axillary
   spreading panicles.
   Fruits: ovoid, glabrous.
   Site: monsoon forest
   subcanopy, scrub; DL.

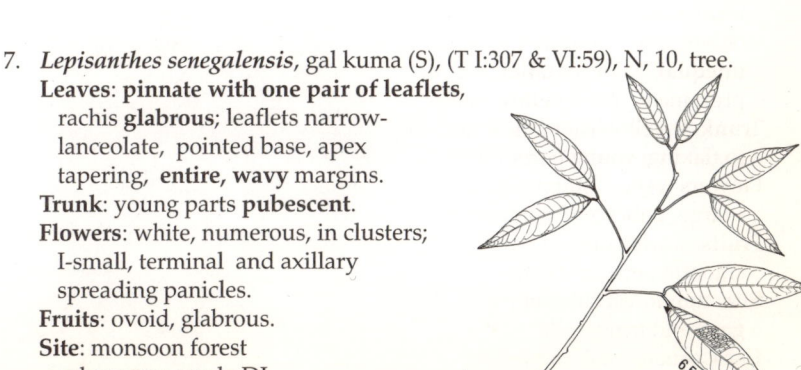

8. *Lepisanthes tetraphylla*, (T I:301 & VI:55), N, 20, tree.
   Leaves: **pinnate**; leaflets nearly opposite,
   lanceolate-oblong, pointed base, apex
   rounded to **notched**, veins **reticulate**.
   Trunk: B-whitish grey; young
   parts **pubescent**.
   Flowers: white, numerous;
   I-lax, clustered, pubescent
   panicles, in axils of fallen leaves.
   Fruit: yellow, oblong-ovoid,
   finely tomentose, pointed.
   Site: monsoon forest canopy; DL.

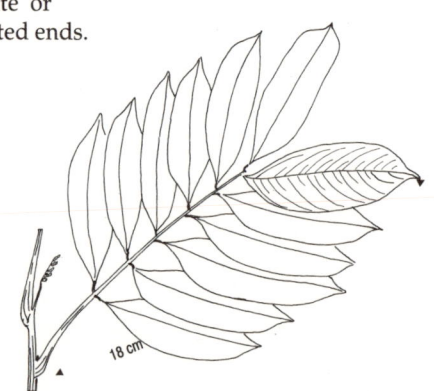

9. *Nephelium lappaceum*, rambutan (E,S,T),
   I (Malesia), 15, small tree.
   Leaves: **pinnate**; leaflets alternate or
   subopposite, lanceolate, pointed ends.
   Trunk: B-brown, smooth.
   Flowers: white, small; I-large,
   spreading, terminal
   and axillary panicles.
   Fruits: **red to green-yellow**,
   softly bristly, globose to
   ovoid; single seed,
   white pulpy aril.
   Site: home gardens; W.
   Uses: fruit-edible.

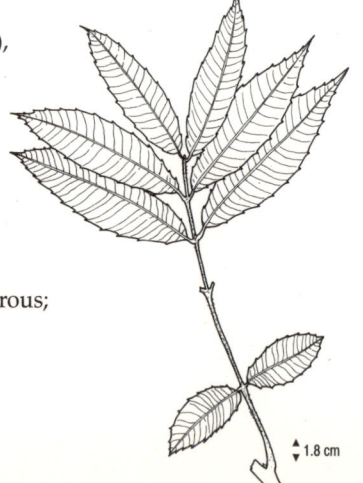

10. *Pometia pinnata*, gal mora (S), (T I:310),
    N, 20, tree.
    Leaves: **pinnate**, **large**, rachis **hairy**;
    leaflets nearly opposite, lanceolate-
    oblong, base **rounded to subcordate,
    unequal**, veins **parallel**,
    prominent; flush **yellow-red**.
    Trunk: B-yellowish to pink, smooth
    to flaking; young parts yellowish hair.
    Flowers: yellowish-brown, small, numerous;
    I-large, pubescent, compact panicles.
    Fruits: scarlet, oblong-ovoid,
    smooth; pericarp fleshy.
    Site: rain forest subcanopy,
    gaps and fringes; W.
    Uses: ornamental; shade tree;
    W-light construction.

11. *Sapindus emarginatus*, penela (S)/panalai (T), (T I:307), N, 15, tree.
 Leaves: **pinnate**; leaflets opposite to subopposite,
  broad, oval-oblong, base pointed to
  rounded, **notched** apex, glabrous
  or pubescent beneath.
 Trunk: much branched; B-whitish,
  rough; W-hard, heavy, yellow.
 Flowers: greenish-white, shortly
  stalked; I-large, terminal,
  spreading panicles.
 Fruits: globose; pericarp
  thick, fleshy.
 Site: monsoon forest canopy; DL.
 Uses: fruit-soap; fruit, root-
  medicinal; W-light and
  medium construction.

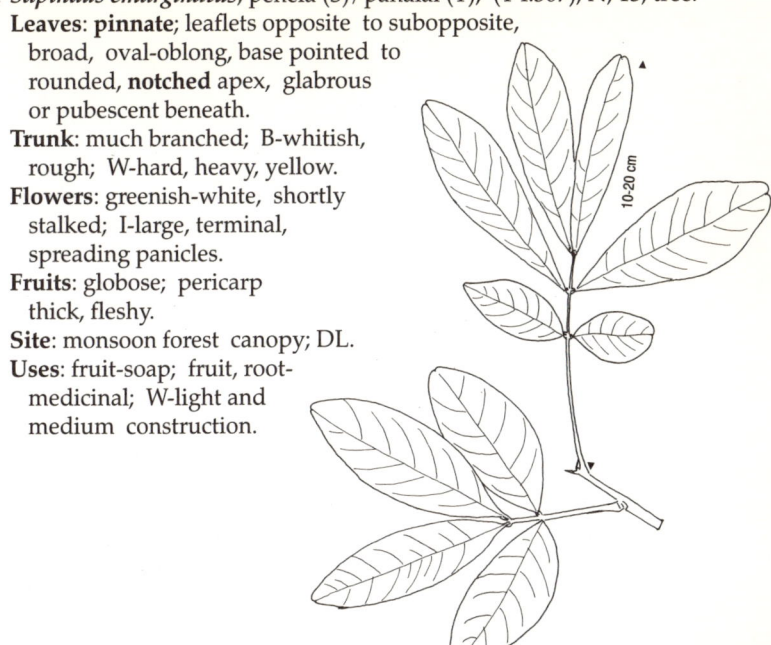

12. *Schleichera oleosa*, kon (S)/puvu (T), (T I:304 and VI:57), N, 20, tree.
 Leaves: **pinnate**, rachis **pubescent, swollen**; leaflets opposite, oblong,
  tapered base, rounded apex, **wavy** margin; buds pubescent.
 Trunk: B-grey, smooth; branches spreading.
 Flowers: green; I-lax, panicles.
 Fruits: ovoid, smooth, sharply
  pointed; aril pulpy.
 Site: monsoon and intermediate
  forest canopy; DL, IN.
 Uses: fruit-edible; seed,
  B-medicinal; seed-macassar
  oil, illuminant, batik
  making, soap; W-light
  construction.

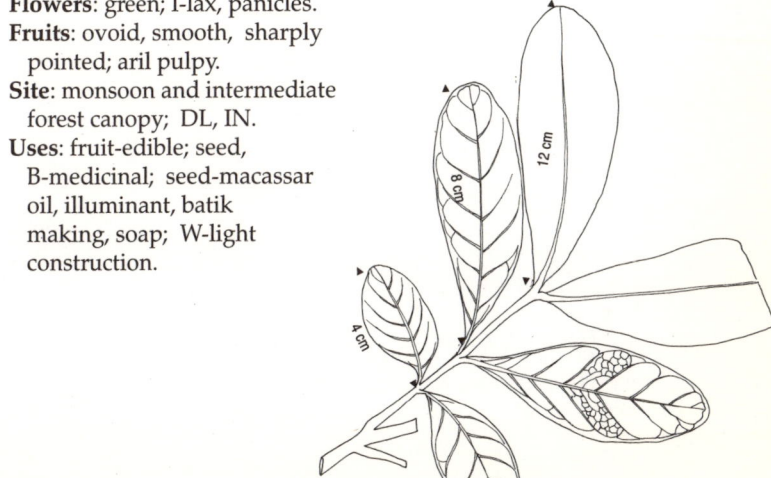

83. SAPOTACEAE

FAMILY DESCRIPTION - Habit: trees or shrubs with milky latex. **Leaves**: usually spiral, rarely opposite, simple, entire. **Stipules**: usually absent. **Flowers**: bisexual, actinomorphic, fasciculate. **Fruits**: berry often with thin leathery to bony outer layer. Seeds with bony, often shining testa, and large broad hilum.

FLOWER PARTS - CALYX: 4-8, rarely 12, free, imbricate, or basally united in 2 whorls of 2, 3 or 4. COROLLA: 4-8, united, lobes imbricate, sometimes with paired appendages. ANDROECIUM: stamens in 1-3 whorls of 4-5 each, but usually only inner whorl fertile, others reduced to staminodes or absent, epipetalous. Filaments free. Anthers 2-locular, dehiscing longitudinally. GYNOECIUM: superior, usually 2-14 carpels and loculi, but rarely upto 30, usually hairy. Each loculus uniovulate, with axile placentation. Style simple, often apically lobed.

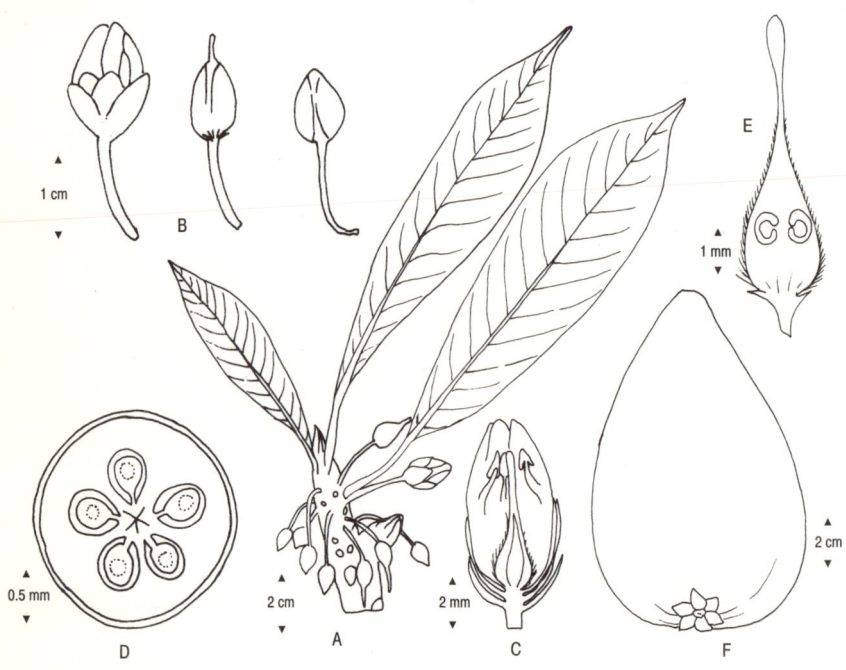

Key: flower clusters (A), buds and full flower (B), half flower (C), transverse section of ovary (D), longitudinal section of gynoecium (E) and mature fruit (F) of *Pouteria campechiana.*

1. *Chrysophyllum cainito*, star apple (E)/rata lawulu (S)/
   seemi pla (T), (W:303), I (C. Amer.), 10, tree.
   Leaves: oval to elliptic, pointed ends, lateral veins many,
      parallel, shiny **green** above, coppery beneath.
   Trunk: B-deeply cracked, greyish brown.
   Flowers: white petals, remaining parts coppery.
   Fruits: purple black, smooth, round,
      cross section star-shaped.
   Site: home gardens; IN, W.
   Uses: ornamental; fruit-edible.

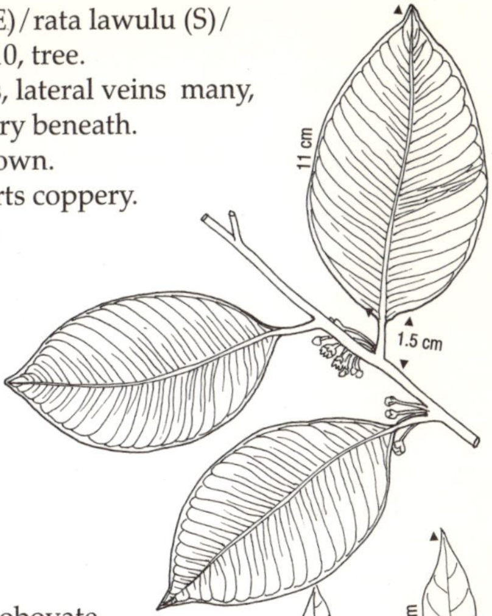

2. *Isonandra compta*, molpadda (S),
   (T III:77), N, 20, tree.
   Leaves: very variable, linear-oval to obovate,
      tapered base, margins often **wavy**, lateral
      veins **distant** and conspicuous, parallel
      intercostals, sometimes hairy.
   Trunk: B-dark brown; W-hard, heavy;
      young parts brownish-pubescent.
   Flowers: pale yellow, small; I-small,
      crowded, axillary clusters.
   Fruits: brilliant scarlet, cylindrical,
      smooth, soft, berry; seeds
      hard and shiny.
   Site: montane and rain
      forest canopy; LM, W.

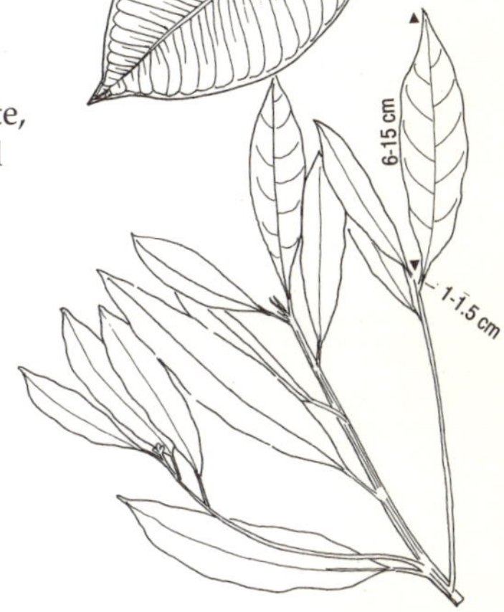

3. *Madhuca fulva*, wana mi (S), (T III:81 & VI:179),
   E, 25, tree.
   Leaves: few, broadly oval to oblong, pointed
      ends, brownish hairy on veins beneath;
      petioles stout, densely **tomentose**.
   Trunk: B-brownish grey; W-hard, heavy,
      yellowish brown; young shoots
      densely **orange tomentose**.
   Flowers: numerous, clusters of
      4-8 in axils of fallen leaves.
   Fruits: brown, ovoid, pointed,
      glabrous berry.
   Site: rain forest canopy; W.
   Uses: stem, leaves-medicinal;
      W-heavy construction.

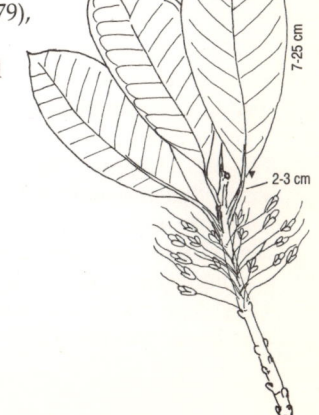

SAPOTACEAE

4. *Madhuca longifolia*, mousey mi (E)/mi (S)/iluppai (T),
   (T III:79 & VI:178), N, 25, tree.
   Leaves: terminally **crowded, linear-lanceolate**, tapering
      ends; flush **yellow to pink**; deciduous.
   Trunk: B-dark yellowish grey, thick, slightly
      furrowed; W-pale reddish, heavy,
      very durable; young parts
      silky pubescent, pinkish white.
   Flowers: pale yellow, solitary, in axils,
      fleshy, sweet, unpleasant mousy odour.
   Fruits: yellow, obliquely ovoid,
      pointed glabrous berry,
      tomentose when young; single
      seed slightly curved with a
      short beak, shiny and hard.
   Site: monsoon and intermediate
      forest canopy; DL, IN.
   Uses: B-tonic; flowers-edible;
      seed-oil, cooking, medicinal;
      W-heavy construction, boat building.

5. *Madhuca microphylla*, (T III:80 &
   VI:179), E, 20, tree.
   Leaves: small, numerous, lanceolate to
      oblong, acute base, rounded apex,
      glabrous, stiff, veins **inconspicuous**;
      buds **yellowish hairy**.
   Trunk: much-branched.
   Flowers: brownish white, small,
      in clusters of 2-4, axillary.
   Fruits: berry.
   Site: rain forest subcanopy; W.

6. *Madhuca neriifolia*, gan mi (S), (T III:80 & VI:179), N, 8, tree.
   Leaves: numerous, **oblong-linear**, tapered apex.
   Trunk: B-dark brown; W-pale reddish, heavy,
      very durable; twigs furrowed.
   Flowers: clusters of 2-6.
   Fruits: linear-ovoid, beaked berry; seed
      solitary and pointed at both ends.
   Site: rain forest understory,
      along streams; W.
   Uses: stem, root-medicinal;
      W-heavy construction.

7. **Manilkara hexandra**, palu (S)/palai (T),
   (T III:86 & VI:179), N, 25, tree.
   Leaves: numerous, small, broadly obovate,
   Tapered base, **truncate to bilobed**
   apex; sapling leaves crowded.
   Trunk: B-blackish grey, deeply
   vertically furrowed;
   W-very heavy, purplish
   brown; often saplings have
   semi-spiny branchlets.
   Flowers: yellow, small, numerous,
   in clusters of 1-3 in leaf axils.
   Fruits: yellow, small, ovoid,
   smooth, berry; seed solitary.
   Site: monsoon and intermediate
   forest canopy; DL, IN.
   Uses: fruit-edible; fruit,
   B-medicinal; W-heavy
   construction.

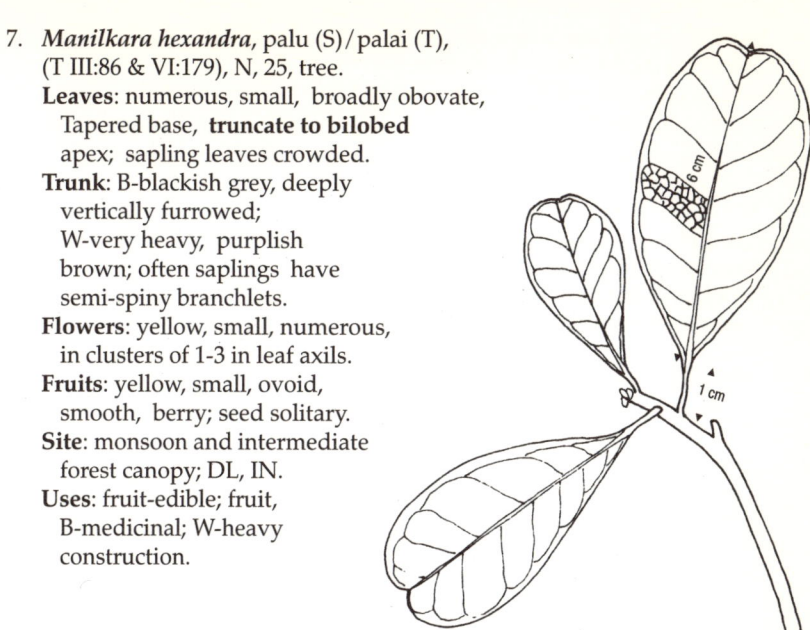

83

SAPOTACEAE

8. **Mimusops elengi**, muna mal (S)/makil (T), (T III:86), N, 25, tree.
   Leaves: numerous, large, oval, rounded base, rounded or
   acute apex, dark green, **wavy margins**,
   lateral veins numerous.
   Trunk: B-reddish brown; W-pinkish red, hard;
   branchlets drooping, young parts silky.
   Flowers: white to cream, fragrant,
   in clusters of 1-4 in leaf axils.
   Fruits: orange-yellow, ovoid,
   pointed berry.
   Site: monsoon and intermediate
   forest canopy; DL, IN.
   Uses: flowers-perfume;
   fruit, flower,
   B-medicinal;
   fruit-edible;
   W-construction.

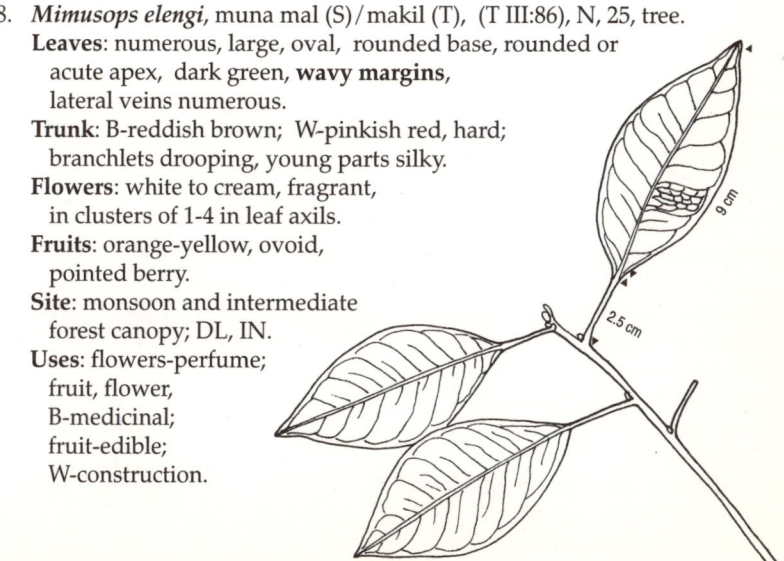

9. **Palaquium canaliculatum**, kiri
   hembiliya (S), (T III:84), E, 25, tree.
   Leaves: scattered, large, oblong-oval,
   rounded to pointed ends.
   Trunk: B-reddish grey.
   Flowers: white, solitary or in
   clusters of 2-6 in leaf axils;
   pedicels slender,
   densely pubescent.
   Fruits: ovoid berry.
   Site: rain forest canopy; W.
   Uses: W-construction,
   plywood.

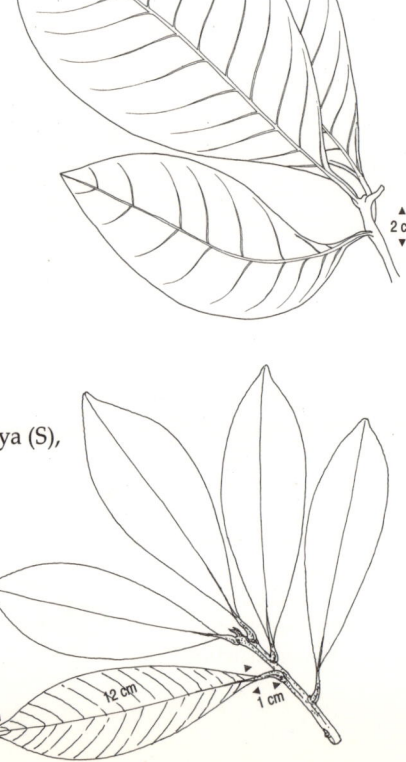

10. **Palaquium grande**, kiri hembiliya (S),
    (T III:82), E, 25, tree.
    Leaves: terminally **crowded**, oval,
    tapered base, apex rounded,
    shortly pointed, lateral veins
    parallel, sometimes
    pubescent beneath.
    Trunk: B-brown, thick,
    prominently scarred.
    Flowers: pale yellow,
    clusters of 2-5 in leaf axils.
    Fruits: purple, nearly globose,
    smooth, berry; 6 ovoid,
    shiny seeds.
    Site: montane and rain
    forest canopy; LM, W.
    Uses: W-construction, plywood.

11. **Palaquium laevifolium**, kiri hembiliya (S),
    (T III:84), E, 25, tree.
    Leaves: numerous, rather small,
    oblong-lanceolate, tapered ends.
    Trunk: B-smooth, grey.
    Flowers: brownish white, few,
    small, in clusters of
    2-4 in leaf axils.
    Fruits: oblong, sharp-pointed.
    Site: rain forest canopy; W.
    Uses: W-construction, plywood.

12. *Palaquium petiolare*, kiri hembiliya (S), (T III:82), E, 30, tree.
    Leaves: terminally **crowded**, oval-oblong, tapered base,
    short-pointed apex, lateral veins parallel,
    distant; flush **pinkish-red**.
    Trunk: B-smooth, grey;
    W-pale, whitish.
    Flowers: greenish pink,
    clusters of 2-5 in axils
    of fallen leaves.
    Fruits: nearly globose,
    pointed berry; supported by
    enlarged woody sepals.
    Site: rain forest canopy; W.
    Uses: W-construction,
    plywood.

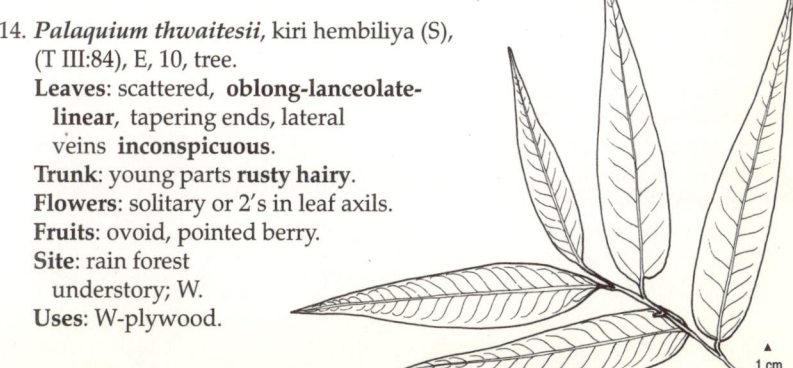

13. *Palaquium rubiginosum*, kiri pedda (S),
    (T III:83), E, 25, tree.
    Leaves: terminally **crowded**, obovate to
    oval, tapered base, rounded apex,
    lateral veins obscure, fine **orange-red
    pubescent** beneath, thick, leathery.
    Trunk: B-dark brown; branchlets
    very stout, leaf scars.
    Flowers: whitish, clusters of 2-7 in leaf axils.
    Fruits: ovoid to globose, pointed berry.
    Site: rain forest canopy; LM, W.

14. *Palaquium thwaitesii*, kiri hembiliya (S),
    (T III:84), E, 10, tree.
    Leaves: scattered, **oblong-lanceolate-
    linear**, tapering ends, lateral
    veins **inconspicuous**.
    Trunk: young parts **rusty hairy**.
    Flowers: solitary or 2's in leaf axils.
    Fruits: ovoid, pointed berry.
    Site: rain forest
    understory; W.
    Uses: W-plywood.

355

15. *Pouteria campechiana*, rata lawulu (S), (T III:76),
     I (Central Amer.), 5, small tree.
   Leaves: numerous, 2-ranked, lanceolate to oblong, somewhat unequal
       tapered base, short-pointed apex; somewhat wavy margins, lateral
       veins many, **parallel**, uniting with an intramarginal vein.
   Trunk: horizontal branches; B-grey, smooth; W-yellowish
       white; young parts **densely yellowish** pubescent.
   Flowers: greenish white, small, numerous,
       in clusters of 2-3.
   Fruits: yelowish orange, ovoid to globose,
       smooth berry; seed brownish
       yellow, hard and shiny.
   Site: home gardens; W.
   Uses: fruit-edible.

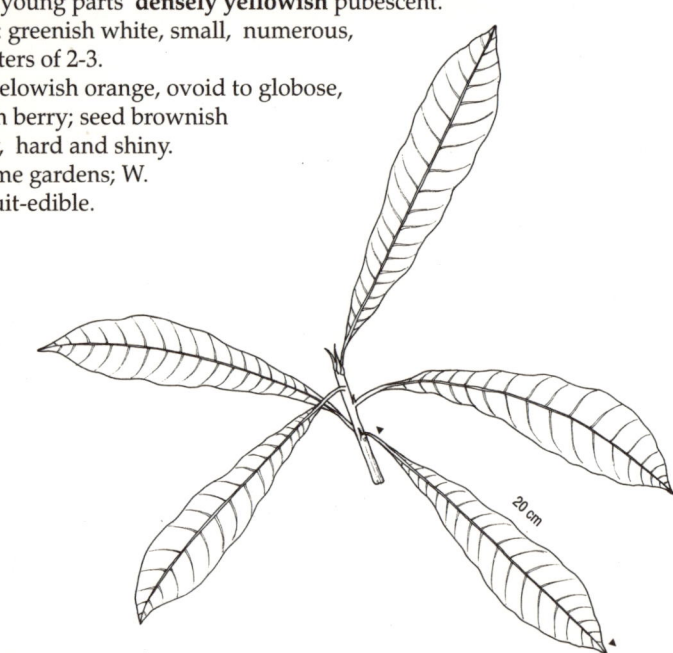

16. *Xantolis tomentosa*, mul makil (T), (T III:77), N, 8, tree.
   Leaves: numerous, oval to obovate, pointed
       ends, stiff; petiole pubescent.
   Trunk: B-cracked; much branched; twigs
       spiny; young parts **rusty pubescent**.
   Flowers: white, small, on stout pubescent
       pedicels; I-small axillary clusters.
   Fruits: pubescent berry becoming
       glabrous green; seeds
       shiny and brown.
   Site: monsoon and
       intermediate forest
       subcanopy; DL, IN.
   Uses: fruit-edible.

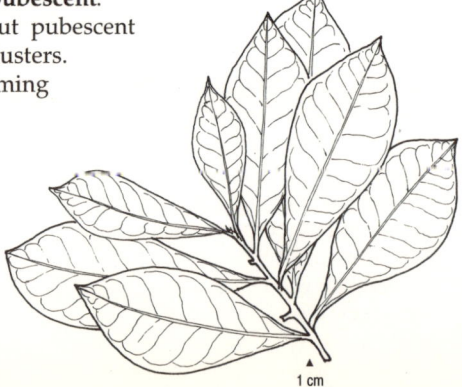

84. SIMAROUBACEAE

FAMILY DESCRIPTION - Habit: trees or shrubs, usually with bitter bark. **Leaves**: spiral, rarely opposite, compound or rarely simple. **Stipules**: absent or very rare. **Flowers**: unisexual often with rudiments of opposite sex, rarely bisexual, actinomorphic, racemes, cymes or thyrses. **Fruits**: capsule or samara, rarely berry or drupe, often schizocarp.

FLOWER PARTS - CALYX: 3-5, basally united into a tube, imbricate or valvate, rarely absent. COROLLA: 3-5, rarely absent, free, imbricate or valvate. ANDROECIUM: stamens equal to or double the number of petals, rarely numerous. Filaments free, often with basal appendages. Anthers 2-locular, opening lengthwise. Disc usually present, sometimes a gynophore or androgynophore. GYNOECIUM: superior, usually 2-5 carpels and loculi, sometimes up to 8, united, rarely free. Ovules solitary or rarely 2 or more per loculus with axile placentation. Styles 2-5.

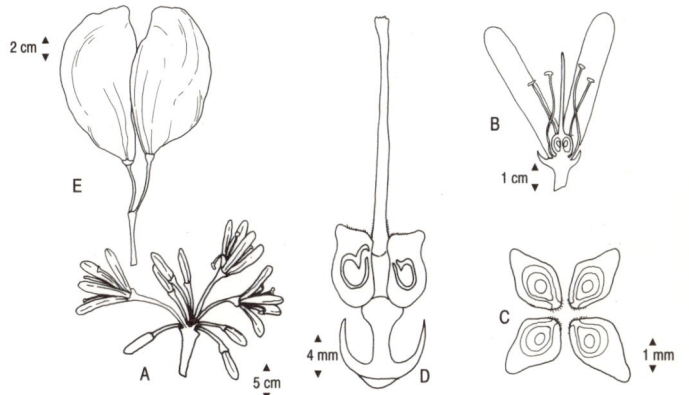

Key: inflorescence (A), half flower (B), transverse section of ovary (C), longitudinal section of the gynoecium (D) and mature fruits (E) of *Quassia indica*.

1. *Quassia indica*, samadera (S), (T I:231), N, 4, small tree.
   Leaves: simple, large, oblong-oval, rounded base, short-pointed apex.
   Trunk: B-transversely cracked; W-light, soft, yellow; young parts glabrous.
   Flowers: pinkish yellow; I-axillary umbels; peduncles red.
   Fruits: flattened, semi-circular, leathery, hard, samara.
   Site: intermediate and rain forest understory; IN, W.
   Uses: leaves-insecticide; whole plant-medicinal.

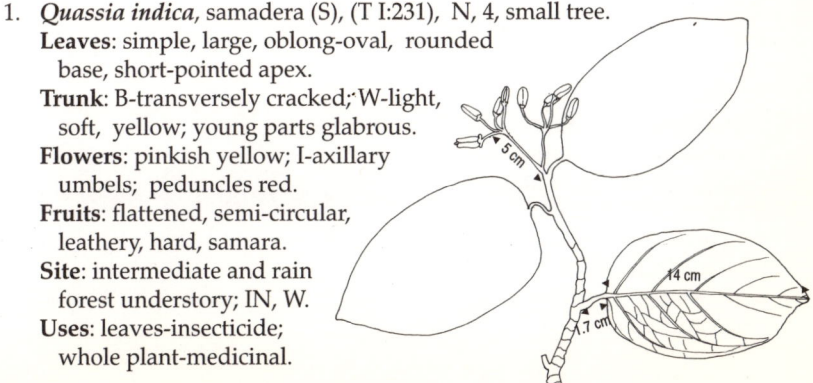

85. SOLANACEAE

FAMILY DESCRIPTION - Habit: trees, shrubs or herbs, sometimes lianas, often spiny. **Leaves**: usually spiral, simple, rarely compound. **Stipules**: absent. **Flowers**: bisexual, actinomorphic, basically cymose inflorescences. **Fruits**: berry or septicidal capsule.

FLOWER PARTS - CALYX: 5, united, persistent in fruit, valvate. COROLLA: 5, rarely 4 or 6, united, sometimes tubular, lobes convolute, imbricate or valvate, variously shaped. ANDROECIUM: stamens 5, epipetalous, usually unequal, sometimes 4 and didynamous or only 2. When less than 5, others often represented by staminodes. Anthers 2-locular, often connivent, with longitudinal slits or terminal pores. Disc around ovary. GYNOECIUM: superior, carpels 2, loculi 2, or 3-5-locular by false septa, rarely unilocular. Placentation axile or free central. Style simple. Stigma 2-lobed.

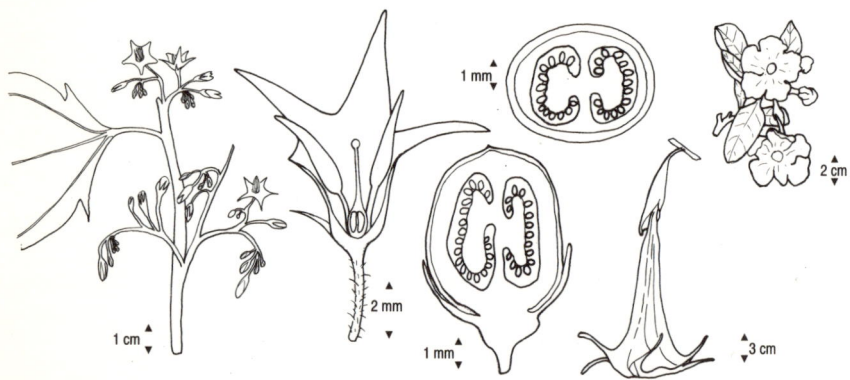

Key: inflorescence (A), half flower (B), young fruit in longitudinal (C) and transverse (D) sections of *Solanum violaceum*. Flowers of *Brugmansia* x *candida* (E) and *Brunfelsia americana* (F).

1. *Brugmansia suaveolens*, angel's trumpet (E)/ rata attana (S),
    (DF VI:401), I (S. Amer.), 3, shrub.
    Leaves: **large**, greyish green,
      thick, **velvety pubescent**.
    Trunk: B-light coloured; W-brittle.
    Flowers: white, **trumpet shaped**,
      large, pendulous, solitary.
    Fruits: fleshy elongated
      capsule.
    Site: roadsides, scrub; M, UW.
    Uses: leaves, seeds-sedative,
      narcotic; ornamental.

2. **Brunfelsia americana**, yesterday, today and tomorrow
(E), (DF VI:366), I (S. Amer.), 2, shrub.
Leaves: dark green, acute to rounded apex.
Trunk: B-light grey.
Flowers: **blue-lavender to white**
when fading, fragrant.
Fruits: yellowish berry; rarely
fruits in Sri Lanka.
Site: home gardens; W.
Uses: hedges; ornamental.

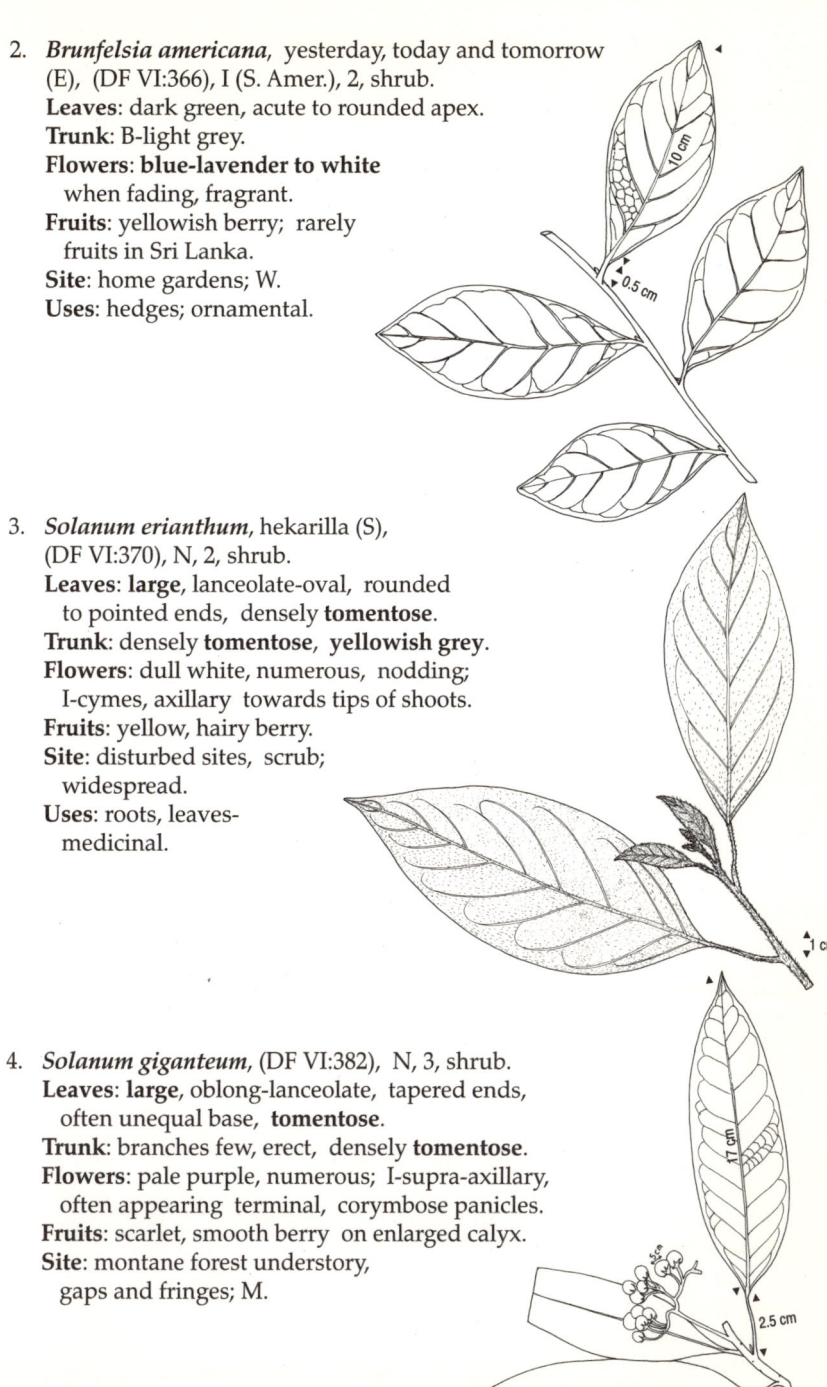

3. **Solanum erianthum**, hekarilla (S),
(DF VI:370), N, 2, shrub.
Leaves: **large**, lanceolate-oval, rounded
to pointed ends, densely **tomentose**.
Trunk: densely **tomentose**, **yellowish grey**.
Flowers: dull white, numerous, nodding;
I-cymes, axillary towards tips of shoots.
Fruits: yellow, hairy berry.
Site: disturbed sites, scrub;
widespread.
Uses: roots, leaves-
medicinal.

4. **Solanum giganteum**, (DF VI:382), N, 3, shrub.
Leaves: **large**, oblong-lanceolate, tapered ends,
often unequal base, **tomentose**.
Trunk: branches few, erect, densely **tomentose**.
Flowers: pale purple, numerous; I-supra-axillary,
often appearing terminal, corymbose panicles.
Fruits: scarlet, smooth berry on enlarged calyx.
Site: montane forest understory,
gaps and fringes; M.

85

SOLANACEAE

86. SONNERATIACEAE

FAMILY DESCRIPTION - Habit: trees. **Leaves**: opposite or whorled, simple, entire. **Stipules**: absent. **Flowers**: bisexual, actinomorphic, rather large, solitary or cymes. Hypanthium present. **Fruits**: berry or capsule, numerous seeds.

FLOWER PARTS - CALYX: 4-8 sepals, free, valvate. COROLLA: 4-8 petals, sometimes absent, free, crumpled in bud, valvate. ANDROECIUM: stamens numerous, in several whorls on hypanthium or clusters. Filaments free. Anthers reniform, versatile, opening lengthwise. GYNOECIUM: superior, 4-20 carpels and loculi. Ovules numerous per loculus, axile placentation. Style long, simple.

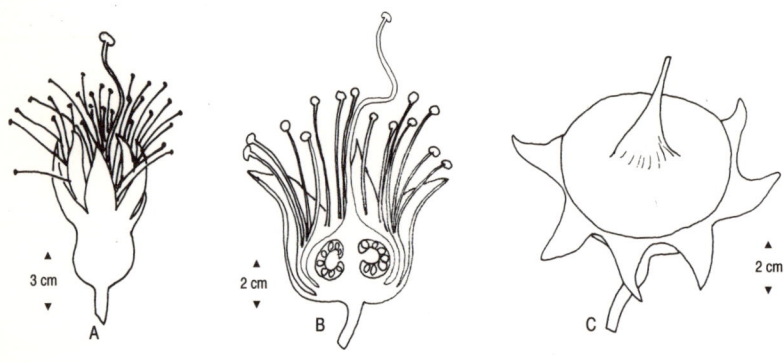

Key: full flower (A), half flower (B), and mature fruit (C) of *Sonneratia caseolaris*.

1. *Sonneratia caseolaris*, kirala (S)/ kinnai (T),
    (DF III:452), N, 12, tree.
    Leaves: oblong-ovate to roundish, base
        tapered, rounded to pointed apex,
        slightly fleshy; nearly sessile.
    Trunk: curious **upwardly growing
        breathing roots**; W-white, soft;
        twigs quadrangular.
    Flowers: red, large, terminal, solitary.
    Fruits: broadly ovoid to subglobose
        berry supported by enlarged
        cup-like calyx.
    Site: estuaries, lagoons,
        mangroves; DC, MC.
    Uses: erect roots-cork
        substitute; fruit-beverage.

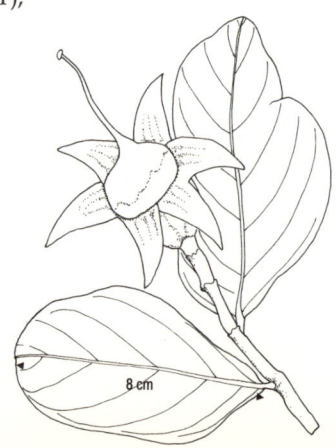

87. STAPHYLEACEAE

FAMILY DESCRIPTION - Habit: trees or shrubs. **Leaves**: opposite or spiral, trifoliolate or pinnate. Leaflets usually toothed. **Stipules**: present or absent. **Flowers**: bisexual or unisexual on dioecious plants, actinomorphic, in racemes or panicles. **Fruits**: head of follicles, drupe or berry, or inflated capsule.

FLOWER PARTS - CALYX: sepals 5, free or sometimes united, often petaloid, imbricate. COROLLA: petals 5, free, inserted on or below a hypogynous disk, imbricate. ANDROECIUM: stamens 5, inserted on hyanthium. Filaments free. Anthers 2-3-locular, opening lengthwise. GYNOECIUM: superior, 2-4 carpels and 2-4 loculi. Each loculus with 6-12 ovules in two rows. Placentation axile. Styles free or united.

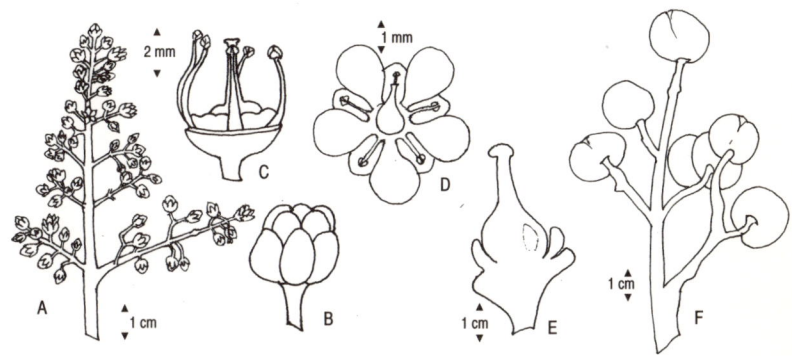

Key: infloresence (A), single flower (B), flower with calyx and corolla removed (C), full flower from above (D), longitudinal section of ovary with nectary (E) and cluster of fruits (F) of *Turpinia malabarica*.

1. *Turpinia malabarica*, eta hiriya (S), (T I:313 & VI:59), N, 4, small tree.
   Leaves: **pinnate**, pinnae 3-9, rachis striate; leaflets paired with terminal, tapered base, long-pointed apex, margins finely serrate; flush **pink-scarlet**.
   Trunk: young parts glabrous.
   Flowers: white, numerous; I-lax, axillary and terminal panicles.
   Fruits: purplish black, globose, smooth; pericarp fleshy.
   Site: montane and rain forest understory, scrub; M, LM, UW.
   Uses: W-mine props.

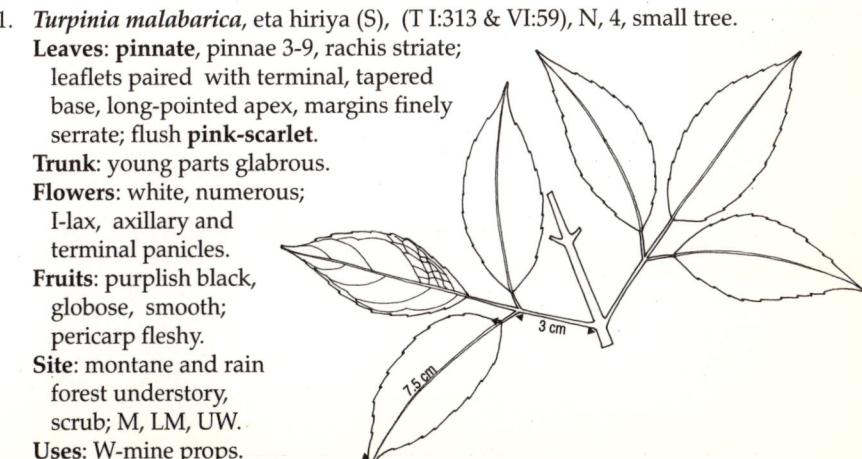

88. STERCULIACEAE

FAMILY DESCRIPTION - Habit: trees, shrubs or lianas, rarely herbs. **Leaves**: spiral or very rarely sub-opposite. Simple or compound. **Stipules**: usually present. **Flowers**: bisexual or unisexual, actinomorphic, often with epicalyx, inflorescences complex. **Fruit**: dehiscent or not, fleshy to leathery or woody, often separating as mericarps. Seeds sometimes arillate.

FLOWER PARTS - CALYX: sepals 3-5, usually basally united, with tufts of glandular hairs acting as nectaries, valvate. COROLLA: petals 5 or absent, free or adnate to staminal tube, usually clawed, contorted or imbricate. ANDROECIUM: stamens 10, 5 often staminodal and petaloid or absent, or bundles of 2-3 each. Filaments united forming a staminal tube around ovary. Anthers 2-locular, opening lengthwise or rarely by apical pores. GYNOECIUM: superior, carpels 2-5 or rarely 10-12, more or less united or reduced to 1. Ovules 2 - several per loculus. Placentation usually axile. Style simple or rarely styles free from the base.

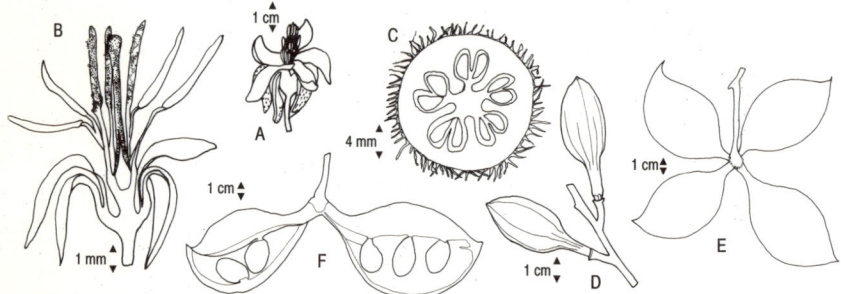

Key: full flower (A), flower in longitudinal section (B), transverse section of ovary (C), and developing fruits (D) of **Pterospermum suberifolium**. Unripe follicle (E), and dehisced follicle (F) of *Sterculia foetida*.

1. *Firmiana colorata*, kenawila (S)/malaiparutti (T), (T I:166 & VI:31), N, 15, tree.
    Leaves: **palmately lobed**; cordate base, variable, lobes long-pointed, margins entire, scattered stellate hairs; deciduous.
    Trunk: branches thick and spreading; B-very **smooth, papery, white to purplish**, shiny.
    Flowers: orange-scarlet, polyga-mous; I-terminal, densely tomentose, racemes.
    Fruits: greenish pink, leaf-like, oblong-oval, follicles, opening before maturity; seed yellow, flattened.
    Site: monsoon forest subcanopy, rock outcrops; DL.
    Uses: iB-fine fibre; ornamental.

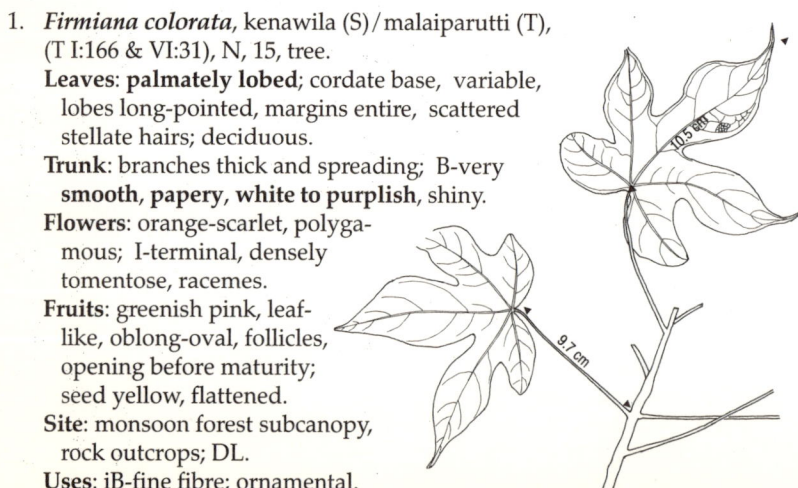

2. *Helicteres isora*, liniya (S)/vallampuri (T), (T I:168),
   N, 3, shrub to small tree.
   Leaves: 2-ranked, rounded to cordate base,
   suddenly long-pointed, sometimes 3-lobed
   apex, slightly hairy, margin **serrate**.
   Trunk: B-finely wrinkled; twigs
   rough, scattered stellate hairy.
   Flowers: crimson to yellowish
   fading blue, large axillary
   clusters, stellate hairy.
   Fruits: tapering to a point,
   stellate-scurfy, cylindrical
   follicles twisting spirally.
   Site: monsoon and intermediate
   forest understory, savannahs,
   scrub; DL, IN, W.
   Uses: iB-tough fibre; whole
   plant-medicinal; ornamental.

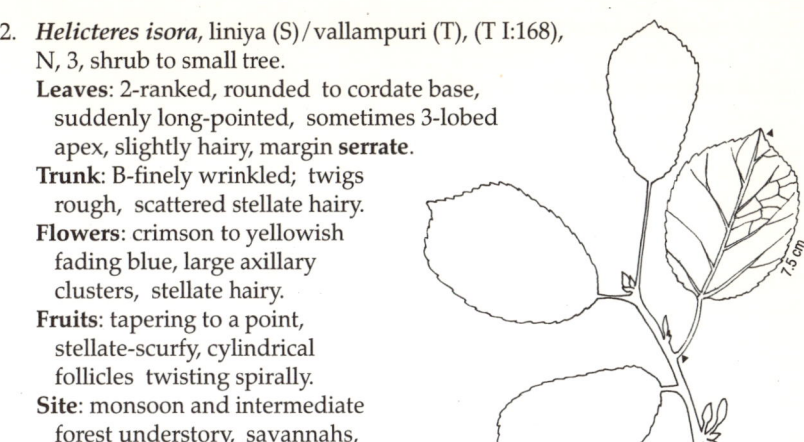

88

STERCULIACEAE

3. *Heritiera littoralis*, etuna (S)/chomuntiri (T), (T I:167), N, 20, tree.
   Leaves: lanceolate-oblong, rounded to pointed base, rounded apex,
   margin **wavy**, **minute silvery scales** beneath.
   Trunk: B-longitudinally furrowed; W-hard, dark red,
   durable; young parts with **peltate scales**.
   Flowers: pale greenish pink, small; I-much-branched,
   drooping, axillary panicles.
   Fruits: pale brown, broadly ovoid, with
   small keel or wing, thick, woody,
   smooth, shiny, indehiscent.
   Site: coastal scrub and forest; DC.
   Uses: W-heavy construction.

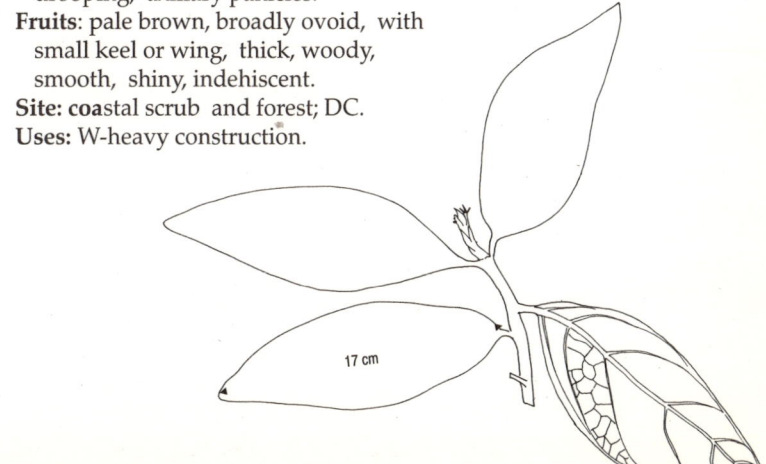

4. ***Pterospermum suberifolium***, welang (S)/vilanku (T), (T I:169 and VI:32), N, 25, tree.

Leaves: 2-ranked, obovate to oblong, rounded base, bluntly long-pointed apex, often **irregularly lobed**, unequal-sided, glabrous above, **yellowish white tomentose** beneath.

Trunk: tendency to coppice; B-thick, longitudinally cracked; W-pale red, hard; young parts tomentose.

Flowers: yellowish white, large, solitary, fragrant, on tomentose, thick pedicel.

Fruits: oblong capsule, tapering to fine point, winged, fine white tomentose.

Site: monsoon and intermediate forest canopy, scrub; DL, IN.

Uses: stem and leaves-medicinal; W-light construction.

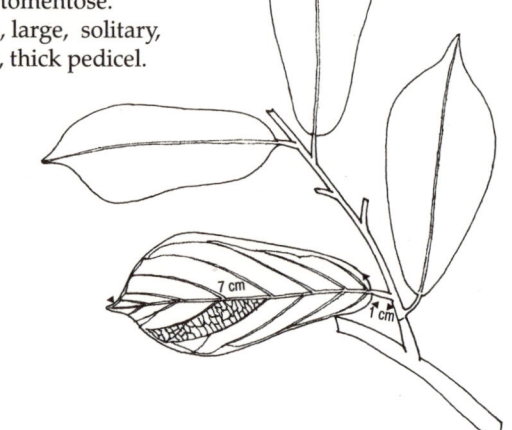

5. ***Sterculia balanghas***, nawa (S), (T I:165), N, 20, tree.

Leaves: **simple**; oval, base rounded, short-pointed apex, glabrous shiny above, paler pubescent beneath.

Trunk: B-whitish; young parts pubescent.

Flowers: green, small, crimson hairs, fragrant; I-stellate hairy panicles, slightly drooping from ends of branches.

Fruits: brilliant orange-scarlet, 4-5 downy follicles, spreading horizontally, oblong-ovoid, leathery, smooth; seeds black.

Site: intermediate forest canopy; IN.

Uses: ornamental; fruit, stem-medicinal.

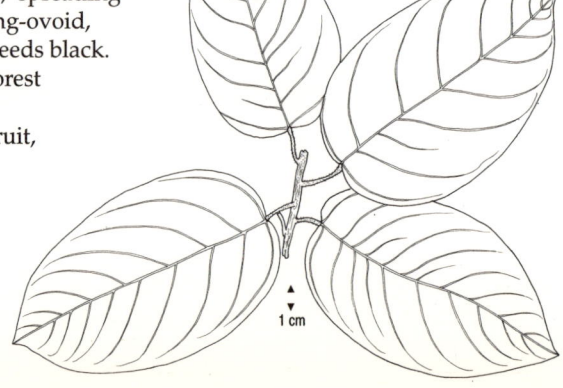

6. *Sterculia foetida,* telambu (S), (T I:164), N, 20, tree.
   Leaves: **palmately lobed**, terminally crowded;
   lobes lanceolate, tapered ends.
   Trunk: whorled, horizontal branches;
   B-whitish, thick, flaky; twigs thick,
   leaf-scarred; W-light yellowish.
   Flowers: dull orange, woolly
   pubescent, offensive odour;
   I-erect racemose panicles.
   Fruits: scarlet, 1-5 (usually 3)
   woody follicles, large,
   pendulous, pear-
   shaped, thick;
   black seeds.
   Site: monsoon and
   intermediate forest
   canopy, scrub; DL, IN.
   Uses: sap-gums, resins;
   seeds-edible.

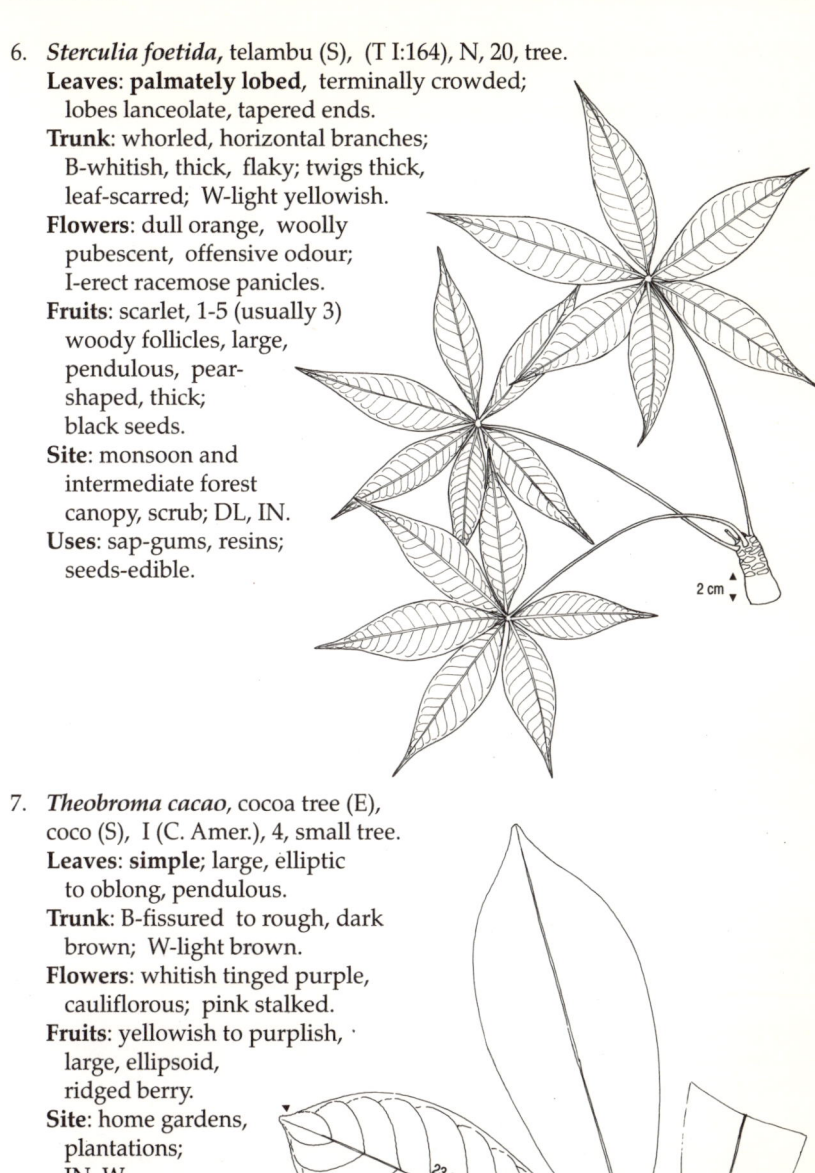

2 cm

7. *Theobroma cacao,* cocoa tree (E),
   coco (S), I (C. Amer.), 4, small tree.
   Leaves: **simple**; large, elliptic
   to oblong, pendulous.
   Trunk: B-fissured to rough, dark
   brown; W-light brown.
   Flowers: whitish tinged purple,
   cauliflorous; pink stalked.
   Fruits: yellowish to purplish,
   large, ellipsoid,
   ridged berry.
   Site: home gardens,
   plantations;
   IN, W.
   Uses: seeds-cocoa
   (beverage, chocolate).

23 cm

2 cm

89. SYMPLOCACEAE

FAMILY DESCRIPTION - **Habit**: trees or shrubs. **Leaves**: spiral, simple. **Stipules**: absent. **Flowers**: bisexual, rarely unisexual on polygamodioecious plants, actinomorphic, racemes or panicles. **Fruits**: usually 1-seeded drupe or berry, topped with a persistent calyx.

FLOWER PARTS - CALYX: sepals 3-5, united, valvate or imbricate. COROLLA: petals 3-11, more or less united, lobes imbricate, sometimes in two rows. ANDROECIUM: stamens 4-many, epipetalous. Filaments free or rarely united. Anthers globose, 2-locular, opening lengthwise. GYNOECIUM: ovary inferior, rarely half- inferior, 2-5 carpels and loculi. Ovules 2-4 per loculus. Placentation axile. Style simple. Stigma capitate, 2-5-lobed.

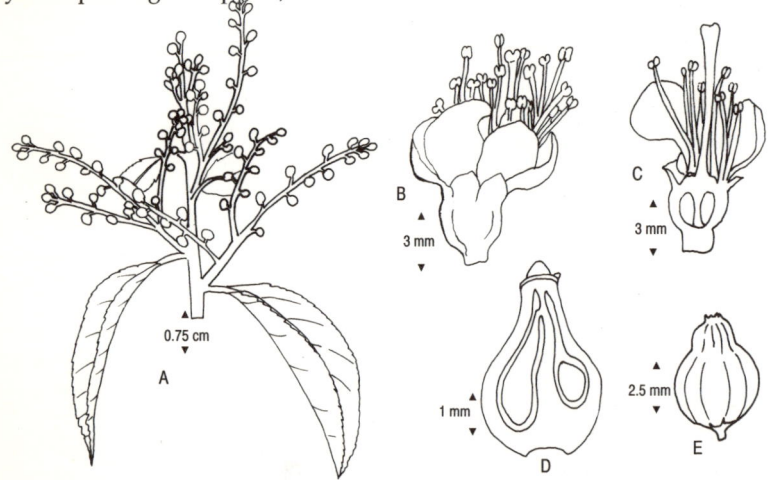

Key: inflorescence (A), full flower (B), half flower (C), longitudinal section of ovary (D) and developing fruit (E) of *Symplocos cochinchinensis*.

1. *Symplocos cochinchinensis*, bombu (S), (DF III:458), N, 5, small tree.
   Leaves: broadly to narrowly elliptic, base tapered, apex
     blunt, margins **dentate to wavy**, lateral
     veins 6-10 pairs, **glabrous**.
   Trunk: greyish, smooth.
   Flowers: white; I-compound
     spike.
   Fruits: purple drupe,
     flask-shaped, vertically
     ribbed when dry.
   Site: montane and rain
     forest understory,
     scrub; M, W.
   Uses: ornamental.

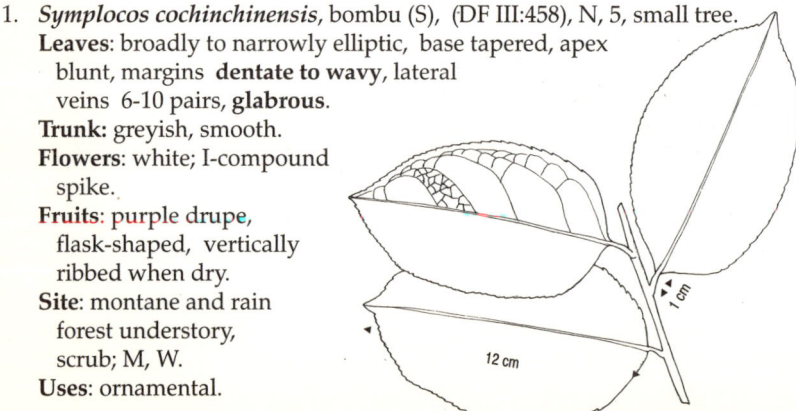

2. ***Symplocos cordifolia***, (DF III:459), E, 6, tree.
   Leaves: elliptic to obovate, base cordate to
     ear-shaped, rounded or pointed apex,
     margin **sharply dentate, strongly rolled
     down**, lateral veins 10-18 pairs.
   Trunk: twigs **glabrous**.
   Flowers: pink; I-many spikes
     terminally crowded.
   Fruits: ovoid to ellipsoid
     with long persistent calyx.
   Site: montane forest
     understory, scrub; M.

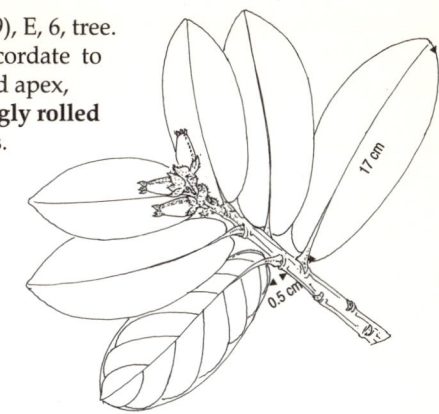

3. ***Symplocos coronata***, ugudu hal (S), (DF III:460), E, 10, tree.
   Leaves: elliptic to obovate, base tapered to rounded, short-
     pointed apex, margin nearly **entire to denticulate**,
     lateral veins 11-16 pairs, tertiary veins **transverse**
     to laterals and midrib, **pubescent** beneath.
   Trunk: twigs sparsely **adpressed
     pilose to long-pubescent**.
   Flowers: pinkish white; I-spikes
     or racemes originating beneath
     uppermost whorl of leaves.
   Fruits: ovoid to cylindrical,
     pubescent drupe.
   Site: rain forest
     understory; W.

4. ***Symplocos cuneata***, (DF III:474), E, 15, tree.
   Leaves: elliptic, tapered base, long-pointed apex, lateral
     veins 6-10 pairs, curved to form intramarginal vein,
     secondaries **transverse; adpressed hairy**.
   Trunk: twigs **adpressed hairy**.
   Flowers: pinkish-white; I-short,
     appressed hairy raceme.
   Fruits: purple, nearly
     cylindrical,
     hairy drupe,
     narrowed
     towards apex.
   Site: rain forest
     understory, gaps
     and fringes; W.

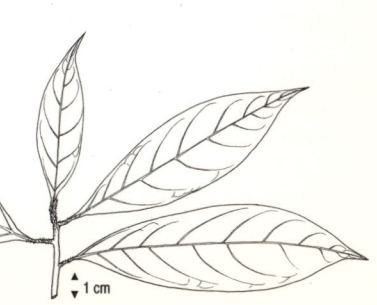

5. **Symplocos elegans** var. *elegans*, (DF III:469), E, 4, shrub to small tree.
   Leaves: ellipitic or ovate, base acute to cordate,
   apex short-pointed, margins **rolled down**,
   hairy on midrib beneath; lateral
   veins 6-10 pairs.
   Trunk: twigs **densely**
   orange tomentose.
   Flowers: white, few, sessile;
   I-small, lax, hairy racemes.
   Fruits: bluish purple
   oblong drupe.
   Site: montane forest
   understory; M.

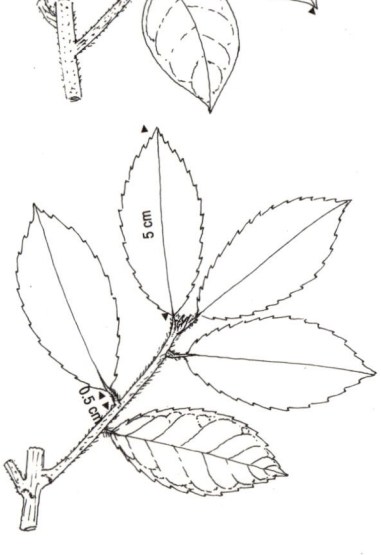

6. **Symplocos elegans** var. *hirsuta*,
   (DF III:470), E, 8, tree.
   Leaves: narrowly elliptic, ends
   pointed, margins **serrate**,
   hairy on midrib beneath,
   lateral veins **6-10** pairs.
   Trunk: twigs with long stiff **hairs**.
   Flowers: white; I-spike or raceme.
   Fruits: bluish purple, ovoid
   to ellipsoid drupe.
   Site: montane forest
   understory; M.

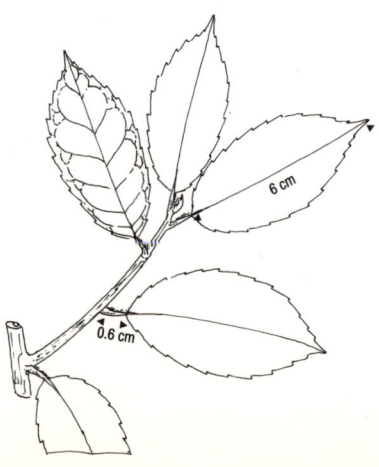

7. **Symplocos elegans** var. *minor*,
   (DF III:471), E, 3, shrub.
   Leaves: elliptic to ovate, tapering to
   cordate base, rounded to pointed
   apex, margins **serrate**, sometimes
   rolled down, lateral veins 7-11
   pairs; **pilose** beneath.
   Trunk: twigs **with long weak** hairs.
   Flowers: white; I-hairy spike.
   Fruits: bluish purple, ovoid
   to ellipsoid drupe.
   Site: montane forest
   understory; M, LM.

8. ***Symplocos macrophylla***, (DF III:477), N, 15, tree.
    Leaves: ovate to elliptic, tapered to slightly
       cordate base, pointed apex, margins often
       rolled down, entire to dentate, lateral
       veins **6-15** pairs, secondaries trans-
       verse, **densely tomentose** beneath.
    Trunk: young twigs densely
       tomentose.
    Flowers: white; I-tomentose
       raceme or spike.
    Fruits: bluish
       purple, ovoid
       to cylindrical,
       glabrous drupe.
    Site: rain forest
       understory; W.

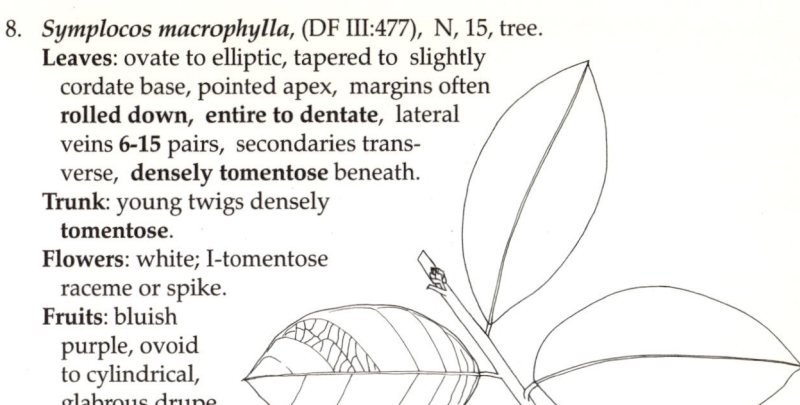

9. ***Symplocos obtusa***, (DF III:465), N, 12, tree.
    Leaves: obovate, rounded to tapered base, rarely
       short-pointed apex, lateral veins **5-10** pairs.
    Trunk: twigs **glabrous**.
    Flowers: white to yellow, fragrant;
       I-long spike or raceme.
    Fruits: bluish purple, ellipsoid
       drupe; conspicuous calyx rim.
    Site: montane forest
       subcanopy; M, LM.

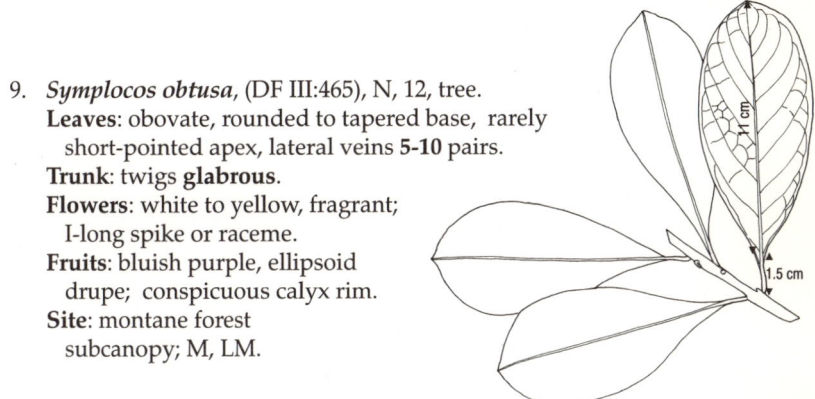

10. ***Symplocos pulchra***, (DF III:473), E, 5, small tree.
    Leaves: broadly ovate to elliptic, base rounded to cordate,
       short-pointed apex, margin **entire to serrate**, lateral
       veins **6-10** pairs; **sparsely hairy** on both surfaces.
    Trunk: young twigs **long-hairy**.
    Flowers: pinkish white; I-raceme.
    Fruits: bluish purple,
       ellipsoid to flask-shaped,
       hairy drupe.
    Site: montane and rain
       forest understory; LM, W.

90. THEACEAE

FAMILY DESCRIPTION - Habit: trees or shrubs, few lianas. **Leaves**: usually spiral, simple, entire to toothed. **Stipules**: absent. **Flowers**: usually bisexual, rarely unisexual and plants dioecious (*Eurya*), actinomorphic, usually solitary or sometimes in axillary or terminal racemes or panicles. **Fruits**: loculicidal capsule or sometimes indehiscent and fleshy berry.

FLOWER PARTS - CALYX: sepals 5, free or shortly united, much imbricate, sometimes persistent. COROLLA: petals 5, free or slightly united, imbricate, sometimes not distinguishable from calyx. ANDROECIUM: stamens numerous in several whorls. Filaments free or basally united into a ring, or in 5 bundles and adnate to corolla, often basally nectariferous. Anthers 2- locular, opening lengthwise or very rarely by a terminal pore. GYNOECIUM: superior, 3-5-carpelled and 3-5-locular. Ovules 2 or more, rarely 1, per loculus. Placentation axile, styles sometimes basally united.

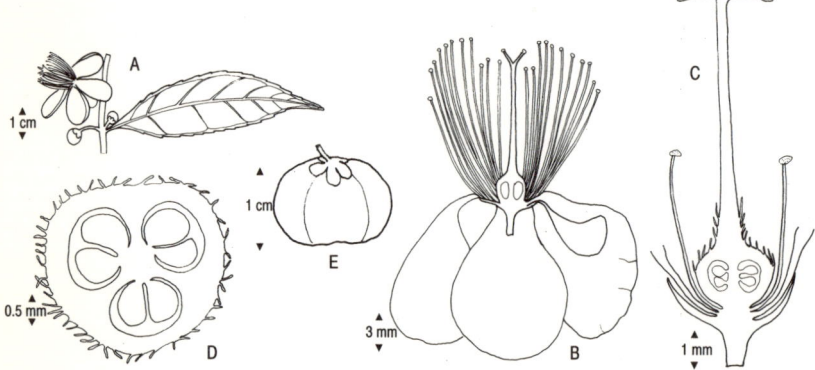

Key: full flower (A), half flower (B), longitudinal section of gynoecium (C), transverse section of ovary (D) and fruit (E) of *Camellia sinensis*.

1. *Adinandra lasiopetala*, ratu mihiriya (S), (T I:108) E, 12, tree.
   Leaves: oblong to lanceolate, base **tapered**, obtuse apex, margins entire to faintly denticulate and rolled down, minutely pubescent beneath.
   Trunk: B-dull brown to mottled grey; W-dark red-brown, hard; young parts minutely pubescent beneath.
   Flowers: white, large, solitary, axillary.
   Fruits: shiny black, globose berry, with pointed, fleshy, persistent sepals; seeds kidney-shaped.
   Site: montane forest canopy; M.
   Uses: W-light construction.

2. ***Camellia sinensis***, tea (E)/the (S,T), (T I:112),
   I (Assam), 3, shrub to small tree.
   Leaves: elliptic, tapered base, long-pointed
     apex, margins **serrate**, slightly
     pubescent beneath when young,
     bright green, stiff, leathery.
   Trunk: mostly **pruned** and
     rarely exceeds 1.5 m.
   Flowers: white to pink, solitary,
     large, conspicuous, axillary.
   Fruits: globular, lobed,
     woody capsule.
   Site: home gardens,
     plantations; IN, M, W.
   Uses: leaves-beverage; prunings-
     firewood; fruit-leech repellent.

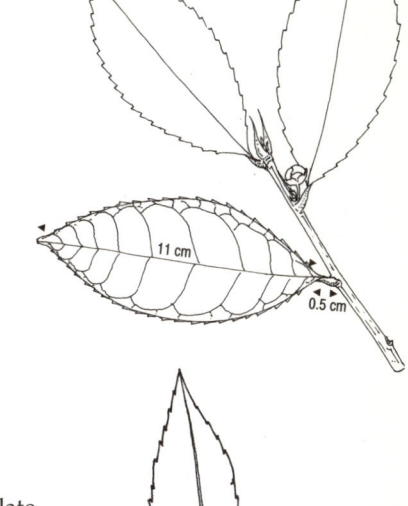

90

3. ***Eurya acuminata***, wana halu (S),
   (T I:110), N, 3, shrub.
   Leaves: narrowly oblong to lanceolate,
     long-pointed, margins serrate;
     buds **silky**.
   Trunk: twigs sparsely hairy.
   Flowers: white, 2-5 together
     in axils, small.
   Fruits: blue-black, ovoid
     berry with persistent
     style.
   Site: rain forest understory,
     fringes; W.

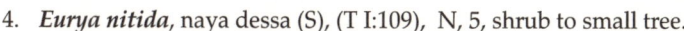

THEACEAE

4. ***Eurya nitida***, naya dessa (S), (T I:109), N, 5, shrub to small tree.
   Leaves: oval to lanceolate, pointed base, long-
     pointed but rounded apex, margin **serrate**,
     veins prominent beneath; **sessile**.
   Trunk: young twigs glabrous or hairy.
   Flowers: white, very
   small, 1-2 in leaf axils.
   Fruits: ovoid berry,
     with persistent style.
   Site: montane forest
     understory, scrub; M.

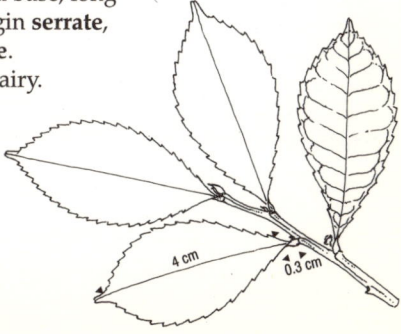

5. *Gordonia ceylanica*, mihiriya (S), (T I:110), E, 12, tree.
   Leaves: oblong to lanceolate, acute to rounded ends, margins
   entire to rolled down, beneath paler and **veinless**.
   Trunk: W-red, smooth, hard; young
   shoots **silky pubescent**.
   Flowers: white, large, axillary, solitary.
   Fruits: capsule, persistent
   sepals; seeds flattened,
   pale brown.
   Site: montane forest
   canopy; M.
   Uses: W-furniture;
   ornamental.

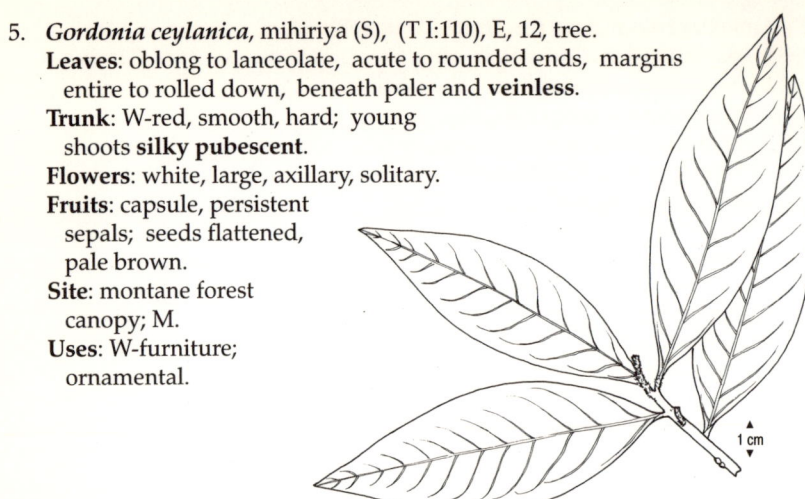

6. *Gordonia speciosa*, ratu mihiriya (S), (T I:111), E, 15, tree.
   Leaves: oblong, thick, base tapered, obtuse apex,
   margins **rolled down, purplish** and
   veinless beneath; **sessile**.
   Trunk: B-smooth; W-red, hard; twigs **glabrous**.
   Flowers: **scarlet and showy**, large, solitary.
   Fruits: ovoid, smooth, five-angled capsule.
   Site: montane and rain forest
   canopy; LM, UW.
   Uses: ornamental; B-medicinal;
   W-light construction.

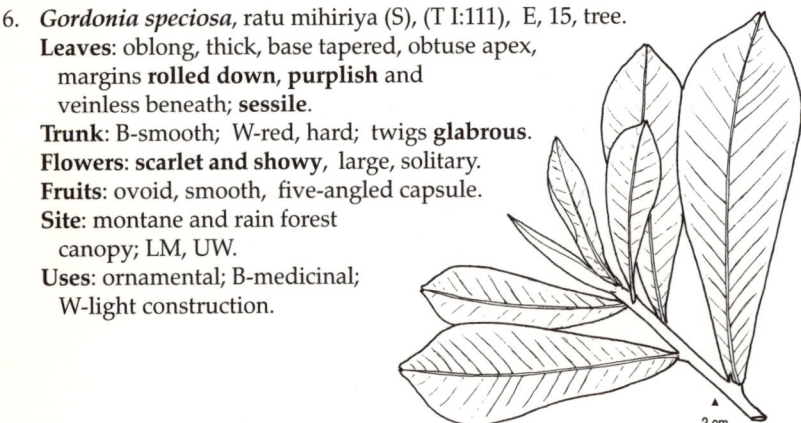

7. *Ternstroemia gymnanthera*, ratatiya (S), (T I:107), N, 15, tree.
   Leaves: closely placed, lanceolate to oval, pointed ends.
   Trunk: B-thick, soft; W-pinkish brown, durable, heavy.
   Flowers: yellow, solitary, in axils of fallen leaves.
   Fruits: ovoid, fleshy berry, persistent style and bracts.
   Site: montane forest canopy; M.
   Uses: B-masticatory; W-furniture,
   turnery.

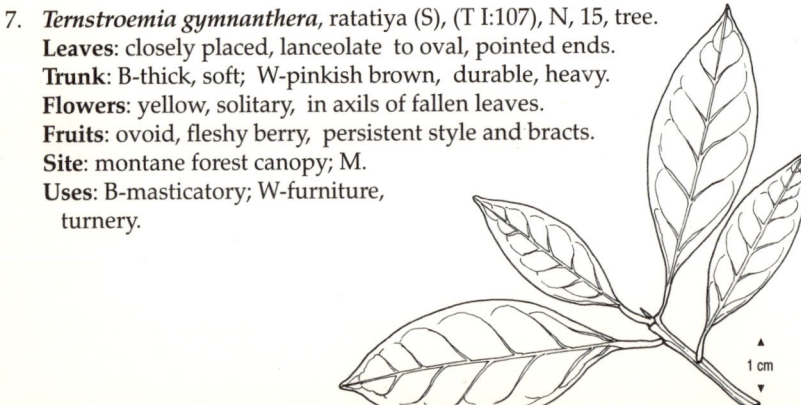

91. THYMELAEACEAE

FAMILY DESCRIPTION - Habit: trees or shrubs, rarely lianas or herbs. **Leaves**: spiral, opposite or whorled, simple, entire. **Stipules**: absent. **Flowers**: bisexual, rarely unisexual, on dioecious plants, actinomorphic or slightly zygomorphic, racemes or heads or solitary. Hypanthium present. **Fruits**: drupe or nut, rarely capsule or berry.

FLOWER PARTS - CALYX: sepals 4 or 5, petaloid, usually united into a tube with spreading lobes, imbricate or valvate. COROLLA: petals 4-12 or absent, scale-like, inserted in throat of hypanthium. ANDROECIUM: stamens 2 to many, mostly same number as calyx, sometimes in 2 whorls, more or less free, arising on hypanthium. Filaments short or absent. Anthers 2-locular, dehiscing longitudinally. Disc present around gynoecium. GYNOECIUM: superior, 1 or 2 carpels, 1 or rarely 2 loculi. Ovule solitary in each loculus. Style present or absent. Stigma discoid or capitate.

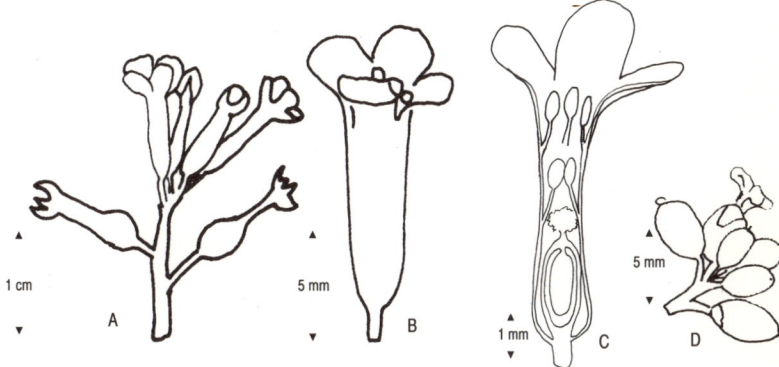

Key: inflorescence (A), full flower (B), half flower (C) and developing fruits (D) of *Wikstroemia indica*.

1. *Gnidia glauca*, naha (S), (DF II:505), N, 4, shrub to small tree.
    Leaves: linear-oblong to oblong-ovate, base tapered, pointed apex, margin entire, lateral veins fine, close, oblique; glabrous (var. *glauca*), dense adpressed hairs (var. *insularis*).
    Trunk: twigs erect, slender; glabrous (var. *glauca*), dense adpressed hairs (var. *insularis*).
    Flowers: bright yellow, drying brown; I-terminal heads, surrounded at base by large, oval, silky bracts.
    Fruits: brown, ellipsoid, adpressed hairy.
    Site: dry patana grasslands: IN.
    Uses: B-fibre, fish poison, medicine.

2. **Gyrinops walla**, walla patta (S), (DF II:510), N, 15, tree.
Leaves: oblong, tapered base, pointed apex, many faint **parallel** lateral veins, densely **adpressed hairy when young**.
Trunk: slender; B-brownish grey, thin, smooth, strongly fibrous; W-white, soft.
Flowers: yellowish white; I-umbel-like terminal heads.
Fruits: reddish brown, obovate, compressed acute capsule; seed smooth, blackish, covered with yellow hairs.
Site: rain forest understory; W.
Uses: B-easily stripped for cordage; W-decorative work; whole plant-medicinal.

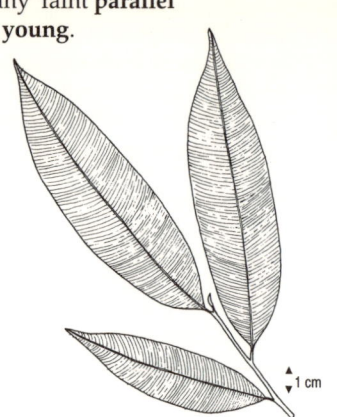

1 cm

92. TILIACEAE

FAMILY DESCRIPTION - Habit: trees or shrubs, rarely herbs. **Leaves**: usually spiral, rarely opposite, sometimes 2-ranked at maturity. Simple, often palmately veined. **Stipules**: present but deciduous. **Flowers**: bisexual or rarely unisexual on monoecious plants, actinomorphic, cymose inflorescences. **Fruits**: fleshy or dry, dehiscent or not, of various types.

FLOWER PARTS - Calyx: sepals 5, rarely 3 or 4, free or sometimes basally united, valvate. Corolla: petals 5, rarely 3 or 4, or absent, free, sometimes like the sepals, contorted or imbricate or valvate. Androecium: stamens 10 to numerous. Filaments free, or rarely united at base or form 5-10 bundles. Anthers 2-locular, opening lengthwise or by a pore. Tufts of glandular hairs serve as nectaries. Gynoecium: superior, 2-10 carpels, 2-10 loculi. Ovules 1 to several per loculus. Placentation axile. Style simple and divided at the apex. Stigmas usually as many as loculi.

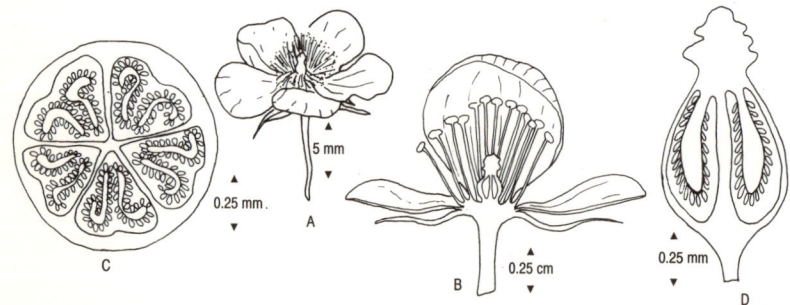

Key: full flower (A), half flower (B), transverse section of ovary (C) and longitudinal section of gynoecium (D) of **Muntingia calabura**.

1. *Berrya cordifolia*, Trincomalee wood (E)/halmilla (S)/
   chavandalai (T), (DF VII: 427), N, 20, tree.
   Leaves: ovate, cordate at base,
      pointed apex, **wavy** margins.
   Trunk: straight, erect;
      B-smooth, pale, young parts
      scaly; W-smooth, pale reddish
      yellow, tough, heavy.
   Flowers: white, pubescent;
      I-numerous, large, lax
      terminal panicles.
   Fruits: globose, pubescent capsule,
      persistent calyx, 6 long papery wings.
   Site: monsoon forest canopy, scrub; DL.
   Uses: W-boat building, heavy
      construction, casks.

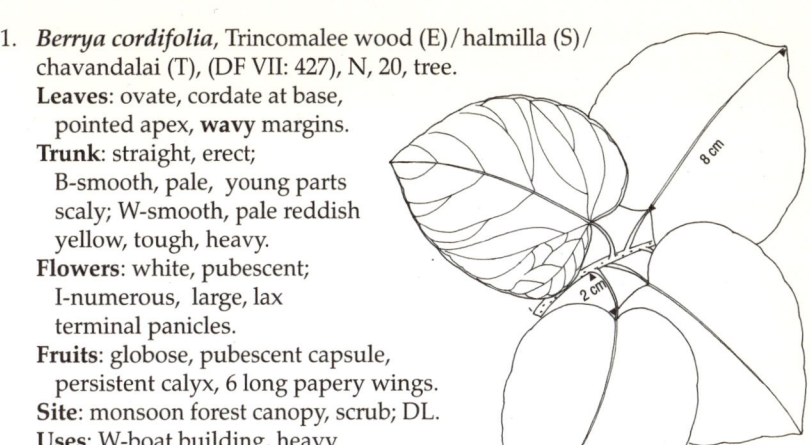

92

2. *Diplodiscus verrucosus*, dik wenna (S)/vidpani (T),
   (DF VII:429), E, 10, tree.
   Leaves: **variable**, usually ovate to oblong,
      subcordate base, rounded apex, **shallowly
      scalloped** margins, stellate-scaly above,
      pubescent and whitish beneath.
   Trunk: B-smooth; W-yellow, strong, heavy;
      Young branches rough with peltate scales.
   Flowers: pink-white, small, numerous,
      I-erect, much branched terminal
      and axillary panicles.
   Fruits: yellow, 5-angled, warty,
      pyriform capsule; persistent calyx.
   Site: monsoon forest subcanopy,
      scrub; DL.
   Uses: W-utensils, handles,
      cart axles, wheels.

TILIACEAE

3. *Grewia damine*, daminiya (S)/chadachchi (T),
   (DF VII: 405), N, 10, tree.
   Leaves: ovate to orbicular, unequal-sided to
      cordate base, pointed to rounded apex,
      serrate margins, lateral veins **3-5** at base;
      stipules ear-shaped with many veins.
   Trunk: B-pale brownish; W,h-hard, heavy, brown.
   Flowers: pale yellow, small, slender pedicels;
      I-stalked, umbellate, axillary clusters of 3.
   Fruits: slightly stellate hairy, 4-lobed.
   Site: monsoon and intermediate forest
      gaps and fringes, scrub; DL, IN.
   Uses: W-tool handles; fruit-edible; IB-fibre.

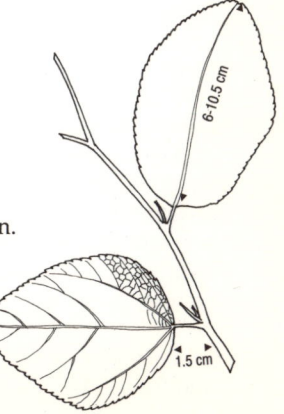

4. *Grewia orientalis*, wel keliya,wel mediya (S)/
   koditaviddai (T), (DF VII:408), N, 3, shrub.
   Leaves: oval, rounded at base, bluntly pointed apex,
   margins **serrate to dentate**, **rough** with stellate hairs.
   Trunk: much-branched, twiggy, young
   parts stellate pubescent.
   Flowers: yellowish white, stellate tomentose
   pedicel; in axillary, or supraaxillary or
   oppositifolious umbellate cymes.
   Fruits: dark yellow, subglobose,
   depressed, somewhat bristly,
   slightly 4-lobed fleshy drupe.
   Site: monsoon forest understory,
   scrub; DL.

5. *Grewia rothii*, bora daminiya (S)/Taviddai (T),
   (DF VII:404), N, 3, shrub.
   Leaves: 2-ranked, oblong-lanceolate, unequal base,
   long-pointed apex, **finely serrate** towards tip,
   3-veined at base, **white pubescent** beneath.
   Trunk: much-branched; B-smooth, whitish;
   W,h-hard, smooth, pale brown; twigs
   slender, young parts pubescent.
   Flowers: small, polygamous;
   I-umbels, 1-4, in axils.
   Fruits: globular, not lobed, hairy.
   Site: monsoon forest understory,
   scrub; DL.
   Uses: B-fibre.

6. *Microcos paniculata*, kohukirilla (S)/kapila (T),
   (DF VII:418), N, 3, shrub.
   Leaves: lanceolate, rounded base, long-pointed
   apex, **3 veined** at base, finely stellate hairy.
   Trunk: young parts pubescent.
   Flowers: white, small, nearly sessile;
   I-crowded in small terminal panicles.
   Fruits: purplish, small, ovoid,
   smooth, fleshy drupe.
   Site: monsoon, intermediate and
   rain forest understory, gaps
   and fringes; DL, IN, W.

7. *Muntingia calabura*, jam tree (E)/jam gas (S)/
   jam maram (T), (W 92), I (S. Amer.), 4, small tree.
   Leaves: **2-ranked**, lanceolate, base unequal, rounded
   to subcordate, long-pointed apex, margins
   serrate to doubly serrate, **3-veined** at
   base, **velvety pubescent** beneath.
   Trunk: much-branched, twiggy; Young
   parts pubescent and leaf-scarred.
   Flowers: white, supraaxillary,
   conspicuous.
   Fruits: red berry, globose,
   persistent stigma.
   Site: home gardens;
   widespread.
   Uses: fruit-edible;
   ornamental.

93. ULMACEAE

FAMILY DESCRIPTION - Habit: trees or shrubs. **Leaves**: usually spiral, simple, distichous. **Stipules**: present, lateral or inter-petiolar. **Flowers**: bisexual or unisexual on monoecious plants, actinomorphic or zygomorphic, solitary and axillary or cymes. **Fruits**: samara, nut or drupe.

FLOWER PARTS - Calyx: sepals 2-9, usually 5, more or less united, bell-shaped, imbricate. Corolla: absent. Androecium: stamens same number as calyx lobes or up to 15. Filaments free. Anthers 2-locular, opening lengthwise. Gynoecium: superior, carpels 2, loculus 1, with a solitary pendulous ovule. Styles 2, divergent.

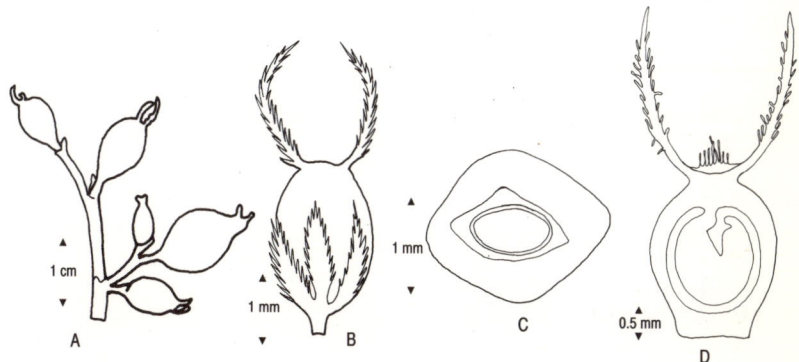

Key: inflorescence (A), female flower in full (B), transverse section of ovary (C) and longitudinal section of gynoecium (D) of *Trema orientalis*.

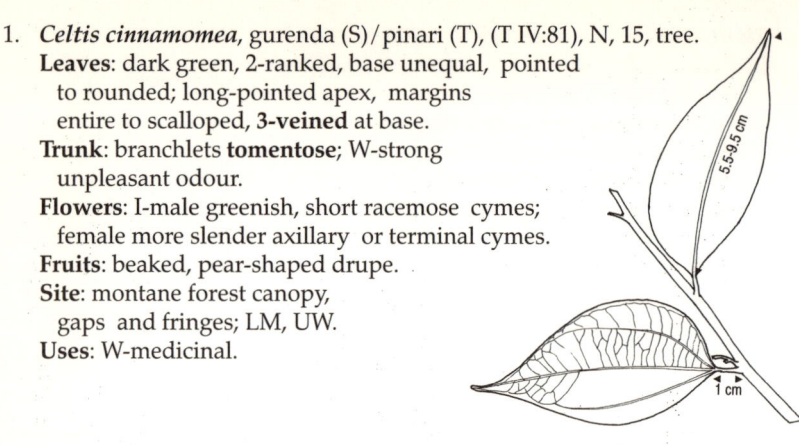

1. *Celtis cinnamomea*, gurenda (S)/pinari (T), (T IV:81), N, 15, tree.
   Leaves: dark green, 2-ranked, base unequal, pointed
        to rounded; long-pointed apex, margins
        entire to scalloped, **3-veined** at base.
   Trunk: branchlets **tomentose**; W-strong
        unpleasant odour.
   Flowers: I-male greenish, short racemose cymes;
        female more slender axillary or terminal cymes.
   Fruits: beaked, pear-shaped drupe.
   Site: montane forest canopy,
        gaps and fringes; LM, UW.
   Uses: W-medicinal.

2. *Celtis wightii*, meditella (S), (T IV:81), N, 15, tree.
   Leaves: oblong-lanceolate, base rounded to pointed,
        long-pointed apex, margins **remotely** to **coarsely**
        toothed, scalloped to **serrate, 3-veined**
        at base. Flush pink; stipules peltate.
   Trunk: young shoots puberulous.
   Flowers: yellow; I-short, hairy cymes.
   Fruits: scarlet, ovoid drupe, with obtuse tip.
   Site: intermediate forest subcanopy; IN.

3. *Gironniera scabrida*, ak mediya (S), (T IV:83), E, 10, tree.
   Leaves: oval-lanceolate to oblanceolate,
        base tapered, long-pointed apex,
        margins **serrate** towards tip,
        lateral veins **up to 10** pairs,
        bristly below.
   Trunk: branchlets scaberulous.
   Flowers: I-female solitary;
        I-male axillary cymes.
   Fruits: bristly, fleshy, ovoid drupe.
   Site: montane and rain forest
        understory; LM, UW, W.

4. *Holoptelea integrifolia,* Indian elm (E)/
   goda kirilla (S)/ ayil (T), (T IV:80),
   N, 20, tree.
   Leaves: oval to oblanceolate-oblong,
   base rounded to cordate, pointed
   apex, margin **serrate when
   young**, lateral veins **5-7** pairs.
   Trunk: branches drooping; B-ash grey,
   pustular; W-heavy, pale, strong.
   Flowers: greenish; I-short racemes
   or clusters in axils of fallen leaves.
   Fruits: broadly ovoid, flat samara,
   wings reticulately veined.
   Site: monsoon forest canopy; DL.
   Uses: ornamental; W-heavy
   construction.

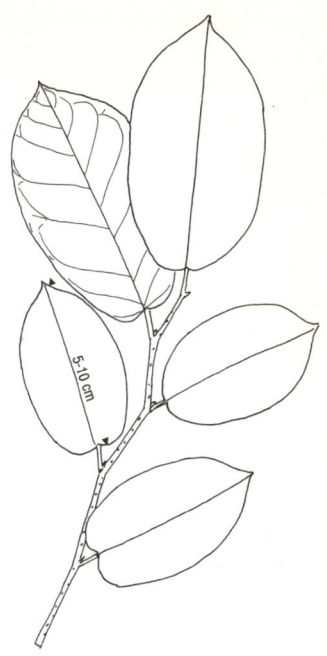
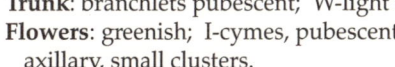

5-10 cm

93

5. *Trema orientalis,* charcoal tree (E)/ gedumba (S)/
   mini (T), (T IV:82), N, 10, tree.
   Leaves: 2-ranked, base unequal, cordate to
   rounded; pointed apex, margin **scalloped**
   to **serrate**, **3-veined** at base, **silvery
   adpressed hairs** beneath.
   Trunk: branchlets pubescent; W-light red.
   Flowers: greenish; I-cymes, pubescent,
   axillary, small clusters.
   Fruits: black, ovoid drupe.
   Site: secondary
   scrub; DL, W.
   Uses: W-charcoal,
   firewood.

ULMACEAE

94. URTICACEAE

FAMILY DESCRIPTION - Habit: shrubs, lianas, herbs and rarely trees. Some with stinging hairs. **Leaves**: spiral or opposite, simple. **Stipules**: present, rarely absent. **Flowers**: unisexual, on dioecious or monoecious plants, actinomorphic, mostly green, axillary cymes, sometimes reduced to 1 flower. **Fruits**: achene, drupe or nut, often enclosed in a persistent perianth.

FLOWER PARTS - PERIANTH: male flowers with 4-5 segments, rarely 3 or 6, united, lobes imbricate or valvate. Female flowers with 4-5 segments, free or united or absent. ANDROECIUM: stamens usually 4, occasionally 3-5. Filaments free, bent inwards in bud and springing back elastically at anthesis, releasing pollen in a sudden burst. Anthers 2-locular, dehiscing longitudinally. Gynoecium vestigial in male flowers. GYNOECIUM: superior or inferior, 1-carpellary and 1- locular. Ovule solitary and basal. Style simple. Stigma often a brush-like tuft. Staminodes often present at base of ovary in female flowers.

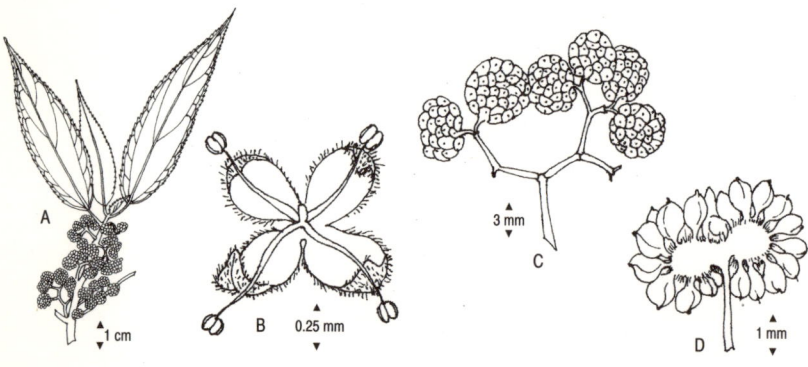

Key: inflorescence-bearing twig (A), male flower from above (B), clusters of fruits in full (C), and in section (D), of *Debregeasia longifolia*.

1. ***Boehmeria nivea***, ramie (E,S), I (India), 3, shrub to small tree.
   Leaves: base rounded, apex tapered, hairs above, densely pubescent beneath.
   Trunk: twigs densely hairy, erect; older parts glabrous.
   Flowers: yellow, small; I-dense, very hairy heads, surrounded by bracts.
   Fruits: drupe.
   Site: grasslands, scrub; IN, W.
   Uses: B-fibre.

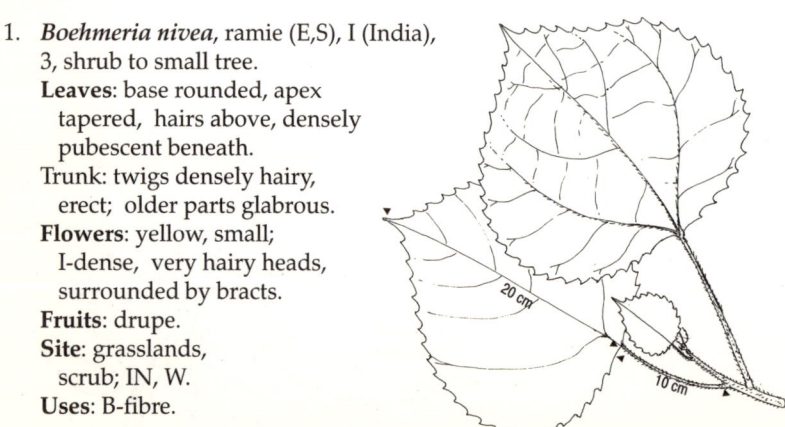

2. *Boehmeria platyphylla*, (T IV:114 & VI:269),
   N, 3, shrub.
   Leaves: mostly opposite; broadly ovate,
   long-pointed apex, upper margin
   coarsely toothed, lateral veins
   3-5 pairs, glabrous or pubescent.
   Trunk: branches soft, glabrous
   or bristly.
   Flowers: minute, pea-shaped
   clusters.
   Fruits: brittle achene, enclosed
   in pubescent perianth.
   Site: montane forest
   understory; M, LM, UW.

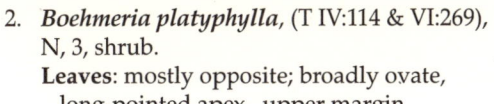

3. *Debregeasia longifolia*, gas dul (S),
   (T IV:119 & VI:271), N, 3, shrub.
   Leaves: oblong-linear-lanceolate, base tapered to
   rounded, long-pointed apex, margin **serrulate**,
   lateral veins 2-3 pairs, 3-veins at base, above
   smooth to rough, white pubescent beneath.
   Trunk: branches pubescent.
   Flowers: orange, I-dichotomous
   cymes with divaricate branches.
   Fruits: bright orange-yellow,
   obovoid to pear shaped.
   Site: montane forest
   understory, gaps and
   fringes; M, UW.
   Uses: B-fibre; ornamental.

4. *Villebrunea integrifolia*, (T IV:118), N, 7, tree.
   Leaves: oval to oblong, base rounded to pointed; long-
   pointed apex, margin entire to scalloped, lateral veins
   5-8 pairs, obliquely ascending, **membranous**.
   Trunk: branches pubescent.
   Flowers: minute, sessile;
   I-crowded, axillary,
   forked cymes.
   Fruits: enclosed in persistent,
   fleshy bracts.
   Site: montane forest
   understory, gaps and
   fringes; LM, UW.
   Uses: B-fibre.

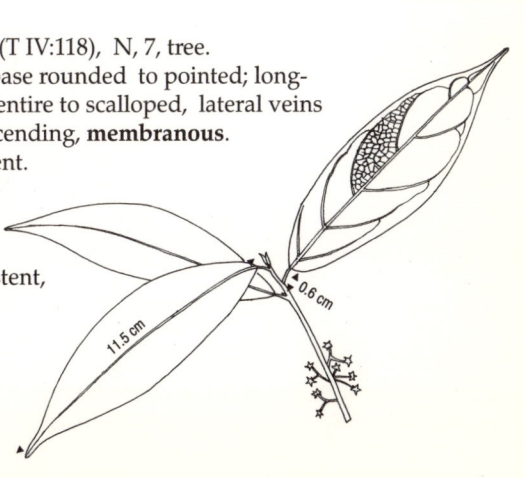

95. VERBENACEAE

FAMILY DESCRIPTION - Habit: trees, shrubs, lianas or herbs. Young twigs often quadrangular. **Leaves**: usually opposite or whorled, simple or compound. **Stipules**: absent. **Flowers**: usually bisexual, more or less zygomorphic, racemes, cymes or heads. **Fruits**: generally a drupe with as many pyrenes as ovules or a head of 1-seeded nutlets or a 4-valved capsule.

FLOWER PARTS - Calyx: typically 5, rarely 4, or 6-8, united, persistent. CorollA: typically 5, rarely 4 or 6-8, united, 2-lipped, often with slender tube, lobes imbricate. ANDROECIUM: stamens 4, didynamous, or rarely 2 or 5, epipetalous. Filaments free. Anthers 2-locular, loculi often divergent, opening lengthwise. Staminodes sometimes present. GYNOECIUM: superior, carpels mostly 2, rarely 4 or 5, loculi as many as carpels or twice as many by false septation. Ovule solitary in each loculus. Placentation axile. Style 1, stigma lobed and as many lobes as carpels.

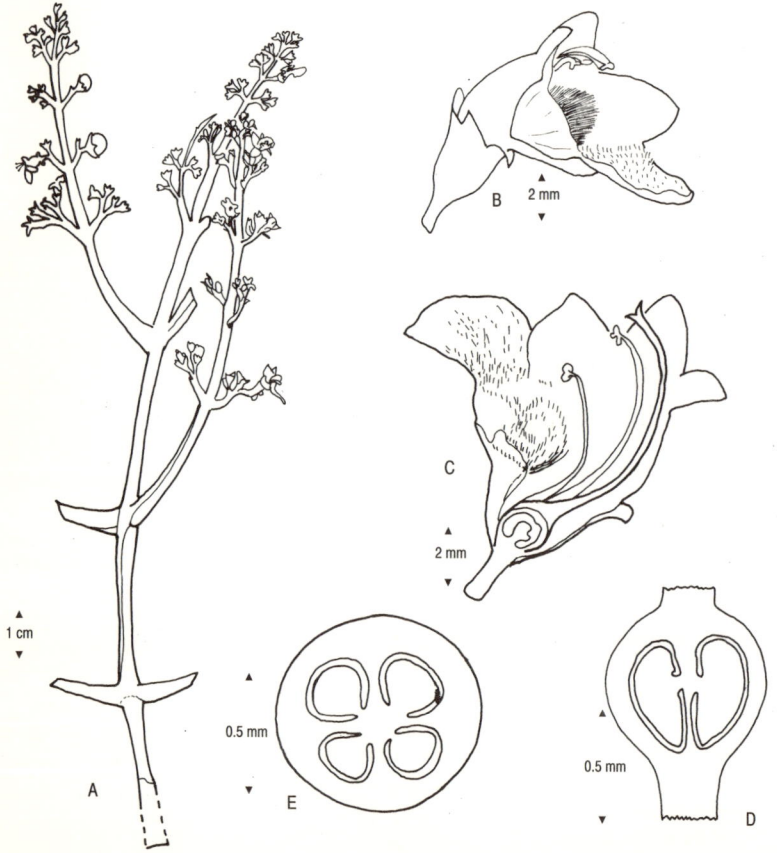

Key: inflorescence (A), full flower (B), half flower (C), ovary in longitudinal (D), and transverse (E) sections of *Vitex altissima*.

1. *Callicarpa tomentosa*, illa (S)/koat komal (T), (DF IV:299),
   N, 5, shrub to small tree.
   Leaves: opposite; crowded, elliptic to lanceolate, base tapered
      to cordate; long-pointed apex, occasionally minutely dentate,
      dark green above, grey below, **tomentose**.
   Trunk: straggly; B-grey to light brown, thin, rough, **corky**,
      aromatic, bitter; W-white to brownish; young
      branches stout, stellate **yellow hair**.
   Flowers: pink to reddish purple, many, tomentose,
      fragrant; I-spreading, dichotomous, axillary cymes.
   Fruits: green to black, globose, smooth, shiny, drupe.
   Site: homegardens, roadsides, scrub; widespread.
   Uses: B-substitute for betel; roots-medicinal;
      W-fuelwood; ornamental.

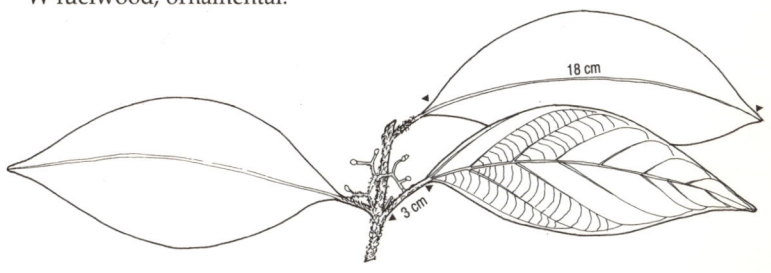

2. *Clerodendrum inerme*, wal gurenda (S)/shen gan kuppi (T),
   (DF IV:447), N, 3, shrub.
   Leaves: opposite, elliptic to lanceolate, pointed
      base, short-pointed to blunt apex,
      often succulent, **aromatic**.
   Trunk: B-dark brown to whitish
      grey, fissured; branchlets often
      angled, arcuate.
   Flowers: white, few to solitary,
      fragrant; I-lax, axillary cymes.
   Fruits: obovate, smooth, dry,
      turning black brown,
      spongy exocarp,
      separating into
      4 corky pyrenes.
   Site: along marshes,
      tidal rivers; MC, DC.
   Uses: leaves, stem, roots,
      seeds-medicinal;
      flowers-perfume;
      twigs-basket making.

3. *Clerodendrum infortunatum*, pinna (S)/
    vata madakki (T), (DF IV:461), N, 2, shrub.
    Leaves: opposite, ovate, base truncate or cordate,
        short-pointed apex, **densely pubescent**.
    Trunk: branchlets slender,
        4-angled, densely pubescent.
    Flowers: white, densely pubescent;
        I-many flowered, terminal,
        paniculate cymes.
    Fruits: violet-black, subglobose,
        shiny, drupes, scarlet pedicels,
        persistent dark red calyx.
    Site: rain forest gaps and
        fringes, scrub; LM, W.
    Uses: whole plant-medicinal.

4. *Clerodendrum paniculatum*, pagoda flower (E)/
    pinna (S), (DF IV:412), N, 2, shrub.
    Leaves: opposite, membranous, ovate, cordate
        base, 3-7-lobed, short-pointed apex,
        margins dentate, scaly beneath.
    Trunk: twigs 4-angled, nodes hairy.
    Flowers: orange-red to scarlet; I-large,
        terminal and axillary panicles.
    Fruits: greenish blue to black drupe.
    Site: scrub; widespread.
    Uses: ornamental; young
        twigs-medicinal.

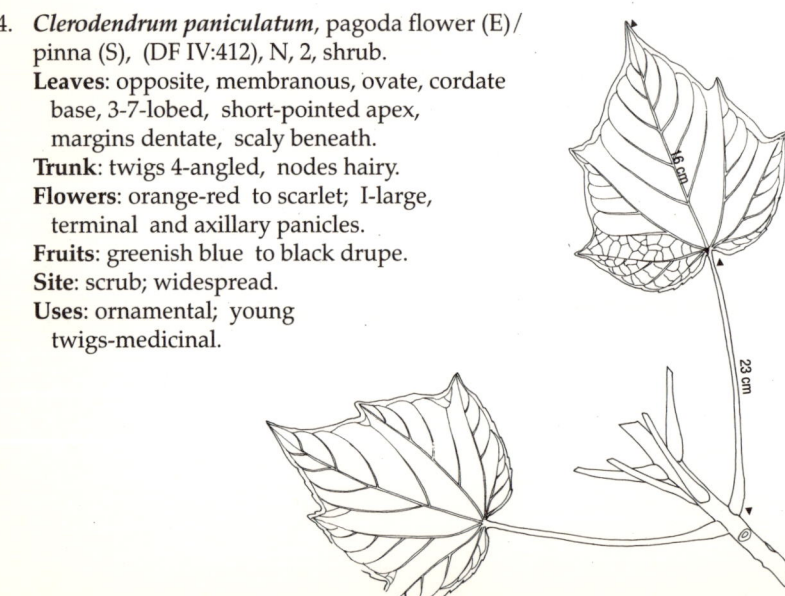

5. *Clerodendrum serratum*, kan henda (S) / siri tekku (T),
   (DF IV:417), N, 4, shrub to small tree.
   Leaves: opposite, elliptic to oblong, short-pointed apex,
      margins **serrate**, **densely white** tomentose.
   Trunk: branches relatively stout, purplish buff.
   Flowers: purplish-pink, hairy; I-terminal
      and axillary cymes.
   Fruits: green to purple, subglobose,
      somewhat succulent, 2-4-lobed,
      drupe; calyx cup-like.
   Site: intermediate and montane forest
      gaps and fringes, grasslands; M, IN.
   Uses: leaves-vegetable; roots,
      seeds-medicinal.

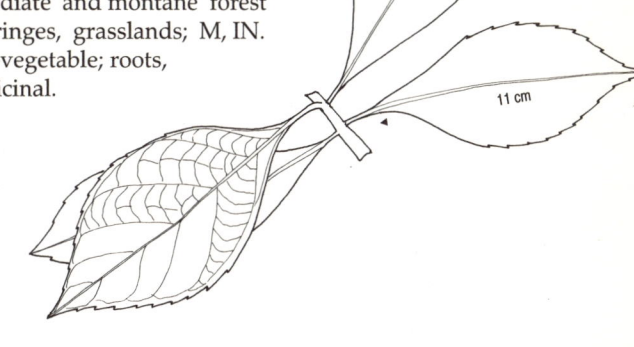

95

VERBENACEAE

6. *Duranta erecta*, pigeon berry (E), (DF IV:278),
   I (C. & S. Amer.), 7, shrub to tree.
   Leaves: opposite; variable in shape and size, ovate
      to elliptic to obovate, base tapered, margins
      occasionally dentate; (variegated in var. **alba**).
   Trunk: extremely variable; branches arching or
      drooping; unarmed or spiny, 4-angled.
   Flowers: blue to violet or white, fragrant,
      many; I-terminal and axillary racemes.
   Fruits: yellow to orange, globose.
   Site: home gardens; widespread.
   Uses: hedges, ornamental;
      whole plant-medicinal.

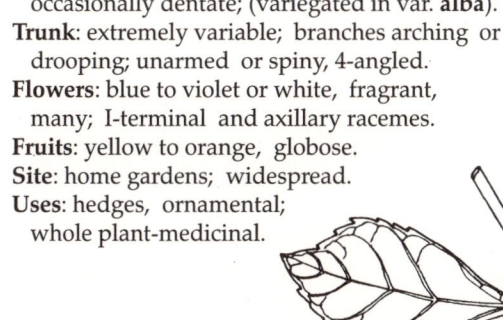

7. *Gmelina arborea*, eth demata (S), umi (T), (DF IV:390), I (S. Asia), 15, tree.
   Leaves: opposite, broadly ovate, base truncate; long-pointed
   apex, margins of young leaves **toothed or lobed**,
   lateral veins **2-10** pairs, densely tomentose.
   Trunk: B-smooth, pale ashy grey-yellow
   with blackish patches, flaking as
   woody plates; IB-pale orange;
   W-light, grey to reddish brown.
   Flowers: yellow to brownish, tomentose;
   I-terminal and axillary cymose panicles,
   tomentose; bracts linear-lanceolate.
   Fruits: ovoid, aromatic drupes;
   exocarp succulent; calyx
   orange-yellow.
   Site: home gardens,
   plantations; DL, IN.
   Uses: fruit-dye, edible;
   roots, stems, leaves-
   medicinal; W-durable
   construction, utensils;
   ornamental.

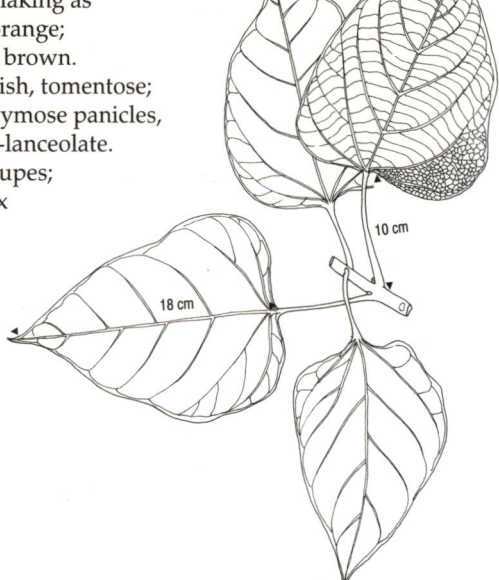

8. *Lantana camara*, lantana (E)/ganda pana (S),
   (DF IV:220), I (W. Indies), 2, shrub.
   Leaves: opposite, ovate to oblong,
   margin serrate to scalloped,
   sparsely brown pilose beneath.
   Trunk: slightly prickly.
   Flowers: yellow to rose
   pink-scarlet; I-axillary heads.
   Fruits: purple to black,
   fleshy drupes.
   Site: naturalised, scrub,
   roadsides, home gardens;
   widespread.
   Uses: ornamental; fruits-edible;
   whole plant-medicinal.

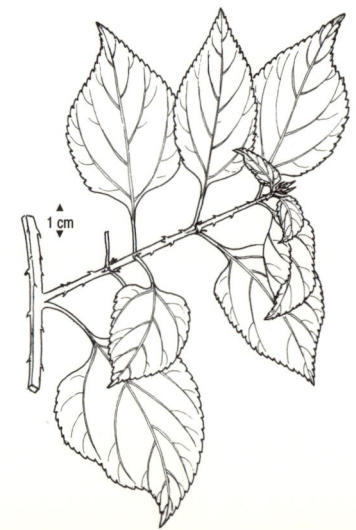

9. **Premna latifolia**, maha midi (S), (DF IV:317), N, 4-5, shrub to small tree.
   Leaves: opposite to alternate; elliptic to lanceolate, cordate to
   rounded base; petioles **rusty pubescent**; deciduous.
   Trunk: woody **spurs**; W-grey with yellow- green-purple
   streaks; twigs rusty pubescent; old
   stems sometimes spiny.
   Flowers: white to creamy-white,
   curry-like odour; I-axillary and
   terminal many-flowered cymes.
   Fruits: subglobose
   fleshy drupes.
   Site: monsoon forest
   understory, scrub;
   DL, IN.
   Uses: leaves-vegetable,
   fodder; W-fuelwood;
   whole plant-medicinal.

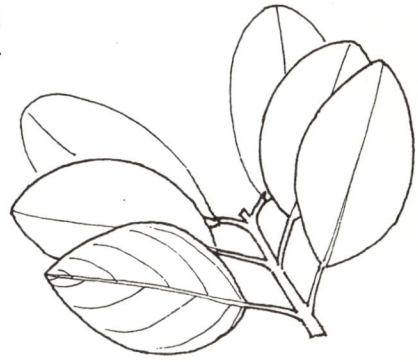

95

VERBENACEAE

10. **Premna serratifolia**, midi (S)/erumaimulla (T),
    (DF IV:334), N, 8, shrub to tree.
    Leaves: opposite; variable, elliptic to subovate,
    margin entire to **scalloped to serrate**,
    lateral veins 3-5 pairs; leaf scars
    prominent, corky; deciduous.
    Trunk: W-very **large medullary rays**.
    Flowers: yellowish white
    to greenish, small,
    unpleasantly aromatic;
    I-terminal panicles.
    Fruits: subglobose drupes.
    Site: scrub, roadsides; DL.
    Uses: whole plant-
    medicinal; W-carving.

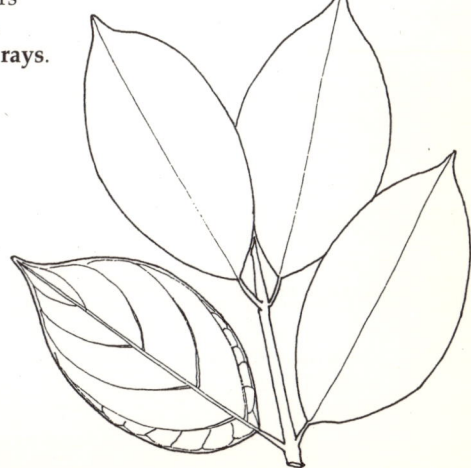

11. *Premna tomentosa*, bu seru (S), koluk kutti (T);
(DF IV:329), N, 8, shrub to tree.

Leaves: opposite, yellow-green, ovate to ovate oblong, base acute, subtruncate or cordate, apex acute to caudate, lateral veins 5-7 pairs, intercostals **numerous and parallel**; deciduous.

Trunk: often crooked, basally **fluted**; B-grey to yellow to light brown, longitudinally fissured, **shaggy and peeling**; W-hard; branchlets densley brownish puberulent, 4-sided.

Flowers: white to greenish yellow, fragrant, stellate tomentose; I-many-flowered terminal and axillary cymes.

Fruits: green to blackish purple, stellate hairy, subglobose drupe; calyx shallow, cup-like.

Site: monsoon and intermediate forest understory, scrub; DL, IN.

Uses: W-turnery, carving, furniture, fuelwood; leaves, roots-medicinal.

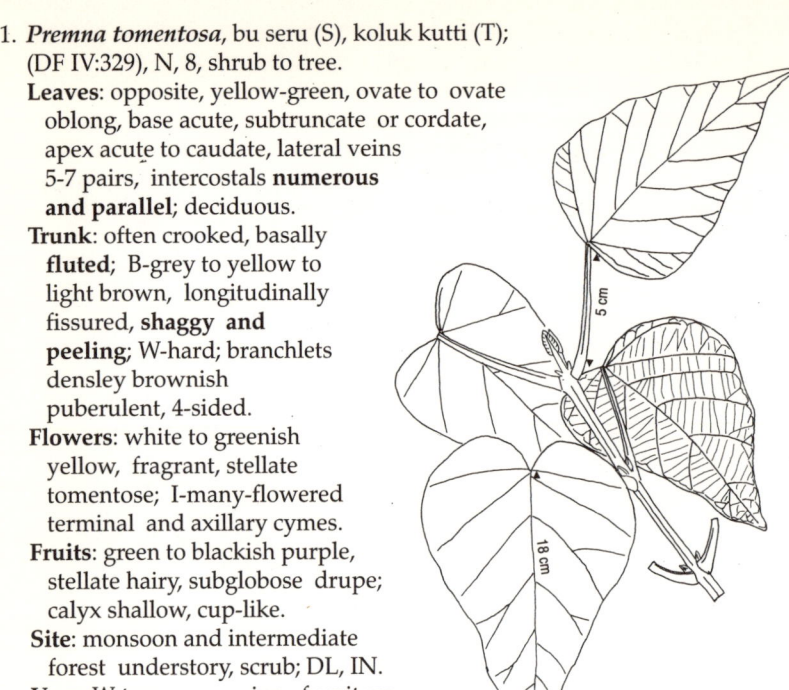

12. *Tectona grandis*, teak (E)/thekka (S)/tekku (T),
(DF IV:304), I (India), 25, tree.

Leaves: opposite, **large**, broadly elliptic, short-pointed, green and shiny above, silvery beneath; deciduous; flush has **orange-red** juice when crushed.

Trunk: B-grey, **fibrous**, peeling; W,s-yellowish; h-yellowish brown, hard, durable; twigs 4-angled.

Flowers: white, minute; I-large axillary and terminal panicles.

Fruits: globular, 4-angled drupe enclosed in calyx.

Site: plantations, home gardens; DL, IN, W.

Uses: W-furniture, heavy construction, veneer; whole plant-medicinal.

13. *Vitex altissima*, milla (S)/kada manakku (T), (DF IV:353), N, 20, tree.
    Leaves: opposite, **trifoliolate**; leaflets subequal, base pointed, elliptic,
    apex long-pointed, pubescent when young; young leaf petiole winged.
    Trunk: W-grey with yellow-brown, hard; branches **drooping**;
    branchlets 4-sided, densely **pubescent**.
    Flowers: white to bluish purple;
    I-axillary to terminal panicles.
    Fruits: greenish to black,
    drupe with enlarged calyx.
    Site: monsoon and interme-
    diate forest canopy and
    subcanopy, rain forest
    gaps and fringes;
    DL, IN, W.
    Uses: B-medicinal;
    W-furniture, turnery,
    construction, resistant
    to termites.

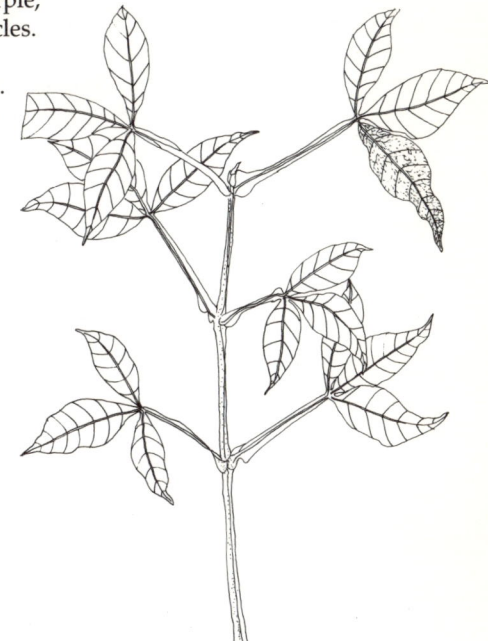

95

VERBENACEAE

14. *Vitex leucoxylon*, nebada (S)/nir nochi (T), (DF IV:365), N, 15, tree.
    Leaves: opposite, leaflets **3-5**, oblong to elliptic, base unequal.
    Trunk: B-white to light grey, smooth, striated; W-dark grey
    to purple-brown, hard, durable; branches twisted, fluted;
    twigs minutely pubescent.
    Flowers: purplish, yellow to brown pubescence; I-lax,
    dichotomous, spreading branches, axillary cymes.
    Fruits: globose to elliptic, green to purple-black drupes.
    Site: monsoon and intermediate forest
    canopy; along waterways
    and tanks; DL, IN.
    Uses: fruit-fish poison;
    leaves, root-
    medicinal;
    W-cart wheels.

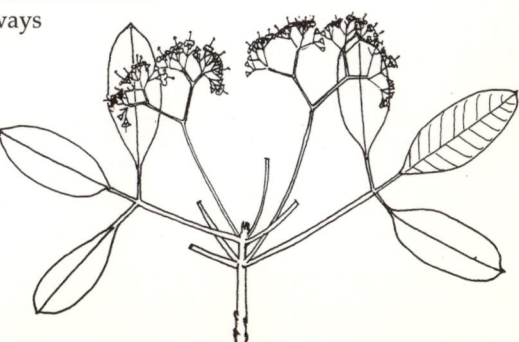

15. *Vitex negundo*, nika (S)/vernochchi (T), (DF IV:373), N, 4, small tree.
   Leaves: opposite, leaflets **3-5**, oblong to elliptic, long-pointed apex, margins entire to wavy, sometimes dentate, **brownish hairy**, **aromatic**; deciduous.
   Trunk: B-pale reddish brown, peeling, papery; W-grey to white, hard.
   Flowers: pale blue, lax, opposite cymes, arranged in terminal panicles.
   Fruits: purple to pink, globose, glabrous, exocarp succulent.
   Site: home gardens; stream banks; DL, IN.
   Uses: leaves-insect repellent; whole plant-medicinal.

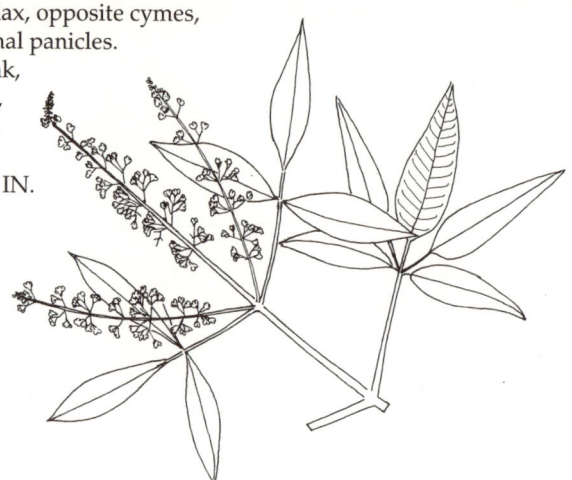

16. *Vitex trifolia*, sudu nika (S)/vettai nochi (T), (DF IV:378), N, 5, small tree.
   Leaves: opposite, usually **trifoliolate**; leaflets subequal, oblong to elliptic, short-pointed; **resinous**, **aromatic**.
   Trunk: B-smooth, brown to light brown; W,s-whitish yellow, soft; h-brownish, hard, brittle; twigs 4-angled.
   Flowers: pale blue to purple; I-cymose panicles with leaf-like bracts.
   Fruits: green-yellow to black, globose to ovoid, densely hairy drupes, rim regularly 5-dentate.
   Site: monsoon and intermediate forest understory, scrub; MC, DL, IN.
   Uses: fruit, leaves-medicinal, insect repellent.

Part III:

Plant uses

Uses of trees in Sri Lanka's history

The forest is a peculiar organism with boundless compassion and endless life-giving qualities that makes no demand for its sustenance and extends generously the products of its life activity; it affords protection to all beings, offering shade even to the axeman who destroys it.—Gautama Buddha.

In Sri Lankan society tradition gives pride of place to trees which epitomize the religious ideal of infinite benevolence towards all beings. Gautama Buddha attained Nirvana under a *Ficus religiosa* tree. In appreciation of the shade that it provided it is said that the Lord Buddha remained in contemplation with his eyes fixed on this sacred tree. He said 'He who worships the Bo tree (*Ficus religiosa*) will receive the same reward as if he worshipped me in person'. The sacred Bo tree was thought to have been brought to Sri Lanka in the 3rd century B.C. as a sapling from the tree under which Gautama Buddha attained enlightenment. It was planted in Maha Meuna Park in Anuradhapura and is believed to be the oldest recorded tree in Sri Lankan history.

Each one of the past twenty four Buddhas attained enlightenment under a tree. Buddhists believe that the next Buddha will attain enlightenment under the Na tree (*Mesua nagassarium*). This is now the national tree of Sri Lanka. Both Buddhists and Hindus have a strong belief that large trees (e.g. Bo, Na) are the abodes of deities or in some instances demons. Tree worship is still continued as a ritual in Sri Lankan society. Favors granted by tree deities are acknowledged by lighting oil lamps around the tree, pouring water or milk at its base, or hanging flags on its branches.

The historical chronicle *Mahawansa* written in the 5th century A.D. described the parks and forests of the ancient cities. Foremost among them was the layout of Anuradhapura which rivalled the modern town planning and landscaping of today. Forests, orchards and shaded roadways were an integral part of royal parks, as described by the Chinese traveler Fa Hsien in the 12th century A.D. The patricidal King Kasyapa (5th century A.D.) built his citadel in Sigiriya and surrounded it with one of the world's most spectacular water gardens. Buddhist monasteries, known as *aramas* (meaning leisure gardens), were also significant places for tree plantings and landscaping. Trees especially connected to temples are not usually cultivated in home gardens as they would be considered unlucky. The timber of certain trees such as Na is used exclusively for temple construction and is taboo for secular use. Also Sri Lankan arts and crafts have a long tradition of tree use. Two examples are Kandyan necklaces made of representations of flower petals or fruits. Beads were made from the dried seeds of *Terminalia belirica* and *Phyllanthus emblica*. It has also been a tradition to associate trees with place names e.g. Meegoda, Horagolla and Nagoda. Tree names are also the surnames of people.

Trees in this guide that have timber uses for various crafts, industries and construction have been listed below. This list provides the family name and number for general location of species description(s) in the field guide. For certain commercial species that have been investigated by the State Timber Corporation a measure of wood density is given in kg/m^3 (kilograms per cubic meter) followed by a class code. Timber class codes for general purposes are given in descending order of rank: SL—super luxury; L—luxury; SC—special class; CI—class I; CII—class II; CIII—class III. In cases where wood has been evaluated by the State Timber Corporation for plywood quality, class codes have also been given, again in descending order of rank: class I—PI; class II—PII; class III—PIII. Lastly code letters are provided for woods with particular known timber uses (see Table 1).

TABLE 1
Code letters for timber uses

Ba - Basketry and wickerwork
Bo - Boat construction and accessories
Ca - Casks, barrels and cart wheels
Cv - Carving and ornament
Fl - Flooring
Fr - Framery of windows and doors
Fu - Furniture and cabinets
Fw - Firewood
Im - Implements, tools and turnery
Lc - Light construction, scaffolding and poles
Mi - Musical instruments
Pa - Panelling and ceiling boards
Pc - Packing cases and chests
Pl - Planking and reepers
Ps - Posts, pit props and bridge beams
Pw - Plywood
Ra - Rafters, beams and joists
Sh - Roof shingles
Sl - Sleepers, carriage and truck bodies
Tp - Toothpicks, pencils and matches
Ut - Shuttles, toys, bobbins and utensils
W - Piling and dockyards

3. ARAUCARIACEAE
Agathis robusta - Fu, Lc, Pc, Pl, Pw, Ut.
Araucaria bidwillii - Lc, Pc, Pl, Pw.
Araucaria cookii - Lc, Pc, Pl, Pw.
Araucaria cunninghamii - Lc, Pc, Pl, Pw.

12. ANACARDIACEAE
Campnosperma zeylanicum - 544; CIII; PIII; Pa.
Lannea coromandelica - 576-768; PIII; Fu, Pl.
Mangifera indica - 624; PIII; Fr, Ra.
Mangifera zeylanica - CIII; PIII; Fr, Lc, Ra.

13. ANNONACEAE
Cyathocalyx zeylanica - PIII; Ut.

14. APOCYNACEAE
Alstonia macrophylla - 640; CIII; PIII; Fr, Pa, Pc.
Alstonia scholaris - 400; CIII; Fu, Fr, Pc.
Pagiantha dichotoma - Cv.

19. AVICENNIACEAE
Avicennia marina - Bo, Fw, W.
Avicennia officinalis - Bo, Fw, W.

20. BIGNONIACEAE
Stereospermum colais - 720; Pl, Ra.
Tabebuia rosea - Fu, Fw, Mi, Ra, Sl.

22. BOMBACACEAE
Bombax ceiba - 320; Pa, Pc, Tp.
Cullenia ceylanica - 512-608; CIII; PIII; Fu, Pa, Pl.
Cullenia rosayroana - 512-608; Fu, Pa,Pl.
Durio zibethinus - 512-608; Pl.

24. BURSERACEAE
Canarium zeylanicum - 432; PIII; Pl.

26. CAPPARIDACEAE
Crateva adansonii - 672; Ut.

28. CASUARINACEAE
Casuarina equisetifolia - 960; Fw, Lc, Ps.

29. CELASTRACEAE
Bhesa zeylanica - 720; PIII; Ca, Fl, Pa, Pl, Ra.
Cassine glauca - 768-928; CI; Ca, Fu, Im, Pl, Ra.
Kokoona zeylanica - Fu, Mi, Pl.
Pleurostylia opposita - 864; CI; Ca, Fu, Im.

31. CLUSIACEAE
Calophyllum bracteatum - 512; CIII; PIII; Bo, Fr.
Calophyllum calaba - 912; CII; Im, Mi, Pl, Pa, Pw,Ra.
Calophyllum inophyllum - 608; Bo, Fu, Pa, Pw, Ra.
Garcinia echinocarpa - 800-880; Ps, Sh.
Garcinia quaesita - 800; Ps.
Garcinia spicata - CII; Ps.
Garcinia terpnophylla - 1248; PIII; Ps.
Mesua ferrea - CII; Fr, Fu.
Mesua nagassarium - 1152; CII; Fr, Fu, Ps.

32. COCHLOSPERMACEAE
Cochlospermum religiosum - 272; Fw.

33. COMBRETACEAE

Anogeissus latifolia - 880; CIII; Fw, Im, Pc, Pl, Ra, Ut.
Terminalia arjuna - 992; CI; Bo, Fu, Im, Ps.
Terminalia bellirica - 748; PIII; Pl, Ps.
Terminalia catappa - 512; Pl, Ps.
Terminalia chebula - 960; CIII; Ra.
Terminalia zeylanica - CIII; PIII; Fu, Ps, Ra.

4. CUPRESSACEAE

Cupressus macrocarpa - Pa, Pc, Ps.
Cupressus torulosa - 512; Pa, Pc, Ps.

38. DATISCACEAE

Tetrameles nudiflora - 448; PIII; Bo, Pc, Tp.

40. DILLENIACEAE

Dillenia indica - 720; CIII; Fr, Ra.
Dillenia retusa - 832; CII; Lc.
Dillenia triquetra - 704; CIII; PIII; Lc.
Schumacheria castaneifolia - Lc.

41. DIPTEROCARPACEAE

Dipterocarpus hispidus - 800-880; CI; Bo, Ca, Fr, Lc.
Dipterocarpus zeylanicus - 720; CI; PI; Bo, Ca,Pa, Pl, Pw, Sl.
Hopea jucunda - Ra.
Shorea congestiflora - 576; CIII; PIII; Fr, Lc, Pa,Pc, Pl, Pw.
Shorea cordifolia - 800; CII; Pw.
Shorea disticha - Fr, Pa, Pw, Ra.
Shorea dyeri - Ra.
Shorea gardneri - 1056; Pa, Pw.
Shorea megistophylla - Fr, Pa, Pw.
Shorea oblongifolia - 960; CII; Ra, Pw.
Shorea stipularis - PIII; Fr, Lc, Pa, Pc, Ps, Pc.
Shorea trapezifolia - 960; CII; PIII; Fr, Lc, Pw,
Shorea worthingtonii - Fr, Pa, Pl, Pw, Ra.
Shorea zeylanica - 1088; Pc, Pw.
Vateria copallifera - 624; CIII; PIII; Lc, Pc, Pl, Pw, Tp.

42. EBENACEAE

Diospyros ebenum - 1120; Cv, Fu, Im.
Diospyros malabarica - 624; Bo, Fu, Ra.
Diospyros melanoxylon - Cv, Fu, Im.
Diospyros oocarpa - 720; CI; Cv, Im.
Diospyros ovalifolia - 912; Lc, Ra.

43. ELAEOCARPACEAE

Elaeocarpus amoenus - Pc.
Elaeocarpus serratus - 528; Lc, Pc.

44. ERICACEAE

Rhododendron arboreum subsp. *zeylanicum* - 672; Fw.

46. EUPHORBIACEAE

Aleurites moluccana - 608; CIII; Pc, Pw.
Aporusa cardiosperma - Ra.
Bridelia moonii - 720-800; CIII; Im, Mi, Pa, Pl, Ra.
Bridelia retusa - 992; Im, Mi, Pa, Pl, Ra.
Chaetocarpus castanocarpus - Fw.
Chaetocarpus coriaceus - Fw.
Dimorphocalyx glabellus - Fw.
Drypetes sepiaria - Fw.
Hevea brasiliensis - Fw.
Macaranga indica - Fw.
Macaranga peltata - Lc.
Mischodon zeylanicus - Pl, Ra.
Phyllanthus emblica - 896; Ra.
Phyllanthus indica - Ra.

47. FLACOURTIACEAE

Homalium zeylanicum - Lc, Ra.
Hydnocarpus venenata - PIII; Ra.

7. GRAMINEAE

Bambusa bambos - Lc, Ut.
Bambusa vulgaris - B, Lc, Ut.
Davidsea attenuatum - B.
Dendrocalamus giganteus - Fu, Lc.
Ochlandra stridula - B
Pseudoxytenanthera monadelpha - B

49. HERNANDIACEAE

Gyrocarpus americanus - Bo.

50. ICACINACEAE

Stemonurus apicalis - 544; CI; Pc, Pw.

51. LAURACEAE

Alseodaphne semecarpifolia - 800; CI; Bo, Fu, Lc.
Cryptocarya wightiana - 1040; Pw, Ra.
Litsea gardneri - CII; PII; Pw.
Litsea glutinosa - 688; CIII; PI; Fu, Im, Pa, Sl.
Litsea quinqueflora - Ra.
Persea macrantha - 560; PIII; Bo, Lc, Pa, Pl.

52. LECYTHIDACEAE

Barringtonia racemosa - 432; Lc, Fw.
Careya arborea - 800; Bo, Ca, Ps.

53. LEGUMINOSAE

Acacia decurrens - Fw.
Acacia leucophloea - 720 CII; Fr, Fw, Im, Ra.
Acacia mangium - Lc, Pl, Ra.
Acacia melanoxylon - 640-800 SC; Fw, Sl.
Acacia planifrons - Fw.
Albizia falcataria - Lc, Pa, Pc, Pl.
Albizia lebbek - 800; Bo, Lc.
Albizia odoratissima - 912 SC; Ca, Fu, Im, Pa, Ra.
Albizia saman - 656; Fr, Pa, Pc.
Bauhinia racemosa - 800; Ps.
Bauhinia tomentosa - Im.
Cassia fistula - ehela 960 CIII; Ps.
Cassia roxburghii - 960; Im, Ps.
Cassia siamea - 848-992; CII; Fu, Fw.
Cassia spectabilis - Im.
Dialium ovoideum - Fu.
Delonix regia - 448; Fw.
Gliricidia sepium - Fw.
Leucaena leucocephala - Fw.
Parkinsonia aculeata - Fw
Peltophorum pterocarpum - Fr, Fu, Fw.
Pericopsis mooniana - 1120 L; Fu.
Pithecellobium dulce— 640; Pc, Pa.
Pongamia pinnata - 672; Ps, Fw.
Pterocarpus indicus - 768; Ca, Cv, Fl, Fr, Fu, Pa, Sl.
Pterocarpus marsupium - 896 SC; Ca, Cv, Fr, Fu.
Saraca asoka - 928; Lc.
Tamarindus indica - 1280 SC; Ca, Cv, Fu.

54. LOGANIACEAE

Strychnos nux-vomica - 912; Ca, Fu, Pa.

55. LYTHRACEAE

Lagerstroemia speciosa - 640; Bo, Ca, Fr, Fu, Im, Pc, Pl, Ps, Ra, Ut.

56. MAGNOLIACEAE

Michelia champaca - 640; CIII; Ca, Fw, Fu, Pa, Pl.
Michelia nilagirica - 656; CIII; Fr, Fu, Pa, Pl.

57. MALVACEAE

Thespesia populnea - 784; Bo, Fu.

58. MELASTOMATACEAE

Lijndenia capitellata - Im.
Memecylon capitellatum - Im.
Memecylon parvifolium - Im, Pos, Ra.
Memecylon rostratum - Im.
Memecylon sylvaticum - Im.
Memecylon umbellatum - Im.

59. MELIACEAE

Azadirachta indica - 752; SC; Ca, Fu, Im, Pa, Ra.
Chukrasia tabularis - 720-800; SC; Ca, Fu, Sl, Pa, Pl.
Dysoxylum binectariferum- 704; Ca, Sl, Pw.
Dysoxylum championii - Ra.
Melia azedarach - 416; Bo, Lc, Pc, Pw.
Swietenia macrophylla - 560 L; Fl, Fu, Mi, Pa, Pl, Ra.
Swietenia mahagoni - 800 L; Fu, Sl, Pa.
Toona ciliata - 496; Bo, Fu, Lc, Pc.
Toona sinensis - 528; CII; PII; Bo, Fu, Lc, Pc.
Walsura piscida - 976; Fr, Lc, Ra.

61. MORACEAE

Artocarpus altilis - 544; Lc.
Artocarpus heterophyllus - 648 SC; Bo, Ca, Fr, Fu, Mi, Pl, Pw, Sl, Ra.
Artocarpus nobilis - 768 CI; PI; Bo, Fr, Fu, Pl, Ra.

63. MYRISTICACEAE

Horsfieldia irya - 640; Pc.
Horsfieldia iryaghedi - Bo, Pc.
Myristica dactyloides - 384; PIII; Bo, Lc, Ra.

65. MYRTACEAE

Eucalyptus camaldulensis - Fw, Ps.
Eucalyptus globulus - 704; CI; Fw, Ps.
Eucalyptus grandis CI; Fr, Fu, Lc, Sl, Ra.
Eucalyptus microcorys - 976; CI; Fr, Fw, Pa, Ps, Sl, Ra.
Eucalyptus pilularis - 880; CI; Fl, Lc, Pl, Ps.
Eucalyptus robusta - 768; CI; Fw, Lc, Pl, Ps.
Syzygium aqueum - 784; Ra.
Syzygium assimile - 816; Ra.
Syzygium cumini - 784-880; CI; Ps, Ra, Sl.
Syzygium gardneri - 1008; CII; Lc, Ra.
Syzygium jambos - B.
Syzygium makul - 784; CII; Fr, Pw, Ra.
Syzygium neesianum - Ps, Sl, Ra.
Syzygium operculatum - 704; CIII; Fr, Fw, Im, Ra.
Syzygium rubicundum - Pw, Ra.
Syzygium umbrosum - 880; Fr, Fw, Ra.
Syzygium zeylanicum - Fw.

68. OCHNACEAE

Ochna lanceolata - 880.

9. PALMAE

Areca catechu - Lc, Ut.
Borassus flabellifer - B, Lc, Ut.
Cocos nucifera - B, Fw, Lc, Im, Ut.

5. PINACEAE
Pinus caribaea - Fr, Fu, Lc, Pa.
Pinus patula - Fr, Fu, Lc, Pa.

72. PROTEACEAE
Grevillea robusta - 640; Fu, Fw, Pc.

75. RHIZOPHORACEAE
Anisophyllea cinnamomoides - 832; CII; Ra, Pl, Pa, Lc.
Bruguiera sexangula 720; Fu, Fw.
Carallia brachiata - 752; CII; Fu, Fw, Im.
Rhizophora mucronata - 1056; Fw, Im, Lc, Ra.

76. ROSACEAE
Prunus ceylanica - 1040; Fw, Lc.
Prunus walkeri - Fw, Lc, Ra.

77. RUBIACEAE
Adina cordifolia - 736; CI; PI; Bo, Ca, Fr, Fu, Im, Mi, Pc, Pl, Sl, Ut.
Canthium dicoccum - 752; Cv, Im, Lc.
Canthium montanum - Ra, Sl.
Mitragyna parvifolia - 752; CI; Fu, Lc, Pl.
Morinda tinctoria - 570-752; Im, Lc, Ra, Ut.
Nauclea orientalis - 608; CIII; Pc, Ra, Ut.
Tarenna asiatica - Lc.
Wendlandia bicuspidata - 880; Fr, Im, Lc, Ra.

78. RUTACEAE
Acronychia pedunculata - 768; CIII; Fw, Lc.
Chloroxylon swietenia - 1024; Cv, Fr, Fu, Im, Ra.
Euodia lunu-ankenda 880; Im, Tp.
Glycosmis pentaphylla - Im, Tp.
Limonia acidissima - 880-960; Im, Ps.
Pamburus missionis - Lc.
Pleiospermium alatum - Im, Ut.

79. SABIACEAE
Meliosma simplicifolia - 560; Fw, Lc.

81. SANTALACEAE
Santalum album - Cv, Fu, Ut.

82. SAPINDACEAE
Allophylus cobbe - 640; Im.
Dimocarpus longan - Lc.
Filicium decipiens - 1024 CI; PI.
Glenniea unijuga - Fr, Lc, Ra.
Nephelium lappaceum - 880; Fw.
Pometia pinnata - 720; Pa, Pc, Pl, Lc, Ra.
Sapindus emarginatus - 1024; CIII; Lc, Sl.
Schleichera oleosa - 1024-1152; C II; Ic, Ra.

83. SAPOTACEAE

Isonandra compta - CIII; Fr, Ra.
Madhuca fulva - 736; CII; Ra, Sl, Ps.
Madhuca longifolia - Bo, Ra, Sl, W.
Madhuca neriifolia - 752; Fu, Ps, Ra, Sl.
Manilkara hexandra - 1120; CI; Ca, Im, Ra, Sl, Ps.
Mimusops elengi - 1008; CI; Bo, Ca, Fu, Fw, Pl, Ps, Ra.
Palaquium canaliculatum - Lc, Pw.
Palaquium grande - Lc, Pw.
Palaquium laevifolium - Lc, Pw.
Palaquium petiolare - 560; CI; PI; Fu, Pa, Pc, Pl, Pw.
Palaquium rubiginosum - 768; CI; Pl; Ps, Ra.
Palaquium thwaitesii - PII; Pw.

84. SIMAROUBACEAE

Quassia indica - 416; Tp.

88. STERCULIACEAE

Firmiana colorata - 384; Ut.
Heritiera littoralis - 1200; Bo, Fr, Ra.
Pterospermum suberifolium - 576; CI; Pa, Pc.
Sterculia foetida - 576; PIII; Bo.

89. SYMPLOCACEAE

Symplocos cochinchinensis - Pl.

90. THEACEAE

Adinandra lasiopetala - 704; Ra.
Gordonia ceylanica - Pa.
Gordonia speciosa - CIII; Pa.
Ternstroemia japonica - 640; Im, Pa, Ra.

92. TILIACEAE

Berrya cordifolia - 784; SL; Sl.
Grewia polygama - 640; CIII; Bo, Ps.
Grewia tiliaefolia - 640; Bo.
Pityranthe verrucosa - Sl.

93. ULMACEAE

Holoptelea integrifolia - 640; PIII; Bo, Cv, Fu, Pa, Pl.

95. VERBENACEAE

Callicarpa tomentosa - Fw.
Clerodendrum inerme - B
Gmelina arborea - 480; CIII;Bo, Cv, Fu, Pa, Pl.
Premna latifolia - Pw.
Premna tomentosa - Cv, Fw, Fu.
Tectona grandis - 720; SL; Bo, Ca, Cv, Fl, Fr.
Vitex altissima - 928; Ca, Fr, Fu, Pa, Pl, Ps, Ra, Sl.
Vitex leucoxylon - 768; Ca, Sl.

Medicinal Plants

The use of medicinal plants
in Sri Lanka

In Sri Lanka, the use of medicinal plants is largely influenced by the Indian ayurvedic system. The Vedic Aryans of India were acquainted with a large number of medicinal plants according to the works of Charaka (1000 BC) and Susruta (800 BC). The early migrants from India to Sri Lanka brought with them the art of healing practised on the mainland at that time.

The earliest book on medicine compiled in Sri Lanka, and written in Sanskrit, was *Sarartha Sangrahaya* by king Buddhadasa, a great physician and surgeon, who ruled Anuradhapura between 330 AD and 400 AD. During the Polonnaruwa period that followed, the traditional system of medicine reached new heights under king Parakramabahu I (1153—1186 AD). The Dambadeni period allowed ayurvedic medicine to blossom further in the hands of monks, since Buddhism proclaims the treatment of the sick and wounded as a virtue. Although several advances were attributed to the Kotte period (1400—1650 AD), the decline of indigenous medicine was thought to have occurred within this period. The decline continued during the Kandyan era and was accelerated by continued foreign invasions. The introduction of the western system of medicine further accelerated the decline to its lowest ebb in the 1960's. Since then it has had a resurgence in popularity particularly in the treatment of minor ailments for which western medicines can be expensive. The efficacy of the traditional system of medicine is well described by Joao Ribero (1640-1688), a captain of the Portuguese army, in his book *The Historic Tragedy of Ceilao*. The same views were experienced by the British sailor, Robert Knox, who was held as a captive by the then Sinhalese king in the outskirts of Kandy, in his book titled *An Historical Relation of Ceylon* (1681).

In traditional society, when one consults an ayurvedic physician, it is customary to offer a sheaf of betel leaves, sometimes with a small present, but not money since the profession is one of prestige and honor. The profession is passed down through a family as a legacy.

Among the more recent books on medicinal plants in Sri Lanka are *Sinhalese Materia Medica* by J. Attygalle, *Vegetable Materia Medica* by E. Roberts, and *Medicinal plants used in Ceylon* by D.M.A. Jayaweera. We have listed below

---

1 It is important to note that the information provided below is for general knowledge only. For all treatments it is essential that advice be sought from a professional ayurvedic.

in alphabetical order the more common ailments and diseases, and the trees and shrubs described in this field guide which are used to treat them. Each species has a reference number for easy access to its description. Most of this information has been condensed from Jayaweera's work and by no means is it an exhaustive list of all plants in this book that have medicinal properties. Instead, we have restricted the plants listed to those used for the most common ailments. It is important that the reader recognize that the information provided below is for general knowledge only. For all treatments it is recommended that advice from a professional ayurvedic be sought.

Code letters for parts of plants—

B   - Bark
BJ   - Bark juice
Fl   - Flowers
FlJ - Flower juice
Fr   - Fruit
FrJ - Fruit juice
FrO - Fruit oils
L   - Leaf
LJ   - Leaf juice
H   - Heartwood
IB   - Inner bark
J   - Plant juice
R   - Roots
RB   - Root bark
RO   - Root oil
S   - Sap
St   - Stem
StO - Stem oil

Antiseptics. Compounds that prevent infection and putrefaction of wounds or sores and destroy bacteria. Plant parts used are either barks or leaves that are pressed to the skin surface (24.1 *Canarium zeylanicum*—B; 31.3 *Calophyllum inophyllum*—B; 59.4 *Azadirachta indica*—L). Also oils or juices can be applied (53.25 *Pongamia pinnata*—StO; 78.6 *Citrus aurantiifolia*—FrJ).

Aphrodisiacs. Compounds that arouse sexual desire. Plant parts are usually pulped and taken as a decoction (22.3 *Ceiba pentandra*—R; 53.14 *Dichrostachys cinerea*—R; 61.10 *Ficus hispida*—Fr).

Asthma. A disease marked by difficulty in breathing due to involuntary constriction of the bronchial pathways often triggered by an allergy. Plant parts used are pulped and taken as a decoction (11.1 *Acanthus ilicifolius*—R, L; 11.5 *Justicia adhatoda*—LJ; 12.5 *Mangifera zeylanica*—LJ; 46.58 *Phyllanthus reticulatus* —RJ; 53.20 *Erythrina variegata*—B, L; 77.31 *Mussaenda frondosa*—Fl; 82.12 *Sapindus trifoliatus*—Fr). Some parts are dried and smoked (11.5 *Justicia adhatoda*—L).

Beri beri. A deficiency disease from a lack of vitamin B1 (thiamine). The individual becomes extremely weak and sometimes paralyzed. *Carica papaya*, (27.1) leaves made into a decoction is a suggested remedy.

Bladder and kidney stones. Stone-like deposits of oxalate crystals that can cause painful blockage of the bladder and kidney. Plant parts are powdered and taken as a decoction (22.3 *Ceiba pentandra*—B; 26.2 *Crateva adansonii*—B; 61.1 *Artocarpus heterophyllus*—L, B; 76.2 *Prunus cerasoides*—nut kernel). Other plants are eaten or taken as a drink (13.5 *Annona muricata*—Fr).

Boils, external ulcers and sores. Hard inflamed suppurating swellings that sometimes secrete pus. Located on the body surface, they are often caused by infected bites or wounds. Some plants are used to hasten suppuration and to draw out the pus. These are usually applied as poultices to the affected area. Leaves and sometimes barks are sometimes applied whole but other parts are usually powdered and made into a paste (12.3 *Lannea coromandelica*—B; 14.3 *Alstonia scholaris*—L; 14.10 *Plumeria obtusa*—L; 18.1 *Calotropis gigantea*—R; 7.1 *Bambusa bambos*—L buds; 24.1 *Canarium zeylanicum*—B boiled with coconut oil; 41.2 *Dipterocarpus zeylanicus*—gum; 45.2 *Erythroxylum zeylanica*—L; 46.54 *Phyllanthus emblica*—Fr pericarp; 46.58 *Phyllanthus reticulatus*—L; 51.15 *Litsea longifolia* B; 53.20 *Erythrina variegata*—L; 54.2 *Strychnos nux-vomica* L; 57.2 *Hibiscus rosa-sinensis*—Fl buds, L; 57.4 *Thespesia populnea*—L, B; 58.5 *Memecylon capitellatum*—L, BO; 61.1 *Artocarpus heterophyllus*—J with vinegar; 61.19 *Streblus asper*—B; 63.1 *Horsfieldia irya*—L; 65.17 *Syzygium caryophyllatum*—L, B; 9.2 *Borassus flabellifer*—toddy and rice flour for gangrenous ulcers; 77.29 *Morinda citrifolia*—L paste with gingelly oil; 77.41 *Tarenna asiatica*—Fr; 78.1 *Acronychia pedunculata*—B; 91.2 *Gyrinops walla*-young L). Other plant parts can be applied as a poultice to prevent suppuration (51.9 *Cinnamomum verum*—B bruised and steamed; 53.47 *Tamarindus indica*—L paste with lime juice; 95.2 *Clerodendrum inerme*—L). To wash and clean ulcers and boils, and to kill maggots within infected areas, juices of plants can be used (46.34 *Flueggea leucopyrus*—LJ; 63.1 *Horsfieldia irya*—J; 77.1 *Adina cordifolia*—LJ).

Bronchial diseases and pneumonia. Constriction and/or inflamation of the lungs. Plant parts used are often made into juices or decoctions that are taken internally (11.5 *Justicia adhatoda*—LJ; 12.4 *Mangifera indica*—LJ; 14.10 *Plumeria obtusa*—L; 7.1 *Bambusa bambos*—camphor in the nodes; 23.3 *Cordia dichotoma*—B; 51.9 *Cinnamomum verum*—B; 53.36 *Cassia fistula*—RB; 74.4 *Zizyphus jujuba*—syrup of dried Fr; 78.3 *Atalantia ceylanica*—LJ; 78.6 *Citrus aurantiifolia* FrJ).

Bruises, sprains and swellings. Injuries to the flesh caused by a blow or pressure that discolours but does not break the skin, or wrenching of the muscle that causes inflammation. Plant parts are applied as poultices to the affected area (12.3 *Lannea coromandelica*,—L, gum; 29.2 *Cassine glauca*—RB; 51.13 *Litsea glutinosa*—R, L; 53. 53.32 *Bauhinia tomentosa*—L; 61.5 *Ficus benghalensis*—J; 78.1 *Acronychia pedunculata*—B; 78.4 *Atalantia monophylla*—L).

Cardiotonics. Compounds that are heart stimulants and can be used to help regulate the heartbeat. Plant parts are taken as a decoction (11.5 *Justicia adhatoda*—RB; 14.9 *Nerium oleander*—whole plant; 78.2 *Aegle marmelos*—RB).

Catarrh and sinusitis. Inflamation of the nasal mucous membrane often caused by an allergy. Plant parts can be taken as a decoction (11.3 *Barleria prionitis*—LJ with honey; 59.4 *Azadiracta indica*—gum exudate; 95.5 *Clerodendrum serratum*—R; 95.15 *Vitex negundo*—LJ), inhaled up the nostrils to bring out the mucus (53.28 *Sesbania grandiflora* L, FlJ), taken as pills (78.3 *Atalantia ceylanica*—LJ), or smoked (95.14 *Vitex leucoxylon*—L). *Vitex negundo* (95.15) leaves stuffed within a pillow can also relieve the patient while sleeping.

Cholera. An infectious and often fatal disease usually transmitted through drinking water with symptoms of continuous diarrhoea and vomiting. Plant parts are taken as a decoction (*Strychnos nux-vomica*—RB) or mixed as a paste with lime juice and made into pills (9.2 *Borassus flabellifer*—R).

Colds and coughs. To soothe sore or itchy throat plant parts can be taken as a decoction (52.1 *Barringtonia acutangula*—Fr; 52.3 *Barringtonia racemosa*—Fr; 52.4 *Careya arborea*—Fl & BJ with honey; 53.11 *Albizia odoratissima*—boiled L and ghee; 78.9 *Citrus limon*—FrJ with sugarcane juice; 78.6 *Citrus aurantiifolia*—FrJ; 83.4 *Madhuca longifolia*—L; 91.15 *Vitex negundo*—R), or as a throat swab (78.9 *Citrus limon*—FrJ with pepper & salt; 82.11 *Sapindus emarginatus*—FrO of pericarp).

Cuts and fresh wounds of the skin. Plant parts are applied as a poultice (11.5 *Justicia adhatoda*—B, L; 53.32 *Bauhinia tomentosa*—seeds are made into a paste), or the juice is used for cleansing (42.6 *Diospyros malabarica*—unripe FrJ; 46.28 *Euphorbia antiquorum*—S kills maggots; 78.6 *Citrus aurantiifolia*—FrJ). Other plant parts can be taken as a decoction to promote healing (74.5 *Zizyphus oenoplia*—B).

Diabetes. A malfunction of the pancreas in which inviduals develop elevated levels of glucose in the blood. Some plant parts can be taken as a decoction to lower blood sugar (12.1 *Anacardium occidentale*—B; 24.1 *Canarium zeylanicum*—B; 29.5 *Kokoona zeylanica*—IB; 53.17 *Butea monosperma*—B; 53.25 *Pongamia pinnata*—Fl; 53.27 *Pterocarpus marsupium*—S; 53.35 *Cassia auriculata*—R; 55.1 *Lagerstroemia speciosa*—L, Fr; 58.13 *Osbeckia octandra*—L; 61.1 *Artocarpus heterophyllus*—L, B; 61.5 *Ficus benghalensis*—B; 61.13 *F. racemosa*—Latex; 65.19 *Syzygium cumini*—Fr).

Diarrhoea and dysentry. Diseases of the bowels often acquired through drinking unclean water and food. Plant parts taken as a decoction (11.5 *Justicia adhatoda*—LJ; 12.1 *Anacardium occidentale*—B; 12.4 *Mangifera indica*—LJ, Fl; 12.12 *Spondias dulcis*—B; 14.3 *Alstonia scholaris*—B; 14.4 *Carissa carandas*—L; 20.4 *Oroxylum indicum*—R; 22.3 *Ceiba pentandra*—R; 31.9 *Garcinia mangostana*—Fr pericarp; 33.4 *Terminalia bellirica*—Fr pericarp; 33.5 *T. catappa*—B; 33.6 *T. chebula*—Fr pericarp with milk; 46.58 *Phyllanthus reticulatus*—LJ; 51.9 *Cinnamomum verum*—B; 51.13 *Litsea glutinosa*—B mucilage; 52.1 *Barringtonia acutangula*—LJ; 52.3 *B. racemosa*—Fr; 53.17 *Butea monosperma*—B; 53.31 *Bauhinia racemosa* -B, Fl, St; 53.32 *Bauhinia tomentosa*—B, Fl, St; 57.3 *Hibiscus tiliaceus*—B mucilage; 59.1 *Aglaia roxburghiana*—BR; 61.19 *Streblus asper*—B; 63.2 *Horsfieldia iryaghedi* Fl, B; 65.10 *Psidium guajava*—B; 78.16 *Syzygium cumini*—B; 73.1 *Punica granatum*—Fr skin; 77.20 *Ixora coccinea*—R; 78.2 *Aegle marmelos* -Fr pulp; 78.16 *Limonia acidissima*—Fr pulp; 81.1 *Santalum album*—H; 82.1 *Allophylus cobbe*—R; 83.8 *Mimusops elengi*—Fr for chronic dysentery; 82.8 *Helicteres isora*—B; 92.5 *Grewia rothii*—R).

Dropsy. A disease whereby watery fluid collects in the body and the individual becomes over-swollen. Plant parts taken as a decoction (12.1 *Anacardium occidentale*—receptacle FrJ; 33.4 *Terminalia bellirica*—Fr pericarp; 33.6 *Terminalia chebula*—Fr pericarp; *Premna latifolia*—L with coriander).

Ear ache. To cure ear aches or to deaden the pain juices of plant parts are often dropped into the ear (12.12 *Spondias dulcis*—LJ; 14.3 *Alstonia scholaris*—LJ; 33.3 *Terminalia arjuna*—LJ; 46.28 *Euphorbia antiquorum* LJ; 53.20 *Erythrina variegata*—LJ).

Emetics. Compounds that cause vomiting and can be used in cases of food poisoning. Plant parts are usually powdered and taken as a decoction (29.2 *Cassine glauca*—R; 57.3 *Hibiscus tiliaceus*—B; 59.14 *Walsura trifoliolata*—Fr; 61.10 *Ficus hispida*—Fr, S, B).

Eye diseases. Several eye ailments can be soothed or cured by plant parts. *Calophyllum inophyllum*, (31.3—B, L & latex steeped in water) soothes sore eyes; *Terminalia bellirica* (33.4—Fr pericarp) and *Terminalia chebula* (33.6—Fr pericarp) can be used to remove white spots on the cornea; *Dichrostachys cinerea* (53.14—bruised L) can cure opthalmia; *Sesbania grandiflora* (53.28—FlJ) improves dim vision; *Ixora coccinea* (77.20—Fl buds) clears reddened eyes.

Fevers. Reaction of the body to a general infection characterized by high temperature and body chills. Plant parts are taken as a decoction (14.3 *Alstonia scholaris*—B; 33.4 *Terminalia bellirica*—Fr pericarp; 40.1 *Dillenia indica*—FrJ; 53.35 *Cassia auriculata*—RB; 53.36 *Cassia fistula*—RB; 61.19 *Streblus asper*—B; 77.28 *Mitragyna parvifolia*—RB; 78.13 *Euodia lunu-ankenda*—L, Fl; 78.15 *Glycosmis pentaphylla*—LJ; 81.1 *Santalum album*—H; 82.11 *Sapindus emarginatus*—S especially for children; 95.2 *Clerodendrum inerme*—LJ, R; 95.3 *Clerodendrum infortunatum*—L & RJ; 95.5 *Clerodendrum serratum*—R; 95.7 *Gmelina arborea*—R, B).

Fractures and dislocations. Breakage of bone or cartilage, and bodily joints that have been forced out of their proper relative position. Plant parts are applied as poultices to the affected area (31.5 *Calophyllum walkerae*—SO; 40.2 *Dillenia retusa*—Fr; 46.30 *Euphorbia tirucalli*—St; 53.46 *Saraca asoca*—B, L; 9.8 *Phoenix zeylanica*—palmheart; 78.1 *Acronychia pedunculata*—B; 82.1 *Allophylus cobbe*—whole plant with oil; 82.2 *Allophylus zeylanicus*—L, B; 83.4 *Madhuca longifolia*—B), and some others are taken internally as a decoction (33.3 *Terminalia arjuna*—B powdered with milk).

Gastric ulcers. Internal sores that line the stomach or intestine. The palm heart of *Caryota urens* (9.3) taken as a cooked vegetable is said to soothe internal ulcers.

Hair growth. Plant parts that supposedly stimulate hair growth are mostly oils applied to the skin surface where hair growth is desired (9.4 *Cocos nucifera*—FrO; 9.3 *Caryota urens*—young Fl; 82.12 *Schleichera oleosa*—SO; 95.12 *Tectona grandis*—FrO).

Headaches. A pain in the head. Plant parts are powdered and used as snuff (29.2 *Cassine glauca*—L; 29.5 *Kokoona zeylanica*—IB), applied as a poultice to the head (44.2 *Rhododendron arboreum*—young L, 46.60 *Ricinus communis*—fresh L), or inhaled through smoking (31.3 *Calophyllum inophyllum*—L; 95.14 *Vitex leucoxylon*—L).

Insecticides and insect bites. Compounds that kill insects and mites and that soothe insect bites on the body surface. Several plant parts can be used to kill insects. *Annona squamosa* (13.7) seeds that are ground to a paste and applied to the hair kill lice. The fruits of *Cassia fistula* (53.36) kill bed bugs. Plants with general insecticidal properties are *Anacardium occidentale* (12.1—FrO) and *Azadiracta indica* (59.4—L). Other plant parts are applied as a poultice to soothe insect bites such as wasp stings (74.5 *Zizyphus oenoplia*—B, LJ; 78.16 *Limonia acidissima*—FrJ & pulp), or tarantula bites (95.15 *Vitex negundo*—bruised L).

Intestinal and stomach upsets. A number of disorders can cause an upset in the stomach or intestine. Most common are bacterial or viral infections from bad food or drinking water. Fruits are often eaten to sooth the stomach (27.1 *Carica papaya*—Fr; 65.15 *Syzygium aromaticum*—Fr). In other plants parts are taken as a decoction to sooth the upset stomach (78.12 *Clausena indica*—St, L, R; 78.16 *Limonia acidissima*—bruised L, especially used for children; 78.18 *Murraya koenigii*—L, BR; 88.2 *Helicteres isora*—Fr, RJ; 95.11 *Premna tomentosa*—RO).

Jaundice and hepatitis. In jaundice the patient is morbid due to the obstruction of bile from the liver and can be characterized as having a yellowish skin. Hepatitis is a viral disease of the liver. For cure plant parts are pulverized and taken as a decoction (11.3 *Barleria prionitis*—L; 18.1 *Calotropis gigantea*—R; 47.1 *Flacourtia indica*—Fr; 53.47 *Tamarindus indica*—Fl; 58.13 *Osbeckia octandra*—L, R; 59.14 *Azadiracta indica*—LJ and honey; 77.31 *Mussaenda frondosa*—sepals; 95.1 *Callicarpa tomentosa*—B).

Laxatives. Drugs that loosen the bowels. Plant parts are taken as a decoction (23.3 *Cordia dichotoma*—R; 2.1 *Cycas circinalis*—L, St; 46.54 *Phyllanthus emblica*—unripe Fr; 46.61 *Ricinis communis*—SO; 53.35 *Cassia auriculata*—S; 53.47 *Tamarindus indica*—Fr pulp; 61.14 *Ficus religiosa*—Fr; 78.2 *Aegle marmelos*—Fr; 78.18 *Murraya koenigii*—L; 88.5 *Sterculia balanghas*—Fr).

Malaria. A fever transmitted by bites of certain mosquitoes. Plant parts that alleviate malarial fever are usually taken as decoctions (52.1 *Barringtonia acutangula*—B,R; 53.36 *Cassia fistula*—RB; 95.3 *Clerodendrum infortunatum*—L, RJ; 95.5 *C. serratum*—RJ).

Menstruation. Monthly period in the estrous cycle when the unfertilzed egg is released along with the superficial lining of the uterus. Irregularities in menstruation can sometimes be corrected by taking a decoction made from the roots of *Memecylon umbellatum* (58.9).

Piles. Swellings of the rectal veins (haemorrhoids). Plant parts can be pulverized and taken as a decoction (12.7 *Semecarpus coriacea*—boiled seeds; 12.8 *S. gardneri*—boiled seeds; 26.1 *Capparis zeylanica*—L; 53.46 *Saraca asoca*—B; 57.4 *Thespesia populnea*—L, B; 61.13 *Ficus racemosa*—S; 78.2 *Aegle marmelos*—Fr pulp). Some can be eaten (33.6 *Terminalia chebula*—Fr pericarp; 2.1 *Cycas circinalis*—cooked L; 47.1 *Flacourtia indica*—Fr; 58.13 *Osbeckia octandra*—L cooked as a porridge).

Pimples and acne. Small swellings within the skin surface that are usually caused by infected pores or hair follicles. Plant parts are usually applied as a paste or lotion to the affected spot (29.5 *Kokoona zeylanica*—IB; 31.10 *Garcinia morella*—St rubbed with water; *Tamarindus indica* (53.47—S) made into a paste with vinegar and lime juice).

Purgatives. Compounds that clean or clear out the body. Plant parts are usually pulverized and made into a decoction (13.7 *Annona squamosa*—RB; 20.4 *Oroxylum indicum*, totilla—St; 31.3 *Calophyllum inophyllum*—BJ; 33.6 *Terminalia chebula*—Fr pericarp; 46.26 *Dimorphocalyx glabellus*—L; 46.28 *Euphorbia antiquorum*—RB; 53.35 *Cassia auriculata*—Fr; 53.36 *C. fistula*—Fr; 61.14 *Ficus religiosa*—L; 62.1 *Moringa oleifera*—LJ; 65.17 *Syzygium caryophyllatum*—St; 78.1 *Acronychia pedunculata*—B).

Rheumatism. An inflammation and pain of the joints. Plant parts can be applied as a poultice, (11.1 *Acanthus ilicifolius*—L; 27.1 *Carica papaya*—bruised L; 95.15 *Vitex negundo*—L), paste (11.3 *Barleria prionitis*—whole plant; 12.12 *Spondias dulcis*—B; 41.2 *Dipterocarpus zeylanicus*, hora—resin paste; 46.46 *Mallotus philippensis*—R; 53.29 *Sesbania grandiflora*—R; 56.1 *Michelia champaca*—Fl), or oil (31.3 *Calophyllum inopyhllum*—SO; 46.19 *Bridelia retusa*—RB liniment with gingelly oil; 53.26 *Pongamia pinnata*—Bruised B & L heated in gingelly oil; 61.5 *Ficus benghalensis*—J; 83.4 *Madhuca longifolia*—SO).

Scalds and burns. An injury to the skin from heat. *Madhuca longifolia* (83.4—L, B) can be applied as a poultice.

Sedatives. Compounds that soothe and often cause drowsiness. Plant parts are macerated and made into a decoction (26.2 *Capparis zeylanica*—RB; 61.1 *Artocarpus heterophyllus*—young L with coconut milk; 77.20 *Ixora coccinea*—Fl).

Sexual (venereal) diseases—Gonorrhea and syphillis. Both are bacterial diseases transmitted through sexual intercourse. Gonorrhoea is the more common of the two and can be characterized as an inflammatory discharge from the urethra. Plant parts that are used for gonorrhoea are *Phyllanthus emblica* (46.55—BJ with honey and tumeric), *Borassus flabellifer* (9.2—R), *Ficus racemosa* (61.13—Fr), *Phoenix zeylanica* (9.8—palm heart). For syphilis a decoction of *Capparis horrida* (26.1—L) or *Capparis zeylanica* (26.2—L) can be used. Other plants can help treat herpes (14.4 *Carissa carandas*—L; 14.10 *Plumeria obtusa*—RB; 23.1 *Carmona retusa*—R; 46.43 *Macaranga peltata*—powdered gum exudate).

Skin diseases. Most skin diseases are fungal. Plant parts are usually made into a paste and applied to the affected area. For eczema *Nerium oleander* (14.8—L, B), *Glycosmis pentaphylla* (78.15—L), or *Tectona grandis* (95.12—H) can be used. *Butea monosperma* (53.17—S), *Jasminum angustifolium* (70.3—R with lime juice), or *Psidium guajava* (65.10—L paste with tumeric) can be used for ringworm. Other plant parts used for more general skin diseases are *Semecarpus coriaecea*, (12.7—Gum, boiled S), *Semecarpus gardneri* (12.8—Gum, boiled S), *Semecarpus subpeltata* (12.10—Gum, boiled S), *Calotropis gigantea* (18.1—B, RJ), *Crateva religiosa* (26.3—B); *Terminalia catappa* (33.5—LJ); *Mallotus philippensis* (46.46—L); *Hydnocarpus venenata* (47.5—SO); *Barringtonia racemosa* (52.3—Fr); *Pongumia pinnata* (53.25—SO) *Cassia fistula* (53.36—B, older L); *Santalum album* (81.1—SO); *Tectona grandis* (95.12—FrO).

Snake bites. Several plants have antidotes to some extent for some of

Sri Lanka's poisonous snakes. *Walidda antidysenterica* (14.16—Fl) can be used for Russell's viper. *Horsfieldia irya* (63.1) roots macerated with lime juice and taken as a decoction can be used for krait. Other plant parts that are used for snake bites are either applied as pastes or poultices (11.1 *Acanthus ilicifolius*—L; 53.32 *Bauhinia tomentosa*—S; 77.29 *Morinda citrifolia*—L; 91.2 *Gyrinops walla*—L), or taken as a decoction (29.2 *Cassine glauca*—R; 29.5 *Kokoona zeylanica*—IB; 33.1 *Anogeissus latifolius*—B; 53.17 *Butea monosperma*—B; 62.1 *Moringa oleifera*—L; 9.3 *Caryota urens*—RB; 76.2 *Prunus cerasoides*—S; 78.6 *Citrus aurantifolia*—FrJ; 78.10 *Citrus medica*—L, B, Fr; 78.16 *Limonia acidissima*—whole plant).

Sore throat. Several plants can be used as gargles to relieve a sore throat (14.16 *Wallida antidysenterica*—whole plant; 41.2 *Dipterocarpus zeylanicus*—gum; 42.6 *Diospyros malabarica*—unripe Fr; 53.17 *Butea monosperma*—gum; 63.3 *Myristica dactyloides*—B, L; 95.12 *Tectona grandis*—L).

Stimulants. Compounds that increase your activity and energy temporarily. Plant parts are taken as a decoction (11.1 *Acanthus ilicifolius*—whole plant; 56.1 *Michelia champaca*—Fl; 59.14 *Walsura trifoliolata*—Fr; 65.33 *Syzygium zeylanicum*—L, R; 9.1 *Areca catechu*—S; 78.16 *Limonia acidissima*—Fr).

Toothache and gum diseases. Many plant parts can be used to relieve tooth ache. Some can be used as a mouthwash to relieve pain (11.3 *Barleria prionitis*—LJ; 61.13 *Ficus religiosa*—B), others can be chewed or applied as a paste/juice to the paining tooth (12.3 *Lannea coromandelica*—B; 14.3 *Alstonia scholaris*—J; 14.11 *Plumeria obtusa*—L; 46.28 *Euphorbia antiquorum*—latex; 46.30 *Euphorbia tirucalli*—Latex; 51.9 *Cinnamomum verum*—B; 53.20 *Erythrina variegata*—LJ; 65.15 *Syzygium aromaticum*—Clove oil; 65.10 *Psidium guajava*—L; 77.33 *Nauclea orientalis*—L). Macerated 91.2 *Gyrinops walla* (L) when put into the offending tooth will loosen it for extraction. *Terminalia chebula*, (33.6—powdered gum of Fr pericarp) can be used to clean teeth. Tooth gums can be strengthened by taking gargles (61.4 *Ficus religiosa*—B; 65.19 *Syzygium cumini*—B; 83.8 *Mimusops elengi*—B) or applied directly (11.3 *Barleria prionitis*—LJ). For bleeding gums a decoction of *Canarium zeylanicum* (24.1—B) can be used. *Carica papaya* (27.1—papain) can be applied to tongue ulcers.

Tonsilitis. Inflammation of the glands at the back of the mouth. Plant parts are applied as ointments to the swollen glands (78.6 *Citrus aurantiifolia*—FrJ with pepper & salt; 82.11 *Sapindus emarginatus*—Fr pericarp).

Urinary diseases. Diseases of the bladder and ducts that pass urine. Plant parts are taken as a decoction (12.7 *Semecarpus coriacea*—boiled seeds; 12.8 *Semecarpus gardneri*—boiled seeds; 12.10 *Semecarpus subpeltata*—boiled seeds; 23.3 *Cordia dichotoma*—Fr mucilage; 26.3 *Crateva adansonii*—B; 53.35 *Cassia auriculata*—R; 61.13 *Ficus racemosa*—B; 9.4 *Cocos nucifera*—nut J; 83.8 *Mimusops elengi*—B).

Warts and corns. Small, hardish, permanent outgrowth on the skin. Ointments and juices of some plants can be applied directly to eradicate the outgrowth (27.1 *Carica papaya*—papain and borax; 46.28 *Euphorbia antiquorum*—latex; 46.30 *Euphorbia tirucalli*—Latex).

Worms. Intestinal worms. Plant parts are powdered and taken as a decoction. *Bambusa bambos* (L bud, young St) and *Mallotus philippensis* (46.46—L) can be used to treat threadworms. *Erythroxylum zeylanica* (45.2—L with rice and honey) and *Areca catechu* (9.1—powdered nut) can be used to treat round worms. Lastly, for tape worms the plant parts of *Mallotus philippensis* (46.46—L) or *Punica granatum* (73.1—RB) can be used.

Medicinal Plants

Part IV:

Index

Index to English names

Index to Sinhala names

Indexes

Index of Tamil names

Indexes

Index to scientific names

NOTES